Handbook of High-speed Machining Technology

OTHER OUTSTANDING VOLUMES IN THE CHAPMAN AND HALL ADVANCED INDUSTRIAL TECHNOLOGY SERIES

V. Daniel Hunt: SMART ROBOTS: A Handbook of Intelligent Robotic Systems

David F. Tver and Roger W. Bolz: ENCYCLOPEDIC DICTIONARY OF INDUSTRIAL TECHNOLOGY: Materials, Processes and Equipment

Roger W. Bolz: MANUFACTURING AUTOMATION MANAGEMENT: A Productivity Handbook

Igor Aleksander: ARTIFICIAL VISION FOR ROBOTS

D. J. Todd: WALKING MACHINES: An Introduction to Legged Robots

Igor Aleksander: COMPUTING TECHNIQUES FOR ROBOTS

Handbook of High-speed Machining Technology

edited by
Robert I. King

NEW YORK LONDON
CHAPMAN AND HALL

This book is dedicated to Tom Vajda, whose inspiration and support in the early 1970s made high-speed machining as it is known today possible.

Acknowledgments

We, the authors, wish to give credit to those who have dedicated years of research to the subject of high-rate metal cutting theory, the long and understanding support of the United States Air Force Materials Laboratory, and the genius of Professor Carl J. Salomon, who developed the initial concepts.

First published 1985
by Chapman and Hall
29 West 35th St., New York, N.Y. 10001

Published in Great Britain by
Chapman and Hall Ltd
11 New Fetter Lane, London EC4P 4EE

Printed in the United States of America

Library of Congress Cataloging in Publication Data
Main entry under title:

Handbook of high speed machining technology.

Includes bibliographies and index.
1. Metal-cutting—Handbooks, manuals, etc.
I. King, Robert I. (Robert Ira), 1924–
TJ1185.H16 1985 671.3′5 85-4113
ISBN 0-412-00811-4

Contents

Preface

The United States now spends approximately $115 billion annually to perform its metal removal tasks using conventional machining technology. Of this total amount, about $14 billion is invested in the aerospace and associated industries. It becomes clear that metal removal technology is a very important candidate for rigorous investigation looking toward improvement of productivity within the manufacturing system. To aid in this endeavor, work has begun to establish a new scientific and technical base that will provide principles upon which manufacturing decisions may be based.

One of the metal removal areas that has the potential for great economic advantages is high-speed machining and related technology. This text is concerned with discussions of ways in which high-speed machining systems can solve immediate problems of profiling, pocketing, slotting, sculpturing, facing, turning, drilling, and thin-walled sectioning. Benefits to many existing programs are provided by aiding in solving a current management production problem, that of efficiently removing large volumes of metal by chip removal.

The injection of new high-rate metal removal techniques into conventional production procedures, which have remained basically unchanged for a century, presents a formidable systems problem, both technically and managerially.The proper solution requires a sophisticated, difficult process whereby management—worker relationships are reassessed, age-old machine designs reevaluated, and a new vista of product/process planning and design admitted. The key to maximum productivity is a "systems approach." The text was structured with this in mind, and the reader can gain the greatest benefit by using the various chapters as building blocks from which an overall production system can be synthesized. The "bottom line" is to increase the overall effectiveness of the factory from whatever source, that is, to obtain the greatest return on investment.

Consider the technical problem of increasing the speed of the cutter through the base material by one magnitude. To realize the benefits of this increase, the table feed must be increased to a compatible rate. This in turn requires lighter inertia tables, more powerful drive motors, and more responsive control systems. As the speed increases, new dynamic ranges are encountered that induce undesirable resonances in the machine and parts being fabricated, requiring additional dampening consideration. Concerning the cutter–material interface, the basic chip morphology changes as new cutting regimes are experienced; hence Taylor's age-old empirical equations no longer hold, since they are not velocity dependent. Even the cutter configuration must now be considered a function of the cutting speed regime, as well as the normal process parameters. The proper incorporation of high-speed machining into factory processes requires the integration of all of the above technical considerations plus many others—a difficult systems problem requiring professional attention.

Part Seven, Management Considerations, is considered the "keystone" of the text and must complement the other parts. High-speed machining should be selectively applied, and only when it is economically justified. This manufacturing procedure is not a panacea for underproductive, high-cost operations; however, if it is used properly, when economics dictate, in a well-loaded and well-balanced factory, the results can be extremely gratifying.

Finally, one must consider the management style required to motivate the employees to accept these new procedures. Keep in mind that the changes suggested in this text are drastic and deviate from practices that have existed for at least a century. Use of the new techniques would be ill advised if the employees are not supportive of them for any reason. Employee involvement and understanding are necessary for success, and fear of the unknown is unacceptable. I am reminded of an excerpt from *The Prince* (1513) by Niccolo Machiavelli:

> It must be considered that there is nothing more difficult to carry out, nor more doubtful of success, nor more dangerous to handle, than to initiate a new order of things. For the reformer has enemies in all those who profit by the old order, and only lukewarm defenders in all those who would profit by the new order, this lukewarmness arising partly from fear of their adversaries, who have the laws in their favour; and partly from the incredulity of mankind, who do not truly believe in anything new until they have had actual experience of it. Thus it arises that on every opportunity for attacking the reformer, his opponents do so with the zeal of partisans, the others only defend him half heartedly, so that between them he runs great dangers.

February 1985
San Jose, California

Robert I. King

Part One

General Theory

CHAPTER 1

Historical Background

Robert I. King
Lockheed Missiles and
Space Company, Inc.

Dr. Carl J. Salomon's Research

The concept of high-speed machining was conceived by Dr. Carl J. Salomon during a series of experiments from 1924 to 1931. This is documented in German patent number 523594 dated 27 April 1931. The patent was based on a series of curves of cutting speeds plotted against generated cutting temperatures. These experiments were performed on nonferrous metals such as aluminum, copper, and bronze. Salomon obtained speeds up to 54,200 surface feet per minute (sfm) [16,500 surface meters per minute (smm)] using helical milling cutters on aluminum. His contention was that the cutting temperature reached a peak at a given cutting speed; however, as the cutting speed was further increased, the temperature decreased. Figure 1.1 is a simplistic presentation of this concept.

As the cutting speed is increased from 0 in the normal mode, V_1, the temperature will increase in a direct relationship until a peak value T_{cr} is achieved. The cutting speed at T_{cr} is commonly called the critical cutting speed, V_{cr}. If the cutting speed is further increased, it was predicted that the cutting temperature would decline. On either side of V_{cr}, Salomon suggested that there was an unworkable regime in which cutters were not able to stand the severe process temperatures and forces. The shape of the curve was thought to be dependent on the exact nature of the base material being cut. When the cutting speed was sufficiently increased, the resulting temperatures were reduced to those of the normal cutting temperatures, and the materials and cutters would once again permit practical cutting procedures. The same cutting

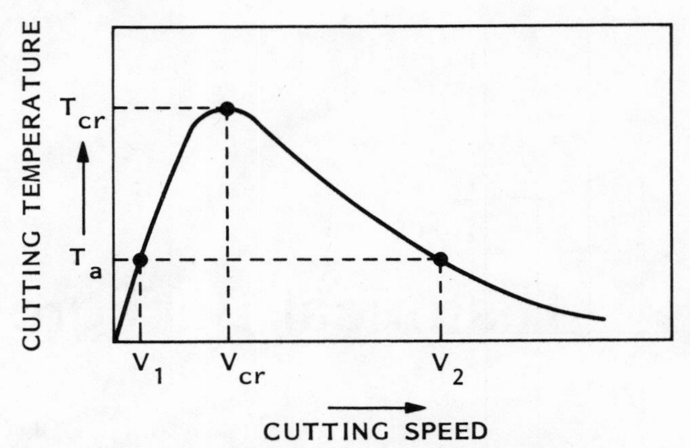

Fig. 1.1 Idealized cutting speed–cutting temperature plot.

temperature, T_a, found in the normal speed range, V_1, could possibly be reproduced in the high-speed range V_2.

There have been many versions of Salomon's curves used as reference by current researchers. Since much of the supporting data were lost during World War II and none of the participants in the research are alive to comment, the exact shape of the curves is left for speculation. However, the most commonly used version cited in most of the recent technical papers is shown in Fig. 1.2. The solid lines represent data that Salomon was supposed to have been developing from experimental results. The broken lines indicate estimated results that were extrapolated by Salomon but not actually verified in the laboratory.

Possibly the most significant contribution of this work is the concept of the bounded nonworkable regime. Cast red brass, for example, is unmachinable with high-speed steel cutters at surface speeds between 200 and 1100 sfm (61 and 335 smm) and with Stellite cutters at speeds from 300 to 985 sfm (91 to 297 smm), according to Salomon's results. A theoretical rationale was not offered, and the assumptions and design of the experiments are not known. However, Dr. Salomon is generally given credit as the father of "high-speed" machining, i.e., machining at speeds higher than those considered in the Taylor equations. His results are now mainly of historical interest since current research is developing more definitive data using more sophisticated test techniques.

Four Periods of Development

The evolution of high-speed machining can be conveniently divided into four periods of time, starting with the original work of Salomon, each subsequent period having a higher level of research activity than the previous period,

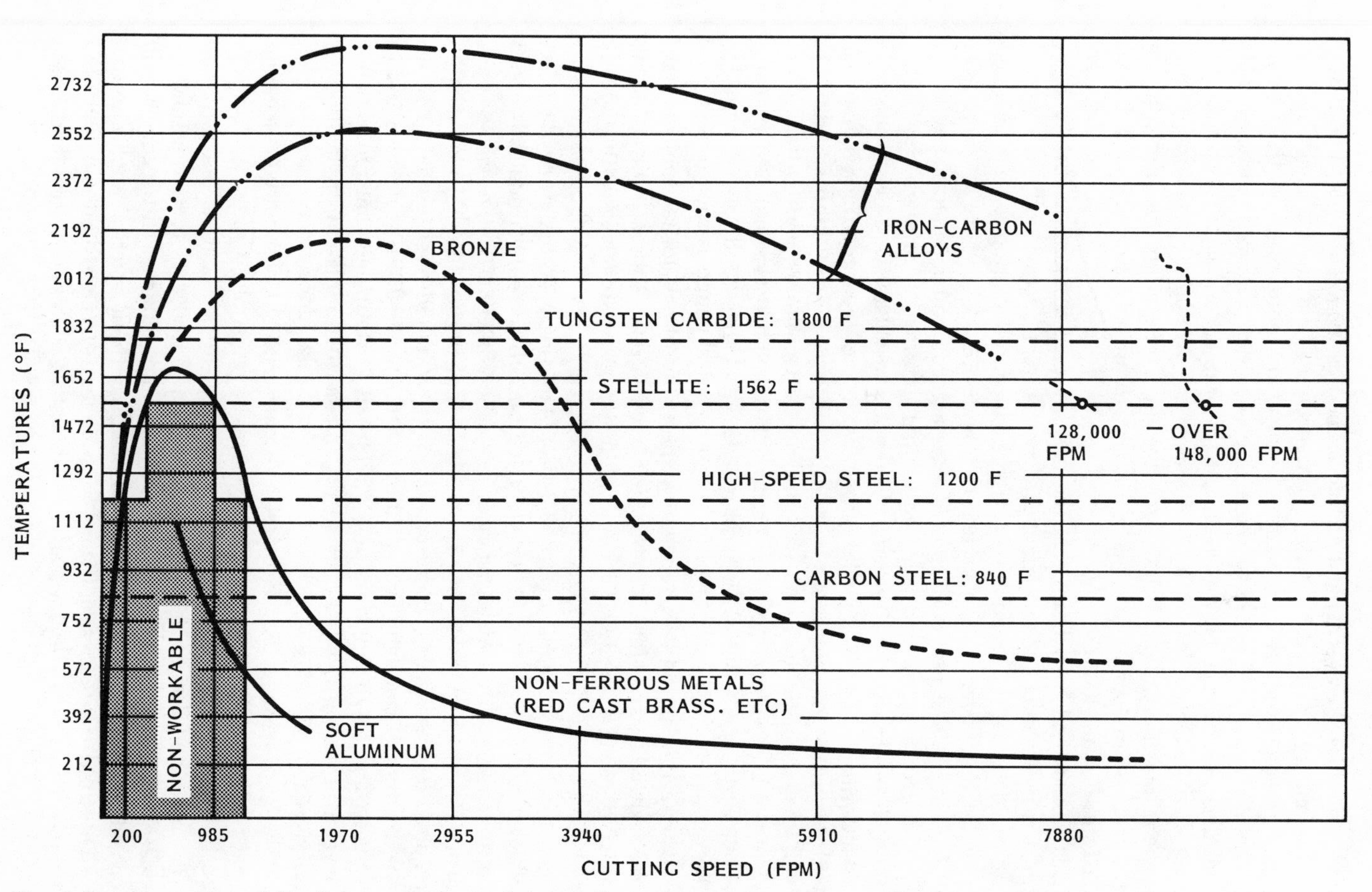

Fig. 1.2 Illustration of Dr. Salomon's theory for the effect of cutting speed on cutting temperature.

and a significant event separating each. The first period includes the 1930s, 1940s, and 1950s. It started with Salomon's research and ended with the first major research project in high-speed machining, sponsored by the United States Air Force, 1958–1961. These three decades produced neither significant data nor interest. However, an article noting a curious phenomenon was published in December 1949, in the *American Machinist.* William Coomly, General Superintendent of the Rice Barton Corporation, reported that at 180 feet (55 m) per minute his planer was drawing 92 hp (69 kW), but at 310 sfm (92 m), it drew only 55 hp (41 kW) in planing semisteel castings. This would tend to indicate that the conventional assumption that "cutting force is independent of cutting speed" is incorrect. Occasionally, this period produced similar statements that a slight reduction in cutting force is obtained when cutting speeds are relatively very high.

During the first period of time, R. L. Vaughn, Lockheed, became aware of Dr. Salomon's patent and acquired limited information through the United States Consul in Berlin, Germany. Salomon's test data were then translated from German to English by Max Kronenberg. The Vaughn group was familiar with the many references on oil well tube perforation at high speeds. It is well known to the art of oil well drilling that explosive perforation cutters were used to perforate oil well casings. These concepts stimulated thinking about very high-speed cuttings of metals at Lockheed and set the stage for the second period.

The second period was initiated by the award of a major research contract to Lockheed by the United States Air Force Materials Laboratory to evaluate the response of a selected number of high-strength materials to cutting speeds as high as 500,000 sfm (152,400 smm). The contract (AF 33[600]36232) was executed from 24 February 1958 to 24 August 1959 under the direction of R. L. Vaughn. The primary objective was to increase producibility and improve the quality and efficiency of fabrication of aircraft and missile components. The experimental procedures used cannons and sleds to obtain the required cutting speeds. Unfortunately, the results did not indicate a reasonable method of translating the data obtained into production methods. Additionally, an analytical model of the high-rate cutting phenomenon was not developed during this contract so that further extrapolation/prediction was not available to industry.

The 1960s was a decade of relative quiescence in high-speed-machining research. Notable exceptions included Arndt in Australia; Coldwell, Quackenbush, and Recht in the United States; Fenton in the United Kingdom; and Okushima and Tanaka in Japan. The second period saw a considerable increase in activity over the first period, although industry and academia still considered the subject an intellectual puzzle without practical application.

The third period was initiated by a series of studies contracted by the United States Navy in the early 1970s with Lockheed Missiles and Space

Company, Inc. The objective of these studies was to determine the feasibility of using high-speed machining in a production mode, initially with aluminum alloys and later with nickel-aluminum-bronze. A team under the direction of R. I. King demonstrated that it was economically feasible to introduce high-speed-machining procedures into the production environment to realize major improvements in productivity. This resulted in a significant increase in overall interest, and a very active period of both experimental and applied research resulted as can be noted in a review of the literature. It soon became clear that some centralized direction was required to resolve some of the conflicts arising out of the results of a myriad of small research studies.

In 1979, the United States Air Force awarded a contract to the General Electric Company (F 33615-79-C-5119) to provide a science base for faster metal removal through high-speed machining and laser-assisted machining. Approximately one year later the Air Force awarded a second contract to General Electric (F 33615-80-C-5057) to evaluate the production implications of the first contract. Both contracts have been supported by a consortium of industrial firms and universities within the United States, integrated by a General Electric team headed by D. G. Flom. These contracts initiate the fourth period, and their results will form the foundation for most of the subsequent chapters.

Industrial Studies

In 1958, Vaughn[148] studied a series of variables involved in traditional machining that become very important in high-speed machining. According to Vaughn:

> The rate at which metal can be machined is affected by: (1) size and type of machine; (2) power available; (3) cutting tool used; (4) material to be cut; and (5) speed, feed, and depth of cut. These five general variables can be broken down further into: (a) rigidity of machine, cutter, and work piece; (b) variations in speed from the slowest to the fastest, depending on machine used; (c) variations in feed and depth of cut from light to heavy and whether cut dry or with the aid of lubricant and/or coolant; (d) type and material of cutting tool; (e) variations in cutter shape and geometry; (f) type and physical characteristics of work material; and (g) specific requirements of desired cutting speed, tool life, surface finish, horsepower required, residual stress, and heat effects.

Recent advances in the development of computer control systems have provided the capability of accurately manipulating high-performance automatic production machines. Progress in the field of bearing design, alternative spindle power sources, automatic tool changing, tool retention devices, and cutter materials have also made contributions toward proving Vaughn's experiments.

The results of Vaughn's research indicate a theoretical limit to improving productivity with several commonly used industrial materials; improvements range from 50 to 1000 times conventional machine performances. "Practical limits will vary with each material, but it is highly unlikely that a productivity improvement of greater than ten is possible with today's technology."[148]

Some of the major conclusions of the Vaughn studies were:

1. High-strength materials can be machined at ultra-high speeds, up to 240,000 sfm.
2. High-speed steel cutters can machine high-strength materials to cutting speeds of 240,000 sfm (73,000 smm).
3. Improved producibility will result through the practical application of ultra-high-speed machining.
4. Brittle (or cleavage) failure of the machined alloy did not occur during high-velocity tests, as suggested by theories of critical impact velocities.
5. Extrapolation of conventional machining curves does not predict results or anticipated phenomena observed during high-speed testing [above 30,000 sfm (9144 smm)].
6. Ultra-high-speed machining improves the surface finish of the cut surfaces over similar surface cut at conventional speeds.
7. Metal removal rates 240 times as great as conventional practice were obtained.

Conclusions on tool wear were:

1. Minimum tool wear resulted at a cutting velocity of 120,000 sfm (36,500 smm) using heat-treated material. Tool wear curves fit a second-degree equation.
2. Cutting velocity had little effect on tool wear when machining annealed material. Tool wear curves fit a second-degree equation.
3. When depths of cut were greater than 0.011 in. (0.28 mm), tool wear was about the same whether machining annealed or heat-treated material.
4. Tool wear per unit metal removed dropped 75% and the ratio of tool wear to metal removal rate dropped 95% by increasing cutting speeds from 30,000 to 150,000 sfm (9144 to 45,700 smm).
5. Less tool wear resulted from machining heat-treated material at 120,000 sfm (36,500 smm) than at 40 sfm (12 smm).
6. 7075-T6 aluminum alloy can be machined at 120,600 sfm (36,560 smm) with no measurable wear.
7. Tool wear patterns (face or flank) differed depending on heat-treated condition of the machined alloy.
8. Total tool wear on long, heavy cuts [48-in. × 150-in. (122-cm × 381-cm) deep] was no greater than tool wear on 6-in. (15.2 cm) long × 0.045-in. (1.1 mm) deep cuts.

Conclusions on cutting forces were:

1. Horizontal and vertical forces are in the range that can be controlled in practical applications, although higher than conventional.
2. In most cases vertical cutting forces were higher than horizontal cutting forces, the inverse of conventional practice.
3. The increase in peak cutting forces over conventional cutting forces was only 33–70% as against a predicted increase of 500%. Use of average force readings would reduce this figure still further.
4. The ratio of cutting force to area of cut was less for deep cuts [0.001 in. (0.025 mm)] than for light cuts [0.011 in. (0.279 mm)].
5. The increase on the shear angle observed leads to the conclusion that lower shearing forces are required to remove metal during ultra-high-speed machining.[151]

King and McDonald[74] set out to confirm and advance Vaughn's research. In doing so, these researchers worked with the Bryant Grinder Corporation and others to equip an existing Sundstrand, five-axis, Model OM-3, Omnimil with a 20,000 revolutions per minute (rpm), 20-horsepower (15-kW) spindle. Based on results obtained with that modified machine, King and McDonald reported a good potential for the high-speed machining of aluminum. They state that, without speculating on the reasons why Vaughn's conclusions were not pursued, it would appear that machine tool technology has progressed to the point that it is now possible to capitalize on the results of his research.

Certainly one of the more profitable types of parts for high-speed milling would be that requiring large quantities of material removal (hog-out parts). It is obvious that if the process were capable of high rates of metal removal, then the more there is to remove, the greater the advantage the process has over conventional techniques. Implicit in this capability for high rates of metal removal is the need to develop trade-off studies for parts normally requiring castings or forgings for economical production. Since the time required for metal removal can be dramatically reduced, it is quite likely that the cost of developing castings/forgings and the associated elaborate holding fixtures will no longer economically offset the machining time of the higher rate cutting methods. King and McDonald state that, "In parts that usually require castings and forgings an additional advantage can be derived from high-speed milling. Often it requires several months (or even years) to develop castings and forgings and to produce sufficient quantities to support production requirements. It is not without precedent that castings and forgings become the pacing milestones in producing a new product. High-speed milling, when economically justified, can significantly reduce the time span between engineering design and production by eliminating castings and forgings".

Thin-walled parts qualify as additional candidates for high-speed milling. Conventional milling techniques with excessive cutter loads cause heat buildup that permanently distorts thin-walled parts. The maintenance of close tolerances under these circumstances may prove to be impossible. This can force the design engineer to develop a less than optimum part configuration to stay within the limitations of the production process or to find an alternative that is more expensive. High-speed milling of thin-walled parts is possible because the cutter loads are greatly reduced or eliminated. Internal stresses are caused chiefly by heat buildup and are dissipated, which allow thin-walled parts of 0.013-in. (0.33 mm) to be obtained.

A Vought[95] (LTV Company) study provided data concerning the effect of cutting speed and cutter geometry on cutting temperature when turning 2014-T652 aluminum. These data indicate that the alloys tested showed that cutting temperature curves tend to peak near the melting point of the aluminum alloys. "It is expected that the cutting temperature curve would plateau at the melting point of the alloy. It would not be expected that the cutting temperature of the material would exceed its melting temperature." When the curves are examined, it would appear "that most of the cutting edge temperature rise occurs at low rather than high cutting speeds and that this is one feature or characteristic which does much toward opening the door to high-speed machining."

An extrapolated, theoretical cutting speed indicates that, where the cutting edge reaches a temperature of 1200°F (650°C) (the melting point of the aluminum alloy), and when the cutting speed is 19,600 sfm (5975 smm), it may be postulated that there is a unique cutting speed at which cutting-edge temperature ceases to rise. If this is the case, several interesting possibilities arise. For one, it would be theoretically possible in this example to continue turning at infinitely higher speeds than 19,600 sfm (5975 smm), because there should be no further rise in cutting-edge temperature and, therefore, no further reduction in cutter-life wear rate. The Vought study, within the cutting speeds tested, did not show a reversing trend as presented on the Salomon curve, but the plateauing effect indicates that infinitely high cutting speeds may be feasible for the machining of aluminum alloys.

Coldwell and Quackenbush,[34] in a study of high-speed milling of titanium, indicated that a cutting chip can be too thin. In their study the chips seemed to be hotter when the theoretical chip thickness was relatively thin. In the Vought (LTV Company) study, the test cutting feeds or chip thicknesses converge near the melting point of aluminum. "Near a cutting temperature of 1200°F (650°C) and a cutting speed of 19,000 ft/min, the theoretical point in this instance at which cutting temperatures cease to rise . . . it would be theoretically possible at cutting speeds beyond 19,000 ft/min to continue turning aluminum at infinitely higher feed rates than 0.0075 inch revolution." It is possible that if such metal-removal rates as these are achieved with no further rise in cutter temperature, there would be no further reduction in cutter life at these speeds and feeds. If these theoretical parameters can be achieved, productivity could progress a quantum jump.

Williamson[166] describes a system of manufacture for small machined components in a paper dated September 1967, "System 24—A New Concept of Manufacture." The system offers advantages far beyond anything presently available. This high-speed machining system, made by Molins Limited, enables the cost of components machined on it to be reduced by a factor between five and ten in comparison with conventional production methods. It also brings large reductions in space, personnel, and surprising capital investment for a given level of manufacture. The most fundamental limitation to increasing the speed of component manufacture is the metal-removal rate. In order to take advantage of high-speed machining rates, the Molins machine uses a turbine-driven cutter spindle that provides high stiffness and long life at speeds up to 30,000 rpm. Mounted on the rear end of the spindle is a 20-bucket Pelton wheel driven by a tangential jet of oil.

King[67] writes that Lockheed Missiles and Space Company, Inc. contracted with the Naval Regional Procurement Office at Philadelphia to provide certain technical services for purposes of evaluating the application of ultra-high-speed machining techniques to the milling of large nickel-aluminum-bronze (Ni-Al-Brz) cast propellers. Significant improvements in metal-removal rates were realized and verified repeatedly even though the tooling used had not been optimized. Cutting rates of four to five times those of the control test rates established during first tests were obtained. Optimization of cutter designs and process parameters should further improve the performance. King[68] stated that Ni-Al-Brz propeller production can be improved by 100% through use of high-speed milling techniques. "Use of solid carbide end mills was extended to include inserted cutters of the type and style used in the actual production processes. The data base has now been increased to include cutter insert performance and additional process efficiency information."

The metal-removal rate in the test operations indicated a significant increase over those rates presently being experienced at shipyards during propeller machining operations. If the controlled cutting of the test series is used as a basis for comparison, the average metal-removal rates during the second test series were three to four times higher, and the maximum rate was over four times higher [i.e., 31.5 in.^{3}/min (516 cm^3/min) vs 6.30 in.^{3}/min (103 cm^3/min)]. It is recognized that maximum rates are indicative only of limiting conditions; however, an average rate improvement from 2.76 to 9.0 in.^{3}/min (45 to 147 cm^3/min) has significant production implications. It should be noted that the cutters used for the first (control) tests were designed for the cutting speeds used, whereas the cutters used in the second tests were not designed for higher cutting rates.

It can be expected that when the cutter designs are finalized for the higher cutting speeds, the possible metal-removal rates should be much greater. It is reasonable to expect eventual test rates of five to ten times the current production metal-removal rates, although other considerations such as cutter wear, spindle horsepower, and cutter force will undoubtedly have a modifying effect on final production process recommendations.

King's evaluation of the various cutter parameters did give some positive guidance for continued cutter development. The most significant results were:

1. Compax-diamond coated inserts gave significantly better overall results than either the carbide VC-2 or VC-7. Of the latter two, the VC-2 performance was somewhat better than VC-7.
2. Considering cutting efficiency only, the square insert gave better results than the triangular insert, which in turn gave better results than the round inserts. However, the round insert displayed the most consistent, stable test result. It is possible that in a more normal (as compared to accelerated) cutting environment, the round insert may turn out to be more desirable.
3. Considering cutter insert wear characteristics only, the triangular shape is marginally better than the square shape, which is appreciably better than the round shape. The best wear-resistant material was the compax-diamond with both grades of carbide somewhat less desirable.

The result is that the triangular compax-diamond came out a resounding first. This program of the high-speed machining of nickel-aluminum-bronze has indicated there are major economic savings to be made in production processes that utilize available cutter materials and high spindle speeds.

A metal removal study summarized by Fenn[35] states that,

> The successful commercial implementation of high-speed machining promises dramatic increases in productivity and significant concomitant reductions in manufacturing costs. One of the more profitable types of operation is where the parts being machined require the removal of large quantities of material. The results demonstrate that cutting speed increases of 500 percent yield a reproducible 300 percent increase in metal removal efficiency regardless of the depth of cut. Machines performing a complex contouring sequence may not reflect all of these savings; however, as cutting speeds increase production, costs decrease.

Okushima et al.[104] adds the following:

> The most important advantages of super-high-speed machining are to improve the productivity of the machine operation and to produce an excellent surface finish and dimensional accuracy. In addition, it is expected by this machining operation to machine those alloys which in missiles and high-performance aircraft are required to withstand high temperatures. On the other hand, super-high-speed machining has disadvantages: rapid wear of cutting tools and vibration of machine tools.

Development of Metal Separation Theory

Researchers in the intervening years have found that machining temperature phenomena may be asymptotic as cutting speeds increase, but the cutting forces tend to reduce at the accelerated speeds. To study these phenomena, a host of researchers have been working on investigations centered in the theories of chip formation, metal fracture, catastrophic slip, and adiabatic shear, as well as chip formation in various materials. To gain some understanding of the fundamentals of the cutting mechanism of chip formation, von Turkovich et al.[157] provide some illumination through their discussion of "New Observations on the Mechanism of Chip Formation When Machining Titanium Alloys." In this presentation, they refer to Professor Shaw's[129] suggestion that "Chip segmentation was due to the onset of instability in the cutting process, resulting from competing thermal softening and strain hardening mechanisms in the primary shear zone." Shaw[130] also stated that "The formation of concentrated shear (also called adiabatic shear) bands was due to the poor thermal properties (low thermal conductivity and low specific specific heat) of these alloys and consequent concentration of thermal energy in those bands."

This instability concept is given further explanation by Recht,[123] who states

> . . . that ductile metals strain harden as they slowly deform plastically. When deformation rate is low, the process is essentially isothermal. Initially, plastic shear strain is restricted to a few weak shear zones within the material. Strain hardening strengthens the weak material in these zones and the burden of strain is distributed through the material. However, if strain hardening did not occur, deformation would remain localized. During rapid plastic deformation, the heat generated locally establishes temperature gradients; maximum temperatures exist at points of maximum heat generation. If the rate of decrease in strength, resulting from the local increase in temperature, equals or exceeds the rate of increase in strength due to the effects of strain hardening, the material will continue to deform locally. This unstable process leads to the catastrophic condition known as "adiabatic" slip.

Recht states further that

> . . . apparently catastrophic shear develops in mild steel at machining velocities near 1300 sfm where apparent shear strength begins to drop. Near the critical velocity, slip planes are close together, spreading farther apart as velocity increases. Fully developed catastrophic slip occurs in the machining geometry when the distance between zones reaches a geometrical maximum. . . . Catastrophic slip reduces the strength in the manner indicated. When the zones are close together the average apparent stress approaches the uniform deformation value. Reduction in strength progresses further when the zones are more widely spaced and thus the average apparent stress is lower.

Recht expresses the following concern:

> Of extreme interest is the fact that a second shear-strength plateau is reached at ultra-high machining speeds. After the average shear stress appears to be independent of strain rate . . . strain rates within the catastrophic shear zone are as high as 10^8 in./in./s. The implication is that, when catastrophic shear is well established, dynamic shear strength tends to be insensitive to strain rate.

Recht concludes that

> . . . heat generated during the dynamic deformation of materials creates temperatures and temperature gradients which can exercise significant influences upon observed dynamic behavior. Certain materials possess thermophysical properties which render them particularly susceptible to catastrophic shear. Catastrophic shear occurs when local temperature gradients offset the strengthening effects of strain hardening; the burden of plastic strain must be supported by a very small portion of the material. . . . The thinness of adiabatic slip zones is helpful for heat-transfer considerations.

Vaughn,[151] when writing about adiabatic shear, states:

> As the velocity of machining increases, an adiabatic condition is approached in which thermal energy is restricted to the preferred slip zone (shear plane—composed of many atomic planes). Because of weakening in the preferred slip zone, additional slip occurs, terminating in complete shear.

Professor von Turkovich[160] states that, according to Arndt,

> The shear zone has a volume consisting originally of solid material. When the speed increases, minute molten regions are generated in the shear zone resulting in a reduction of solid material volume. The shear zone is resolved into planes of infinitesimal thickness which are parallel to the shear plane.

In his conclusions, von Turkovich states,[160] "The solution of the energy balance equation in the shear layer indicates that an adiabatic process may take place leading to a rather sharp (thin) transition layer."

Rogers[125] writes that

> All of the adiabatic shearing phenomena are based on two facts: Approximately 90 percent of the work of plastic deformation is converted to heat, and, the flow stress of most metals is quite sensitive to temperature, decreasing as the temperature increases. That localized temperature increases and strain concentration plays a major part in high-speed deformation of metals was recognized by Zener and Hollomon in 1944. The phenomenon is most clearly identified in most steels, in which heating above the transformation temperature causes the transformation of ferrite to austenite. On rapid cooling the austenite retransforms to a product that etches with difficulty and appears as a white band

against the dark background of the remainder of the etched steel. These materials thus retain evidence of adiabaticity of the deformation, while in most metals the evidence is significantly less definite.

Rogers[125] cautions,

> . . . the use of the term "adiabatic deformation" is obviously an oversimplification in the sense that some heat always transfers out of any deforming region. Moreover, to categorize one situation as "adiabatic" and another not is in many instances equivalent to labeling shades of gray as black or white. It will nevertheless be used herein for convenience, recognizing these limitations.

Rogers continues:

> . . . all the characteristics described above for highly localized deformation through adiabatic shearing instabilities are also found in high-speed machining operations. In fact, because of the tool-workpiece geometry in orthogonal cutting, there are two zones of intense shear. The primary shear zone results from the initial deformation of the layer to be removed. The geometry of the metal chip flow is such that, after shearing, the metal in the forming chip is forced to flow normally to the surface of the solid being machined. This flow is also parallel to the rake face of the tool against which the chip is forced with considerable pressure. Frictional heating causes the chip to seize on the rake face. The differential flow between the dead metal at the rake face and the rapidly moving chip takes place over the narrow zone of secondary shear.

Wright and Bagchi[171] showed that even in low carbon iron, temperatures as high as 1832°F (1000°C) could be generated in this zone.

More significant from the adiabatic shearing standpoint are the studies by Laenaire and Backofen[88] of discontinuous chip formation, and the study by Recht[187] of catastrophic shear zone formation in chips during machining. In the former study, the separation of discontinuous chips was shown to take place through one of three possible mechanisms. The light etching bands can be seen in the type 2 and 3 schemes for chip formation observed in this 18% nickel steel as the cutting speed increases. From the horizontal cutting force–time curves, the authors were able to correlate discontinuous drop in force with the formation of the bands. Furthermore, the author's analysis shows that the only way to obtain the temperature sufficient to produce these bands is to release the elastic energy stored in the machine into the locally adiabatically deforming zone.

According to Arndt,[10] in a model of the cutting process at very high speed, ". . . the shear zone has a volume consisting originally of solid material. When the speed increases, minute molten regions are generated in the shear zone. . . . The shear zone is resolved into planes of infinitesimal thickness which are parallel to the shear plane."

Almost 20 years ago, Tanaka et al.140 stated the following: (1) The cutting mechanism is affected mainly by temperature and its distribution in shear and tool-chip contact zones. (2) With increase in cutting speeds, shear angles increase, for the decreasing of tool face friction force is achieved by temperature rise on the tool-chip interface up to the softening or melting point of the tool or work materials. (3) With cutting speed, the metal strength of shear plane does not drop, but rather increases in appearance. "The cause may be in the following: In shear zone, temperature gradient in direction normal to the shear plane becomes steeper with increasing cutting speed; it is then expected at higher speeds that the strength of the next coming shear plane may not be affected by heat due to present shear processes. As a result, apart from the strain rate effects, a positive tendency in V_c relationship exists. (4) Except for the size effect, specific cutting force and shear energy remain constant irrespective of cutting speed. Those are proper to each of the work materials."

Nachtman,[102] addressing the Machining with High Speeds and Feed Clinic of the Society of Manufacturing Engineers, concisely summarizes the problem and possible solutions, as well as the benefits of high-speed cutting. He suggests that

> Almost all of the energy of metal cutting is used for plastic deforming of the chips and in overcoming friction between chip, tool, and workpiece. All of these actions result in heat, and that fraction dissipated in the tool causes softening and reduced resistance to abrasion, thus limiting tool life. Consequently, the crux of the problem of super-high-speed machining is to concentrate the heat in the chips and minimize the quantity of heat transfer to the tool.
>
> It is postulated that the heat generated per unit volume of metal removed should be lower at very high speeds because ductility of metal decreases with increasing strain rates. In addition, if the heat accompanying deformation is confined to the chips, chip temperatures will be raised and their strength reduced. Unfortunately, the available data do not permit quantitative estimates of importance of temperature and strain rate on the energies required for deforming chips at excessively high speeds.
>
> In addition to the points discussed, high-speed metal cutting seems feasible on other grounds. It might reasonably be expected that the tool would not reach a high temperature, even when the heat liberated in the chip is higher per unit of time, because the time available for conduction is limited by the high speed.

References

The following is a listing of publications considered by the author as the most significant generic high-speed machining documents prior to 1980. There are numerous others which are either replications or were considered not particularly germane to the subject. Additional references both prior to and after 1980 which address specific topics within this text can be found at the end of the respective chapters. This listing is intended to be an aid and guide for the engineer who wishes to gain a greater understanding of the overall subject.

1. Akiyama, T., et al., "Study of the Orthogonal Cutting Mechanism by Controlled Shear Angle Experiments," Asahikawa Tech. Coll., Japan, Men Fac. Eng. Hokkaido University, Sapporo, Japan, Vol. 14, No. 1, March 1975, pp. 13–20.

2. *American Machinist*, "High Speed Machining," Special Report 710, March 1979, pp. 115–130.

3. Anonymous, "Penetration of Metal Plate by Projectiles," and "Adiabatic Shear Bands in Steel," *D.S.L. Annual Report*, 1968–69, Maribyrnong, Victoria, Australia also see earlier reports.

4. Armarego, E. J. A. and R. H. Brown, *The Machining of Metals*, Englewood Cliffs, N.J., Prentice-Hall, 1964.

5. Arndt, G., "Ballistically Induced Ultra-High-Speed Machining," Ph.D. Thesis, Monash University, Melbourne, Australia, 1971.

6. Arndt, G., "Further Considerations of Ultra-High-Speed Machining," *Proc. 6th Intern. Conf. High Energy Rate Fabrication*, Essen, W. Germany, September 1977 (in English).

7. Arndt, G., "On the Study of Metal-Cutting and Deformation at Ultra-High Speeds," *Proc. Harold Armstrong Conf. Prod. Sci. Industry*, Vol. 30, Monash University, 1971.

8. Arndt, G., "Temperature Distributions in Orthogonal Machining," M. Eng. Sc. Thesis, University of Melbourne, Australia, 1964.

9. Arndt, G., "The Development of Higher Machining Speeds: Part I, Historical," *Prod. Engnr.*, 1970, Vol. 49, p. 470; "Part I, Present Practice and Theory," *Prod Engnr.*, 1970, Vol. 49, p. 517.

10. Arndt, G., "Ultra-High-Speed Machining: A Review and an Analysis of Cutting Forces," *Proc. Inst. Mech. Engrs.*, London, Vol. 187, 1973, pp. 625–634.

11. Arndt, G., "Ultra-High-Speed Machining: Notes on Metal Cutting at Speeds up to 7,300 ft/sec," *15th Proc. Intern. Mach. Tool Des. and Res. Conf.*, September 1974, pp. 203–208.

12. Arndt, G., "On the Temperature Distribution in Orthogonal Machining," *Int. J. MTDR*, 1967, Vol. 7, pp. 39–53.

13. Arndt, G., "Design and Preliminary Results From an Experimental Machine Tool Cutting Metals at up to 8,000 ft/sec," *Proc. 13th Intern. MTDR Conf.*, New York, Macmillan, 1972, pp. 217–223.

14. Arndt, G., "Ultra-High-Speed Machining," *Ann CIRP*, Vol. 21, No. 1, Monash University, Victoria, Australia, 1972, pp. 3–6.

15. Arndt, G. and J. T. McHenry, "A Computerized Internal Ballistic Analysis of Conventional Gun System With Muzzle Velocities of up to 8,000 ft/sec," *Explosivst.*, 1970, Vol. 18, p. 253.

16. Backofen, W. A., *Deformation Processing*, Reading, Mass. Addison-Wesley, 1972, p. 271.

17. Bailey, J. A. and D. G. Bhanvadia, "Correlation of Flow Stress With Strain Rate and Temperature During Machining," *J. Eng. Materials and Technol.*, Vol. 95H, 1973, p. 94.

18. Barash, M. M., "Mechanical State of the Sublayer of a Surface Generated by Chip-Removal Process—Cutting With a Sharp Tool," *Trans. ASME,* Paper No. 75-WA/Prod-9 for Meeting 30 Nov–5 Dec 1975.

19. Bhattacharyya, B. and R. R. Scrutton, "Plastic Flow at the Chip-Tool Interface During Hot Machining," *ASME,* Paper No. 70-WA/Prod-1, 1970.

20. Bitans, K., "Investigation of the Stress-Strain Characteristics of Materials at High Rates of Strain," Ph.D. Thesis, University of Melbourne, 1970.

21. Black, J. T., "Flow Stress Model in Metal Cutting," *Trans. ASME,* Paper No. 78-WA/Prod-27, to appear in *J. Eng. Ind.*

22. Black, J. T., "On the Fundamental Mechanism of Large Strain Plastic Deformation," *Trans. ASME J. Eng. for Ind.*, 1971.

23. Black, P. H., *Theory of Metal Cutting,* New York, McGraw-Hill, 1961.

24. Boothroyd, G., "Fundamentals of Metal Machining and Machine Tools," *Scripta Met.*, Washington, D.C., 1975.

25. Boston, O. W., *Bibliography on the Cutting of Metals, 1864–1943,* New York, ASME, 1954.

26. Bredendick, F., "Die Massenkrafte beim Zerspanvorgang," *Werkstart Betr.*, 1959, Vol. 92, No. 10, p. 739.

27. Brunton, J. H., et al., *Metals for the Space Age—Plansee Proceedings,* 1964, 1965, Vol. 137, ed. F. Benesovsky, Berlin, Springer-Verlag.

28. Campbell, J. D. and S. G. Ferguson, "The Temperature and Strain-Rate Dependence of the Shear Strength of Mild Steel," *Phil. Mag.*, Vol. 21, No. 169, 1970, p. 63.

29. Carrington, W. E. and M. I. V. Gayler, "The Use of Flat-ended Projectiles for Determining Dynamic Yield Stress, III: Changes in Microstructure Caused by Deformation Under Impact at High Striking Velocities," *Proc. R. Soc.*, 1948, Vol. 194A, p. 323.

30. Chakrabartz, J., "New Slipline Field Solution for the Orthogonal Machining of Metals," *Proc. Intern. Conf. Prob. Eng.*, 27th New Delhi, India, 27 August–4 September 1977, Inst. of Eng., India, Calcutta, 1977, Vol. 1, 8 pp.

31. Chao, B. T. and K. J. Trigger, "An Analytical Evaluation of Metal Cutting Temperatures," *Trans. Am. Soc. Mech. Engrs.*, 1951, Vol. 73, p. 57.

32. Choudry, A. and P. J. Gielissi, "Dynamic Elastic Model of Ceramic Removal," Univ. of R.I., Kingston, Inst. *Symp. on Spec. Top in Ceramic Proc.*, Alfred Univ., N.Y., Plenum, 27–29 Aug 1973 (Mater. Res., No. 7: Surfaces and Interfaces of Glass and Ceramic), New York, 1974, pp. 149–166.

33. Coldwell, L., J. Mazur, and J. Angell, "Diagnostic Sensing in Machining Operations," Dept. of Mech. Eng., The University of Michigan, Report No. 320357, January 1975.

34. Coldwell, L. and L. Quackenbush, "A Study of High-Speed Milling," Office of Research Administration, The University of Michigan, Report No. 05038-1 and 2-F, Parts I and II, December 1962.

35. Committee on Science Base for Materials Processing, National Materials Advisory Board, *Science Base for Materials Processing—Selected Topics,* Contract No. MDA-903-78-C-0038 Final Report, National Materials Advisory Board, National Academy of Sciences, 2101 Constitution Ave., NW, Washington, D.C., 20418, November 1979.

36. Craig, J. V. and T. A. C. Stock, "Microstructural Damage Adjacent to Bullet Holes in 70-30 Brass," *J. Aust. Inst. Metals*, 1970, Vol. 15, No. 1, p. 1.

37. Crerar, J., "Metal Cutting Bibliography 1943–1956," *Am. Soc. of Tool and Mfg. Engrs.*, Detroit, Mich., Boston, O. W., Vol. 1, 1960.

38. Datsko, J., "Material Properties and Manufacturing Processes," New York, Wiley, 1966.

39. Dean, R. N., "The Effect of Temperature on Young's Modules," No. NRL-M-2886, Naval Research Lab., November 1946.

40. DeGroat, G. H. and A. Ashburn, "Ultra-High-Speed Machining," *American Machinist/Metalworking Manufacturing*, Special Report No. 484, 22 February 1960, pp. 111–126.

41. Dhosi, J. M., et al., "High Temperature Deformation and Fracture Behavior of Metals Under High Strain Rate Conditions," NOW-63-0502C, New England Materials Lab., Inc., NEMLAB-0502-FR, October 1964, p. 20.

42. Dorn, R. S., F. Hauser, and J. E. Dorn, "Theoretical Prediction of Strain Distribution Under Impact Loading," *Proc., Joint Meeting Univ. of NM and ASTM*, Sep 1962.

43. Earles, S. W. E. and M. J. Kadhim, "Friction and Wear of Unlubricated Steel Surfaces at Speeds up to 655 ft/sec," *Proc. Instn. Mech. Engrs.*, 1965–1966, Vol. 180, Part 1, pp. 531–548.

44. Ernst, H. and M. E. Merchant, "Chip Formation: Friction and Finish," Cincinatti Milling Co., OH, 1941.

45. Ernst, H., "Machining of Metals," *Trans. AMEE*, 1938, p. 24.

46. Ernst, H., "Physics of Metal Cutting," *Am. Soc. Met.*, 1938.

47. Fenton, R. B. and P. L. Oxley, "Mechanics of Orthogonal Machining: Predicting Chip Geometry and Cutting Forces from Work-Material Properties and Cutting Conditions," *Proc. Inst. Mech. Engrs.*, London, Vol. 184, Part 1, 1969–1970. p. 417.

48. Fenton, R. G. and W. L. Cleghorn, "Mechanics of Machining: Strain Rate in the Primary Zone," *Proc. 3rd NAMRC*, Pittsburgh, 1975, Carnegie-Mellon University, Pittsburg, Pa., p. 661.

49. Fenton, R. G. and P. L. B. Oxley, "Predicting Cutting Forces at Super-High Cutting Speeds From Work Material Properties and Cutting Conditions," *Proc. 8th MTDR. Conf.*, Oxford, Pergamon Press, 1967, pp. 247–258.

50. Findley, W. M. and R. M. Reed, "The Influence of Extreme Speeds and Rake Angles in Metal Cutting," *Trans. ASME*, Vol. 85, No. 2, 1963, pp. 49–67.

51. Flom, D. G., et al., "Advanced Machining Research Program (AMRP)," General Electric Co., Schenectady, NY, Air Force Systems Command, Air Force Wright Aeronautical Laboratories/MLTM, Wright-Patterson Air Force Base, Dayton, Ohio, February 1980, p. 13.

52. Gane, N., "Chip Fracture During Metal Machining," CSIRO, Melbourne, Australia, *Mech. Eng. Trans. Inst. Eng. Aust.*, ME 3, 1978, pp. 5–8.

53. Gane, N., "Chip Fracture During the Machining of Brass," *Fracture at Work* (Proc. Conf.), Melbourne, Australia, 1979.

54. Gilbert, W. W., "Economics of Machining," Machining—Theory and Practice, *Trans. Am. Soc. Met.*, 1950, pp. 465–485.

55. Gilman, J. J., "Dislocation Dynamics and the Response of Materials to Impact," *Appl. Mech. Rev.*, 1968, Vol. 21, No. 8, p. 767.

56. Glass, C. M., G. M. Moss, and S. K. Golaski, "Response of Metals to High-Velocity Deformation," eds. P. G. Shewman and V. F. Zackey, New York, Interscience Publishers, 1961.

57. Groat, H. G. and A. Ashburn, eds., "Ultra-High-Speed Machining," *Am. Mach.*, Vol. 104, 1960, p. 111.

58. Holloman, J. H. and J. D. Lubahn, "Flow of Metals at Elevated Temperatures," *Gen. Electric Rev.*, 1947, Vol. 50, No. 2, p. 28, No. 4, p. 44.

59. Iwata, K. and K. Veda, "Significance of Dynamic Crack Behavior in Chip Formation," Kobe University, Japan, *Ann. CIRP*, Vol. 25, No. 1, 1976, pp. 65–70.

60. Kahles, J. F., M. Field, and S. M. Harvey, "High-Speed Machining—Possibilities and Needs," *Ann. CIRP*, Vol. 27, No. 2, 1978.

61. Kauzmanm, W., "Flow of Solid Metals From the Standpoint of Chemical Rate Theory," *Trans. AIME*, 1941, Vol. 143, p. 57.

62. Kececioglu, D., "Shear Strain Rate in Metal Cutting and Its Effect on Shear Flow Stress," *Trans. ASME*, Vol. 80, 1958, p. 158.

63. Kellock, B., "High-Speed Machining of Alloy Road Wheels," *Mach. Prod. Eng.*, N3253, Vol. 126, April 1975, pp. 330–337.

64. Kronenberg, M., "A New Approach to Some Relationships in the Theory of Metal Cutting," ASTME Paper No. 86, 1958.

65. Kienzle, O., "Die Best immung von Kraften und Leistungion und spanenden Werkzeugen und Werkzeugmaschinen," VDI 94, No. 11/12, 1952.

66. King, R. I., "High-Speed Production Milling of Non-Ferrous Materials," *Proc. 4th NAMRC Conf.*, May 1976, pp. 334–338.

67. King, R. I., "Phase IIA Summary Technical Report of the Feasibility Study for High-Speed Machining of Ships Propellers," Contract No. 00140-79-C-0326, Lockheed Missiles and Space Company, Inc., Sunnyvale, CA, Nov 1979.

68. King, R. I., "Results of Some Recent Research of the High-Speed Machining of Nickel Aluminum Bronze," *1980 Intern. Conf. on Tooling Applications and Materials for the 80's*, Purdue University, Soc. Carbide and Tool Eng., June 1980, pp. 181–197.

69. King, R. I., "The Economics of Ultra-High-Speed Machining," *Tool Prod.*, January 1978, Vol. 43, No. 10, pp. 92–95.

70. King, R. I., "The Economics of Ultra-High-Speed Machining," *Proc. 41st Westinghouse Tool Forum*, June 1977.

71. King, R. I., "Ultra-High-Speed Machining Offers Benefits," *Man. Tech. J.*, Vol. 2, No. 4, pp. 22–27.

72. King, R. I., "Ultra-High-Speed Machining of Nonferrous Metals," *Proc. CIRP Conf.*, Vol. 2, August 1977, p. 10.

73. King, R. I., "Update of Ultra-High-Speed Machining," *Proc. 42nd Westinghouse Tool Forum*, June 1978.

74. King, R. I. and J. McDonald, "Product Design Implications of New High-Speed Milling Techniques," *Trans. ASME*, November 1976, pp. 1170–1175.

75. Komanduri, R., "The Mechanics of Chip Segmentation," Ph.D. Thesis, Monash University, Melbourne, Australia, 1972.

76. Komanduri, R. and R. H. Brown, "The Formation of Microcracks in Machining a Low Carbon Steel," *Metals and Materials*, December 1972, p. 531.

77. Koontz, J. and W. Mitchell, "Ultra-High-Speed Machining," *Am. Mach.*, June 1977, pp. 135–139.

78. Krabacher, E. J. and M. E. Merchant, "Basic Factors in Hot-Machining of Metals," *Trans. ASME*, Vol. 73, 1951, pp. 761–769.

79. Krabacher, E. J. and M. E. Merchant, "Zweiter Bericht uber die Vevielfachung heute ubliches Schnittgeschwindigkeiten," *Werkstattstechnik*, Vol. 51, No. 133, 1961.

80. Kumar, S. and V. Chandra, "A New Force System in High-Speed Machining," *Proc. Ann. CIRP*, New Delhi, Vol. 1, 1977, Calcutta, India, Institution of Engineers.

81. Kumar, S. and C. Mishra, "Hydrodynamic Action and Tool Wear at Tool–Chip Interface During High-Speed Machining," *J. Inst. Engrs.*, India, Mech. Eng. Div., Vol. 54, May 1974, pp. 191–197.

82. Kuznetsov, V. D., G. D. Polosatkin, and M. P. Kalashnikova, "Investigation of the Cutting Process at Ultra-High Speeds," *Fiz. Metallov Metallovedenie*, Vol. 10, September 1960. (Translation by Air Information Div., Report, pp. 60–109.)

83. Kuznetsov, V. D., G. D. Polosatkin, and M. P. Kalashnikova, "The Study of Cutting Processes at Very-High Speeds," *Fiz. Metallov Metallovedenie*, 1960, Vol. 10, No. 3, pp. 425–434; Transl. pp. 107–116.

84. Kuznetsov, V. D., "Super-High-Speed Cutting of Metals," *Iron Age*, 1945, Vol. 155, pp. 66–69.

85. Larsen, R. J. and J. V. Barks, "Machines Keep Finding New Ways to Cut It," *Iron Age*, 23 April 1979, p. 108.

86. Larsen, R. J., et al., "The Shape of Things to Come," *Metalcutting*, Vol. 222, No. 47, 17 December 1979, pp. 64–67 and 70–74.

87. Lee, E. H., "Wave Propagation in Anelastic Materials," *Deformation and Flow of Solids Colloquium*, Madrid, Spain, Proceedings, 1955, p. 129.

88. Laenaire, J. C. and W. A. Backofen, *Metall. Trans.* 1972, Vol. 3, No. 4, pp.77–82.

89. Lira, F. and E. G. Thomsen, "Metal Cutting as a Property Test," *Trans. Am. Mech. Engrs., Engng. Ind.*, 1967, Vol. 89, No. 3, p. 489.

90. Anonymous, "Longer Tool Life, Higher Removal Rates Offered by Polycrystalline Diamond Tooling," *Cutting Tool Eng.*, Vol. 26, No. 3–4, March–April 1974, pp. 8–9.

91. MacGregor, C. W. and J. C. Fisher, "A Velocity-Modified Temperature for the Plastic Flow of Metals," *Trans. ASME, J. Appl. Mech.*, 1946, p. 68.

92. Malvern, L. E. "The Propagation of Longitudinal Waves of Plastic Deformation in a Bar of Material Exhibiting a Strain-Rate Effect," 1951, Vol. 18, *Trans. ASME*, Vol. 73, *J. Appl. Mech.*, 1951, p. 203.

93. Manion, S. A. and T. A. C. Stock, "The Measurement of Strain in Adiabatic Shear Bands," *J. Aust. Inst. Metals*, 1969, Vol. 14, No. 3, p. 190.

94. McGee, F. J., "An Assessment of High-Speed Machining," Vought Corp. (Pamphlet) Society of Mfg. Engineers, SME Tech., Paper No. MR 78-648, p. 22.

95. McGee, F. J., "Final Technical Report for Manufacturing Methods for High-Speed Machining of Aluminum," Vought Corp. LTV Co., Tech. Reg. No. 6089,

Mfg. Methods and Tech. Branch (DRDMI-EAT), U.S. Army Missile Research and Development Command, Redstone Arsenal, Al., 1 February 1978.

96. McGee, F. J., P. Albrecht, and H. N. McCalla, "Development of Cutter Geometry Based on Material Properties," Technical Report No. AFML-TR-68-350, Air Force Systems Command, December 1968.

97. McLellan, D. L. and T. W. Eichenberger, "Strain Rate Effects on the Compressive Behavior of Pure Aluminum," High Speed Testing, Vol. VI, "The Rheology of Solids," *J. Appl. Polym. Sci.* 1969, Vol. 11, Interscience, New York, Wiley, pp. 185–204.

98. Merchant, M. E., "Basic Mechanics of the Metal-Cutting Process," *J. Appl. Mech.*, September 1944, pp. A-168–175.

99. *Metal Cutting Bibliography 1943–1956*, 1960 NY, ASME.

100. Monarch Machine Tool Co., *Speeds and Feeds for Better Turning Results*, 1957, pp. 37–38.

101. Moss, G. and C. M. Glass, "Some Microscopic Observations of Cracks Developed in Metal by Very Intense Stress Waves," Ballistic Research Labs., BRL 1312, Aberdeen Proving Grounds, N.D., April 1960, p. 21.

102. Nachtman, E., "High-Speed Machining," Machining With High Speeds and Feeds Clinic, 13–15 September 1977, Hotel Sonesta, Hartford, Conn.

103. National Twist Drill and Tool Co., "Some Effects of Flute Helix and Rake Angles on Milling Cutter Performance," *Metal Cuttings*, Vol. II, No. 3, July 1963.

104. Okushima, K., et al., "A Fundamental Study of Super-High-Speed Machining," *Bull. Japan Soc. Mech. Engrs.*, Vol. 8, No. 32, 1965, p. 702.

105. Okushima, K., K. Hitomi, and S. Sto, "A Study of Super-High-Speed Machining," *Ann. CIRP*, Vol. 13, MS. 72/8, 1966, Great Britain, pp. 399–410.

106. Olberts, D. R., "A Study of the Effects of Tool Flank on Tool Chip Interface Temperatures," *Trans. ASME*, Vol. 81, May 1959, pp. 152–158.

107. "100th Anniversary Issue of *American Machinist*," Section J., November 1977.

108. Osborn, C. J. and N. E. Ryan "Metallography of Powder-powered Fastenings in Mild Steel," *J. Aust. Inst. Metals*, 1957, Vol. 11, No. 2, pp. 48–53.

109. Osina, V., "Metallumformung mil hohen Geschwindigkeiten and Energien," (Abstract), *Industrie-Anzeiger*, 1967, Vol. 89, No. 30, pp. 588–589; also *Metal Treatment*, 1966, Vol. 33, No. 248, 193; Original *Strojierenstvi*, 1964, Vol. 14, No. 9, p. 667.

110. Ostafiev, V. A. and S. Kobayashi, "Stress, Strain and Strain Rate in Metal Cutting," Proc. 7th MTDR Conf., 1966, p. 479, Oxford, Pergamon Press.

111. Oxley, P. L. B., "Applied Research in Plastic Deformation," *Aust. Mach., Prod. Engng.*, 1968, Vol. 21, No. 233, pp. 12–18.

112. Oxley, P. L. B., "Rate of Strain Effect in Metal Cutting," *Trans. Am. Soc. Mech. Engrs., J. Engng. Ind.*, 1963, Vol. 85, pp. 335–338.

113. Oxley, P. L. B. and M. G. Stevenson, "Measuring Stress/Strain Properties at Very High Strain Rates Using a Machining Test," *J. Inst. Metals*, 1967, Vol. 95, pp. 308–313.

114. Pauls, F. E., "Introduction to the Rotary Cutter," SME Tech. Paper Ser. MR for ESTEC Conf., Los Angeles, 14–17 March 1977, Book 1, Paper No. MR 77–208, Cutters Unlimited Company, p. 12.

115. Polosatkin, G. D., "Rezaniye metallov so skorostyami ot 100-700 m/sek," *Sibirsk Phys. Tekhn. Inst., Sci. Rep.*, 1948.

116. Polosatkin, G. D., et al., "Cutting and Grinding at Ultra-High Speeds," *Isvestiya Uchebuykh Zavedenii, Fiz.*, 1967, Vol. 5, No. 10, pp. 93–101.

117. Polsatkin, G. D. and A. N. Khludkova, "Determination of the Specific Work Expended in Plastic Deformation in the Ultra-High-Speed Cutting of Metals," *Isvestiya Vyschikh Uchebuykh Zavedenii, Fiz.*, 1967, Vol. 6, No. 7, pp. 81–83.

118. Polsatkin, G. D. and V. B. Titov, "On the Question of Tool Wear at Cutting Speeds of 200–600 m/sec," *Isvestiya Vyschikh Uchebuykh Zavedenii, Fiz.*, 1967, Vol. 5, No. 3, pp. 124–125.

119. Prevey, P. S. and M. Field, "Variation in Surface Stress Due to Metal Removal," *Ann. CIRP*, Vol. 24, No. 1., 1975, pp. 497–501.

120. Pugh, H. D., "Mechanics of the Cutting Process," *Proc. IME Conf. Tech. Eng. Mfr.*, Inst. Mech. (London), 1958, p. 237.

121. Read, H. E., et al., "Dislocation Dynamics and the Formulation of Constitutive Equations for Rate-Dependent Plastic Flow in Metals," Final Report DASH-01-70-C-0055, December 1970.

122. Recht, R. F., "The Feasibility of Ultra-High-Speed Machining," M.S. Thesis, University of Denver, 1960.

123. Recht, R. F., "Catastrophic Thermoplastic Shear," *Trans. ASME, J. Appl. Mech.*, Vol. 31, June 1964, pp. 189–193.

124. Robichand, R. L. R., "Introduction to Rotary Cutters," *Can. Mach. Metalwork*, Vol. 89, No. 11, pp. 36–37.

125. Rogers, H. C., "Adiabatic Plastic Deformation," *Ann. Revised Material Sci.*, Vol. 9, 1979, p. 283.

126. Salomon, C., "Process for the Machining of Metals or Similar Acting Materials When Being Worked by Cutting Tools," German Patent No. 523594, April 1931.

127. Salomonovich, E. D., "Investigation of Temperature in the Case of Super-High Cutting Rates," *Vestn. Mashinostroeniya*, 1954, Vol. 34, No. 9, pp. 45–46.

128. Schmidt, A. O., "Ultra-High-Speed Machining . . . Panacea or Pipedream," *The Tool Engr.*, November 1958, pp. 105–109.

129. Shaw, M. C., et al., "Machining Titanium," Cambridge, Mass., M.I.T., 1954.

130. Shaw, M. C., "Machinability," 151 Special Report 9/4, The Iron and Steel Institute, London, U.K., p. 1, 1967.

131. Shaw, M. C., *Metal Cutting, Principles*, 3rd ed., Cambridge, Mass., M.I.T. Press, Vol. 1, 1957.

132. Shewman, P. G. and V. F. Zackay, "Response of Metals to High Velocity Deformation," American Inst. of Mining, *Metallurgical Conference on Response to Metals to High Velocity Deformation*, Estes Park, 1960 and New York, Interscience, 1961.

133. Siekmann, H. J., "High-Speed Cutting With Ceramic Tools," *The Tool Engr.*, April 1958, pp. 85–88.

134. Stanford, J. E., "New Tools From New Materials," *Iron Age*, 13 April 1977, p. 209.

135. Stevenson, M. G. and P. L. B. Oxley, "An Experimental Investigation of the Influence of Speed and Scale on the Strain-Rate in a Zone of Intense Plastic Deformation," *Proc. Inst. Mech. Engrs.*, 1969–1970, Vol. 184, Pt. 1, pp. 561–576.

136. Stevenson, M. G. and P. L. B. Oxley, "An Experimental Investigation of the

Influence of Strain Rate and Temperature in the Flow Stress Properties of a Low Carbon Steel Using a Machining Test," *Proc. Inst. Mech. Engrs.*, 1970–1971, London, Vol. 185(55/71), p. 741.

137. Stevenson, M. G., and P. L. B. Oxley, "High Temperature Stress-Strain Properties of Low Carbon Steel From Hot Machining Tests," *Proc. Inst. Mech. Engrs.*, London, Vol. 187(23/73), p. 263.

138. Stock, T. A. C. and K. R. I. Thompson, "Penetration of Aluminum Alloys by Projectiles," *Metall. Trans.*, 1970, Vol. 1, pp. 219–224.

139. Takeyama, H., T. Murai, and H. Usui, "Speed Effect on Metal Machining," *J. Mech. Lab., Japan*, 1955, No. 2, pp. 59–61.

140. Tanaka, Y., H. Tsuwa, and M. Kitano, "Cutting Mechanism in Ultra-High-Speed Machining," ASME Paper No. 67-Prod-14, 1967.

141. Tanaka, Y. and M. Kitano, "Metal Cutting with Extremely High Speeds," *Technology Reports of the Osaka University*, Vol. 16, No. 670, 1965, pp. 305–314.

142. Tangerman, E. J., "Are We Slow-pokes at Machining?" *American Machinist*, Vol. 93, December 29, 1949, p. 55–57.

143. Taylor, F. W., "On the Art of Cutting Metals," *Trans. ASME*, Vol. 28, No. 1907, p. 31.

144. Thompson, K. R. L., T. A. C. Stock, and B. H. McConnoll, "Evidence for Melting of a Low-Melting-Point Alloy During High-Velocity Impact," *J. Aust. Inst. Metals*, 1970, Vol. 15, No. 4, p. 26.

145. Tlusty, J., "Analysis of the State of Research in Cutting Dynamics" (McMaster Univ., Hamilton, Ontario), *Ann. CIRP*, Vol. 27, No. 2, 1978, pp. 583–589.

146. Trent, E. M., *Metal Cutting*, London, Butterworth, 1977.

147. Vaughn, R. L., "A Theoretical Approach to the Solution of Machining Problems," ASTME Technical Paper No. 164, September 1958.

148. Vaughn, R. L., "Ultra-High-Speed Machining," Interim Engineering Report No. 1, Air Force Contract AF 33 600 36232, Production Engineering Department, Lockheed Aircraft Corp., Burbank, Calif., May 1958.

149. Vaughn, R. L., "Ultra-High-Speed Machining," Interim Engineering Report No. 4, Air Force Contract AF 33 600 36232, Production Engineering Department, Lockheed Aircraft Corp., Burbank, Calif., February 1959.

150. Vaughn, R. L., "Ultra-High-Speed Machining (Feasibility Study)," Final Technical Engineering Report (Phase 1), AMC Tech. Report 60-7-635 (1), AMC Aeronautical Systems Center, USAF, Wright-Patterson AFB, June 1960.

151. Vaughn, R. L., "Ultra-High-Speed Machining," *Am. Mach.*, Vol. 104, No. 4, 22 February 1960, pp. 111–126.

152. Vaughn, R. L., "Ultra-High-Speed Machining—Solution to Producibility Problems," *The Tool Engr.*, October 1958, pp. 71–76.

153. Vaughn, R. L. and R. R. Krueck, "Recent Developments in Ultra-High-Speed Machining," ASTME Technical Paper No. 255, April 1960.

154. Vaughn, R. L., L. J. Quackenbush, and L. V. Coldwell, "Shock Waves and Vibration in High-Speed Milling," ASTME Technical Paper No. 62-WA-282, November 1962.

155. Venkatesh, V. C., "High-Speed Machining of Cast Iron and Steel," *Ann. CIRP*, Vol. 15, 1967, pp. 387–391.

156. Venkatesh, V. C. and P. K. Philip, "Investigation of Deformation in High-Speed Orthogonal Machining of a Plain Carbon Steel Using a Ballistic Set," Indian Inst. of Technology, Madras, *Ann. CIRP*, Vol. 21, No. 1, 1972, pp. 9–14.

157. Von Turkovich, B. F., "Dislocation Theory of Shear Stress and Strain Rate in Metal Cutting," *Proc. 8th MTDR. Conf.*, 1967, pp. 531–542, Oxford, Pergamon Press.

158. Von Turkovich, B. F., "High Velocity Machining," *Proceedings of the Ken Trigger Symposium on Metal Cutting and Manufacturing*, University of Illinois at Urbana, Champaign–Urbana, Ill. April 1977.

159. Von Turkovich, B. F., "Influence of Very-High Cutting Speed on Chip Formation Mechanics," *VII North American Metalworking Research Conference Proceedings*, 13–16 May 1979, SME Tech., 1979.

160. Von Turkovich, B. F., "On a Class of Thermomechanical Processes During Rapid Plastic Deformation" (with special reference to metal cutting), *Ann. CIRP*, Vol. 21, No. 1, 1972, p. 15.

161. Von Turkovich, B. F., "Deformation Mechanics During Adiabatic Shear," *Proc. 2nd North American Metalworking Research Conference*, Madison, Wis., 1974, Supplement, p. 682.

162. Vukelja, D., "Thermodynamics of Cutting," Monografije Iama, 1970, Vol. 2, Institute of Metal Cutting, Belgrade.

163. Weill, R., "Contribution à l'Étude des Outils Céramiques," *Microtechnic*, Vol. 12, No. 2, 1958.

164. Williams, J. E., "Observations of Deformation Occurring in the Cutting Process Related to a Three-Zone Model of Machining," *Proc. 3rd North American Metalworking Research Conf.*, Pittsburgh, 1975.

165. Williams, J. E., "Some Aspects of a Three-Zone Model of Machining," *Ware*, Vol. 48, No. 1, May 1978, pp. 55–77.

166. Williamson, D. T. N., "System 24–A New Concept of Manufacture," *8th Intern Machine Tool Des. Res. Conf.*, University of Manchester, September 1967.

167. Williamson, D. T. N., "New Wave in Manufacturing," *Am. Mach.* (Spec. Report, No. 607), 1967, Vol. 3, No. 19, 11 September, pp. 143–154.

168. Wingrove, A. L., "A Note on the Structure of Adiabatic Shear Bands in Steel," *J. Aust. Inst. Metals*, 1971, Vol. 16, No. 1, pp. 67–70.

169. Wolak, J. and I. Finnie, "A Comparison of Stress-Strain Behavior in Cutting and High Strain-Rate Compression Tests," *Proc. 8th MTDR Conf.*, Oxford, Pergamon Press, pp. 233–246, 1967.

170. Wright, P. K., "Metallurgical Effects at High Strain Rates in the Secondary Shear Zone of the Machining Operation," Dept. of Sci. and Ind. Res., Auckland, New Zealand; *Metal Eff. at High Strain Rates*, Conf., Proc., Paper and Discussion, Albuquerque, NM, New York, Plenum Press (Metall., Soc. of AIME Proc.), 1973, 5–8 February 1973, pp. 547–558.

171. Wright, P. K. and A. Bagchi, "Tool Wear Processes in High-Speed Machining," *Proc. 8th NAVRC*, Rolla, M., 1980, p. 277.

172. Wright, P. K., J. G. Horne, and D. Tabor, "Boundary Conditions at the Chip Tool Interface in Machining: Comparison Between Seizure and Sliding Friction, Wear," June 1979, Vol. 54, No. 2, pp. 371–390 (in English).

173. Wright, P. K. and K. C. Mannie, "Strain Hardening, Strain Rate and Temperature Effects in Metal Cutting," *Fracture at Work (Proc. Conf.)*, Melbourne, Australia, 12–14 February 1979, University of Melbourne, Melbourne, Australia, 1979.

174. Wright, P. K. and S. P. McCormick, "Effect of Rake Trace Design on Cutting Tool Temperature Distribution," *J. Eng. for Ind. Trans. ASME*, Paper No. 79-WA/Prod-3, 1979.

175. Wright, P. K. and J. L. Robinson, "Material Behavior in Deformation Zones of Machining," *J. Metals Tech.*, Vol. 4, 1977, p. 240.

176. Yamada, K. and N. Nakayama, "Ultra-High-Speed Machining and Its Technique," *Sci. Mach.*, Vol. 13, 1961, pp. 779–782 and 911–916.

177. Yamamoto, A. and S. Nakamura "Study on Chip Formation in Ultra-High-Speed Cutting: Cutting of Photoelastic Materials at Speeds Over Elastic Distortion Wave Propagation Velocity," *Bull. Japan Soc. Precision Engr.*, September 1971, Vol. 5, No. 3, pp. 67–72 (in English).

178. Zener, C., "The Micro-Mechanism of Fracture, Fracturing of Metals," *Trans. Am. Soc. Metals*, 1948, pp. 3–31.

179. Zener, C. and J. H. Holloman, "Plastic Flow and Rupture of Metals," *Trans. Am. Soc. Metals*, 1944, Vol. 33, pp. 163–235.

The following publications provide additional overview of the theory of metal cutting.

180. Barrow, G., *Tribology*, February 1972, p. 22.

181. Bhattacharyya, A. and I. Ham, *Trans. ASME, J. Eng. for Ind.*, August 1969, p. 790.

182. Cook, N. H., *Trans. ASME, J. Eng. Ind.*, November 1973, p. 931.

183. Ham, I. and N. Narutaki, *ASME, J. Eng. for Ind.*, November 1973, p. 951.

184. Hoggatt, C. R. and R. F. Recht, *J. Appl. Phys.*, 1968, Vol. 39, pp. 1856–1862.

185. Hsu, T. C., *Trans. ASME, J. of Eng. for Ind.*, August 1969, p. 652.

186. Lemaire, J. C. and W. A. Backofen, *1972 Metall. Trans.*, Vol. 3, pp. 477–482.

187. Recht, R. F., *1964 Trans. ASME*, Vol. 86 (Ser. E), pp. 189–193.

188. Siekmann, H. J., "The Use of an Ultra-High-Speed, 150 Horsepower Lathe for Machinability Studies," ASTE, Vol. 58, No. 82, May 1958, p. 8.

189. Stock, T. A. C. and K. R. L. Thompson, "Penetration of Aluminum Alloys by Projectiles," *1970 Metall. Trans.*, Vol. 1, pp. 219–224.

190. Stock, T. A. C. and A. L. Wingrove, *J. Mech. Eng. Sci.*, 1971, Vol. 13, pp. 110–115.

191. Wright, P. K. 1973 (see pp. 547–558), M. E. Beckman, S. A. Finnegan. In *Metallurgical Effects at High Strain Rates*, eds. R. W. Rohde, B. M. Butcher, J. R. Holland, C. H. Karnes, pp. 531–543, New York, Plenum, 1973, p. 699.

192. Wright, P. K. and E. M. Trent, *Met. Tech.*, Vol. 1, January 1974, p. 13.

193. Zener, C. and J. H. Holloman, *J. Appl. Phys.*, Vol. 15, 1944, pp. 22–32.

CHAPTER 2

Cutting Theory and Chip Morphology

B. F. von Turkovich
University of Vermont

Introduction

Machining is a mechanical process where excess material from a workpiece can be removed by cutting action to produce a part of specified geometrical shape and surface finish. Machining can be performed on virtually all solid materials even though the term commonly applies to the cutting of metals and alloys, less frequently to plastics and wood, and rarely to other solid substances (rocks, composites, etc.).

One of the characteristics of the process is the smallness of the deformation (cutting) volume wherein the chip is formed in comparison with the dimensions of the workpiece. In this relatively small deformation volume, a strong energy conversion takes place; i.e., the mechanical work supplied by the machine tool power drive is converted almost completely into heat as the chip emerges and the new surface is formed on the workpiece.

The theory of machining is concerned with the various features of the cutting process including the forces, strain and strain rates, temperatures, and wear of tools. Even though the process resides at any given time in only a very small portion of the metal being machined, it is nevertheless fairly complex, and in order to provide an engineering basis for its control and exploitation, substantial simplifications are introduced to render the computation manageable.

The process can be represented, in the simplest terms, as a force traveling on the surface of the workpiece. This force changes in time and space and so do its components, related to a reference frame attached to the workpiece. It is therefore advantageous for the first level analysis to establish how these cutting-force components vary with the speed, the cross-sectional area of the nascent chip, and the tool geometry, expressed initially only by the rake angle. In the second level of analysis the problems of strain and of stress distributions are addressed, including the strain rate and the temperature. At this level the theory of machining makes a connection with the metallurgical theory of metal deformation through the formulation of appropriate constitutive equations. These equations are either theoretical and/or empirical formulas relating the stress, strain, strain rate, temperature in, and the structure and state of a material to each other in a particular range of deformation. A somewhat special subject is the study of tool wear and the estimation of tool life under particular cutting conditions.

Since machining is an important component in the overall manufacturing activity, many additional features of the process have been and are still actively studied. For instance, because of the variability of all cutting-force components as functions of chip load and speed, very severe forced and self-excited vibrations can be observed, which are detrimental to the tool life, workpiece geometry, finish, and, finally, the machine tool itself. Chip control and the measures to reduce the tool wear by design or by the use of cutting fluids are also active areas of research and development. Many actual cutting operations are quite complex owing to the tool shape and because of special requirements regarding the final form of the workpiece such as threads, bores, or sharp corners. Moreover, the operations are separated into rough and finishing kinds, and for each, separate tools may be used. The process is also subdivided into subgroups: milling, turning, drilling, threading, etc.

A subject of great practical importance is the surface integrity. Since all cutting operations involve a large specific power consumption, which means that the local forces are sufficiently large to cause permanent deformation on the new surface as it is being generated, on one hand, and on the other hand, that the temperatures are also high enough to permit metallurgical changes on the new surface, the evaluation of the surface state (i.e., its integrity) is very important for the manufacture of parts subject to fatigue and corrosion. There is also a large activity in the economic analysis of machining operations, where the various costs are taken into account, such as the tool life, machined surface of a part, cost of various tool materials and fluids, and are frequently related to the most economical material-removal rates.

The present chapter is a brief review of all the above subjects. Its primary thrust is to develop the elementary features of the theory and to show how they may be applied in more complex cutting situations.

A Simple Model of Cutting Process

All cutting operations require a three-dimensional coordinate frame and time for the full description. However, a two-dimensional model has been developed over the years which retains almost all essential variables. In this model it is assumed that the cutting action proceeds at a constant cutting speed. Figure 2.1 is the schematic representation of the model.

The following assumptions are made regarding this model:

1. The cutting speed is constant.
2. The width of cut b_1 is much larger than the feed t_1, and both are constant.
3. The tool is perfectly sharp.

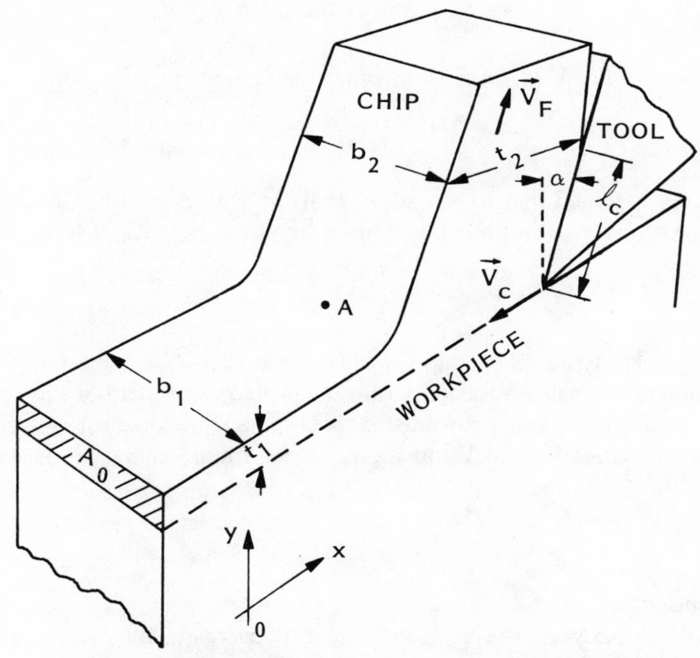

t_1 = FEED (DEPTH OF CUT)

b_1 = WIDTH OF CUT

t_2 = CHIP THICKNESS

b_2 = WIDTH OF CHIP

α = RAKE ANGLE

$\vec{V}_c$ = CUTTING VELOCITY

$\vec{V}_F$ = CHIP VELOCITY

ℓ_c = LENGTH OF CONTACT

Fig. 2.1 Basic dimensions of model.

4. The chip is a continuous ribbon.
5. The cutting velocity vector $\mathbf{V}_c$ is normal to the cutting edge.
6. The workpiece material is a homogeneous, isotropic, incompressible solid.
7. The workpiece is at room temperature.
8. The cutting is performed in air with no liquid coolants.
9. There is no wear on the tool.
10. The tool is a rigid body, and its width, b_t, is larger than b_1.
11. The steady state of cutting has been reached, i.e., there is no effect arising from the incipient cutting regime.

From the general continuity equation, where ρ is the material density and $\mathbf{V}$ is the velocity vector,

$$\frac{\partial \rho}{\partial t} + \operatorname{div}(\rho \mathbf{V}) = 0, \tag{2.1}$$

which reduces to div $\mathbf{V} = 0$ for an incompressible material, it is derived that:

$$V_c t_1 b_1 = V_F t_2 b_2. \tag{2.2}$$

See Fig. 2.1.

If it is now assumed that $b_1 = b_2$, i.e., that the chip does not spread laterally because of deformation due to cutting, then from Eq. (2.2) it follows that

$$\frac{V_F}{V_c} = \frac{t_1}{t_2} = r_c. \tag{2.3}$$

The ratio r_c is known as the "chip thickness ratio."

Since the chip remains attached to the workpiece and increases in length only, any point A in its interior must have a velocity $\mathbf{V}_s$, which is a vector sum of the velocities $\mathbf{V}_c$ and $\mathbf{V}_F$ in the reference frame (O, x, y) indicated in Fig. 2.1.

Thus

$$\mathbf{V}_s = \mathbf{V}_c + \mathbf{V}_F \tag{2.4}$$

which is shown in Fig. 2.2.

The angle ϕ between the vectors $\mathbf{V}_c$ and $\mathbf{V}_s$ is called the "shear angle," and is determined from the expression

$$\tan \phi = \frac{V_F \cos \alpha}{V_c - V_F \sin \alpha} = \frac{r_c \cos \alpha}{1 - r_c \sin \alpha}. \tag{2.5}$$

Thus it is sufficient to measure the chip thickness t_2 and using Eq. (2.3) determine the value of ϕ.

The procedure so far presented is known in the technical literature as the "orthogonal cutting model."

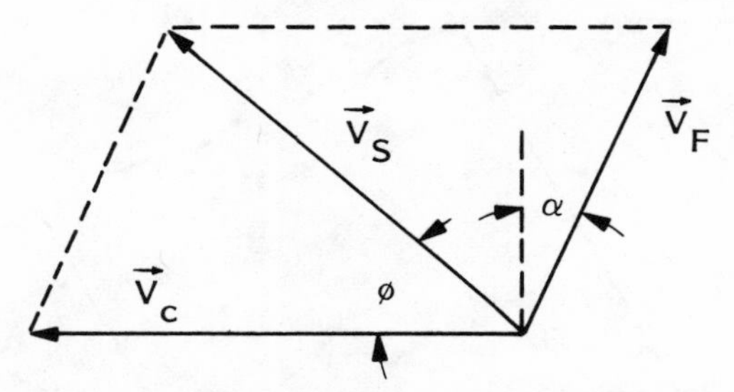

Fig. 2.2 The velocity diagrams (hodograph).

The cutting force **R**, which the tool applies to the chip and through it to the workpiece, is the vector sum of a force normal to the tool face, **N**, and a tangential force, **F**, parallel to the tool face. These forces are distributed along the length of contact l_c in Fig. 2.3. The contact length can be obtained by measurement.

In Fig. 2.3, the direction of the cutting force **R** is shown to pass through the point *C*, which is situated in the line *AB*. This is the line of shortest distance between the tool edge, *A*, and the free surface, *B*.

If it is now assumed that the resultant reaction force of the workpiece passes through *C*, i.e., the cutting force **R** and the reaction **R**′ are collinear and equal but of opposite sign, there is no moment acting on the chip. It is in static equilibrium as a rigid body.

The resolution of the resultant cutting force **R** with the initially perpendicular components, $\mathbf{F}_c$ and $\mathbf{F}_T$, is a convenient way to study the effects of cutting speed, feed, and rake angle on the power consumption. The component $\mathbf{F}_c$, called the "cutting force," is parallel to the cutting velocity vector, so that

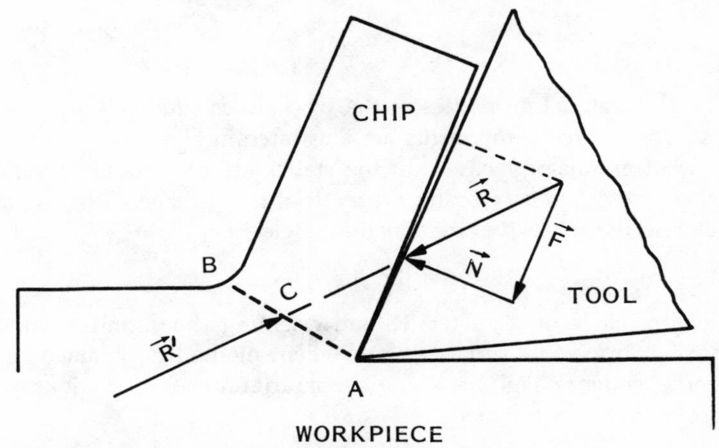

Fig. 2.3 Basic tool forces.

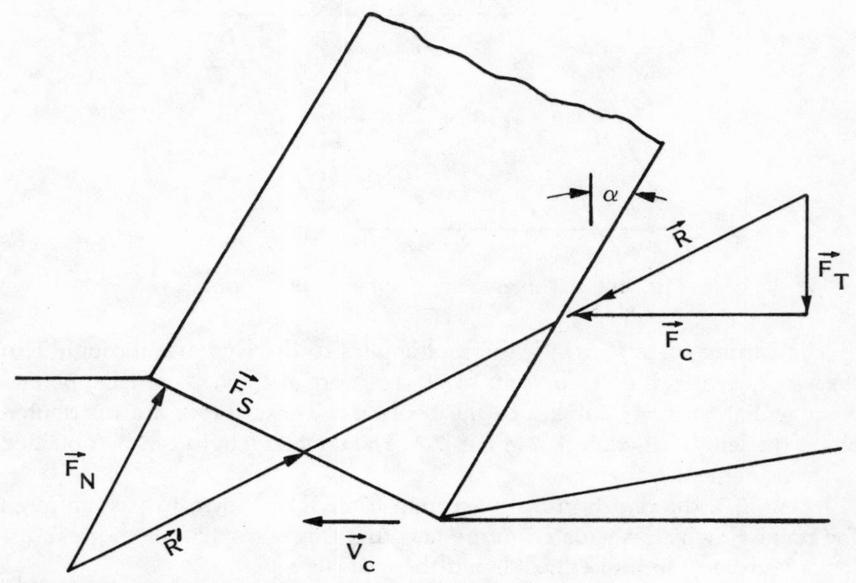

Fig. 2.4 Forces on tool and shear plane.

the product $F_c V_c$ represents the power used in cutting. The component normal to $\mathbf{V}_c$ and thus to $\mathbf{F}_c$ is called the "thrust force" and is frequently taken as an indicator of cutting stiffness. The resultant for $\mathbf{R}'$ in the shear plane has two orthogonal components, the shear force $\mathbf{F}_S$ in the shear plane AB and the component $\mathbf{F}_N$, which is normal to the shear plane. The components $\mathbf{F}_S$ and $\mathbf{F}_N$ can be translated into the average stresses acting in the shear plane, i.e.,

$$\tau = F_S/A_S, \quad \sigma = F_N/A_S, \tag{2.6}$$

which are the material properties of the workpieces and $A_S = b_1 t_1/\sin\phi = A_0/\sin\phi$. These force components are illustrated in Fig. 2.4.

In view of an analogy based on the statics of solid bodies, the ratio of the cutting forces $\mathbf{F}$ and $\mathbf{N}$ acting on the tool face is called "the friction coefficient"; and the angle β is the "friction angle," i.e.,

$$\mu = \tan\beta = F/N. \tag{2.7}$$

By drawing a circle of diameter $|\mathbf{R}|$ and orienting the various components of $\mathbf{R}$ by their respective surface, a convenient method is obtained to show their interdependence, Fig. 2.5. Using simple trigonometric relations, the formulas are obtained as follows:

$$F_c = R\cos(\beta - \alpha), \tag{2.8}$$

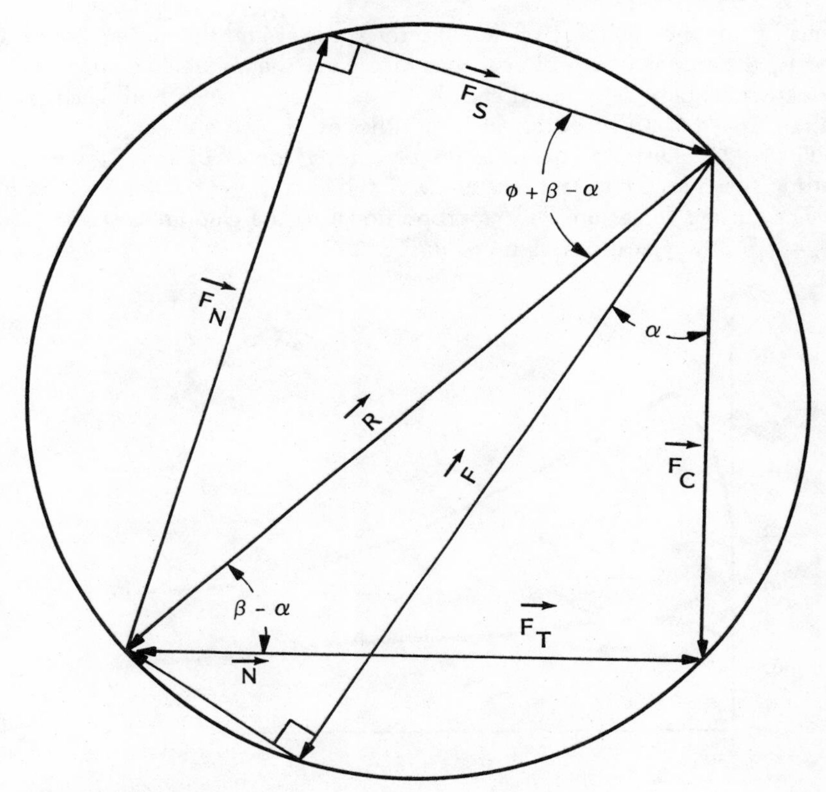

Fig. 2.5 The force diagram.

$$F_T = R \sin(\beta - \alpha), \tag{2.9}$$

$$N = R \cos \beta = F_c \cos \alpha - F_T \sin \alpha, \tag{2.10}$$

$$F = R \sin \beta = F_c \sin \alpha + F_T \cos \alpha, \tag{2.11}$$

$$F_N = R' \sin(\phi + \beta - \alpha) = F_T \cos \phi + F_c \sin \phi, \tag{2.12}$$

$$F_S = R' \cos(\phi + \beta - \alpha) = F_c \cos \phi - F_T \sin \phi, \tag{2.13}$$

$$\mu = \frac{F_T + F_c \tan \alpha}{F_c - F_T \tan \alpha}, \tag{2.14}$$

$$\tau = \frac{F_c \cos \phi - F_T \sin \phi}{t_1 b_1} \sin \phi, \tag{2.15}$$

$$\sigma = \frac{F_c \sin \phi + F_T \cos \phi}{t_1 b_1} \sin \phi. \tag{2.16}$$

Thus by the measurement of the chip thickness t_2 and the cutting forces $\mathbf{F}_c$ and $\mathbf{F}_T$ (by means of a tool post dynamometer), the variables ϕ, μ, τ, σ can be experimentally determined and shown how they change with changes of cutting speed V_c, the rake angle α, and the feed t_1.

Figure 2.6 illustrates the changes of cutting forces F_c and F_T with the cutting speed for a constant rake angle, α, the feed t_1 and the depth of cut b_1.

The cutting forces are always proportional to the chip cross-section area $A_0 = t_1 b_1$. A typical case is given in Fig. 2.7.

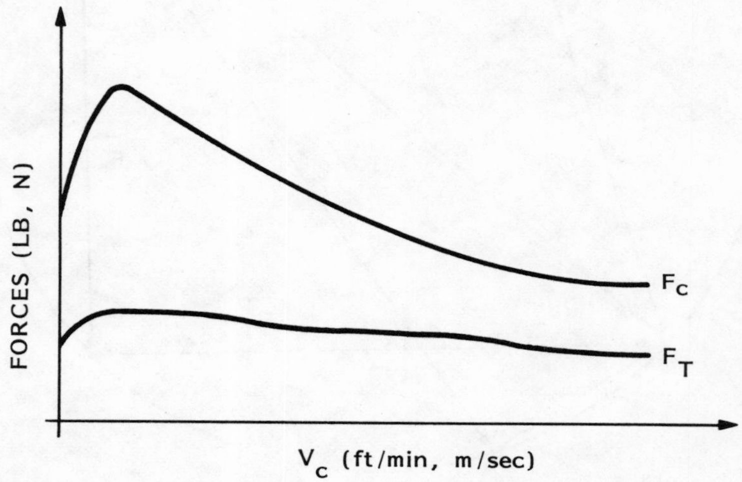

Fig. 2.6 The cutting forces versus cutting speed.

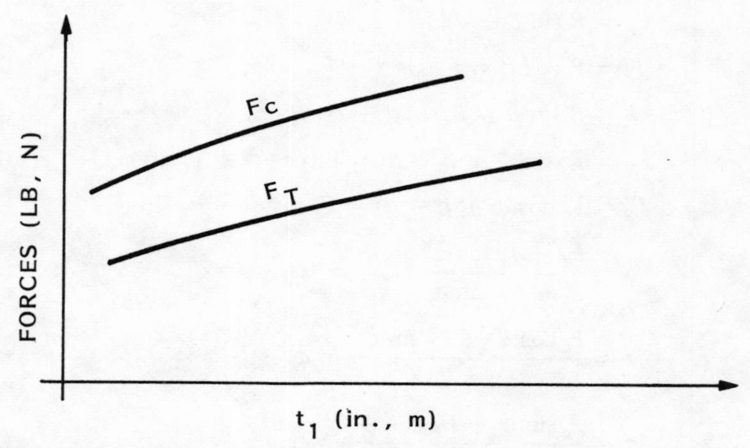

Fig. 2.7 The cutting forces versus feed.

The chip thickness t_2 decreases with the cutting speed, as shown in Fig. 2.8.

The rake angle has a strong influence on the cutting forces, as illustrated in Fig. 2.9.

The shear angle ϕ generally increases with the cutting speed and the positive rake angles. The coefficient of friction μ, however, tends to decrease with the

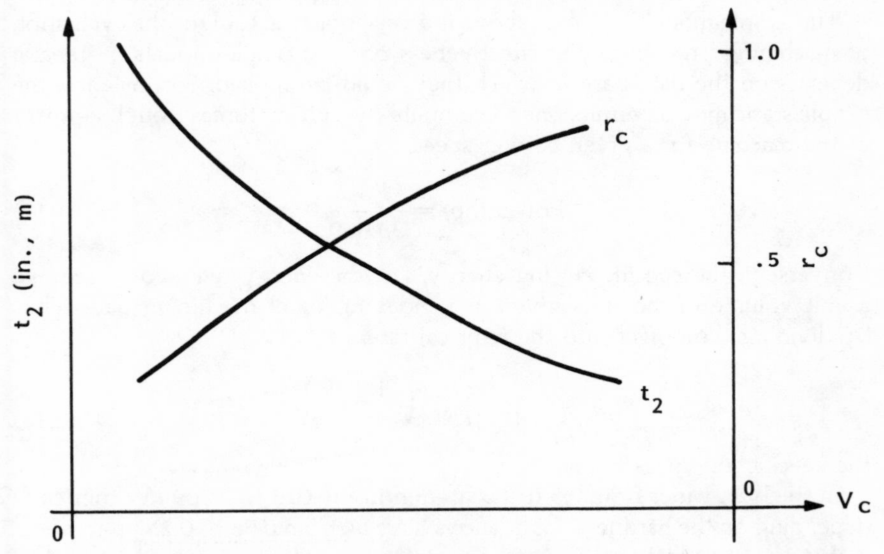

Fig. 2.8 Chip thickness t_2 and cutting ratio r_c versus cutting speed.

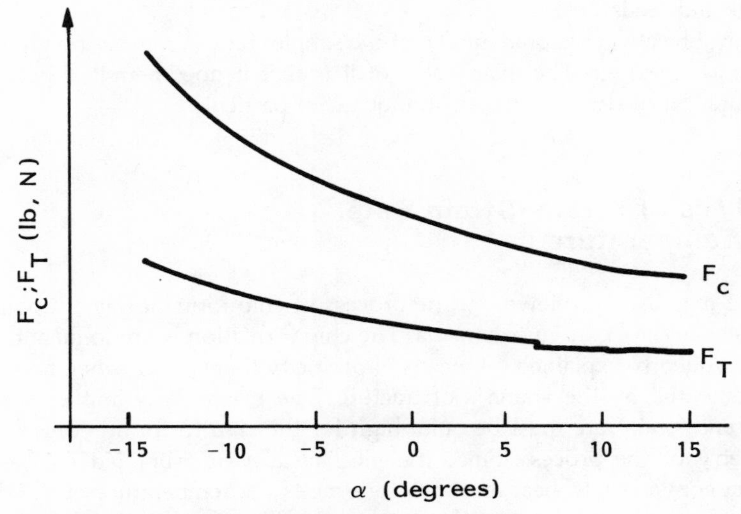

Fig. 2.9 Cutting forces versus rake angle.

increasing cutting speed. Each workpiece–tool material pair has a specific value of μ, which, in view of Eq. (2.14), depends also on the rake angle. The frictional coefficient μ is a concept based on the static analysis of forces, and for this reason it is not really a basic variable of the process. The physically significant force is the friction force. This force can be viewed as the integral of a variable shear stress distributed over the chip–tool contact area.

The simple model described above is a very practical tool for the evaluation of machining processes. The effectiveness of very simple models in general depends on the data base in which they could be applied. For instance, the simple static model permits one to compute the cutting forces from the power of the machine for a given cutting speed, i.e.,

$$\text{Power (hp)} = \frac{F_c V_c}{33{,}000}. \tag{2.17}$$

Conversely, the specific cutting energy, i.e., the energy required to remove a unit volume of metal, is given in various tables of machining data. Thus by dividing the power into the removal rate, i.e.,

$$w_c = \frac{F_c V_c}{12 b_1 t_1 V_c} \left(\frac{\text{hp-min}}{\text{in.}^3} \right), \tag{2.18}$$

the specific energy is obtained.

In steels w_c varies from 0.5 to 1.5 hp-min/in.^{3} (0.023 to 0.069 kW-min/cm^3) depending on the hardness. Light alloys have w_c from 0.08 to 0.25 hp-min/in.^{3} (0.004 to 0.012 kW-min/cm^3). If a metal-cutting dynamometer is available, the machine tool efficiency can be very precisely determined for various speeds and feeds.

It must be always kept in mind that the simple static model has no predictive power whatsoever. The main value of it is that it does permit a systematic presentation of data, for the continuous chip particularly.

Analysis of Strain, Strain Rate, and Temperature

From a macroscopic viewpoint the process of chip formation is a problem in the mechanics of continuous media. The chip formation is predominantly and most frequently explained by means of plasticity theory and, when necessary, supplemented by the analysis of fracture. The plastic flow implies that the strain and strain rate must be quite high for the chip to form because of the geometry of the process. Since the mechanical work supplied by the tool motion converts into heat and therefore raises the temperature of the deformation zone, the temperature has a very specific effect on the entire operation. The process is thus properly viewed as a thermomechanical problem that requires for its solution a complete set of equations:

the continuity equation,
the equation of motion,
the energy balance equation,
Clausius–Duhem inequality, and
the constitutive equation of the material.

In order to obtain the solutions of these equations it is necessary to specify the boundary and initial conditions. In the case of orthogonal cutting (the two-dimensional model), a special procedure is employed for the materials which can be approximated by the rigid, ideally plastic medium. This procedure is the slip-line method. The slip-line method can be also extended to work-hardening materials, but it leads to substantially more difficult analysis. The fully general case has not as yet yielded an analytical solution. In principle it is possible that a complete solution may be obtainable by numerical techniques using computers.

The engineering approach to these problems must therefore be guided by various approximations. The first approximation is to consider the kinematical and thermal aspects of the process separately.

Analysis of Strain

In Fig. 2.3 the line *AB* represents the boundary between the chip and the workpiece. Below this line the workpiece material is not deformed, and above it the chip is fully formed. It is, therefore, convenient to imagine that this line is a thin layer, Fig. 2.10. This layer *AA'B'B* is deformed by shear into the layer *ACDB*, as the tool progresses from the right to the left in the figure. The layer *AA''B''B* is the next layer to be deformed. The shear strain is defined as

$$\gamma = \frac{A'C}{AE} = \frac{A'E}{AE} + \frac{CE}{AE} = \cot\phi + \tan(\phi - \alpha). \tag{2.19}$$

In Eq. (2.5) the angle ϕ is defined in terms of velocities and, also, the chip thickness ratio, so that the shear strain can alternatively be given as

$$\gamma = \frac{r_c^2 - 2r_c \sin\alpha + 1}{r_c \cos\alpha} = \frac{\cos\alpha}{\sin\phi\cos(\phi - \alpha)}. \tag{2.20}$$

Thus, it is sufficient to measure the chip thickness to obtain the shear strain.

The strain has a minimum value when $r_c = 1$, which corresponds to the frictionless sliding of the chip over the tool rake face. Then $\phi = \pi/4 + \alpha/2$ is the corresponding shear angle.

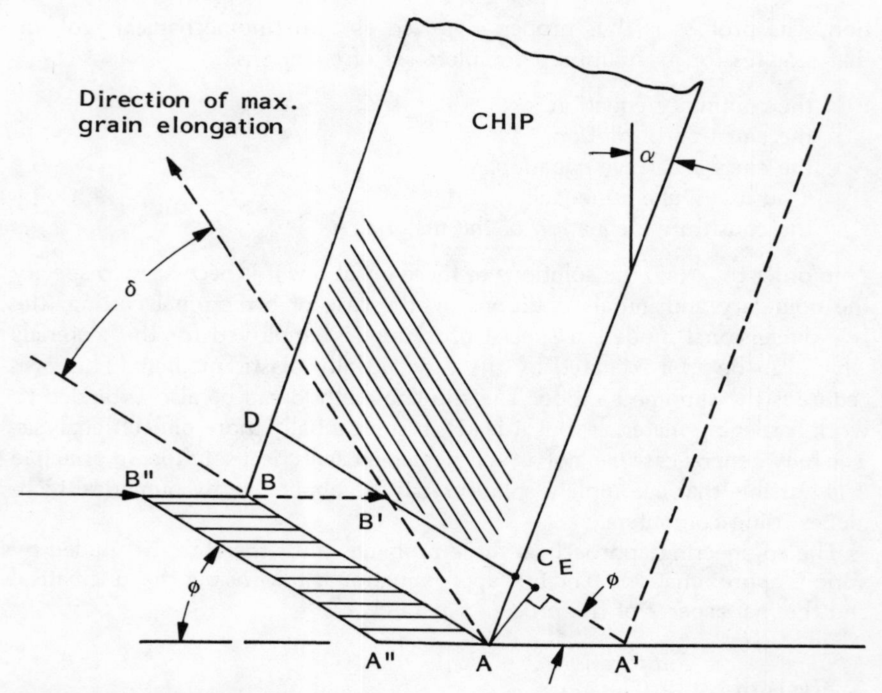

Fig. 2.10 Strain in chip formation.

The strain γ [Eq. (2.20)] is always larger when friction is taken into account. It can be represented as a sum of the frictionless part and the redundant strain due to friction, i.e.,

$$\gamma = \frac{2(1 - \sin \alpha)}{\cos \alpha} + \frac{r^2 - 2r_c + 1}{r_c \cos \alpha}. \tag{2.21}$$

In the quick-stop experiments, the actual shear angle can be determined directly. If the material is composed of equiaxed (spherical) grains, the maximum elongation of the grains forms an angle θ with the shear plane, so that (Fig. 2.10)

$$\gamma = 2 \cot 2\theta. \tag{2.22}$$

Analysis of Strain Rate

The strain rate is determined from the shearing velocity, V_S, and the thickness of the shear zone, the distance AE in Fig. 2.10:

$$\dot{\gamma} = V_S/d. \tag{2.23}$$

Since the shear strain γ does not depend on the thickness of the deformation (shear) zone or on its shape but exclusively on the chip thickness ratio and the rake angle (for the continuous chip only), a continuous particle path can be visualized within the shear zone. Then, considering that

$$\dot{\gamma} = \frac{d}{dt}\left(\frac{V_S}{V_N}\right) \tag{2.24}$$

where $\mathbf{V}_N$ is the velocity normal to the shear velocity, the average strain rate is given by

$$\dot{\gamma}_{\mathrm{av}} = \frac{1}{t_d}\int_0^{t_d} \dot{\gamma}\, dt = \frac{\gamma}{t_d},$$

where t_d is the time required for the particle to cross the shear zone at the constant normal velocity V_N, i.e., $t_d = d/V_N$. Then

$$\dot{\gamma}_{\mathrm{av}} = \frac{V_N}{d}\frac{V_S}{V_N} = \frac{V_S}{d},$$

which is Eq. (2.23).

In practical computations the velocity $\mathbf{V}_S$ is very close to $\mathbf{V}_c$ and d is somewhat smaller than the feed, t_1, so that a good approximation for the average strain rate is obtained from Eq. (2.25):

$$\bar{\dot{\gamma}}_{\mathrm{av}} = V_c/t_1. \tag{2.25}$$

Thus for the cutting speed of 120 in./sec (600 ft/min) and the feed of $t_1 =$ 0.006 ipr, the approximate strain rate is

$$\bar{\dot{\gamma}}_{\mathrm{av}} = \frac{120}{6 \times 10^{-3}} = 2 \times 10^4\ \mathrm{sec}^{-1}.$$

At very high cutting speeds and very shallow feeds, the strain rate can be very large, on the order of $10^7\ \mathrm{sec}^{-1}$. As the chip slides along the tool rake face, further deformation is imparted to it. This secondary deformation zone has a great influence on the chip formation process. If the flow along the tool face is not restricted by either a chip breaker or another obstacle, the chip slides over a length l_c (Fig. 2.11). The thickness of the secondary zone is usually a small percentage of the chip thickness t_2. The grain deformation pattern is extremely dense and shows strains much larger than those in the bulk. The strain rate $\dot{\gamma}_F$ is obtained from the sliding velocity $\mathbf{V}_F$ and the thickness of the secondary layer δ, i.e., $\dot{\gamma}_F = V_F/\delta$. Considering that $V_c = V_F r_c$, and that the contact length $l_c = nt_1$, where $n = 2$ to 5, the

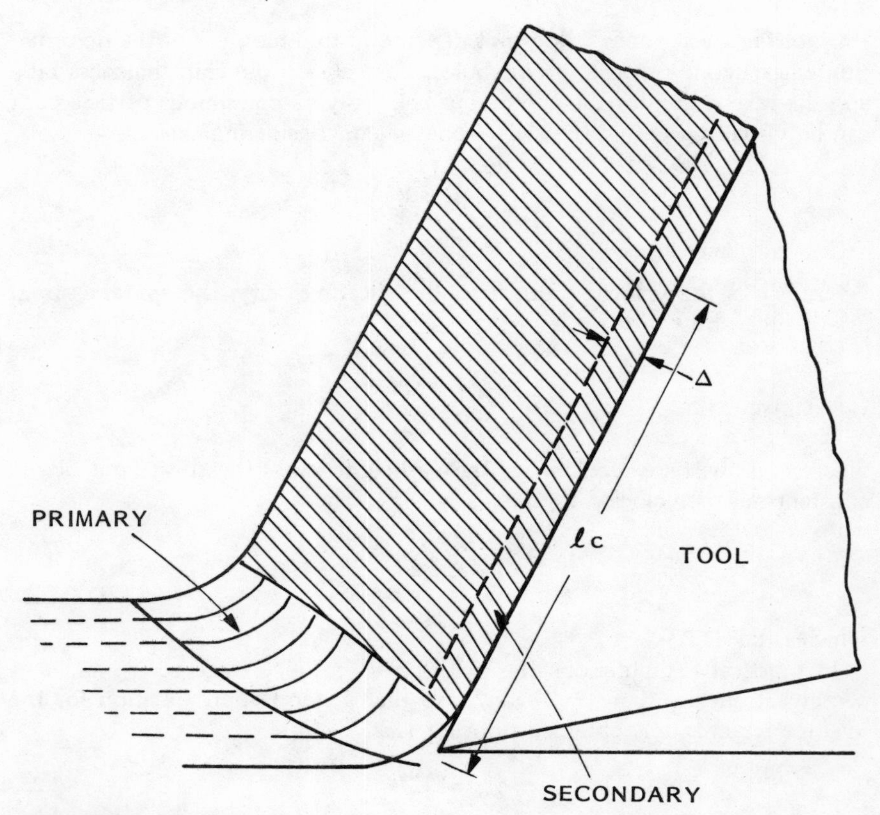

Fig. 2.11 Deformation zones.

average shear strain is then

$$\gamma_F = \int_0^t \dot{\gamma}_F \, dt = \dot{\gamma}_F \frac{l_c}{V_F} = \frac{l_c}{\delta} = \frac{nr_c}{\xi}, \tag{2.26}$$

where $\xi = 0.1$ to 0.3. Thus the typical values of γ_F are between 7 and 15.

Analysis of Temperature

The total power input in cutting, $\dot{w}_T$, is composed of several parts:

$$\dot{w}_T = \dot{w}_S + \dot{w}_F + \dot{w}_A + \dot{w}_M \tag{2.27}$$

where

$\dot{w}_S$ = the power consumed in the primary (shear) zone,
$\dot{w}_F$ = the power consumed in the secondary zone,
$\dot{w}_A$ = the power consumed to generate two new surfaces,
$\dot{w}_M$ = the power consumed in changing the chip momentum.

The specific energies $w_S = \tau\gamma$ and $w_F = \tau_F\gamma_F$ are dissipated almost entirely by conversion to heat. The specific energies w_A and w_M are relatively small and are neglected in the analysis except that (i) w_A is an appreciable portion of the total energy at very small feeds, $t_1 \approx 0.0001$ in. (0.0025 mm) or less, and (ii) w_M gains importance at very high cutting speeds [usually above $V_c =$ 15,000 ft/min. (4,570 m/min)].

The heat evolution in the primary and secondary shear zones gives rise to the chip temperature. The problem of temperature determination is a moving-heat-source problem. The heat evolving in the primary shear zone flows into the chip, and a small part into the workpiece. Then, using the method of moving heat sources, the shear zone temperature is given by

$$T_S = \frac{\eta\tau\gamma}{JC_1\rho_1[1 + 1.328\sqrt{K_1\gamma/V_c t_1}]} + T_0, \tag{2.28}$$

where

T_S = chip temperature due to shear, °F;
τ = shear stress on the shear plane [Eq. (2.15)], lb/in.^{2};
γ = shear strain [Eq. (2.19)];
C_1 = specific heat of workpiece, Btu/lb-°F;
J = mechanical equivalent of heat, 9339 in.-lb/Btu;
ρ_1 = specific weight of workpiece material, lb/in.^{3};
K_1 = workpiece thermal diffusivity, in.^{2}/sec;
V_c = cutting speed, in./sec;
t_1 = feed, in./rev;
T_0 = initial workpiece temperature, °F;
η = heat conversion factor $\approx$ 0.9.

The average temperature rise due to the chip tool friction at the contact area can be computed by the expression

$$\Delta T_F = \frac{0.377 l_c}{k_2\sqrt{L_2}} R_2 \frac{V_F F}{J l_c b_1} \tag{2.29}$$

where

T = temperature rise due to friction, °F;
l_c = length of chip–tool contact, in.;
k_2 = chip thermal conductivity at resultant temperature T
$= T_S + \Delta T_F$, Btu/in.^{2}-sec-°F/in.;
$L_2 = V_F l_c / 4k_2$, a dimensionless number;
K_2 = thermal diffusivity of chip at resultant temperature T, in.^{2}/sec;
R_2 = fraction of (w_F) going to chip;
F = friction force, lb;
V_F = chip velocity, in./sec;
J = mechanical equivalent of heat, 9339 in.-lb/Btu;
b_1 = the width of chip, in.

A typical result of these computations is given for AISI 1015 steel, cutting with carbide and ceramic tools (Fig. 2.12).

By using the numerical procedure the temperature distribution in chip formation, including the tool, can be obtained.

General Equations of Metal Cutting

The chip-formation process is, as it has been indicated above, a problem in the mechanics of continuous media. It is a thermomechanical problem and the general equations are the following:

the continuity equation,

$$\frac{\partial \rho}{\partial t} + (\rho v_i)_{,i} = 0; \tag{2.30}$$

the equations of motion,

$$\sigma_{ij,j} + x_i = \rho \frac{\partial v_i}{\partial t}; \tag{2.31}$$

the energy balance equation,

$$c\rho\left(\frac{\partial T}{\partial t} + T_{,i}v_i\right) = s_{ij}\dot{e}_{ij} - q_{i,i}; \tag{2.32}$$

the Clausius–Duhem inequality,

$$\begin{aligned} T\rho\dot{\delta}_{\mathbf{local}} &\equiv s_{ij}\dot{e} \geq 0, \\ T\rho\dot{\delta}_{\mathbf{conduction}} &\equiv -\frac{1}{T} q_i T_{,i} \geq 0; \end{aligned} \tag{2.33}$$

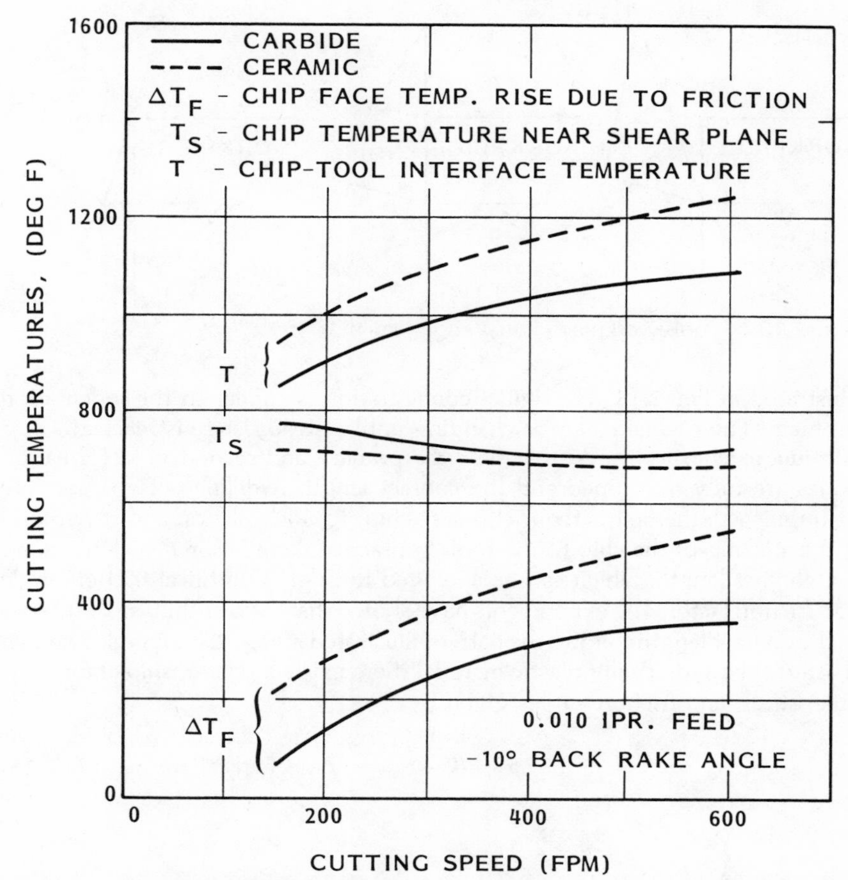

Fig. 2.12 Cutting temperature at various cutting speeds for ceramic and carbide cutting materials.

the constitutive equation,

$$\sigma_{ij} = f(\varepsilon_{ij}; \dot{\varepsilon}_{ij}; T); \tag{2.34}$$

where ρ is the density, v_i are the velocity vectors, x_i are the body forces, c is the heat capacity, s_{ij} is the deviatoric stress tensor, e_{ij} is the deviation strain rate tensor, q_i is the heat flux, and $\dot{\delta}$ is the internal entropy production rate per unit mass.

The relationship between the strain tensor ε_{ij} and the strain deviation is given by $\varepsilon_{ij} = e_{ij} - \frac{1}{3}\varepsilon_{\kappa\kappa}\delta'_{ij}$. A plasticity condition (yield criterion) is added for the plastically deforming materials. The chip formation is a partially constrained problem, i.e., its geometry is not governed by a complete enclosure by tool surfaces. The initial configuration of tool–workpiece engagements is

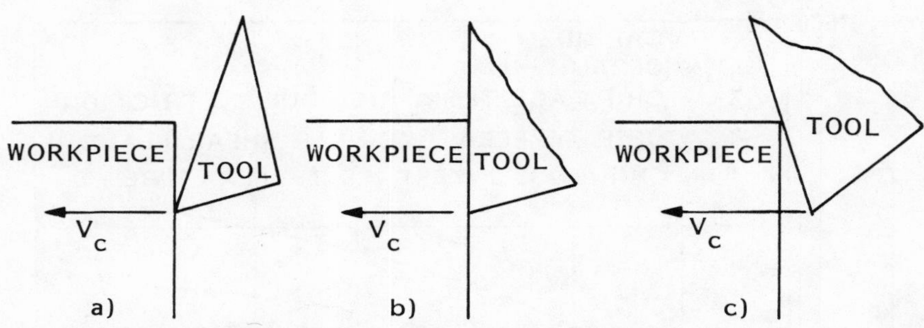

Fig. 2.13 Tool–workpiece initial engagement.

illustrated in Fig. 2.13. This initial configuration is similar to the indentation problem. The geometry keeps changing until a steady state is reached, i.e., a continuous chip forms. At that time the primary and secondary deformation zones are fully developed and the contact length is defined. Two cases are distinguished: the first is the frictionless chip–tool interaction and the second is the sticking of the chip to the tool. In practice there is some sticking along the contact length, which can be averaged through an artifical friction angle or a friction factor. The experiments have shown that the distribution of stresses in the contact length follows the pattern illustrated in Fig. 2.14. It is also known that for the rigid, ideally plastic material there exists a relationship among the rake, shear, and friction angles given by

$$\phi + \beta - \alpha = \frac{\pi}{4}. \tag{2.35}$$

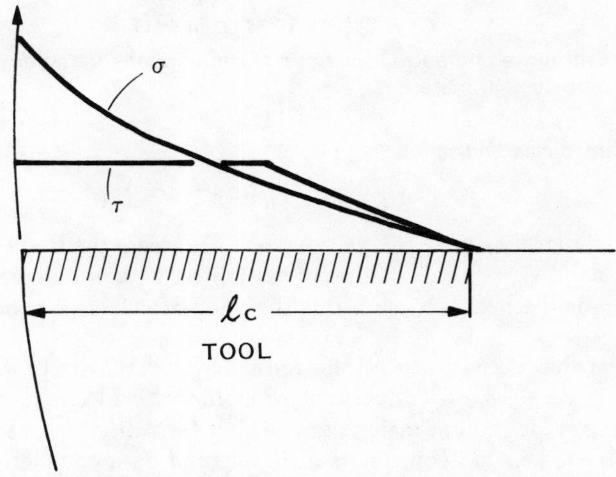

Fig. 2.14 Stress distribution along tool rake face.

If the minimum-cutting-force concept is used to show this angle relationship, then

$$\phi + \frac{\beta - 2}{2} = \frac{\pi}{4}. \tag{2.36}$$

Once the fully developed steady state is achieved, each of the deformation zones can be treated separately.

The equation of motion for each zone is

$$\rho \frac{\partial u}{\partial t} = \frac{\partial \sigma_x}{\partial x} + \frac{\partial \tau_{xy}}{\partial y}, \tag{2.37}$$

$$\rho \frac{\partial v}{\partial t} = \frac{\partial \tau_{xy}}{\partial x} + \frac{\partial \sigma_y}{\partial y}, \tag{2.38}$$

where $u(x, y)$ is the velocity parallel to the deformation zone and $v(x, y)$ is the velocity perpendicular to it. Differentiating Eq. (2.37) with respect to y and Eq. (2.38) with respect to x and summing results in

$$\rho \frac{\partial^2 \gamma}{\partial t^2} = \left(\frac{\partial^2}{\partial x^2} + \frac{\partial^2}{\partial y^2} \right) \tau_{xy} + \frac{\partial^2}{\partial x \, \partial y} (\sigma_x + \sigma_y). \tag{2.39}$$

Since both zones are thin in comparison to their length, and the deformation occurs only in one direction, i.e., parallel to the length, but it may have a gradient in the other, Eq. (2.39) simplifies to

$$\rho \frac{\partial^2 \gamma}{\partial t^2} = \frac{\partial^2 \tau_{xy}}{\partial y^2} \tag{2.40}$$

under the assumption that the variation of stresses are much steeper in the y direction (perpendicular to the zone).

The energy balance equation [Eq. (2.32)] is transformed as follows: $w_S = \eta \tau \gamma$ is the heat produced by plastic flow in the zone. The term $s_{ij} e_{ij}$ is expressed as

$$s_{ij} \dot{e}_{ij} = \eta \left(\frac{\partial w_S}{\partial t} + u \frac{\partial w_S}{\partial x} + v \frac{\partial w_S}{\partial y} \right)$$
$$= \eta \tau \left(\frac{\partial \gamma}{\partial t} + u \frac{\partial \gamma}{\partial x} + v \frac{\partial \gamma}{\partial y} \right).$$

Since

$$q_{i,i} = -k \left(\frac{\partial^2 T}{\partial x^2} + \frac{\partial^2 T}{\partial y^2} \right),$$

the energy equation becomes

$$\eta\tau\left(\frac{\partial\gamma}{\partial t}+u\frac{\partial\gamma}{\partial x}+v\frac{\partial\gamma}{\partial y}\right)=c\rho\left(\frac{\partial T}{\partial t}+u\frac{\partial T}{\partial x}+v\frac{\partial T}{\partial y}\right)-k\left(\frac{\partial^2 T}{\partial x^2}+\frac{\partial^2 T}{\partial y^2}\right). \tag{2.41}$$

Again the gradients in the y direction are much larger than in the x, and, owing to incompressibility, $V = 0$, one obtains finally

$$\eta\tau\frac{\partial\gamma}{\partial t}=c\rho\frac{\partial T}{\partial t}-k\frac{\partial^2 T}{\partial y^2}. \tag{2.42}$$

The constitutive equation is of the form

$$\tau=f\left(\gamma,\dot{\gamma},T;\,T=\int\tau\,d\gamma\right). \tag{2.43}$$

Equations (2.40) and (2.41) are difficult to solve, even if the stationary form Eq. (2.42) is used.

The main interest in Eqs. (2.40) and (2.42) arises from the stability analysis. It is possible to show that these equations predict the onset of unstable chip formation. The instability in the primary deformation zone leads to segmental chip, and the instability in the secondary zone causes oscillations of the shear planes and, therefore, a wavy chip form.

Considering that the constitutive equation expresses the material behavior in either zone, the question arises, what happens if the strain required to produce chip γ is larger than the strain which the material can sustain, i.e., it fractures or shows a negative slope of the stress–strain curve?

This behavior can be exhibited by considering the constitutive law of the form

$$\tau=\tau_y\gamma^n\left(\frac{\dot{\gamma}}{\dot{\gamma}_0}\right)^m\exp\left(\frac{A}{T}\right). \tag{2.44}$$

The strain at which $\partial\tau/\partial\gamma\to 0$ is then given by

$$\gamma^*=n\left[\frac{Jc\rho}{\tau_y A\eta}T^2\exp\left(-\frac{A}{T}\right)\right]^{1/(n+1)}\left(\frac{\dot{\gamma}}{\dot{\gamma}_0}\right)^{-m/(n+1)} \tag{2.45}$$

where T is evaluated as

$$dT=\frac{\eta}{Jc\rho}\tau\,d\gamma,$$

and n, m, A, τ_y are independent of the strain, strain rate, and temperature. Equation (2.45) is equivalent to Eq. (2.46):

$$\gamma^*=\frac{nJc\rho}{\tau A\eta}T^2, \tag{2.46}$$

which is applicable at all speeds. If γ^* is smaller than γ, then the continuous chip cannot form.

The specifics of the discontinuous chip formation are currently under active study. Since the process differs significantly from the continuous-type chip mechanics, it will be necessary to reexamine classical models and develop new procedures for the computation of cutting temperatures and, to some extent, the cutting forces.

References

Hodge, Jr., P. G., *Continuum Mechanics*, New York, McGraw-Hill, 1970.

Shaw, M. C., *Metal Cutting Principles*, 3rd ed., Cambridge, Mass., MIT, 1959.

Zorev, N. N., *Metal Cutting Mechanics*, New York, Pergamon Press, 1966.

Armarego, E. J. A. and R. H. Brown, *The Machining of Metals*, Englewood Cliffs, N.J., Prentice-Hall, 1969.

CHAPTER 3

Machine Dynamics

J. Tlusty
McMaster University

Introduction

The text presented here, in its major Part A, is a comprehensive discussion of self-excited vibrations in machining with special emphasis in its illustrations and conclusions on milling and, mainly, on ultra-high-speed end milling. It is heavily based on a number of previous research papers of the author, which were mostly coauthored by F. Ismail, whose contributions are gratefully acknowledged. He worked for the past seven years, first as a Ph.D. student and subsequently as a research engineer, in the author's laboratory. The sections on the effect of spindle speed and on special cutters is based on the work of W. Zaton, who was another research collaborator of the author. In Part B the problems of deflections and of forced vibrations of end mills and their effect on accuracy of machining are discussed. These problems are significant in end milling of aerospace structures and, in this instance, also in ultra-high-speed milling.

Credit is also to be given to the Lawrence Livermore National Laboratory for the permission to use significant parts of the text which the author wrote for the Machine Tool Task Force, which is quoted as reference 2.

In Part A the general theory of chatter (self-excited) vibrations is first presented and, subsequently, ways of increasing stability of machining against chatter are discussed in three categories. In the first category the methods of measurement and of computations of the machine tool structure are indicated which lead to improved stability. In the second category the choice of optimum spindle speeds is explained, and in the third category milling cutters with nonuniform tooth pitch and with alternating helix as well as with serrated or undulated edges are described and their potentials are evaluated. In Part B the problems of deflections of end mills and their effect on accuracy of the machined surface are discussed.

Summary

Extensive research was devoted in the past to the problems of chatter in machining and all the three previously mentioned categories of measures for stability improvement have been explored. Reference 1 summarizes this research work. However, most of this work was based on theories in which the teeth of a milling cutter were represented by one "mean" direction of the cut. This assumption, which permits a more general theoretical treatment, is used here partly, too. However, it is recognized that it leads to errors in the calculated values of limit width of chip and, mainly, in the calculated efficiency of the optimum speed and of the special cutters. In the past three years most of these inaccuracies have been investigated and corrected by the application of time domain simulations in the work carried out at McMaster University, which was reported in references 3, 4, 5, and 13. The results of this work are included here and, in this respect, this is significantly the most up-to-date summary of the Dynamics of Metal Cutting.

In the sections dealing with the *theory of chatter* the limit width of chip b_{lim} is expressed for cutters with uniform tooth pitch and for the most unfavorable spindle speed. This is the basic, or borderline, stability. The most common and most significant case of systems with two mutually perpendicular degrees of freedom is explored, detailed, and the effects of directional orientation of cutting with respect to the system are evaluated. For single-point tool operations like turning or boring these effects may be practically utilized for obtaining best stability. In milling, however, owing to the different orientation of each tooth and the great variety of process orientations in up and down milling and slotting, which may all be combined in one operation, these measures do not have practical significance. In addition, it is shown that the closer the two natural frequencies are, the lower is the stability, and it is worst for the case of two equal frequencies, which is the case where a flexible spindle or a flexible tool, like a long end mill, has the dominant flexibility.

The part on the *design of the machine tool structure* is rather brief because it is not of a great interest to the user of machine tools, whereas the builders of machine tools need a deeper knowledge than could be presented in this text and its limited space. For the latter we recommend references 1, 6, and 7. However, enough examples are introduced, especially with reference to Figs. 3.42–3.47, and the most pertinent case for ultra-high-speed milling of optimizing the design of a spindle carrying an end mill is briefly explained with reference to Fig. 3.51.

The problem that is very important for the user of machine tools is the one of *optimum spindle speed.* It is discussed in Part A. It is shown that for ultra-high-speed milling significant increase in stability (up to 10 times) can be achieved by choosing the right speed. However, it has to be chosen rather accurately because this effect diminishes very quickly if the speed differs from the optimum one; see Figs. 3.56 and 3.57. It is therefore recommended to determine, by dynamic excitation measurements, the transfer functions of the machine tool used with all the cutters involved and to determine and know the exact optimum speeds. The phenomenon of damping in the cutting process is described, and it is shown that it leads to significant stability increase for very low cutting speeds. While this is generally important for the practice of some operations, like broaching, it still has a special meaning for end milling with the common cutting speeds of 82–197 ft/min (25–60 m/min) involving spindle speeds of 1000–3000 rev/min for end mill diameters from 0.39 to 1.97 in. (10 to 50 mm). At these speeds, which are generally used for machining steel and titanium but also aluminum in those cases where higher spindle speeds are not available, the very flexible mode of the long and slender end mill cannot participate in self-excitation because its rather high frequency (1600–2000 Hz) would lead to very short surface undulations [below 0.04 in. (1 mm)] which cannot reproduce well; see Figs. 3.58 and 3.61. An unexpected but theoretically well understandable feature is encountered when longer end mills vibrate much less than short end mills. This is illustrated in Fig. 3.60, and it is explained by the fact that the mode involved is one of spindle bending which has a much lower frequency (500–600 Hz), and the end mill flexibility alternates the force variation of this frequency. However, once higher spindle speeds are used, generally beyond 3000 rev/min, see Table 3.2, the flexible mode of long end mills becomes active. This is the case of most of the advanced aluminum machining, and the only methods to prevent chatter are the use of the exact optimum spindle speed and the use of cutters with special tooth forms.

Special cutters that improve stability are also discussed in Part A. Three types of special cutters are considered: with nonuniform tooth pitch, with alternating helix, and with serrated or undulated edges. It is shown that all of them represent rather powerful means of increasing stability of milling. However, in order to achieve their best efficiency, their design parameters have to be matched to a given range of applications. Significant efficiency can be obtained with cutters with coarse pitch nonuniformity and especially with cutters with alternating helix in the ultra-high-speed milling applications. All this is explained in reference to Figs. 3.73, 3.75, 3.77 and, ultimately, Fig. 3.78.

In Part B the deflections between tool and workpiece, which are produced by the cutting force in end milling, are analyzed as well as their effect on the errors of the geometry of the machined surface. The cutting force has a mean component and a component that varies with a basic period of the passage of one cutter tooth. The mean component varies slowly as the stock to remove varies, which is significant mainly in milling internal corners of pockets. The variable component may have a rather high frequency, from 100 to 2000 Hz, the latter in ultra-high-speed milling. It might, therefore, be in resonance with one of the natural frequencies of the system. It is essential to avoid the resonance situation by knowing the dynamic characteristics of the machine and avoiding the speeds at which the tooth frequency would equal one of the natural frequencies. The nature of force variations is described in the text, and it is shown that it is possible to choose the helix angle versus axial depth of cut so that the variable force component becomes zero for any radial depth of cut; see Fig. 3.73.

The relationship between cutter and workpiece deflections and the errors of geometry of the machined surface is not simple and depends on the number of teeth of the cutter, the helix angle, and the mode of milling (up or down). Although it might not be expected or commonly known, the fewest errors are obtained with two-fluted cutters with straight teeth (helix angle zero). However, these cutters give the biggest variable force component and could cause serious forced vibrations; therefore, they are not generally recommended. The geometric errors of the machined surfaces may be rather large, of the order of 0.004–0.02 in. (0.1–0.5 mm) on a web which is itself only 0.2 in. (5 mm) thick. A typical illustration is provided in Fig. 3.76.

In order to obtain small and acceptable errors it is possible to use computations to evaluate the best cutter geometries and suitable cutting pass distributions and to develop suitable "cornering routines."

Conclusions

Theoretical and experimental analysis of chatter in milling and of end mill deflections and their effect on accuracy of milling has reached a rather advanced state in which it is possible to compute the limit of stability and the accuracy of machining.

To increase stability of chatter various approaches are possible in three categories: the machine tool structure, optimum spindle speed, and optimum cutter design. For the first category the methods to be used are indicated in reference to Fig. 3.52. The two latter categories are the concern of the machine tool user. In order to utilize the potential increases in stability, which are considerable in high-speed end milling, it is necessary to determine experimentally the transfer functions at the various cutters to be used on the particular machine tool and to develop and suitably package corresponding software for choosing optimum speed and optimum cutter design and for evaluating the effects obtainable. Such software is not presently ready but it can be provided for the individual users by a number of research establishments. The computational methods are described here and examples of results of pertinent computations are given in reference to Figs. 3.56 and 3.57 for regular pitch cutters and in reference to Figs. 3.73, 3.75, 3.77, and 3.78 for special cutters.

To obtain satisfactory accuracy in end milling it is necessary to minimize force variations, to avoid resonances, and to use suitable cutting pass distribution and cornering routines.

A. CHATTER VIBRATIONS

Theory of Chatter

General Characteristics of Chatter in Machining

Chatter is recognized by characteristic noise, by chatter marks, and by the undulated or dissected chip. In Fig. 3.1 photographs of chatter marks are

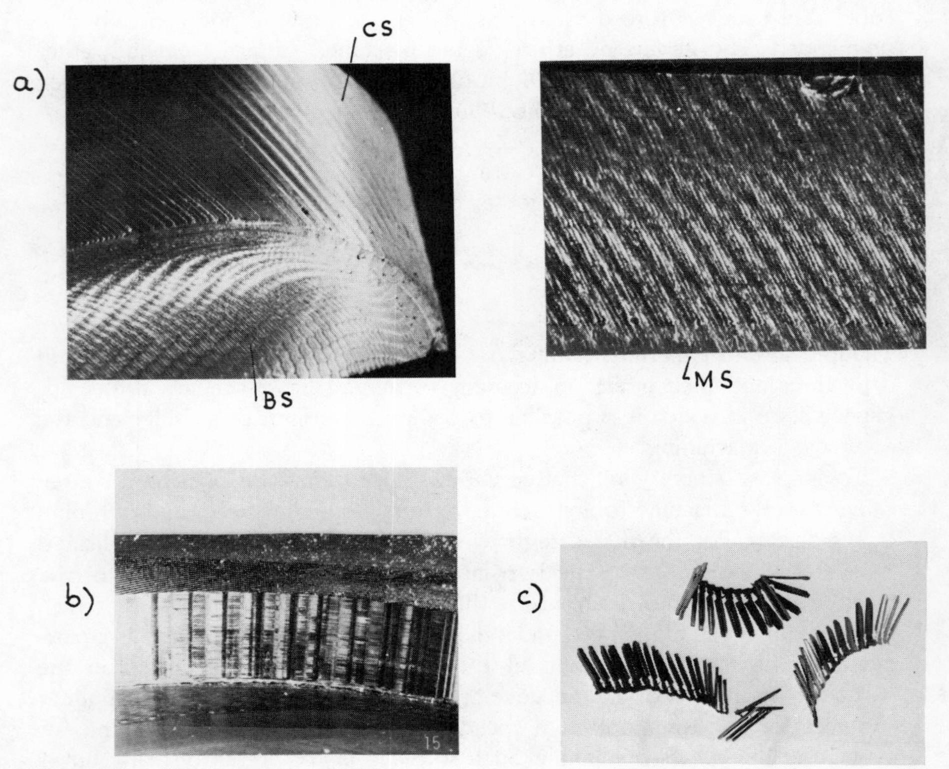

Fig. 3.1 Typical chatter marks.

shown as obtained in end milling of aluminum (Fig. 3.1a) and in carbide face milling of steel (Fig. 3.1b). In Fig. 3.1c typical dissected chips of the operation shown in Fig. 3.1b are shown. The notations CS, BS, MS coincide with those in Fig. 3.2. As will be explained later, the frequency of chatter vibrations is equal to a dominant mode of the natural vibrations of the mechanical structure involved. Typically, this frequency f may be about 200 Hz in turning shafts, about 550 Hz in face milling and in end milling steel or titanium, and about 1500–2000 Hz in milling aluminum. The rather loud noise arising from chatter has the corresponding pitch. The chatter marks on a turned surface and on the surface of the arc being cut (CS) by the teeth of a milling cutter, see Fig. 3.2, will have a wavelength (pitch) w, which depends on the frequency f and on the cutting speed v:

$$w = v/f \text{ (in consistent units).}$$

If using the units v(ft/min), f(Hz), w(in.), it is

$$w = (12/60)v/f = 0.2v/f; \tag{3.1}$$

if v(m/min), f(Hz), w(mm), then $w = 16.7v/f$.

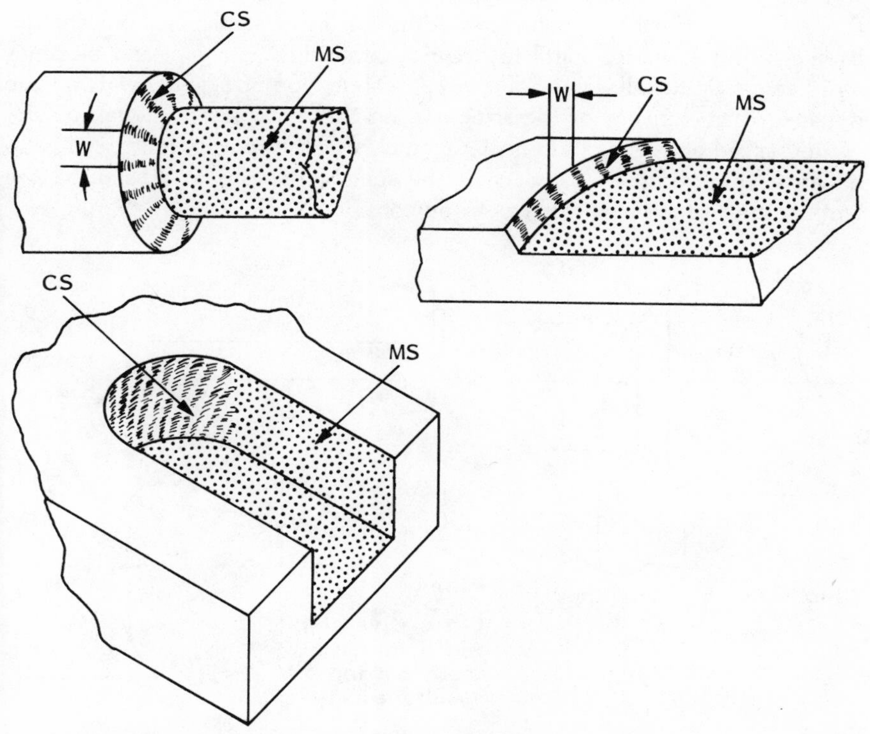

Fig. 3.2 Marks on the cut and on the machined surfaces.

Correspondingly, the chatter marks on the cut surface CS will have a long pitch at high cutting speeds and low frequencies and a short pitch at the reverse conditions. Typically,

carbide turning of steel, $v = 500$ ft/min (152.4 m/min), $f = 200$ Hz, $w = 0.5$ in. (12.7 mm);
carbide face milling of steel, $v = 500$ ft/min (152.4 m/min), $f = 550$ Hz, $w = 0.2$ in. (5.1 mm);
HSS milling of steel, $v = 90$ ft/min (27.4 m/min), $f = 550$ Hz, $w = 0.032$ in. (0.813 mm);
end milling of aluminum, $v = 1000$ ft/min (304.8 m/min), $f = 1800$ Hz, $w = 0.110$ in. (2.794 mm).

However, chatter marks on the surface MS machined in milling, see Fig. 3.2, will be obtained as a combination of the feed/tooth marks with the previously described chatter marks. They are best recognized by showing the helix angle of the cutter, while regular feed marks obtained in the absence of chatter are parallel with the cutter axis.

It is recognized that the most significant cutting parameter, which is decisive for the generation of chatter, is the *width of cut* (*width of chip*) b; see Fig. 3.3. For sufficiently small chip widths cutting is stable, without chatter. By increasing b chatter starts to occur at a certain width b_{lim} and becomes more energetic for all values of $b > b_{lim}$. The value of b_{lim} depends on the dynamic characteristics of the structure, on the workpiece material, on the cutting speed and feed, and on the geometry of the tool. In milling, the cumulative chip width b_{cum}, which is the sum of the chip widths of all the teeth cutting simultaneously, has to be considered.

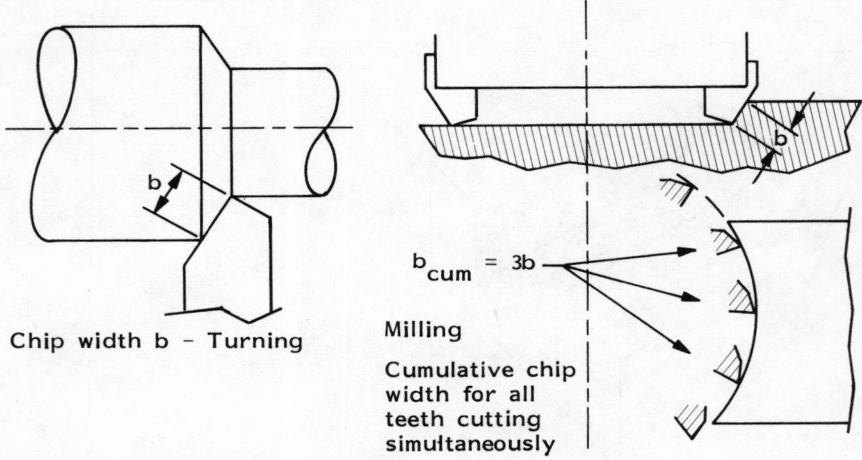

Fig. 3.3 Width of chip b is the most important parameter.[2]

There are two main sources of self-excitation in metal cutting: (a) mode coupling and (b) regeneration of waviness.

Mode coupling is a mechanism of self-excitation that can only be associated with situations where the relative vibration between the tool and the workpiece can exist simultaneously in at least two directions in the plane of the orthogonal cut. This is symbolically expressed in Fig. 3.4, where the tool is shown attached to a mass that is suspended on two sets of springs perpendicular to each other. This graph and some of the others in this section are reproduced from ref. 8. Simultaneous vibration in the directions X_1 and X_2 with the same frequency and a phase shift between the two results in an elliptical motion as shown. The workpiece may be moving with a steady cutting speed v. Assume that the tool moves on the elliptical path in the direction of the arrows. The cutting force F has the direction as shown. For the part of the periodic motion of the tool from A to B the force acts against this motion and takes energy away. During the motion from B to A the force drives the tool and imparts energy to its motion. Because motion $B \rightarrow A$ is located deeper in the cut, the force F is larger than during the motion $A \rightarrow B$ and the energy delivered by the force F to the periodic motion in section $B \rightarrow A$ is larger than the energy taken away in the section $A \rightarrow B$. Periodically, there is surplus of energy sustaining the vibrations against damping losses. The conditions under which this process of self-excitation is possible are derived further on.

Regeneration of waviness, see Fig. 3.5, is possible because in almost all machining operations the tool removes the chip from a surface which was produced by the tool in the preceding pass, i.e., the surface produced in turning during the preceding revolution or, in milling, by the preceding tooth of the cutter. If there is relative vibration between tool and workpiece, waviness is generated on the cut surface. The tool in the next pass (next revolution in turning, next tooth in milling) encounters a wavy surface and removes a chip

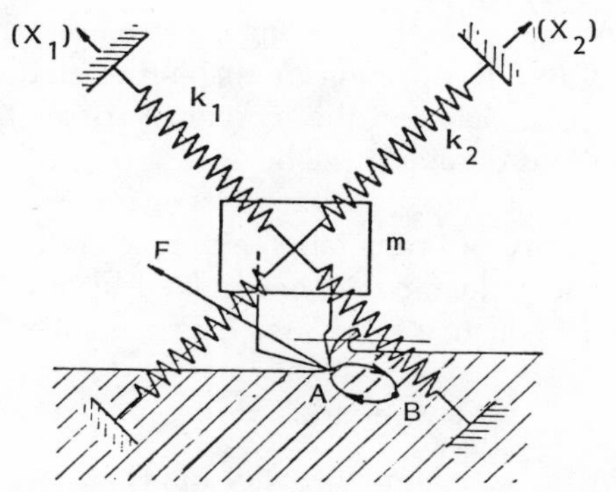

Fig. 3.4 Mechanism of mode coupling.[8]

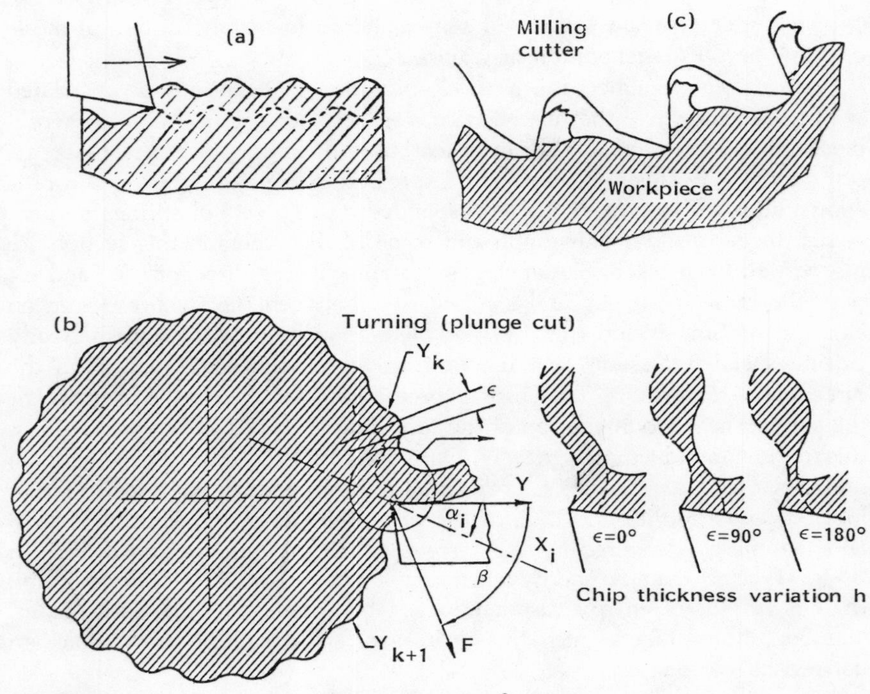

Fig. 3.5 Regeneration of surface waviness.[2]

with periodically variable thickness. The cutting force is periodically variable. This produces vibrations and, depending on conditions derived further on, these vibrations may be at least as large as in the preceding pass. The newly created surface is again wavy, and, in this way, the waviness is continually regenerated.

Regeneration of waviness is influenced by the geometry of the operation, which imposes a constraint on the phasing between undulations produced in subsequent passes. Only in a planing or shaping operation is there no such constraint; see Fig. 3.5a. Between the individual passes the tool exits from the cut and its vibrations decay. At the beginning of the next cut its vibration may freely adjust its phase for maximum regeneration. In turning, see Fig. 3.5b, where for simplicity of presentation a plunge cut is depicted, the phase ε between subsequent undulations is determined by a relationship between spindle speed n and frequency of chatter f. The number of waves between subsequent cuts is

$$N + \varepsilon/2\pi = f/n, \tag{3.2}$$

where f is frequency in Hz, n is spindle speed in rev/sec, N is the largest possible integer such that $\varepsilon/2\pi < 1$. In other words, there are N full waves and a fraction $\varepsilon/2\pi$ of a wave on the circumference of the workpiece. Depending on ε the chip thickness variation $(Y_0 - Y)$, see Fig. 3.5c, can be either zero for $\varepsilon = 0$ or maximum for $\varepsilon = \pi$. Obviously, there would be no chatter with $\varepsilon = 0$ because no periodic self-excitation force would be generated. It will be shown later that for maximum self-excitation the phasing is close to $\varepsilon = -\pi/2$. If the just-described geometric constraint led to a strongly different ε, the system would adjust frequency to achieve the most favorable phasing. In turning, even a small change of frequency leads to a large change of ε; the change of the number N by 1 produces full 2π change of ε. If, e.g., there are $N = 50$ waves on the circumference of the workpiece, a change of 2% in frequency produces 2π change of ε. Correspondingly, the geometric constraint on phase ε has little effect on the limit of stability in turning. It should be noted that 2% change of spindle speed has the same effect on ε.

In milling, see Fig. 3.5c, the geometric constraint is much more important. Regeneration of waviness on the surface is not done in subsequent spindle revolutions as in turning but by subsequent teeth of the cutter. Equation (3.2) is now changed to

$$N + \varepsilon/2\pi = f/nz, \tag{3.3}$$

where z is the number of teeth of the cutter. It may easily be that the number N is two or three; i.e., there are two and a fraction (three and a fraction) waves between subsequent teeth. Now, a change of ε by 2π, for $N = 2$, requires 50% change of frequency or of cutting speed. The geometric condition has an important influence on stability and this has great practical significance to be discussed in later sections.

At this stage, we will not consider any geometric constraint on the phasing of subsequent undulations in regenerative chatter. This is equivalent to saying that we will be dealing with chatter as it would develop under the most favorable phasing, which in reality can always occur at some spindle speed in milling and which is almost always obtained in turning. In other words, it is a realistic simplification for turning, and it is the worst possible case in milling. It is sometimes called the borderline stability. We will now derive expressions for the limit of stability for regenerative and nonregenerative (mode-coupling) borderline chatter.

Limit of Stability for Regenerative Chatter

We will now derive the condition for the limit of stability for the usual case of machining which would involve the regeneration of waviness of the cut surface. Let us consider the diagram in Fig. 3.6. The structure of the machine tool is shown as a frame M, which carries the tool on one end and the workpiece on the other end and it implies a relative cutting motion in the direction v between the two. Vibrations between the two are possible simultaneously

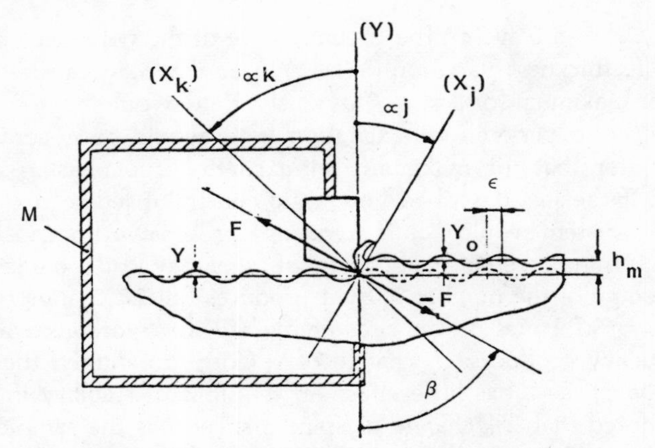

Fig. 3.6 Diagram for deriving limit of stability.[8]

in several "modes of natural vibrations" (discussed subsequently). The directions X_k and X_j of two such modes are indicated, and they are measured by the angles α_k, α_j from the direction Y, which is normal to the cut. The components of the X vibrations projected into Y produce waviness of the surface and modulate the chip thickness. If, in a preceding pass, there was vibration

$$y_0 = Y_0 \sin \omega t \tag{3.4}$$

and, in the current pass, the vibration is

$$y = Y \sin(\omega t + 2\pi N + \varepsilon) \tag{3.5}$$

and the "mean chip thickness" is h_m, the chip thickness is

$$h = h_m + y_0 - y = h_m + h_v$$

where the variable component of chip thickness is

$$h_v = y_0 - y. \tag{3.6}$$

The cutting force F has the direction inclined by β from Y and its magnitude is assumed proportional to chip width b (normal to the plane of paper) and to chip thickness h:

$$F = Cbh = F_m + F_v,$$

where the variable component is

$$F_v = Cbh_v = Cb(y_0 - y). \tag{3.7}$$

The "cutting force coefficient" C will, for now, be considered a real number. This means that, with b being real anyway, the force is in phase with

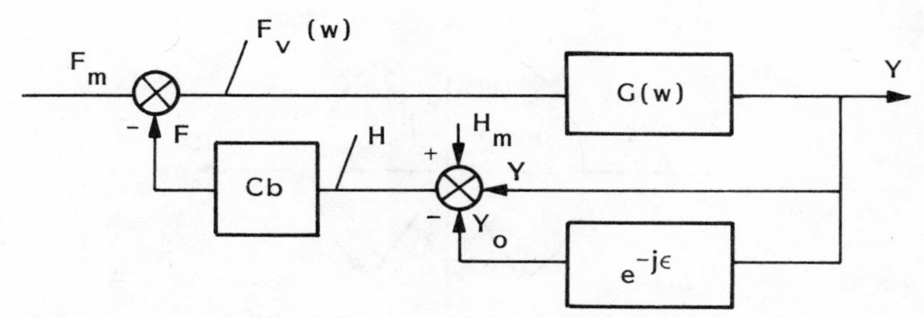

Fig. 3.7 Block diagram of vibration in regenerative cutting.

the chip thickness variation $(y_0 - y)$. Later, the possibility of a complex C will be discussed with the implication of a damping in the cutting process.

The process of self-excitation is represented by the closed loop in Fig. 3.7. The variable force F_v excites the structure. Vibration, relative between tool and workpiece, is produced through the corresponding transfer function $G(\omega)$ of the structure. The difference between Y and vibration Y_0 of the preceding pass, which produced the waviness of the surface from which the cut is taken and which was phase shifted by ε, is the chip thickness h. When multiplied by Cb it produces the force F, which when compared with the mean force F_m leaves the variable force F_v that excites the structure.

At the *limit of stability* any vibration would remain constant without decaying or increasing. The absolute values of amplitudes of vibrations in subsequent passes are equal:

$$|Y| = |Y_0|. \tag{3.8}$$

This will be obtained, according to the Nyquist criterion, if the open-loop transfer function has the value -1:

$$CbG(\omega)(1 - e^{-j\varepsilon}) = -1. \tag{3.9}$$

The value of the chip width b at the limit of stability is

$$b_{\text{lim}} = -1/CG(1 - e^{-j\varepsilon}). \tag{3.10}$$

Since the coefficient C was chosen to be real and b is obviously always real, the condition (3.10) is only satisfied if

$$G(1 - e^{-j\varepsilon}) = G - Ge^{-j\varepsilon} \text{ is real.} \tag{3.11}$$

Because $e^{-j\varepsilon}$ is a unit vector,

$$|G| = |Ge^{-j\varepsilon}|,$$

condition (3.11) is only satisfied (see Fig. 3.8) if:

$$G(1 - e^{-j\varepsilon}) = 2\,\text{Re}(G) \tag{3.12}$$

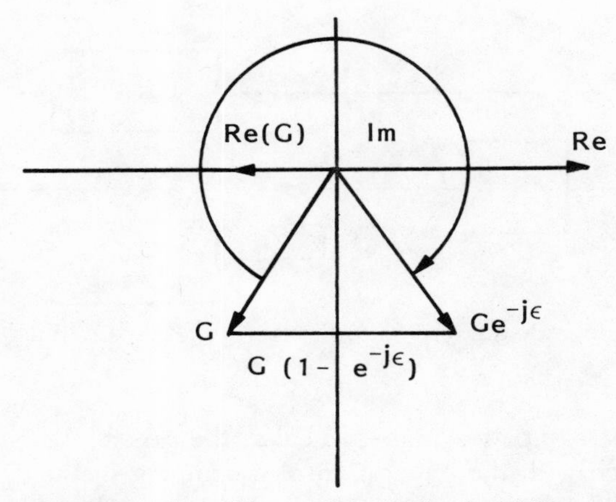

Fig. 3.8 Conditions at the limit of stability.

and Re(G) < 0. Consequently, at the limit of stability, it is

$$b_{\text{lim}} = -1/2C\,\text{Re}(G). \tag{3.13}$$

However, if the phase shift ε is left free to accept any value, Eq. (3.13) does not determine a unique value of b_{lim} and chatter may occur at many frequencies at many values of b_{lim}. Of all of them there will be one which will be minimum, and this one represents the borderline stability, which is

$$\begin{aligned}(b_{\text{lim}})_{\text{min}} &= \{-1/2C\,\text{Re}(G)\}_{\text{min}},\\ (b_{\text{lim}})_{\text{min}} &= 1/2C\,\text{Re}(G)_{\text{lim}},\end{aligned} \tag{3.14}$$

where $\text{Re}(G)_{\text{lim}}$ is the absolutely maximum negative value of Re(G).

Let us illustrate all this by using the example of a single degree of freedom system, Fig. 3.9. The system can only vibrate in the direction X, and has mass m, stiffness k and damping coefficient c. It is usual to express the modal parameters of each structural mode as they correspond to the *direct transfer function* (DTF) of the mode. The DTF is the ratio of vibration in X (relative between tool and workpiece) over a force F_x acting in the direction X, between tool and workpiece. We will denote it G_d:

$$G_d = X(\omega)/F_x(\omega) = (1/k)\omega_n^2/(\omega_n^2 - \omega^2 + 2j\zeta\omega_n\omega), \tag{3.15}$$

where the natural frequency ω_n is

$$\omega_n\,(\text{rad/sec}) = (k/m)^{0.5}$$

and the damping ratio ζ is:

$$\zeta = (c/2)/(km)^{0.5}.$$

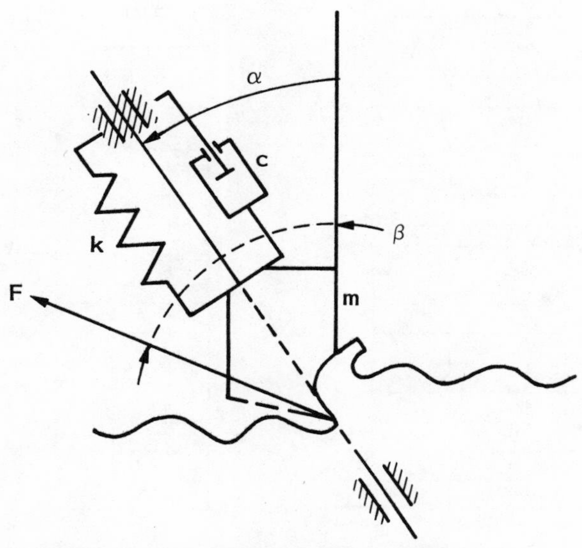

Fig. 3.9 Single degree-of-freedom system.

The DTF can be graphically expressed as a function of ω, as in Fig. 3.10a, by its absolute value $A = |X/F_x|$ and phase ϕ between F_x and X:

$$F_x = \sin \omega t,$$

$$X = A(\sin \omega t + \phi),$$

where $-\pi < \phi < 0$; or, as in Fig. 3.10b by its real and imaginary parts:

$$\begin{aligned}\mathrm{Re}(G_d) &= \mathrm{Re}(X/F_x)\\ &= (1/k)\omega_n^2(\omega_n^2 - \omega^2)/[(\omega_n^2 - \omega^2)^2 + (2\zeta\omega_n\omega)^2]. \quad (3.16)\end{aligned}$$

The minimum $G_{d,\mathrm{min}}$, i.e., the maximum negative point of Re(G), is

$$\mathrm{Re}(G_d)_{\mathrm{min}} = -1/4k\zeta(1 + \zeta) \quad (3.17)$$

at $\omega_{\mathrm{min}} = \omega_n(1 + \zeta)$,

$$\mathrm{Im}(G_d) = \mathrm{Im}(X/F_x) = -(2/k)\zeta\omega_n^3\omega/[(\omega_n^2 - \omega^2)^2 + (2\zeta\omega_n\omega)^2] \quad (3.18)$$

and its negative maximum is reached at $\omega = \omega_n$ and it is

$$\mathrm{Im}(G_d)_{\mathrm{min}} = -1/2k\zeta; \quad (3.19)$$

or, as in Fig. 3.10c by a polar plot in the phase plane where the frequency values are marked along the curve that represents the complex values of G. The points 1 and 2 represent the maximum negative imaginary and the maximum negative real parts corresponding to expressions (3.17) and (3.19). At

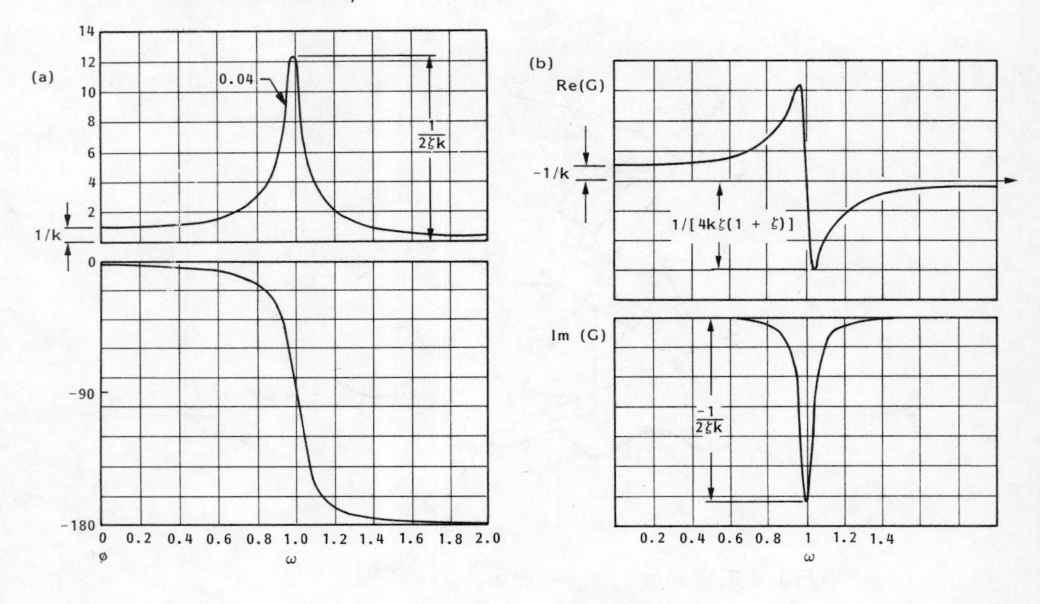

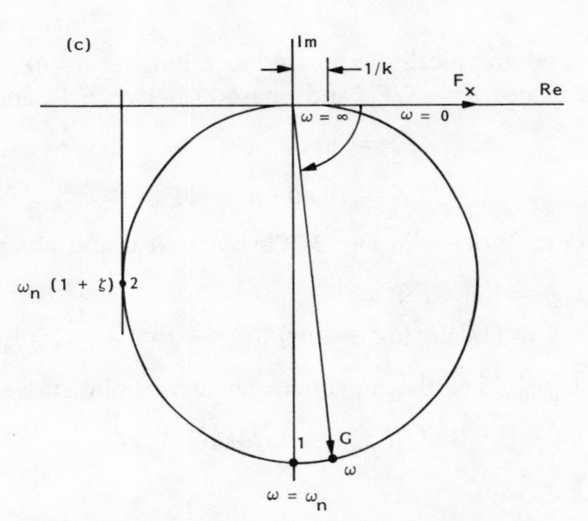

Fig. 3.10 Various ways of expressing the transfer function.

any frequency ω the length of the vector represents the magnitude of G and its phase ϕ, the phase angle between the force and the vibration. In this graph it is customary to draw the force on the real axis.

The conditions at the limit of stability are best explained using the representation as in Fig. 3.10c. However, it is first necessary to realize that the transfer function G which was used in Fig. 3.7 and in Eqs. (3.9)—(3.14) represented the response in the direction Y to a force acting in the direction of the cutting force. We will call it the *oriented transfer function* (OTF) and denote it simply as G.

It is obvious from Fig. 3.9 that it is only the projection $F \cos(\beta - \alpha)$ of the cutting force which excites the mode in the direction X and that it is the projection $X \cos \alpha$ of the vibration of the mode into the direction Y which will modulate the chip:

$$G(\omega) = Y(\omega)/F(\omega) = [X(\omega)/F_x(\omega)] \cos(\beta - \alpha) \cos \alpha = uG_d, \quad (3.20)$$

where $u = \cos(\beta - \alpha) \cos \alpha$ is the directional factor of the mode.

Now, the OTF is plotted in Fig. 3.11a. The cutting force F excites vibration Y. At the limit of stability of this vibration will have the same magnitude as the vibration Y_0 in the preceding pass, and it is (see Figs. 3.7 and 3.8):

$$Y_0 = Ye^{-j\varepsilon}$$

and

$$F = Cb_{\text{lim}}(Y_0 - Y) \quad \text{[see Eq. (3.7)]},$$

and, because Cb_{lim} is real, $(Y_0 - Y)$ is in phase with F, and it is shown as real. It is, for $F = 1$,

$$1/Cb_{\text{lim}} = -2 \operatorname{Re}(G)$$

and

$$b_{\text{lim}} = -1/2C \operatorname{Re}(G),$$

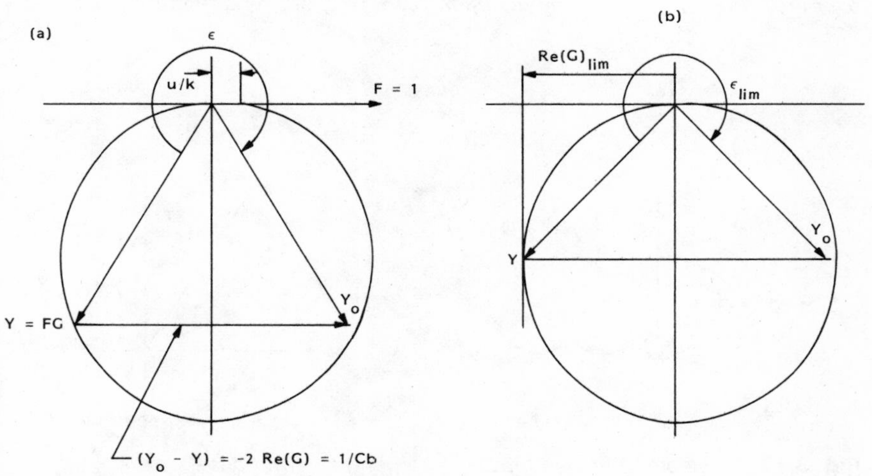

Fig. 3.11 Determining the limit of stability.

which is the expression as obtained in Eq. (3.13). If there is no constraint on ε, the minimum b_{lim} can be obtained as indicated in Fig. 3.11b,

$$(b_{lim})_{min} = -1/[2C\ \mathrm{Re}(G)]_{lim},$$

which was derived in Eq. (3.14).

The cases where there is no constraint on ε will be called basic or borderline stability and instead of writing $(b_{lim})_{min}$ we will write $b_{b,lim}$ or, simply, b_{lim} if it is understood from the context that it is the minimum value that is being considered.

It may be seen that the value ε_{lim}, which is the one that gives the borderline stability, is approximately 1.5π and, for a single-degree-of-freedom system, considering Eqs. (3.17) and (3.20):

$$b_{b,lim} = 1/4k\omega\zeta(1 + \zeta). \tag{3.21}$$

It is also useful to consider the graph in Fig. 3.10b and measure the magnitude of the minimum (negative maximum) of G to use in Eq. (3.14); in this way, Eq. (3.21) is again obtained. This equation is very important to remember.

For cases of systems with more than one degree of freedom it is best to use the representation as in Fig. 3.10b because the real part of the OTF of a complex system is simply a sum of the OTF's of the individual modes involved. This will now be shown for a system with two degrees of freedom, such as Fig. 3.4. Later, it will be shown that systems with two mutually perpendicular modes of vibration are very frequent and important in practice.

Referring to Fig. 3.12, there are two mutually perpendicular modes, $\alpha_2 = \alpha_1 + \pi/2$, with parameters k_i, m_i, c_i, $i = 1, 2$. The directional factors are

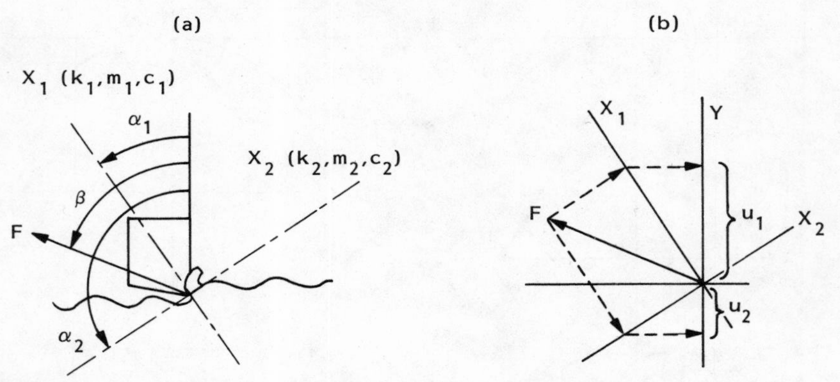

Fig. 3.12 System with two mutually perpendicular modes.

shown in Fig. 3.12b as geometric constructions by taking a unit vector F and projecting it onto X_1 and X_2 and further onto Y. Assuming, e.g., $\beta = 70°$, $\alpha_1 = 35°$,

$$u_1 = \cos(\beta - \alpha_1) \cos \alpha_1 = 0.671,$$

$$u_2 = \cos(\beta - \alpha_2) \cos \alpha_2 = -0.329.$$

The directional factor for mode 2 is negative. Choosing $m_1 = 48.7$ kg, $m_2 = 39.5$ kg, $k_1 = k_2 = 1 \times 10^4$ N/m, $\zeta_1 = \zeta_2 = 0.04$, we have $f_{n1} = \omega_{n1}/2\pi = 90$ Hz, $f_{n2} = \omega_{n2}/2\pi = 100$ Hz, and

$$G = u_1 G_{d1} + u_2 G_{d2}. \tag{3.22}$$

For the case of $\alpha_1 = 35°$, the OTF of the two modes as well as their sum G are shown in Fig. 3.13a. The "opposite" case in which mode 2 with the higher frequency is at $\alpha_1 = 35°$ (the two modes switched directions) is illustrated in Fig. 13b. It is seen that the resulting values of G_{lim} differ very much. By changing the directional orientation of a system with two perpendicular modes with close frequencies, the degree of stability can be very strongly influenced. Assuming $C = 2000$ N/mm^2, the limit depths of cut $b_{b,\text{lim}}$ of the two cases are

$$(b_{b,\text{lim}})_a = -1/2C \, \text{Re}(G)_{\text{min}} = 0.42 \text{ mm},$$

$$(b_{b,\text{lim}})_b = -1/2C \, \text{Re}(G)_{\text{min}} = 0.79 \text{ mm}.$$

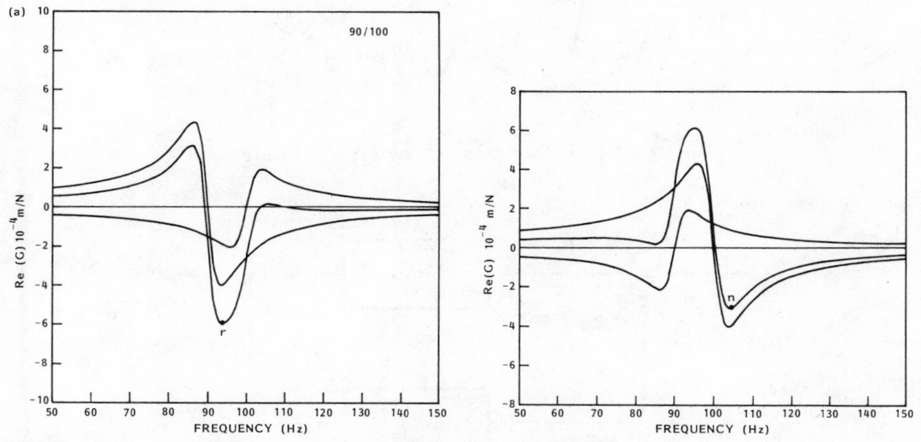

Fig. 3.13 Graphs for orientations with highest and lowest stability.

Nonregenerative Chatter

It was stated previously that in all practical machining operations the tool removes the chip from a surface that was generated in the preceding pass. Correspondingly, all practical situations are regenerative. One could arrange a very special case shown in Fig. 3.14 in which the top of a flat thread is turned and in which the tool does not cut into the surface generated in the preceding revolution of the workpiece; regeneration is not possible, at least not in one pass. Self-excitation can only be based on mode coupling, as explained in reference to Fig. 3.4.

This very special case is not practical and would not be a justification for discussing the limit of stability of nonregenerative chatter. However, it will be shown later that regeneration is hindered if the phase ε of subsequent undulations is constrained to $\varepsilon = 0°$, or if milling is done with special tools like those with alternating helix. In those cases chatter may only exist in the nonregenerative way. It is therefore useful to discuss briefly the difference in stability between regenerative and nonregenerative situations.

For nonregenerative chatter the block diagram of Fig. 3.7 is replaced by the one in Fig. 3.15. The cutting force equation is obtained from Eq. (3.7) by

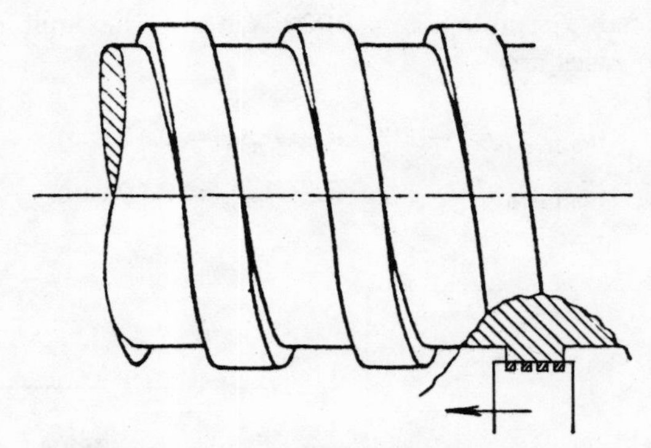

Fig. 3.14 Arrangement for nonregenerative cutting.[8]

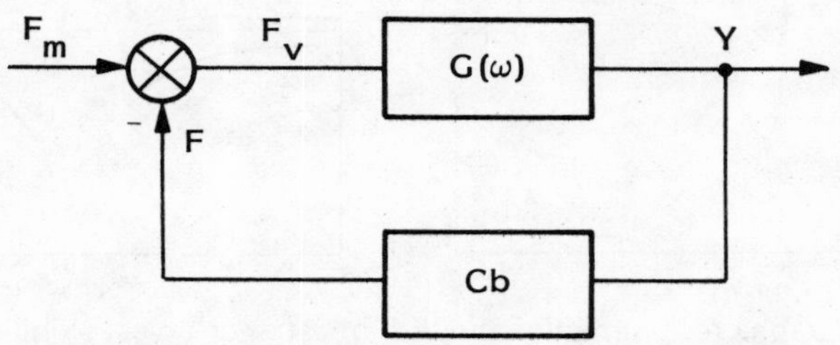

Fig. 3.15 Block diagram for nonregenerative cutting.

setting $y_0 = 0$,

$$F = F_m - Cby. \tag{3.23}$$

For the limit of stability the open-loop transfer function is set equal to -1:

$$GCb = -1. \tag{3.24}$$

Condition (3.24) is only satisfied if $G(\omega)$ is purely real, i.e., if $\mathrm{Im}(G) = 0$. Checking with Fig. 3.10c it is seen that for a single degree of freedom this is so only for $\omega = 0$ and $\omega = \infty$; the corresponding values of G are $G = 0$ and this would lead to $b_{\mathrm{lim}} = \infty$. For a single-degree-of-freedom system, nonregenerative chatter cannot exist. This was, in any case, already stated in the description of Fig. 3.4 where an elliptical motion resulting from the combination of two vibratory motions was postulated.

For such systems where $\mathrm{Re}(G)$ has a final value G_{lim} at such frequency at which $\mathrm{Im}(G) = 0$, the limit of nonregenerative instability is obtained for

$$b_{b,\mathrm{lim}} = -1/CG_{\mathrm{lim}}, \qquad \text{when } \mathrm{Im}(G)_{\mathrm{lim}} = 0. \tag{3.25}$$

This result is illustrated in Fig. 3.16a for the same system with two degrees of freedom as in the case of Fig. 3.13a. The same diagram of $\mathrm{Re}(G)$ is reproduced again with $\mathrm{Im}(G)$ included. The frequency is found for which $\mathrm{Im}(G) = 0$ and the corresponding G_{lim} is determined. It is, in this case,

$$(b_{b,\mathrm{lim}})_{a,\mathrm{nonregenerative}} = -1/CG_{\mathrm{lim}} = 0.87 \text{ mm}.$$

In the graph of $\mathrm{Re}(G)$ the value of G_{lim} for regenerative chatter is simply marked by r and the value for nonregenerative chatter by n. The two values are not far apart, but, since there is a factor of 2 in Eq. (3.14) which is not in Eq. (3.25), the limit width of chip of nonregenerative chatter is approximately two times greater than for regenerative chatter:

$$(b_{b,\mathrm{lim}})_n \simeq 2(b_{b,\mathrm{lim}})_r. \tag{3.26}$$

In graphs 3.16b the $\mathrm{Re}(G)$ of Fig. 3.13b is reproduced again, together with $\mathrm{Im}(G)$; it is seen that in this case for $\mathrm{Im}(G) = 0$ the corresponding $\mathrm{Re}(G)$ value is positive. No b_{lim} value can then satisfy Eq. (3.25) and, for the nonregenerative situation, the case would be infinitely stable. In the orientation in which the mode with the positive directional factor has higher natural frequency than the mode perpendicular to it, no nonregenerative chatter can occur. In the physical interpretation of Fig. 3.4 this orientation would force the elliptical motion in the direction opposite to the arrows shown, and the cutting force would actually be consuming the vibratory energy.

Both the regenerative and nonregenerative limits of stability for systems with two mutually perpendicular degrees of freedom are now recapitulated. The case of a system with both modes exactly equal, with the two natural frequencies $f_{n1} = f_{n2} = 100$ Hz, is taken as reference. The damping ratios

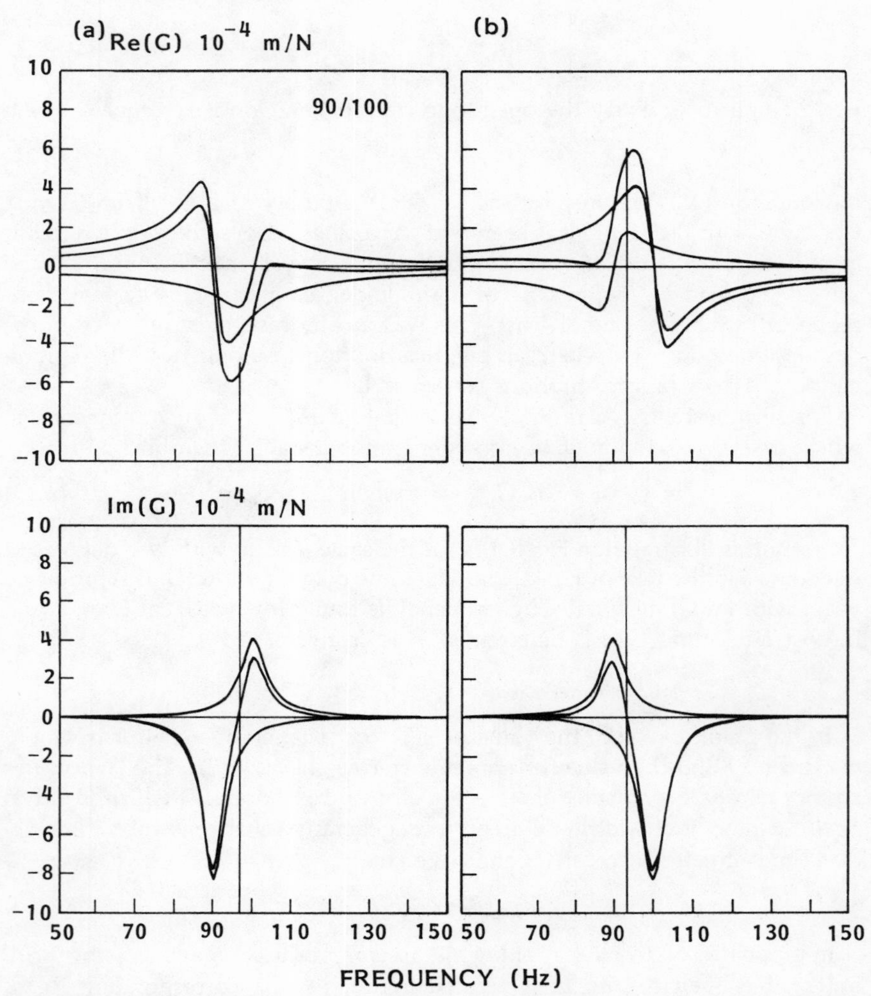

Fig. 3.16 Comparison of stability limits for regenerative and nonregenerative cutting.

are $\zeta_1 = \zeta_2 = 0.04$ and the masses, $m_1 = m_2$, as well as the two stiffnesses, $k_1 = k_2$, are equal. This case will correspond to the diagrammatical situation as shown in Fig. 3.17, where a tool is attached to the end of a boring bar with a round cross section. Such a bar possesses two mutually perpendicular degrees of freedom with equal frequencies and their orientation can be freely chosen. In the diagram it is chosen so that X_1 is normal to the cut, $X_1 = Y$ and X_2 is parallel with the cutting speed. The angle β of the cutting force is $\beta = 70°$. The directional factor $u_2 = 0$ because the vibration X_2 has no component in

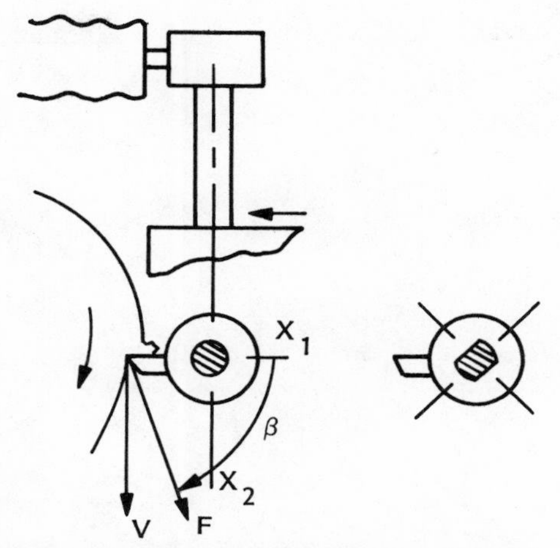

Fig. 3.17 Two degree-of-freedom "boring bar" system.[2]

Y. It is sufficient to consider the single mode in X_1. The value b_{lim} is

$$b_{\text{lim}} = -1/2CG_{\text{lim}} = 4k\zeta(1 + \zeta)/2Cu, \tag{3.27}$$

where $u = \cos\beta = 0.342$; and with $\zeta = 0.04$,

$$(b_{\text{lim}})_{\text{ref}} = 0.243k/C. \tag{3.28}$$

This result is very simple and easy to remember. One can say that with the common amount of damping ($\zeta = 0.04$) the round boring bar gives a limit width of chip which is about 25% of the ratio k/C of the spring stiffness to "cutting stiffness," the cutting stiffness C being defined as the increment in the magnitude of the cutting force per unit increment of the depth of cut, at unit chip width. For instance, for cutting steel the value of C is commonly 2000 N/mm^2; a system with $k = 10{,}000$ N/mm (10 N/μm) would give $b_{\text{lim}} = 1.22$ mm.

All the cases for which $f_{n1} \neq f_{n2}$ could be represented as in Fig. 3.17b with a tool at the end of a bar with a rectangular cross section such that $k_1 < k_2$ and $f_{n1} < f_{n2}$, and the angle α_1 denotes the direction of the mode with the lower frequency. In reality, the two modes may arise from many different design configurations and their stiffnesses may be nearly equal while the natural frequencies differ. Correspondingly, we will keep the two damping values $\zeta_1 = \zeta_2$ and the two stiffness values $k_1 = k_2$ equal to those of the reference case and vary the mass m_1 so as to achieve $f_{n1} < f_{n2}$, and we will try different ratios f_{n1}/f_{n2} while keeping $f_{n2} = 100$ Hz. We will vary the angle α_1 through 360° and check the variation of b_{lim} for both the regenerative (r) and nonregenerative (n) cases as related to the b_{lim} of the reference case, which will be

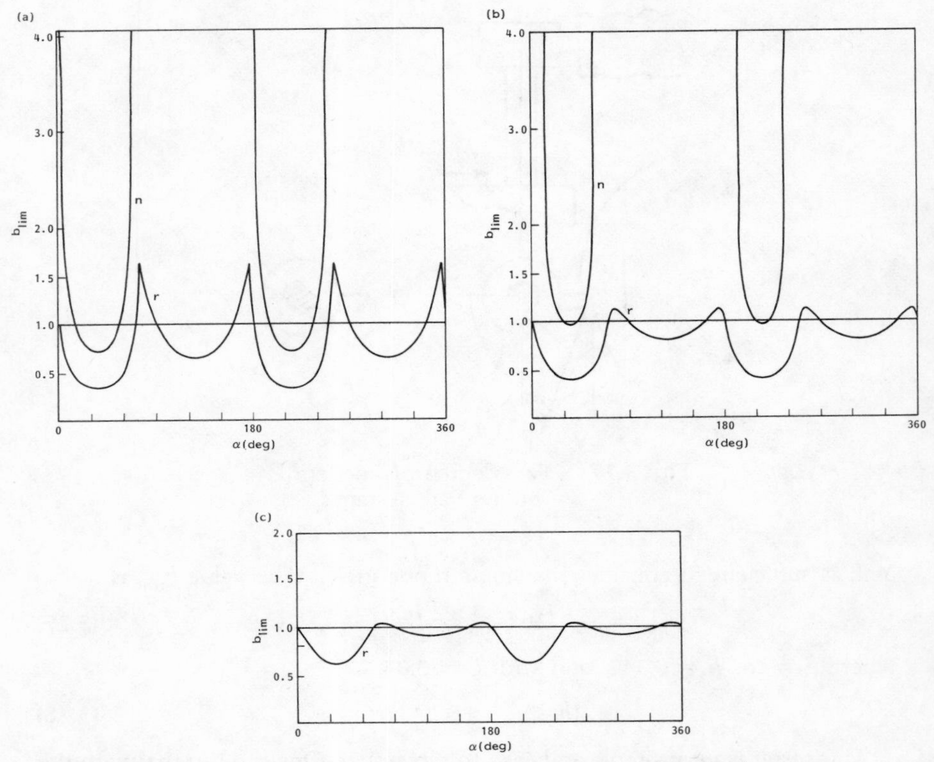

Fig. 3.18 Effect of directional orientation on stability of the boring bar system.

set as

$$(b_{\text{lim}})_{\text{ref}} = 1.0.$$

The results of this investigation are plotted in Fig. 3.18 for three f_{n1}/f_{n2} ratios with (a) $f_{n1} = 90$ Hz, (b) $f_{n1} = 96$ Hz, (c) $f_{n1} = 98$ Hz. It is seen that non-regenerative chatter exists for only a narrow range of orientations in Fig. 3.18a, and this range is narrowing as the two frequencies come closer, and it practically disappears in Fig. 3.18c. The regenerative chatter is also strongly influenced by the orientation of the system. For instance, in Fig. 3.18a the value of b_{lim} in the best orientation, $\alpha_1 = 75°$, is five times higher than in the worst orientation, $\alpha_1 = 37.5°$.

The significance of the directional orientation for two mutually perpendicular modes with close natural frequencies can be illustrated using the example of a horizontal knee-type milling machine.[8] Four significant modes were measured on this machine, as are shown in Fig. 3.19. Two of them had the direction X of relative vibrations between tool (overarm) and table, which is horizontal and parallel with the table axis. Their natural frequencies were

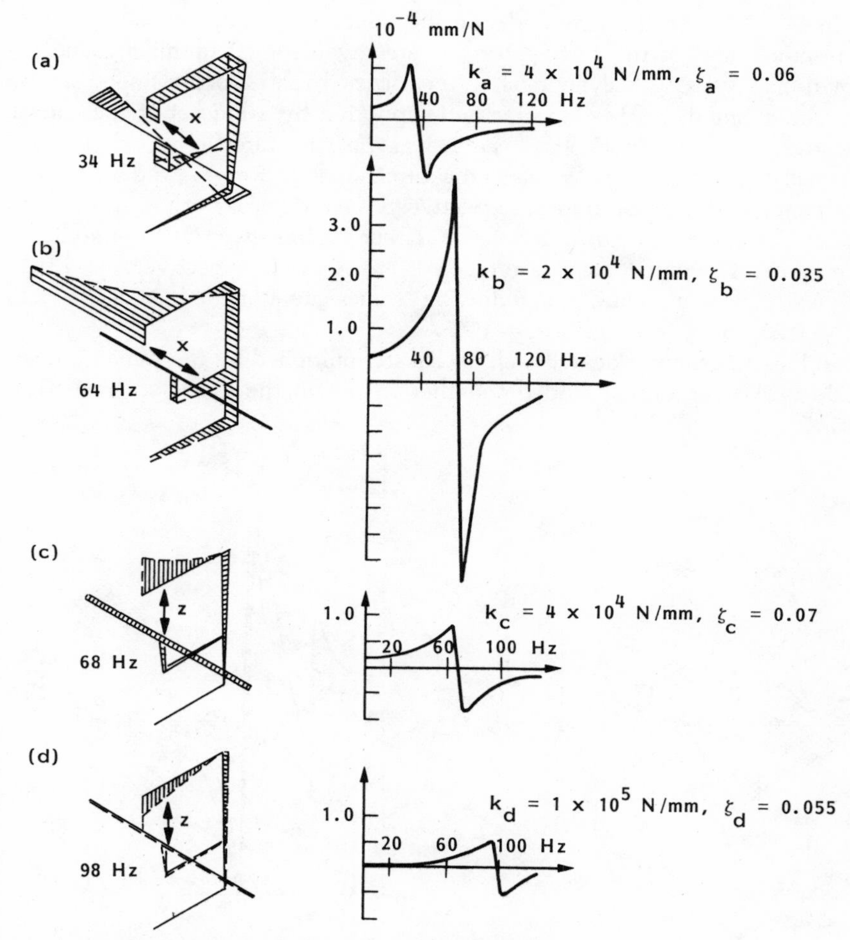

Fig. 3.19 Mode shapes of a milling machine.[8]

34 and 64 Hz. Relative vibrations between the tool and table in two other modes had the vertical direction Z. The mode in Fig. 3.18a consisted essentially of the column with overarm vibrating against the knee and table. The mode in Fig. 3.18b was characterized by horizontal vibration of the overarm against the column. The mode in Fig. 3.18c was essentially based on the first vertical mode of the overarm and column, and the mode in Fig. 3.18d was based on the second higher shape of these two structural elements.

The modal DTFs of each mode are presented in Fig. 3.19 by their real parts, and modal stiffnesses and damping ratios are also given. They will now be used to construct the OTF's for two cases of milling with different directional orientations.

In Fig. 3.20, on the left-hand side the directions of the normal Y to the cut surface and of the cutting force F are given for (a) up-milling and (b) down-milling using a cylindrical cutter carried by an arbor clamped in the horizontal spindle of the machine and supported by a bracket at the end of the overarm. The directions Y are equal, but the directions F differ substantially. Correspondingly, also, the directional factors u_x and u_z of the horizontal and vertical modes, respectively, have different values in the two cases. The geometric constructions of u_x and u_z are given in the middle of the figure. The unit force F is projected into X or Z, respectively, and subsequently into Y. Thus, the following values are obtained: (a) $u_x = 0.25$, $u_z = 0.25$; (b) $u_x = -0.18$, $u_z = 0.68$.

In Fig. 3.20a, coincidentally, all DTF's are multiplied by the same number, 0.25 (broken lines), and added together to obtain the OTF (full line). The

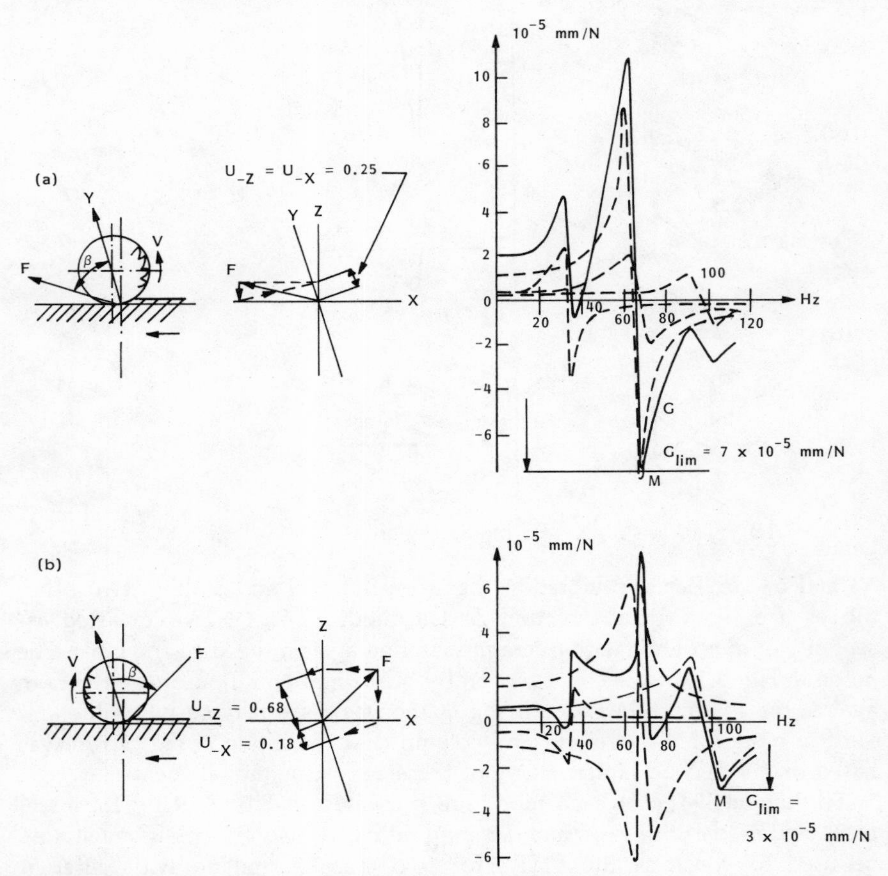

Fig. 3.20 Oriented transfer function and stability for up- and down-milling.[8]

mode 3.19b being predominant determines the minimum, $G_{min} = 8.74 \times 10^{-5}$ mm/N. In Fig. 3.20b, the DTF of Fig. 3.19b is multiplied by -0.18 while Fig. 3.19c, which has a very similar frequency, is multiplied by a four times higher directional factor. The two curves almost cancel each other, and it is now the much stiffer mode Fig. 3.19d that determines the $G_{min} = 3.77 \times 10^{-5}$ mm/N. Correspondingly, the value of the limit width of milling b_{lim} is $8.77/3.77 = 2.3$ times greater for down-milling than for up-milling.

So far, all our analysis was based on cutting with a single-point tool (turning, boring). For milling, where not only are there several teeth cutting simultaneously but also where chip thickness varies continuously, and where the cutting process on each tooth has a different directional orientation with respect to the vibratory system of the structure of the machine tool, the theory as presented can only give an approximate assessment. Experience has shown that such an approximation is not satisfactory for determining the effects of spindle speed and of special cutters on chatter. Therefore, an analysis that is more realistic is presented in the following section; it is based on time-domain simulations.

Theory of Chatter by Time-Domain Simulation

The theoretical aspect of chatter can be well illustrated and understood by using digital simulation. This was first explained in Ref. 3, which is the basis for this Section. This approach not only permits a good insight into the behavior of the vibrating system but also makes it possible to correctly take into account the basic nonlinearity of the process and the cross effects of teeth of milling cutters. These latter aspects will be explained as we proceed. However, right now, let us indicate that the nonlinearity of machining chatter is mainly due to the fact that when vibration grows larger, the tool jumps out of the cut for a part of the vibrational period and the cutting force disappears for this time instant. The cross effects in milling arise from the different directions of the normals to the cut surface at each tooth; e.g., a vibration that is tangential to the cutting of one tooth and does not, therefore, modulate its chip thickness may be most efficient in modulating the chip thickness of a tooth 90° apart on the cutter circumference for which it is normal to the cut surface, and vice versa.

Let us first look at the case of turning. The model used for simulation is shown in Fig. 3.21. A system with two mutually perpendicular degrees of freedom, X_1 and X_2, is used. A tool is attached to the mass of the system, and it cuts an undulated surface and leaves another undulated surface behind. The normal to the cut surface is denoted Z.

The parameters of the simulated system are chosen so that stiffnesses of both modes are equal, $k = 18{,}000$ N/mm (10^6 lbf/in.), but the frequencies

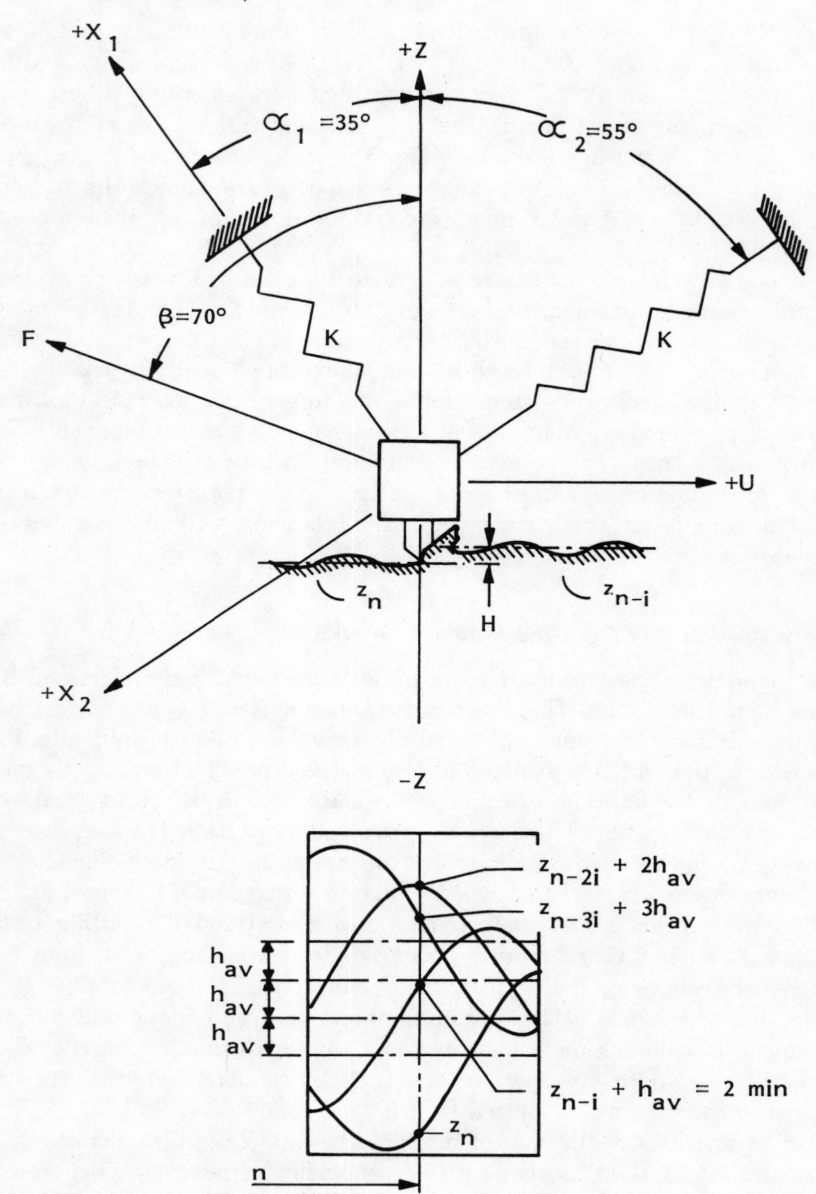

Fig. 3.21 Model for simulation of turning.

differ, $f_{n1} < f_{n2}$. Various values of the ratio $R = f_{n1}/f_{n2}$ will be used. The directional orientation is given by $\alpha = 35°$, $\beta = 70°$. Damping ratios of both modes are taken to be $\zeta_1 = \zeta_2 = 0.05$. Cutting stiffness is chosen as $C = 2000\ \text{N/mm}^2$ (2.9×10^5 psi). The simulation is carried out in small time steps dt.

Displacements x_1, x_2, z are subscripted by n. Subsequent cuts follow one another after i steps of the computation (after i computational loops), i.e., after the time $T = i\,dt$. The time T corresponds to one revolution of the workpiece. The average chip thickness is h_{av}. In our examples it is chosen $h_{av} = 0.15$ mm (0.006 in.). The instantaneous chip thickness h is usually taken as the difference between two consecutive passes:

$$h = h_{av} + z_{n-i} - z_n$$

but, as the vibrations grow, more than one preceding cut may be involved. Looking at the bottom of Fig. 3.21, at any time instant n we may check which of the cuts preceding the nth cut reached lowest into the workpiece material and denote such a position z_{min}:

$$z_{min} \text{ is lowest of } z_{n-i} + h_{av}, \quad z_{n-2i} + 2h_{av}, \quad z_{n-3i} + 3h_{av}, \tag{3.29}$$

and the instantaneous chip thickness is

$$h = z_{min} - z_n. \tag{3.30}$$

The cutting force is taken proportional to chip width b and to chip thickness h:

$$F_n = Cbh, \tag{3.31}$$

but, if $F_n < 0$,

$$F_n \leftarrow 0 \tag{3.32}$$

where Eq. (3.32) expresses the basic nonlinearity. If the force becomes formally negative, it means that the tool is out of cut and the force must actually be set to zero. The cutting force is never less than zero.

The force excites vibrations x_1 and x_2 by its components

$$\begin{aligned} F_1 &= F\cos(\beta - \alpha_1), \\ F_2 &= F\cos[\pi - (\beta - \alpha_2)]. \end{aligned} \tag{3.33}$$

These forces excite vibrations according to:

$$m_j\ddot{x}_j + c_j\dot{x}_j + k_jx_j = F_j, \quad j = 1, 2, \tag{3.34}$$

and the motion z which modulates the chip thickness is obtained from the components of the two vibrations

$$z = x_1 \cos\alpha_1 - x_2 \cos\alpha_2. \tag{3.35}$$

In the simulation, in each step n the acceleration $\ddot{x}_{jn}$ is determined from Eq. (3.34) and the displacement $x_{j,n+1}$ is obtained by double integration. Thus, e.g., for x_1:

$$\begin{aligned} \ddot{x}_{1,n} &= (F_{1,n} - c_1\dot{x}_{1,n} - k_1x_{1,n})/m_1, \\ \dot{x}_{1,n+1} &= \dot{x}_{1,n} + \ddot{x}_{1,n}\,dt, \\ x_{1,n+1} &= x_{1,n} + \dot{x}_{1,n+1}\,dt. \end{aligned} \tag{3.36}$$

Similarly $x_{2,n+1}$ and $\dot{z}_{n+1}$ are obtained.

This sytem is linear as long as the tool does not leave the cut. Therefore, for incipient vibration the classical equation (3.14) for the "basic" limit of stability width of chip b_{lim} applies:

$$b_{\text{lim}} = -1/2CG_{\text{min}}$$

where G_{min} is the real part of the "oriented" transfer function between vibration z and force F and most favorable phasing is considered between vibrations in subsequent cuts. This latter condition is practically fulfilled if the number of waves between subsequent cuts is large; in our example it is 16.6.

With a system of two degrees of freedom oriented like the one in Fig. 3.21 the value of b_{lim} depends on the ratio R of the two frequencies. The values of b_{lim} versus this ratio as established from Eq. (3.14) and using the abovementioned parameters of the system are plotted in Fig. 3.22. It is seen that b_{lim} is minimum for $R = f_{n1}/f_{n2} = 0.94$ and that for $R = 1.0$ the value of

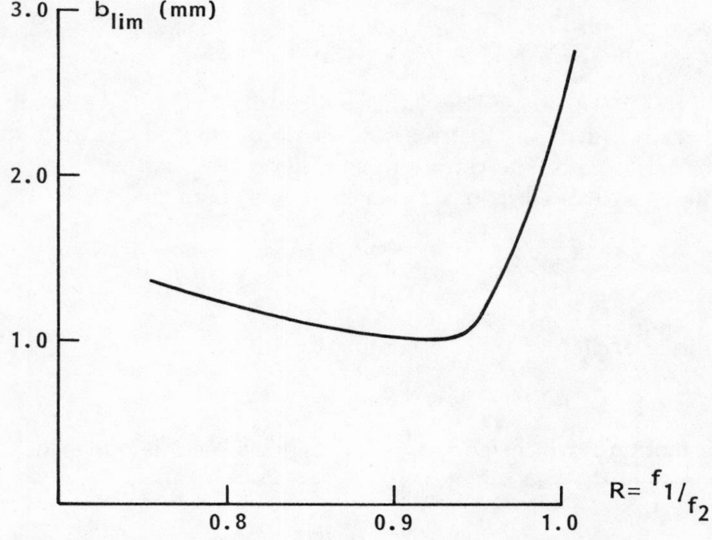

Fig. 3.22 Limit of stability versus tuning of two modes.

b_{lim} is 2.5 times larger than that. The case of $R = 0.94$ is selected here for simulation.

In Fig. 3.23a vibration z during the first three cuts (passes) is plotted versus time $t = n\,dt$, where $dt = 0.0002$ sec. The chip width is chosen as $b = 1.212$ mm, which is slightly over b_{lim}. The length of one cut corresponds to one revolution of workpiece (in turning) per 0.1 sec. This means that the number of simulation steps between consecutive cuts is $i = 500$. The individual workpiece revolutions are correspondingly indicated as i, $2i$, $3i$, etc. It is seen that at the entry of the tool into the first cut the natural decaying vibration of the system is produced, which leaves undulations on the surface, and these begin to reproduce during the second and third cut. Between the second and third cut a growth is seen indicating instability. However, already the first cut is not purely natural vibration because of "mode coupling." In Fig. 3.23b simulation of a case is presented with larger chip width $b = 2.2b_{lim} = 2.222$ mm. The vertical scale of the graph is 70 times larger to accommodate the much faster rate of increase of vibration. Here it is seen that there is practically no decay at the beginning. This is due to a larger self-excitation energy arising

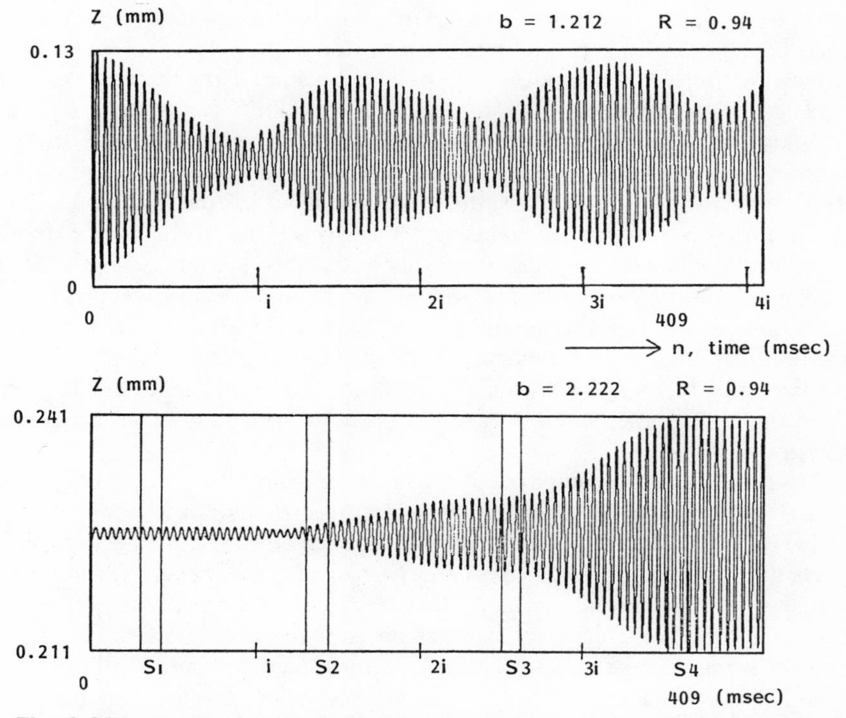

Fig. 3.23 Development of chatter in turning at two different chip widths.[3]

from the "mode coupling" due to the larger chip width. Also "wave regeneration" is much more active and in the fourth revolution the vibration z reaches an amplitude 0.22 mm, which is 1.5 times the average chip thickness h.

The details of the process are shown in Figs. 3.24 and 3.25 which expand on sections of Fig. 3.23b.

In Fig. 3.24a vibrations z are plotted for sections S1, S2, S3, S4 of Fig. 3.23b. These sections follows each other exactly always after one revolution, and they represent subsequent undulations on one and the same section of the circumference of the workpiece. It is seen that in the fourth cut, already, the tool jumps out of the cut for a part of the vibration cycle. The shaded area represents the material removed in this cut. The horizontal coordinate in this case is the angle of workpiece rotation in degrees. There are about 17 waves per circumference of the workpiece. In Fig. 3.24b the cutting force F is shown for the last cut, z_n, of Fig. 3.24a. It is seen that, repeatedly, the force becomes zero for a part of the cycle. In Fig. 3.24c vibration z_n of Fig. 3.24a is plotted versus vibration u_n; see Fig. 3.21. The graph shows the actual motion of the tip of the tool in the Z, U coordinates. The tool moves on this elliptical path in the direction of the arrow, and during the motion from point A to point B it moves outside of the cut. The cutting force F has the direction as indicated at one arbitrary point along the path. During the motion from point 1 to point 2 the force has a positive component into the direction of the velocity of motion—it drives the tool. During motion from point 2 to point 1 the force component acts against the motion—the force brakes the motion. Even without the tool jumping out of the cut, with the direction of motion as shown here, the 1–2 motion is carried out deeper in the cut than 2–1, and, therefore, during the 1–2 part the force is greater and the work supplied is greater than during 2–1 and a surplus of self-excitation energy results. This is the mechanism of "mode coupling" in metal-cutting chatter as will be described in a subsequent section. This mechanism acts here in addition to the "regeneration of waviness" mechanism and increases the intensity of self-excitation. For a single-degree-of-freedom system the tool moves on a straight line and mode coupling is not active. Similarly, for $f_{n1} = f_{n2}$ the tool moves on a straight line (both vibrations have the same phase) and therefore the case of $R = 1$ has higher stability in the graph, Fig. 3.22.

Now, with the tool jumping out of cut the braking action works only from B to 1 and there is still more energy left for self-excitation.

In Fig. 3.25a vibration z is shown in sections over the same part of workpiece revolutions numbers 99–103 after the start of the cut. Therefore, this graph actually follows the one in Fig. 3.24a after 98 revolutions of workpiece. By this time vibration z has stabilized at an amplitude of 0.35 mm (0.014 in.) (2.3 times the average chip thickness). The tool leaves the cut for a little more than half the vibratory cycle. The shaded area shows the material cut during each cycle. In Fig. 3.25b the tool motion is shown in coordinates Z, U. Again,

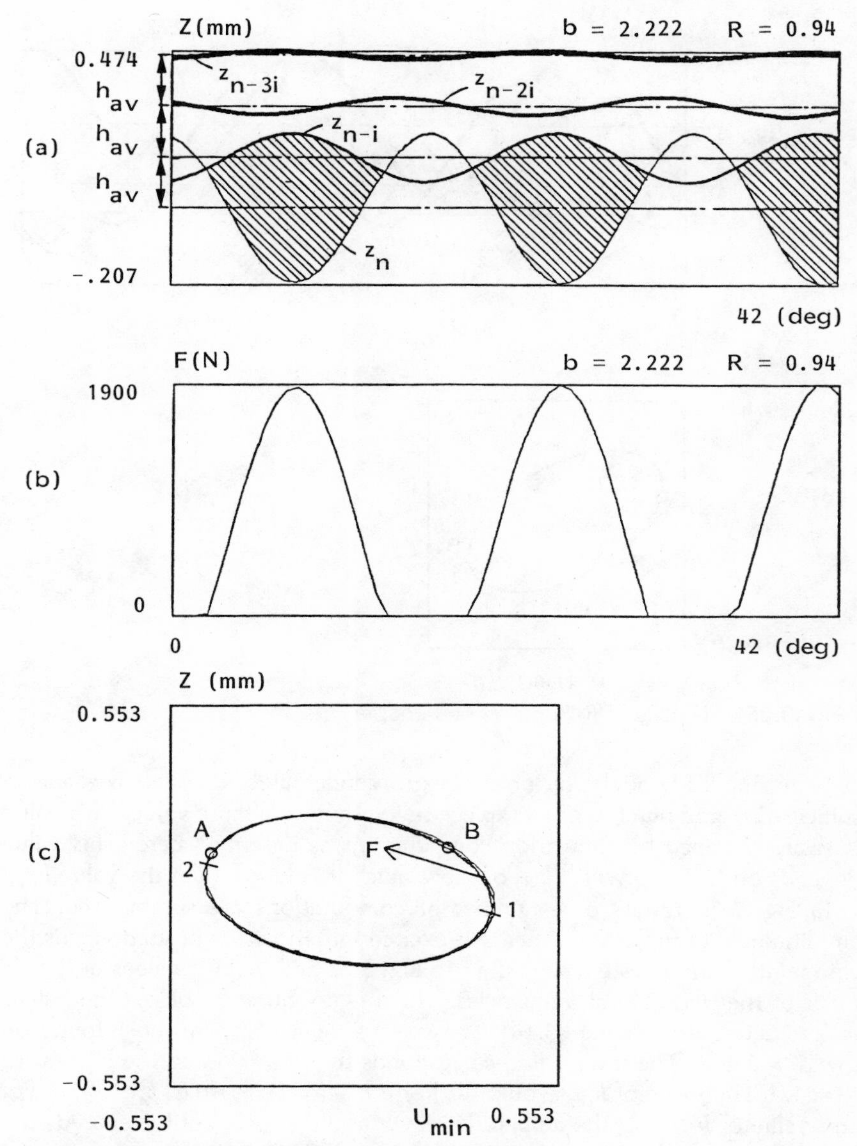

Fig. 3.24 Details of the developing chatter.

during motion 1–2 the force drives the tool. Between A and B the tool does not cut. Thus, there is no force to brake the tool along the path 2–1.

The diagrams in Figs. 3.23, 3.24, and 3.25 illustrate very clearly the process of self-excitation in turning. Owing to the nonlinearity of the tool leaving the

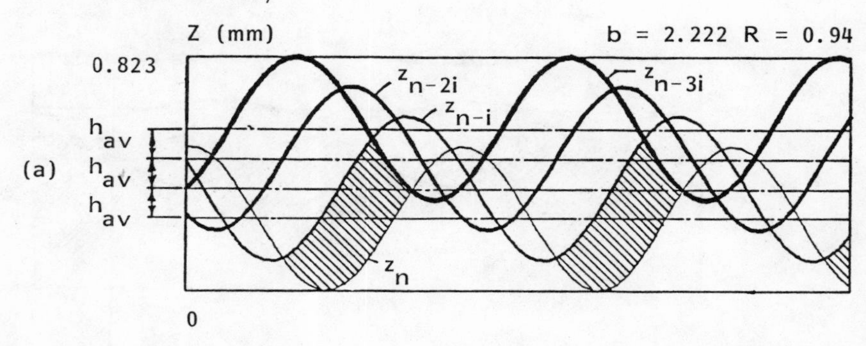

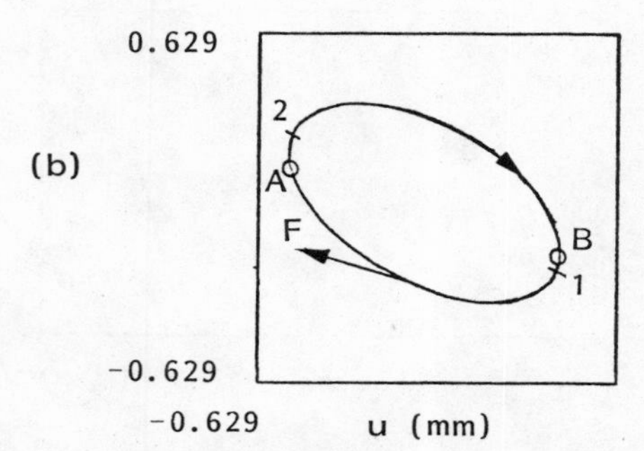

Fig. 3.25 Details of fully developed chatter.

cut, the amplitude of chatter does not grow indefinitely but stabilizes after a sufficiently large number of workpiece revolutions at a finite value. This value is easily obtained by simulation computations as described here. This value depends on the chip width b—on how much it is larger than the value b_{lim}.

In Fig. 3.26 results of a number of computations are summarized. The amplitudes Z of fully established self-excited vibrations are plotted versus the chip width b for a system according to Fig. 3.21 and various values of the ratio R of the two natural frequencies. There is no vibration below the values b_{lim} as determined by Eq. (3.14). The various values b_{lim} are those found on the $Z = 0$ axis. The lowest b_{lim} corresponds to $R = 0.94$ and the highest to $R = 1.0$. The value of b_{lim} would further grow for $R > 1$ (i.e., $f_{n1} > f_{n2}$). For any value of $b > b_{lim}$ the amplitude of vibrations grows until it exceeds the average chip thickness h_{av} (the feed per revolution). In our example $h_{av} =$ 0.15 mm (0.006 in.). Then the tool starts jumping out of the cut and the amplitude stabilizes on a value that increases with any further increase in b.

However, the vibration with an amplitude larger than h_{av} is usually considered unacceptable. Correspondingly, one may conclude that the boundary between stable and unstable machining with continuous cuts (turning, boring) is very sharp.

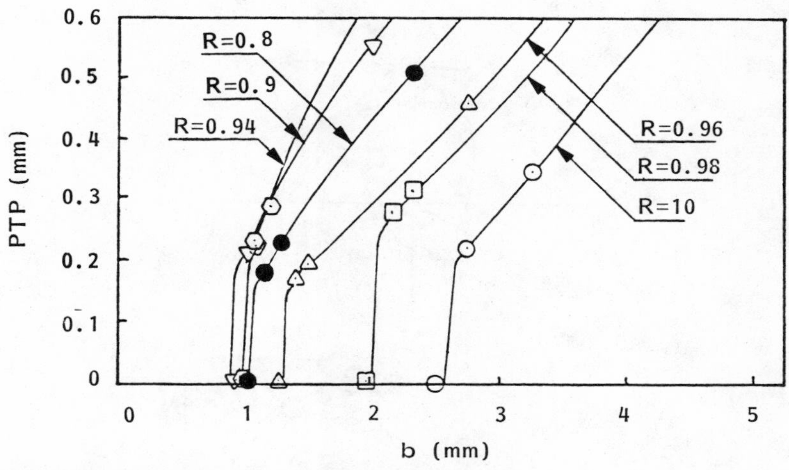

Fig. 3.26 Amplitudes of chatter in turning.

Chatter in Milling The geometry of the milling operation is depicted in Fig. 3.27, where three cases are selected, which will be further analyzed: (a) half-immersion (radial depth of cut equals half cutter diameter) up-milling; (b) half-immersion down-milling; (c) full immersion milling, i.e., slotting. In slotting the chip thickness h varies from zero to a maximum of f_t at $\phi = 90°$ and again to zero. The value f_t is feed per tooth and is $h = f_t \sin \phi$. Cases a and b represent the first and second halves of case c, respectively. We consider an end milling cutter with four teeth, and, for simplicity, straight (non-helical) teeth are assumed.

Two modes of vibrations are considered, one in direction X and the other in direction Y. They have the same stiffness $k = 2 \times 10^4$ N/mm (1.14×10^5 lb/in.) and damping ratio $\zeta = 0.05$ but different natural frequencies, f_x, f_y.

In the following we will be discussing five different arrangements of the vibratory system:

(a) $f_x = f_y = 660$ Hz,
(b) $f_x = 633$ Hz $< f_y = 660$ Hz,
(b′) $f_x = 660$ Hz $> f_y = 633$ Hz,
(c) $f_x = 660$ Hz $< f_y = 960$ Hz,
(c′) $f_x = 960$ Hz $> f_x = 660$ Hz.

Cases a, b, and c differ by the ratio of the natural frequencies of the two modes of vibration. In all these three cases the direction of feed is X, as shown in Fig. 3.27. Cases b′ and c′ are the reverses of b and c, respectively. They will be indicated as milling with feed Y, because they act as b and c if f_x and f_y are interchanged.

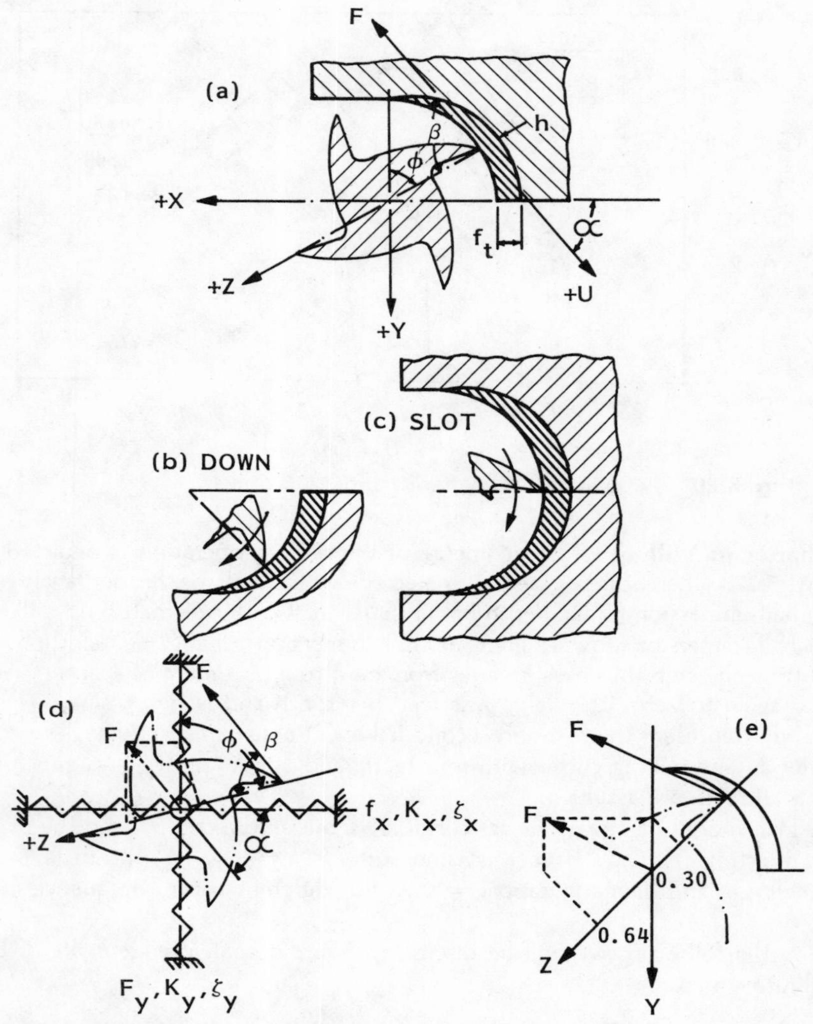

Fig. 3.27 Model for simulation of milling.

The cutting force F is assumed simply proportional to chip thickness. Its direction varies with the rotation of the cutter. In the common terms in discussing chatter in machining, the "directional orientation" of the cut varies; see Fig. 3.27d.

The cutting force F is translated into the milling cutter axis. It excites the X mode by its projection $F \cos \alpha$ and the Y mode by its projection $F \sin \alpha$. Each of these vibrations participates in the motion Z modulating the chip thickness:

$$z = x \sin \phi + y \cos \phi. \tag{3.37}$$

Correspondingly, the directional factors

$$u_x = \cos\alpha \sin\phi = \cos[\beta + \phi - (\pi/2)] \sin\phi$$
$$= \sin(\beta + \phi) \sin\phi$$

and

$$u_y = -\sin\alpha \cos\phi = \cos(\beta + \phi) \cos\phi$$

act as gain factors of the involvement of the X and Y modes in regeneration of surface waviness.

In analyzing chatter in milling it has, so far, always been done so that instead of the variable directional factors [Eq. (3.7)] ones were used that correspond to the mean directional orientation. This is illustrated for the up-milling case in Fig. 3.27e. They would correspond to (a) $\phi = 45°$, (b) $\phi = 135°$, and (c) $\phi = 90°$, and for $\beta = 70°$ their values would be:

(a) $u_x = 0.64$, $u_y = -0.03$,
(b) $u_x = -0.03$, $u_y = 0.64$,
(c) $u_x = 0.34$, $u_y = 0$.

This is, of course, an oversimplification.

If the variation of the directional factors is appreciated, it is realized that during the cut either the X or the Y mode takes over and the frequency of vibration may change.

If excitation were an instantaneous phenomenon, then the limit of stability against chatter would vary during the rotation of the cutter tooth as the directional factors and the combination of modes vary. However, excitation cannot vary instantaneously, because vibrations decay or increase slowly; time is needed for transitions, especially since chatter cannot start with one frequency and change instantaneously into another frequency. Obviously in each of the presented cases there is some "average" value of $b_{\lim}$.

Furthermore, the variation of chip thickness also has strong implications for the "saturation" of the force as expressed by the nonlinear condition (3.32). At the beginning of the cut in case a, at the end of cut in case b, and both at the beginning and end of cut in case c, the average chip thickness is zero and, therefore, the tool will jump out of the cut at even the smallest vibration.

The cutting force F and its components F_y and F_x are also periodically variable in the case of stable cutting, without chatter. This variation has a basic period equal to the interval between teeth, which we will choose to be 1/80 sec, and it has a number of harmonic components. We will not discuss the form of the variation of the forces; it will be well illustrated in the following plots. It is, however, necessary to point out that in cases a and b the sudden changes at the exit and entry, respectively, of the cut periodically excite decaying natural vibrations in both X and Y (more so in Y). In the case c, where two teeth cut simultaneously, the sum of their forces is constant and it has a constant direction (if the cutter has no run-out). Therefore, in this case, the cutting force itself does not excite any vibration unless self-excitation occurs. All these aspects—the varying directional orientation, the varying chip thickness, and the forced and transient natural vibrations due to the variation of the cutting force—make the analysis of milling much more complicated than that of turning. The only way it can be done properly is by time-domain simulation.

The simulation follows the procedure as presented in Eqs. (3.29)–(3.36), except that now instead of Eqs. (3.33), we have

$$F_x = F \cos \alpha = F \sin(\phi + \beta),$$
$$F_y = F \sin \alpha = F \cos(\phi + \beta), \tag{3.38}$$

and Eq. (3.35) is replaced by Eq. (3.37), and all the variables F_x, F_y, x, y, z are also functions of the angle ϕ of the rotation of the cutter tooth. If there are two or more teeth cutting simultaneously, then each of them generates corresponding F_x and F_y components (angles ϕ differ for the different teeth) and all the F_x and F_y components are added, respectively.

The transition between stable and unstable cutting is "blurred" by the transient natural vibrations excited every tooth period occurring below the limit of stability. This is illustrated in Figs. 3.28 and 3.29. In Fig. 3.28

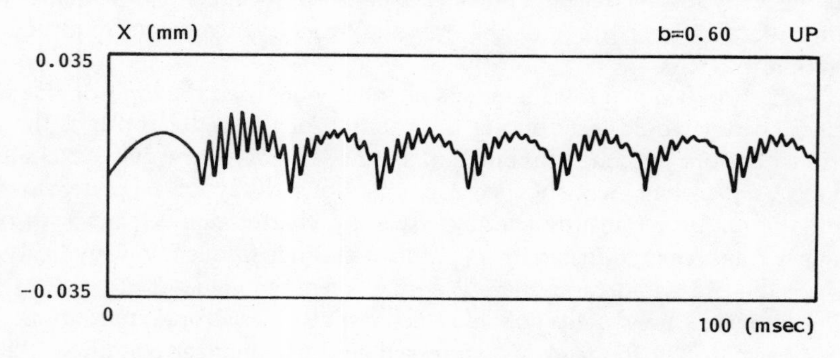

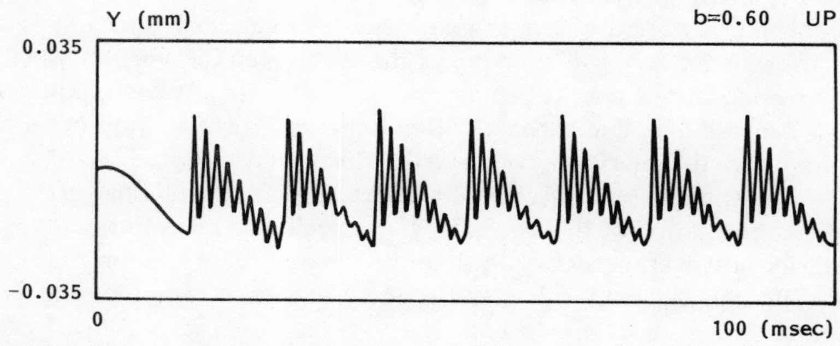

Fig. 3.28 Vibrations in stable up-milling.

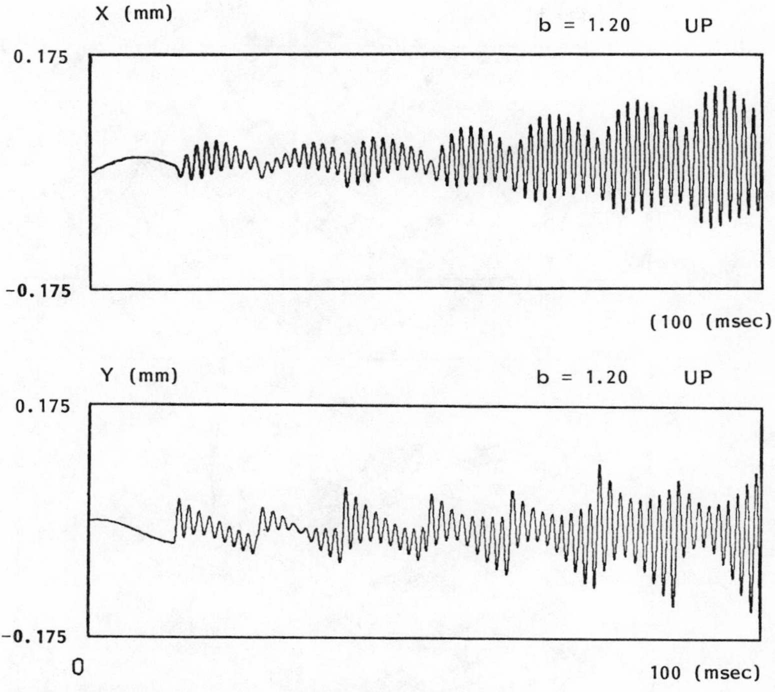

Fig. 3.29 Vibrations in unstable up-milling.

the X and Y vibrations are plotted (as simulated on the computer) for the first eight tooth periods after the beginning of cutting for case a, $f_x = f_y$, up-milling, for chip width $b = 0.6$ mm (0.024 in.) which is slightly below the limit of stability [as shown later, $b_{\text{lim}} = 0.7$ mm (0.028 in.) for this case]. The basic periodicity per cutter tooth is clearly visible as well as the transient vibrations with the natural frequency of the system. In Fig. 3.29 the X and Y vibrations of the same case are shown, but a larger chip width $b = 1.2$ mm (0.47 in.) is used. This is already beyond the limit of stability, and vibrations are increasing every tooth period. The vertical scale of this graph is five times larger than that of Fig. 3.29.

Some insight into the self-excitation process in milling is provided in Figs. 3.30 through 3.34. In Fig. 3.30a vibration in direction Z (normal to the cut—radial to the cutter) in four subsequent cuts (subsequent teeth) numbers 51 to 54 (after 50 cuts from the beginning) is shown for the $0 < \phi < 90°$ up-milling cut of case a and chip width $b = 1.2$ mm (0.47 in.). This is obviously well-established chatter. The shaded area shows the actually removed material in cut number 54. Figure 3.30b shows the tool-tip trajectory in the (Z, U) coordinates. It is interesting to see that now, unlike in turning, even

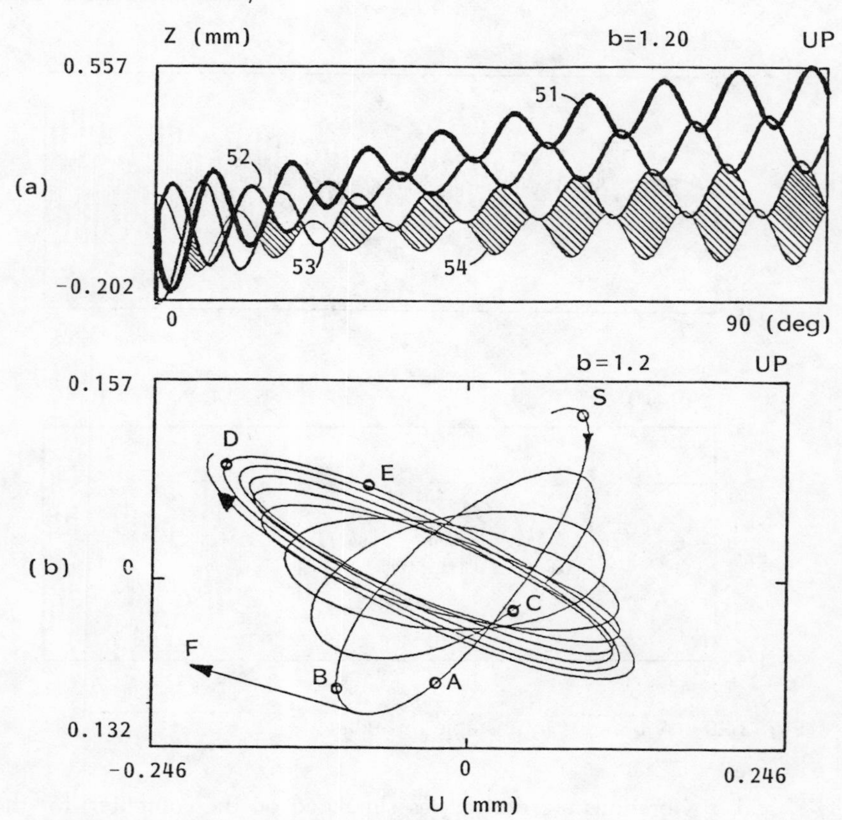

Fig. 3.30 Details of chatter in up-milling.

for $f_x = f_y$ the trajectory is not a straight line but an (evolving) ellipse and "mode coupling" is active. The arrow shows the direction of motion. For two of the cycles the points *A, B, C* and *D, E* are indicated. From the start *S* to *A* the tool moves outside of the cut, from *A* to *B* it cuts, from *B* to *C* it does not cut, etc.; from *D* to *E* it does not cut, etc.

In a way similar to Fig. 3.30a the case of down-milling is shown in Fig. 3.31 for cuts 100 to 103 and in Fig. 3.32 the case of slotting [$b = 0.4$ mm (0.0157 in.)] for cuts 100 to 103. Again, in both these cases chatter is fully established and vibration does not vary from cut to cut (tooth after tooth).

In Figs. 3.33 and 3.34 the cases b of $f_x = 633$ Hz and $f_y = 660$ Hz and b′ of $f_x = 660$ Hz and $f_y = 633$ Hz are shown, respectively, for up-milling and $b = 1.2$ mm (0.47 in.). In Fig. 3.33a the vibration Z is shown for cuts 50 to 53 and in 3.33b the trajectory of the tool tip is shown in the *Z, U* coordinates. It is seen that chatter is fully developed and the mode-coupling

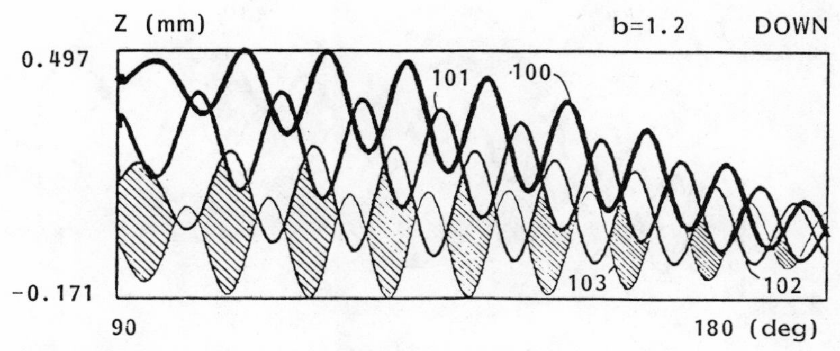

Fig. 3.31 Details of chatter in down-milling.

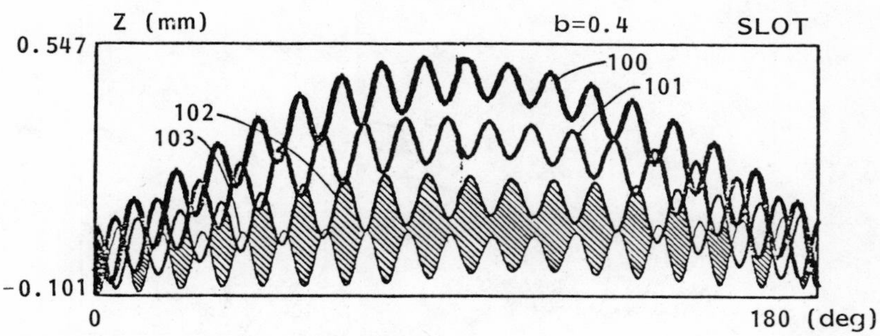

Fig. 3.32 Details of chatter in slotting.

mechanism is strongly active: the ellipses are wide and the tool moves in the right direction for mode-coupling self-excitation; note the direction of the cutting force *F*.

In Fig. 3.34 the case of the reversed frequencies is shown ($f_x > f_y$, milling in direction *X*). In Fig. 3.34a vibration *Z* is shown for cuts 50 to 53 and in Fig. 3.34b the corresponding tool-tip trajectory is shown. The direction of the motion is correct for mode-coupling self-excitation but the ellipses become very flat in the later part of the cut, and this mode of self-excitation is, therefore, weak. Although there is sustained chatter that is fully developed, its amplitude is much smaller than in the case of Fig. 3.33.

Computations of the various cases were carried out for a number of values of chip width *b* and sufficient cycles carried out to reach established vibrations. The summary of the results is presented in Fig. 3.35. The three diagrams correspond to slotting (S), down-milling (D) and up-milling (U), respectively.

In each graph the a, b, c and b′, c′ cases of the various frequency combinations 660/660, 633/660, 660/690 are included always with $f_x < f_y$, while

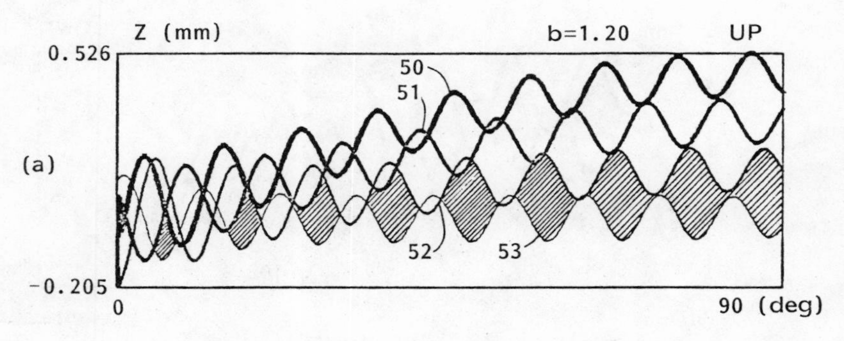

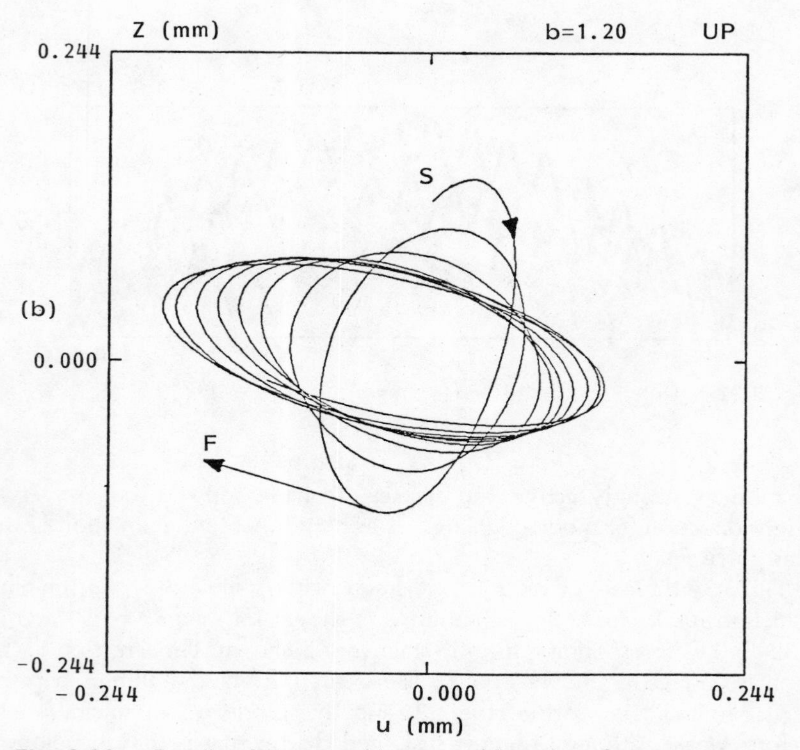

Fig. 3.33 Details of chatter with unequal but close frequencies.

X means milling in direction X (cases a, b, c of the previous classification) and Y means milling in direction Y (cases b′, c′ of X with reversed frequencies).

The following observations can be made:

1. In slotting, the transition between stable and unstable machining is rather sharp and for any $b > b_{\text{lim}}$ the amplitudes of chatter increase fast. In

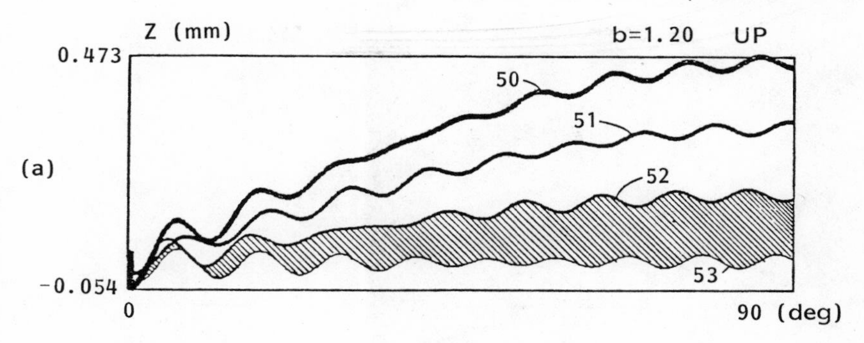

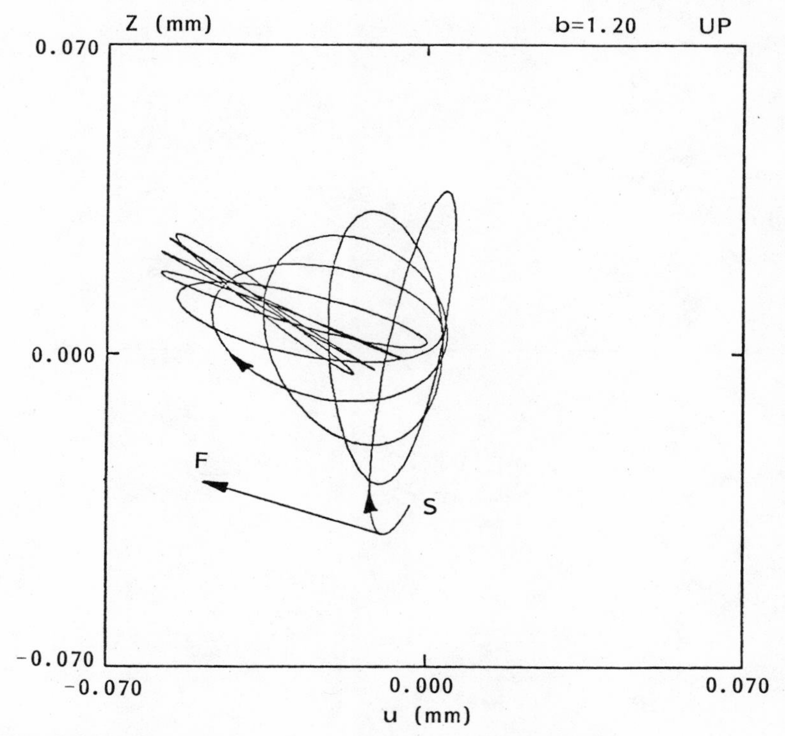

Fig. 3.34 Details of chatter with frequencies reversed.

up-milling and down-milling (half-immersion milling) the boundary between stable and unstable cutting is wide; with chip width b increasing, the amplitude of established vibration increases, first, rather gradually. It is not easy to clearly define the limit of stability b_{lim}. We suggest

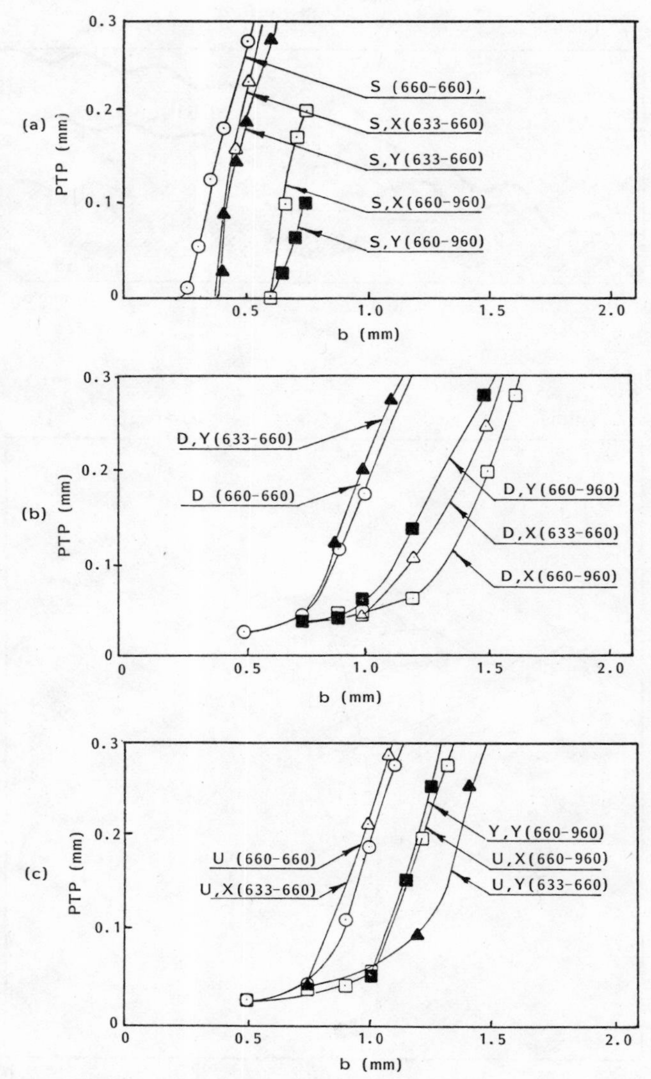

Fig. 3.35 Amplitudes of chatter in milling.

extrapolating the steep part of the individual lines down to PTP = 0 and taking the intersect as $b_{\lim}$.

2. In slotting $b_{\lim}$ is on the average about one-half of that in the half-immersion cuts. This is due to the fact that twice as many teeth cut simultaneously. However, for $f_x = f_y$ it is less than half and it is the lowest stability of all the cases. (This is different from turning where

the case of $R = 0.94$, which corresponds here to 633/660 Hz, was much less stable than $R = 1.0$). This is due to the rotation of the cutting force. This effect is more pronounced the closer the chatter frequency is to the tooth frequency (here they were far apart, in a ratio of 8:1).

3. In slotting the direction of the cut does not matter. Stability of case c with frequencies f_x and f_y far apart is much higher than for $R = 1$ and $R = 0.95$. The same applies to up- and down-milling except for the cases b and b′ with frequencies f_x and f_y not equal but close: $R = 0.95$, $R = 1.05$. These are the cases with strong mode coupling. In these cases the direction of the feed has a strong effect.

The final recapitulation is presented in Fig. 3.36 where just the b_{lim} values are assembled (black bars). The same main conclusions can be repeated as for Fig. 3.35.

Up- and down-milling are about the same and feed direction does not matter if f_x and f_y are either equal or far apart; but they differ and feed direction

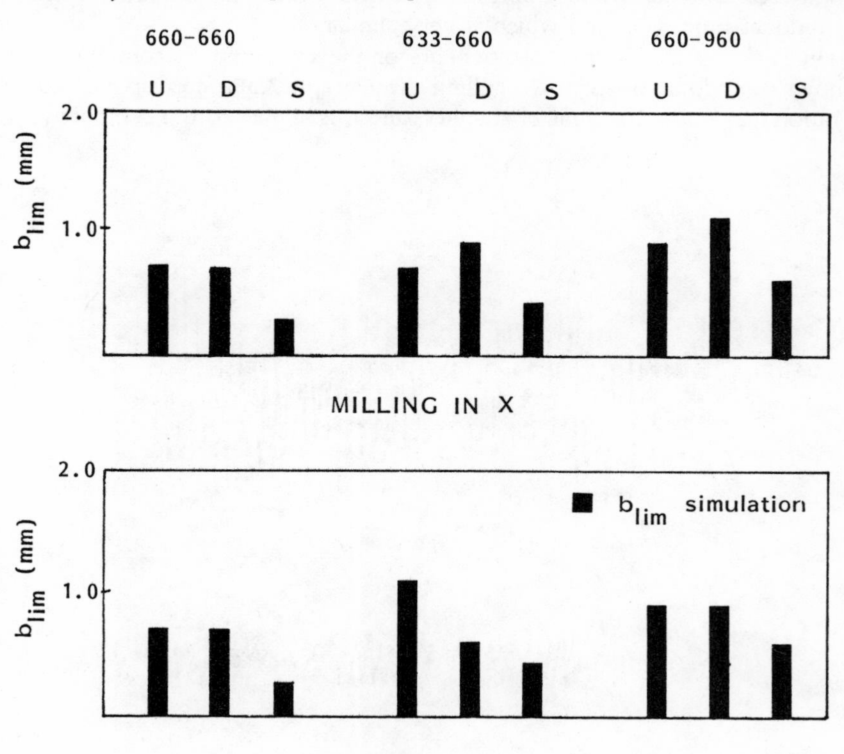

Fig. 3.36 Comparison of limits of stability calculated with constant and with varying directional orientation.

has an effect if f_x and f_y are different but close (end milling). Slotting is about half as stable as half-immersion except for $f_x = f_y$ where it is still less stable.

Experimental Results Experiments were carried out on chatter in milling steel with end mills and face milling cutters. One of the greatest difficulties in such experiments is how to assess the vibration at the end of the tool. If a vibration measurement is taken at the quill, it does not represent vibration of the tool, especially in the case of the end mills.

We consider it best to use a good dynamometer with a reasonably flat frequency characteristic and measure the F_x and F_y force components. The cutting force is a reasonably true picture of the relative vibration between tool and workpiece.

In Fig. 3.37 a record of F_x and F_y is reproduced for down-milling in direction Y. The F_y component is smaller and contains two frequencies, while the F_x component is almost purely 600 Hz. This is a picture of fully developed chatter and it corresponds to the graph in Fig. 3.38a, which was obtained by computational simulation and which is very similar.

In Fig. 3.39 the Fourier components of the measured F_x component are shown for vibration in down-milling. The graph 3.38a corresponds to vibration just below the limit of stability, and it is similar to the computed Fig.

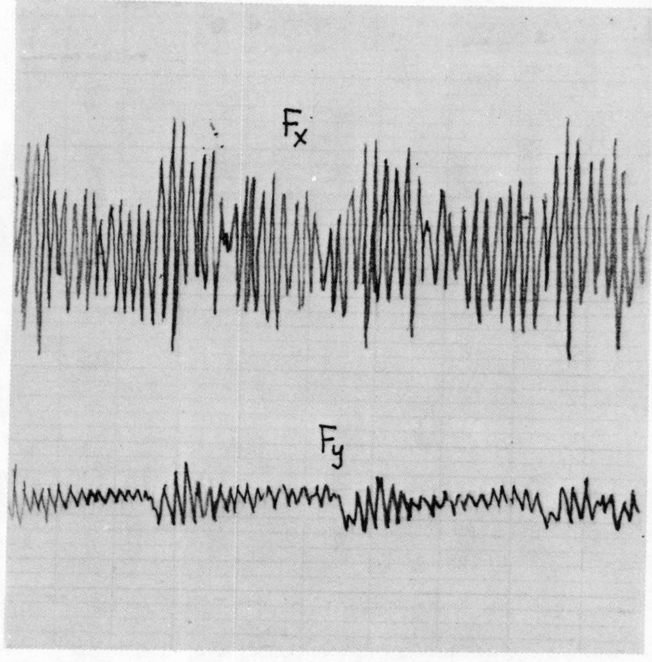

Fig. 3.37 Record of chatter vibration.

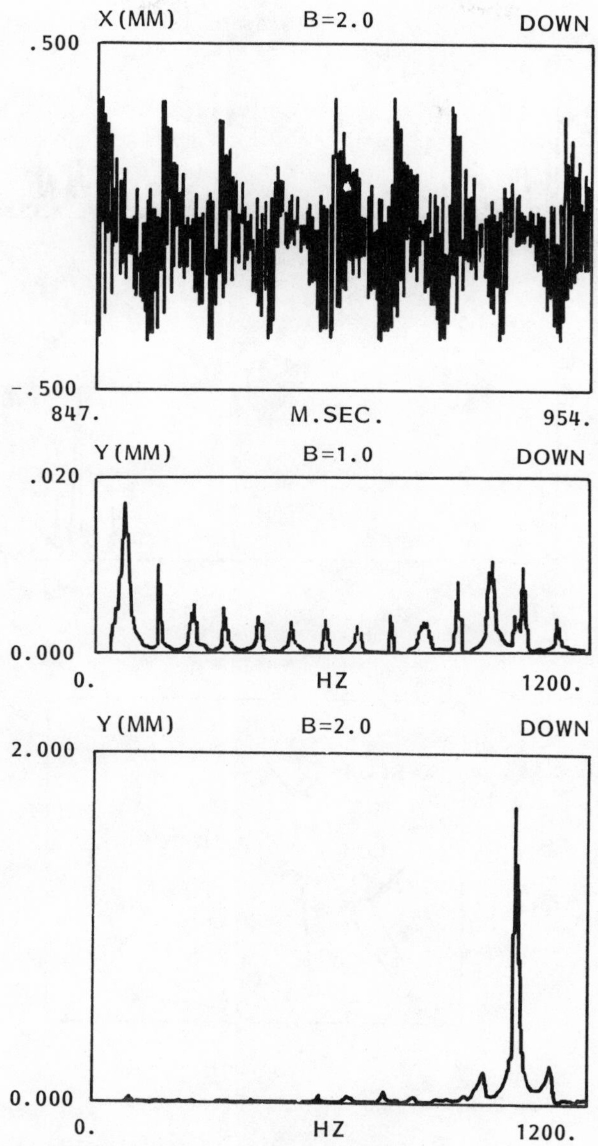

Fig. 3.38 Vibration simulation for down-milling.

3.38b. Graph 3.38b corresponds to fully developed chatter, and it is similar to the computed Fig. 3.38c.

In Fig. 3.40 the measured amplitudes of chatter are plotted versus depth of cut b for end milling steel. The tool had a 25.4 mm (1 in.) diameter, was

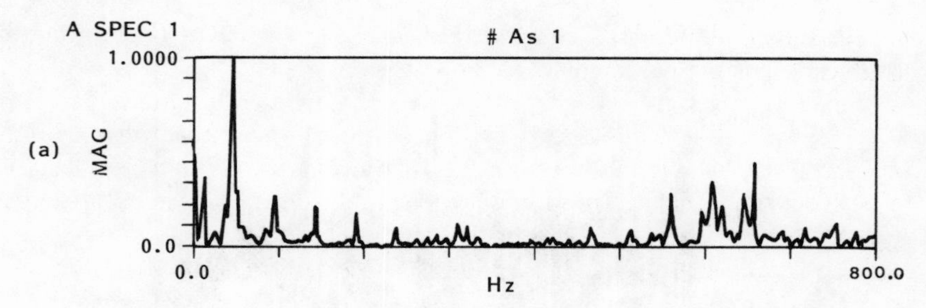

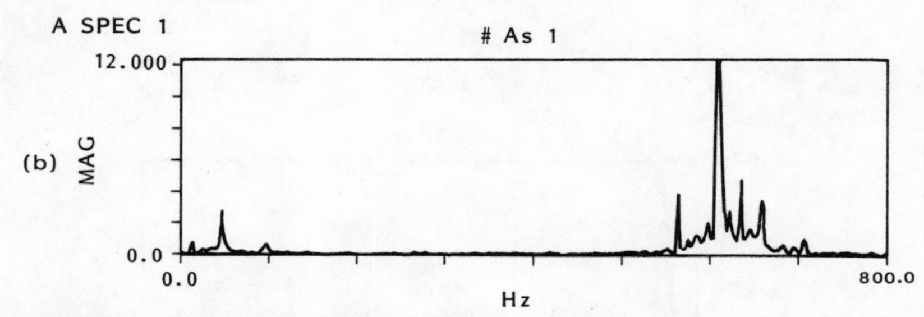

Fig. 3.39 Fourier components of measured vibrations.

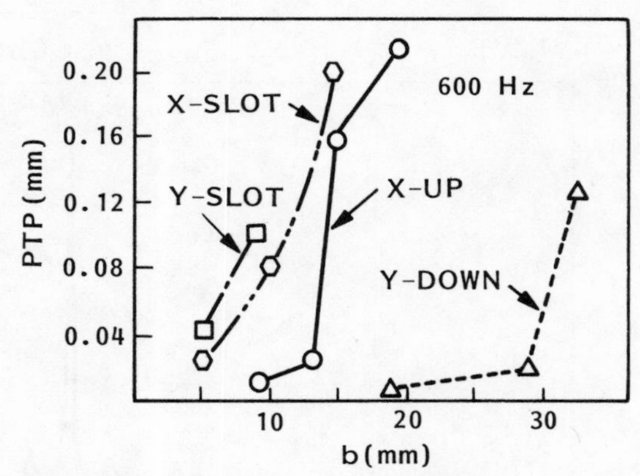

Fig. 3.40 Established amplitudes of vibration (measured).

70 mm (2.76 in.) long, and had four teeth. Cutting speed was 56 m/min (184 fpm), feed per tooth $f_t = 0.1$ mm (0.004 in.). The results and relations between up- and down-milling and slotting are very similar to the computed ones shown in Fig. 3.35. There is, of course, a difference in the scale of b because the stiffness of the modes in the computation was nominally chosen

low. But the amplitude of chatter with its saturation at about the maximum chip thickness is in agreement with the computations.

In Fig. 3.41 the measurements of F_x, F_y and of the quill vibration are recorded for three depths of cut of a slotting operation. They show the transition from stable to unstable machining in a way similar to the computed results.

The comparison of measured vibrations in milling with those obtained by computer simulation shows very good agreement.

The theory of chatter as presented in this chapter in both basic ways, one in a general, simplified way using the frequency domain and the other as obtained by digital-computer time-domain simulations, is the basis for all the practical approaches to suppressing chatter as will be treated in subsequent chapters.

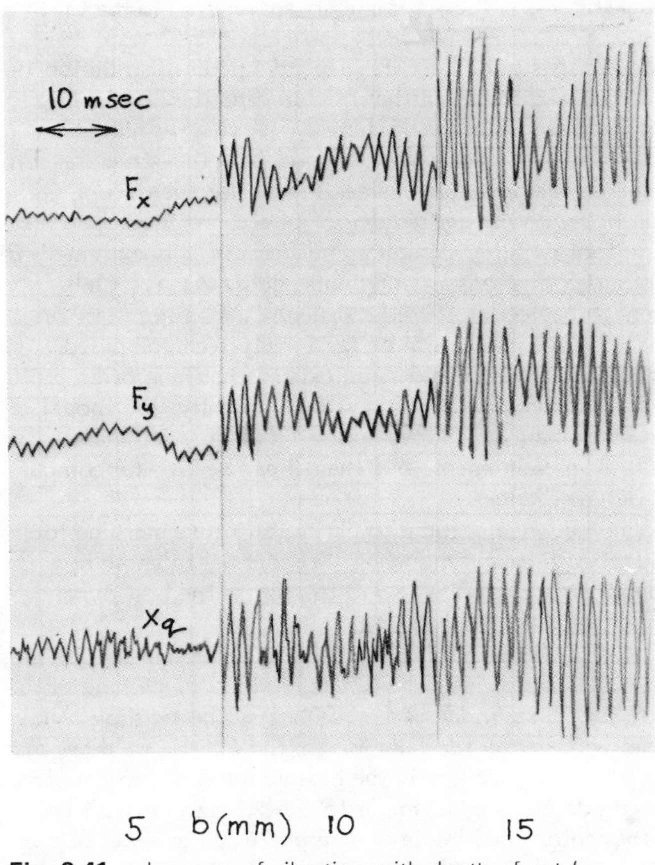

Fig. 3.41 Increase of vibration with depth of cut *b*.

The Role of the Structure of the Machine Tool

In the section, "Limit of Stability for Regenerative Chatter," it was shown that the structure of the machine tool is represented in the process of chatter vibrations by the oriented transfer function (OTF) between the tool and the workpiece. It is oriented so that it expresses the vibration between tool and workpiece, in the direction normal to the cut surface (i.e., normal to both the cutting speed and the cutting edge), as it is excited by a harmonic force acting between the tool and workpiece, in the direction of the cutting force. This concept was introduced in relation to Eq. (3.20) and illustrated in the example based on Figs. 3.12 and 3.13. The OTF is considered as the sum of responses of the individual normal modes of vibration of the structure. Each mode is characterized by its modal parameters, natural frequency f_n, damping ratio ζ, and modal stiffness k (these parameters are referred to the OTF), and by its "mode shape."

The mode shapes and the OTF depend on the distribution of stiffness, damping, and mass throughout the structure, and they could, ideally, be computed from the drawings of the machine using either the finite-element method or, in a more approximative way, a model of the structure consisting of "lumped" springs, dampers, and masses. Practically, such a computation cannot be accomplished with the desired degree of accuracy because the structure of the machine tool is rather complex and because it is extremely difficult to compute local deformations around joints and guideways. Only exceptionally is it sufficient to neglect most of the structure and concentrate on a subgroup like that of a spindle, which can be reasonably well computed.

In most instances one considers an existing machine or its prototype and measures its structural characteristics. Subsequently, the model of the machine as an assemblage of lumped springs, dampers, and masses can be identified from these measurements and, then, it can be used for computing results of various design changes.

In the following we are going to illustrate various machine tool structures and their identification and modeling. We will try to show how these structures are designed and modified so as to possess the best dynamic characteristics for high stability of machining.

In Fig. 3.42a, a sketch of a light milling machine and its five most prominent mode shapes are shown. The 22-Hz mode is the rocking of the machine on the floor; the 75-Hz mode is represented by the twisting of the overarm, and the head with the motor is its inertia; the 125-Hz mode is characterized by the rotation of the overarm in the flexible turret at the top of the column; the 350-Hz mode is a higher mode to the 75-Hz one with the top of the head with the motor vibrating in counterphase to the lower part of the head; and the 570-Hz mode is the higher mode to the 125-Hz one with prominent

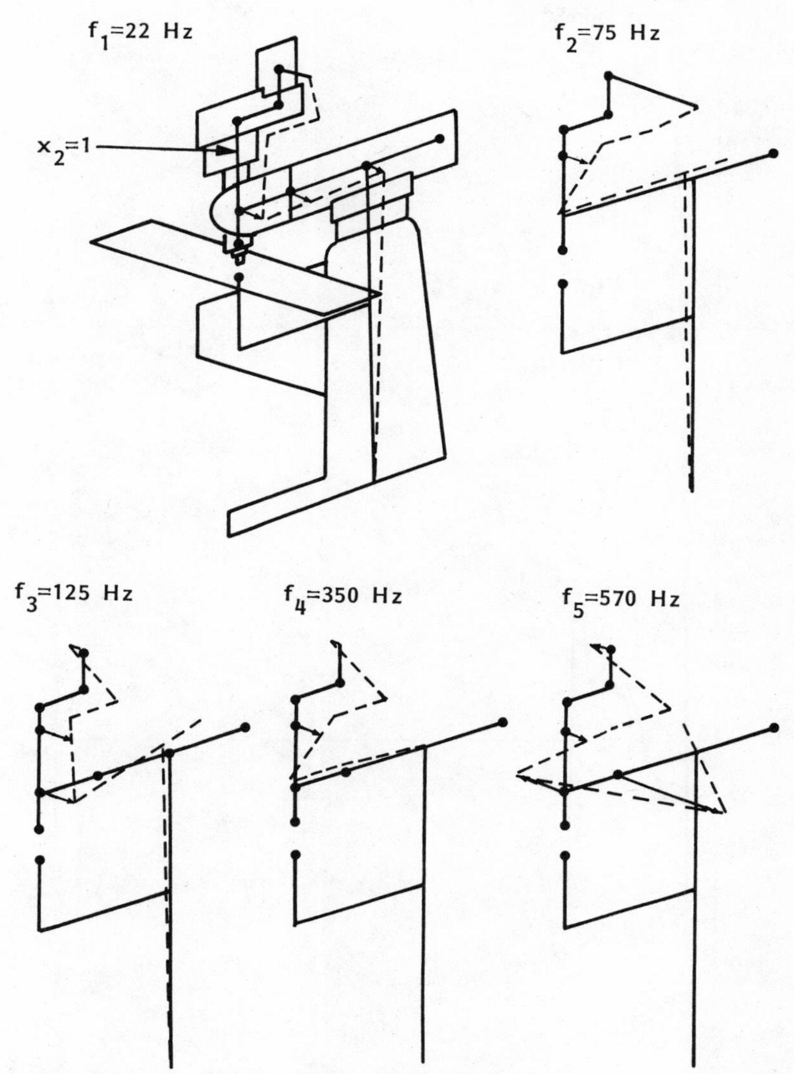

Fig. 3.42 Mode shapes of the milling machine.

flexibility at the base of the U-type end of the overarm. Two transfer functions of the machine are shown in Fig. 3.43.

The transfer function in Fig. 3.43a is relative between tool and workpiece (the real part), and it indicates that the 125-Hz mode is most responsible for chatter on this machine. The graph in Fig. 3.43b is an absolute transfer function on the point x_2 of Fig. 3.42 and it serves to determine the modal parameters of the structure. A lumped spring–mass model of the machine is shown

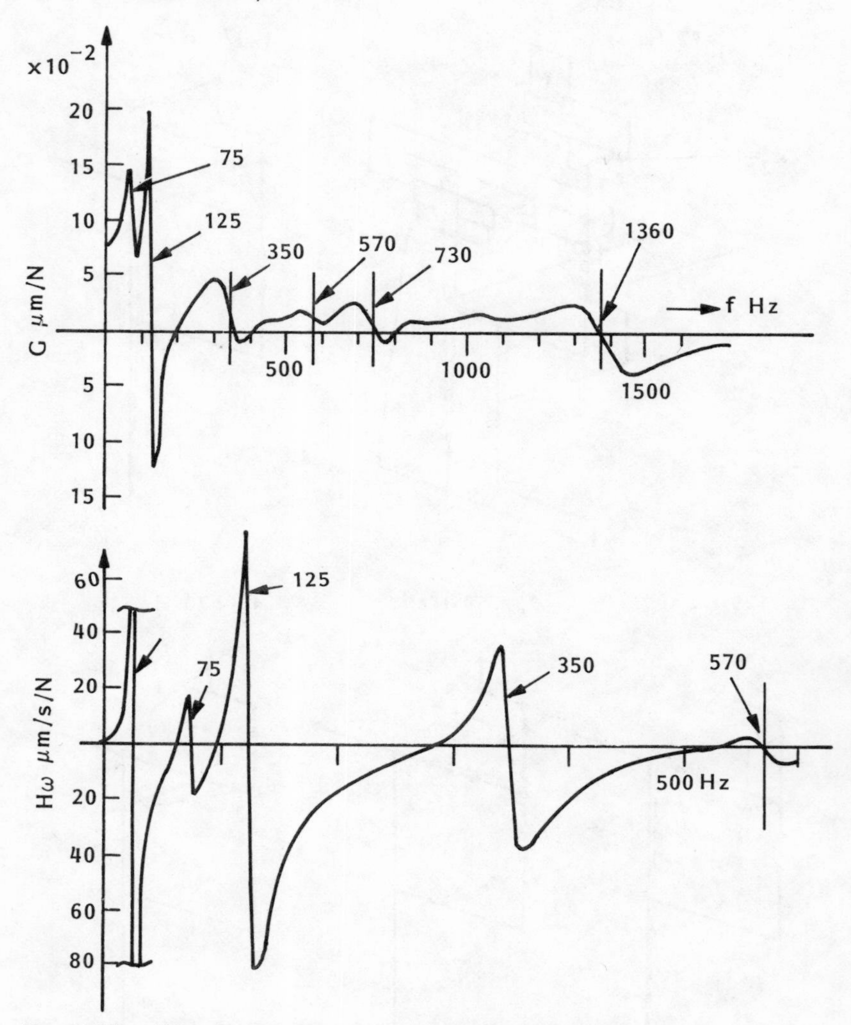

Fig. 3.43 Transfer functions of the milling machine.

in Fig. 3.44. The individual concentrated masses are indicated by black circles. The double lines represent rigid links. Five springs denote the flexibilities γ_1 to γ_5. There are five basic coordinates x_1 to x_5 of the model, corresponding to the five degrees of freedom as expressed by the five modes visible in the transfer function in Fig. 3.43b. In a project carried out in our past experience the "local" parameters of this model were "identified" (see ref. 6), and subsequently various changes tried so as to minimize the negative peak of the transfer function in Fig. 3.43a.

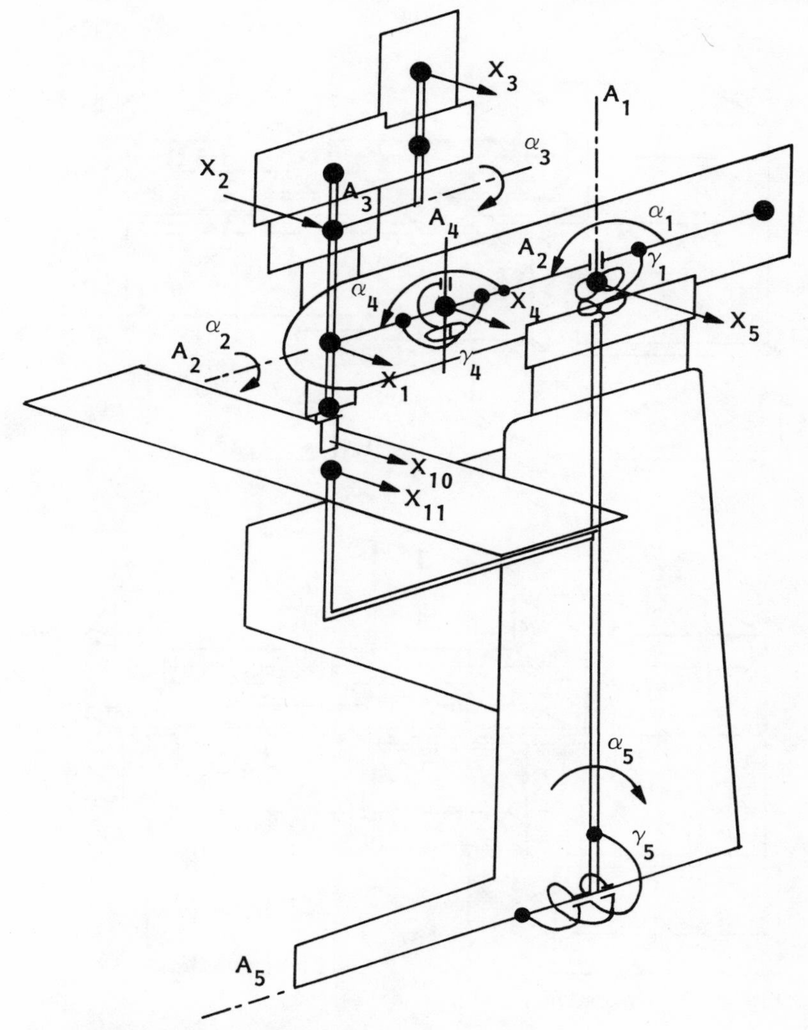

Fig. 3.44 Model of the milling machine.

Another example of mode shapes is in Fig. 3.45. The two basic modes of a lathe with a heavy shaft-type workpiece are such that the workpiece is the main inertia in both, and the spindle and, mainly, the tailstock quill and center are the most prominent springs. The frequencies of the two modes are not far apart, and their directions between the tool and the workpiece are approximately perpendicular to each other.

In the case of the horizontal milling machine described in reference to Fig. 3.19, there were also two prominent modes with frequencies close to one

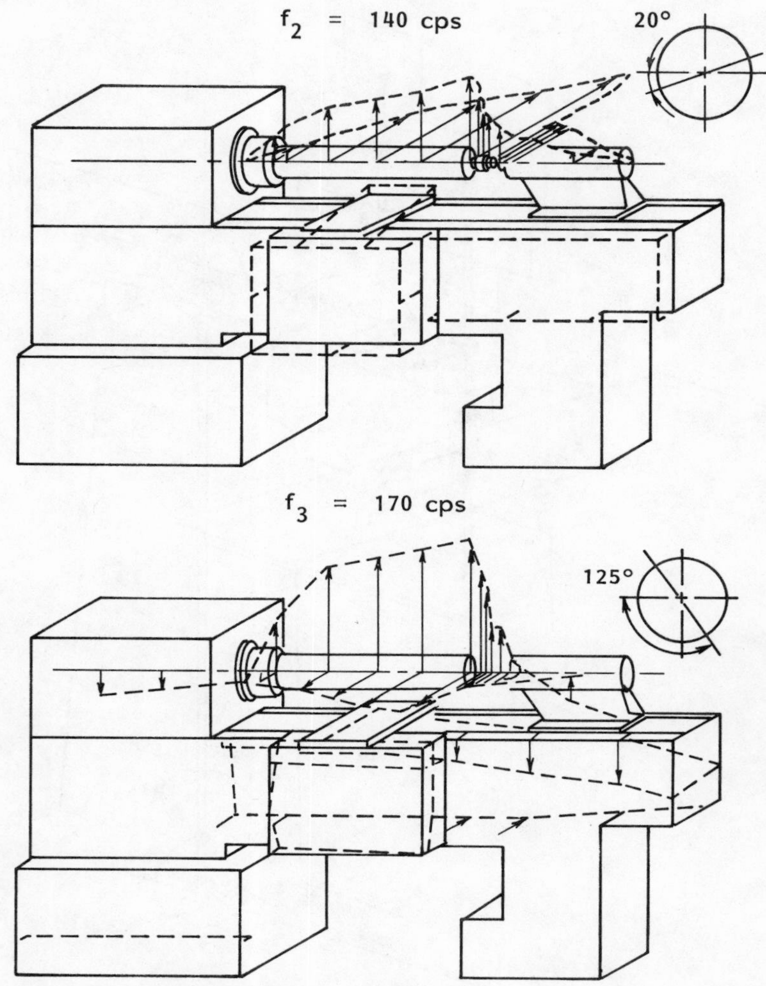

Fig. 3.45 Mode shapes of a lathe.

another and directions mutually perpendicular. This is often the case because the most flexible part of the structure is like a beam (the overarm, the shaft workpiece on the lathe, a spindle, etc.) vibrating in two directions perpendicular to its axis. We refer here to the basic diagram of mode coupling, Fig. 3.4, where we used two mutually perpendicular modes with close frequencies as well as to the cases referred to in Figs. 3.12, 3.17, 3.21, 3.27.

However, in some instances there is one mode so prominent (like case a in Fig. 3.20) that all the other modes can be neglected. This is represented diagrammatically in Fig. 3.46a and the corresponding value of G_{min} is indicated,

$$G_{min} = -\tfrac{1}{4}k\zeta(1 + \zeta), \tag{3.39}$$

and the corresponding b_{lim} is the same as it was derived in Eqs. (3.26) and (3.27). This is the simplest possible case and we will call it the *very basic* case.

A case that acts like the one in Fig. 3.46a is obtained with a round bar as in Fig. 3.46b. This situation can be thought of as having two degrees of freedom, X_1 and X_2, with identical stiffness and identical frequencies; we may choose X_1 and X_2 in any direction, e.g., $\alpha_1 = 0$, $\alpha_2 = 90°$. Then X_2 has a zero-directional factor, and the case is equivalent to the one in Fig. 3.46a.

Let us now take four cases having two identical degrees of freedom in two mutually perpendicular directions and treat them as *very basic* cases. Although in reality their limits of stability will differ from what we are going

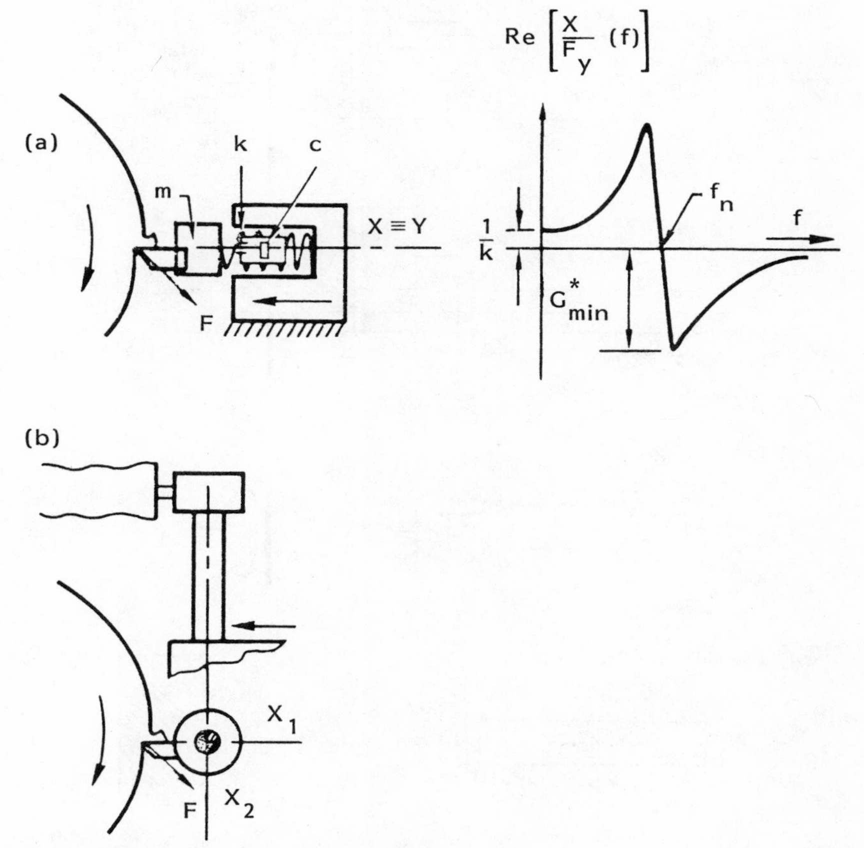

Fig. 3.46 Single degree-of-freedom system.[8]

to establish in this simplified way, it will at least give us some orientation. The four cases considered are illustrated in Fig. 3.47. The data given here are based on computed modes of vibration considering absolutely rigid clamping of the boring bars and absolutely rigid housing of the spindles. The values of flexibilities obtained agree generally with values measured on good designs.

In each case a damping ratio of $\zeta = 0.035$ was assumed. This is an average value encountered in practice, where damping ratios between 0.02 and 0.06 are usual. Using a damper may double to quadruple the damping ratio. All cases relate to roughing operations in the machining of steel with a Brinell hardness of about 250. The cutting stiffness is taken as $C = 1764h^{-0.15}$ N/mm^2 ($2.6 \times 10^5 h^{-0.15}$ psi).

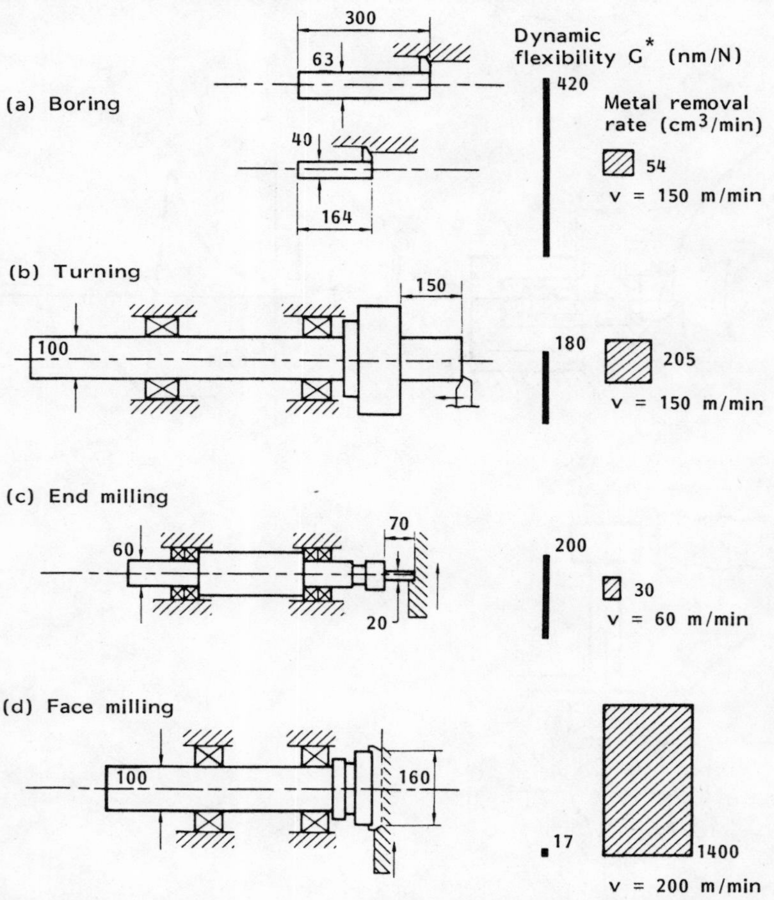

Fig. 3.47 Limits of stability of four "overhang" machining cases.[8]

(a) *Boring* (see Fig. 3.47a). The two illustrated boring bars, $d = 63$ mm (2.48 in.), $l = 300$ mm (11.8 in.) and $d = 40$ mm (1.57 in.), $l = 164$ mm (6.46 in.), have the same stiffness. The minimum of the real part of the direct transfer function is $G_{min} = 0.42$ μm/N (6.1×10^{-5} ft/lb). Assumed cutting data: $\sigma = 30°$, $f = 0.2$ mm/rev (0.008 in./rev), $v = 150$ m/min (522 ft/min), $C = 2295$ N/mm^2 (3.33×10^5 psi). Limit depth of cut results as $a_{lim} = 1.8$ mm (0.071 in.) and the corresponding metal removal rate MRR = 54 cm^3/min (3.3 in.^{3}/min). If a plunge cut ($\alpha = 90°$) should be taken, the limit width would be $b_{lim} = 0.9$ mm (0.04 in.).

(b) *Turning* (see Fig. 3.47b). Overhang turning is considered at the end of a workpiece 1.5 times longer than the diameter of the spindle bearings. Bearings of a very stiff type [$d = 100$ mm (3.94 in.), double roller] are considered with a well-balanced spindle design. Inevitably the point of machining is rather distant from the front bearing. For this case $G_{min} = 0.18$ μm/N (2.63×10^{-6} ft/lb). Cutting data: $\sigma = 30°$, $f = 0.3$ mm (0.012 in.), $v =$ 150 m/min (522 ft/min); $C = 2113$ N/mm^2 (3.1×10 psi). Limit depth of cut results as $a_{lim} = 4.56$ mm (0.18 in.) with corresponding MRR = 205 cm^3/min (12.5 in.^{3}/min) and power 6 kW. If a plunge cut should be taken, limit width would be $b_{lim} = 2.28$ mm (0.09 in.).

(c) *End milling* (see Fig. 3.47c). A case of an end mill of diameter $d = 20$ mm (0.79 in.) and length 70 mm (2.78 in.) is considered, using a spindle with angular contact bearings $d = 60$ mm (2.38 in.). For this case $G_{min} = 0.2$ μm/N (2.9×10^{-6} psi). Cutting data: $\sigma = 0°$, feed per tooth $f_t = 0.2$ mm (0.008 in.); $C = 2245$ N/mm^2 (3.3×10^5 psi), HSS cutter with four teeth, $v = 60$ m/min (209 ft/min), radial depth of cut $a_r = 10$ mm (0.4 in.). The limit axial depth of cut is $a_{lim} = 4$ mm (0.16 in.), corresponding to MRR = 30 cm^3/min (1.9 in.^{3}/min).

(d) *Face milling* (see Fig. 3.47d). A spindle with double roller bearings $d = 100$ mm (4 in.) is considered in a very compact design with the minimum possible distance of the cutter teeth from the front bearing. $G_{min} = 0.017$ μm/N (2.5×10^{-7} ft/lb) for this case. Cutting data: $d =$ 160 mm (6.36 in.) with 16 carbide tools, $f_t = 0.25$ mm (0.01 in.), $\sigma = 30°$, $v = 200$ m/min (696 ft/min), width of cut $B = 80$ mm (3.18 in.); $C =$ 2220 N/mm^2 (3.22×10^5 psi). The limit depth of cut results as $a_{lim} = 11$ mm (0.44 in.) with corresponding MRR = 1400 cm^3/min (85.4 in.^{3}/min) and corresponding power 35 kW.

The comparison of cases b and d illustrates the significance of the distance from the front bearing for the "overhang" systems. In many instances the effect of the clamping of the boring bar or the flexibility of the headstock of the milling cutter would lead to smaller limit depths of cut and smaller limit MRR values. Also, there are instances where a much larger limit width of cut is required than derived for these four cases. On a roll lathe, a tool with a very large side cutting edge angle of, say, $\sigma = 85°$ is often used to spread the load over a wide cutting edge, and may give a width of cut $b = 80$ mm (3.18 in.). This requires static stiffness of the corresponding mode to be $k_{st} = 1 \times 10^6$ N/mm (5.71×10^6 lb/in.), which is extremely difficult to achieve. As another example, in broaching the "Christmas tree slots" in turbine disks, the total chip width may be 150 mm (5.96 in.)—requiring an impossible stiffness of the holding fixture. Obviously, other factors have to be taken into account than those considered so far in the very basic cases.

In reality, structures have more than one degree of freedom. There are numerous cases of two fairly prominent modes with frequencies close to one another. All the cases of Fig. 3.47 have two prominent modes, due either to the nonsymmetry of the clamping of the boring bar, to a key slot in a boring spindle, or to different stiffnesses of the spindle housing in two perpendicular directions. The existence of a second degree of freedom may substantially influence the limit of stability. Let us illustrate this by means of a simple example. This example will be presented symbolically as a boring bar, but it represents a general case of a system with two mutually perpendicular degrees of freedom.

Figure 3.48a shows a boring bar with circular cross section, similar to that in Fig. 3.46c. In Fig. 3.48b a bar with rectangular cross section is chosen (from many possible rectangles) so that its moment of inertia I_A about the axis A is the same as that of the round bar, while I_B is increased by a factor of 6.

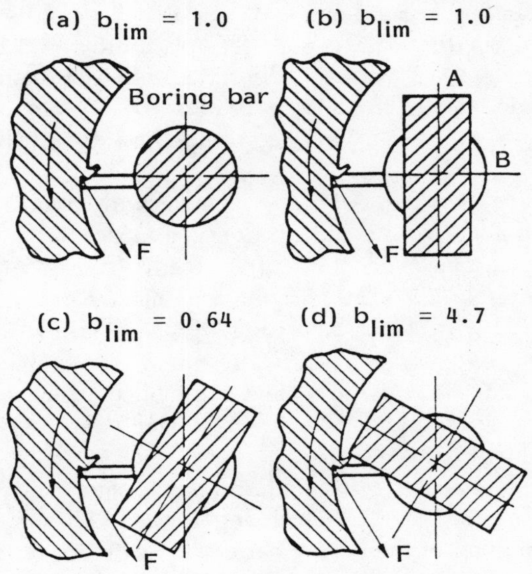

Fig. 3.48 System with two mutually perpendicular degrees of freedom; limit of stability is dependent on orientation.[8]

The two cases 3.48a and 3.48b give the same limit of stability. Cases 3.48c and 3.48d show two other orientations of the same rectangular bar that result in limits of stability 0.64 and 4.7 times that of case 3.48a, respectively. This example shows that a system with two degrees of freedom, one of them as stiff as the original single degree of freedom and the other six times stiffer, may give either less stability or much more of it, depending on the orientation of the two modes.

In multi-degree-of-freedom systems various combinations of modal participation in the resulting oriented transfer function arise. The real part of the OTF may have a shape like the one shown in Fig. 3.49a. The OTF denoted by G is a sum of the oriented transfer functions G_i corresponding to the individual modes. Some of them, like G_3, are "inverted" because the corresponding directional factor u_3 has a negative value. Quite often one mode is dominant, as in the stability diagram in Fig. 3.49b. Such a case approximates the single-mode case of Fig. 3.46 except that it may have a different directional orientation. Also, cases are frequent where there are two modes with close frequencies and not very different stiffnesses. In such cases, depending on the orientation, the two modes can amplify each other as in Fig. 3.49c or subtract from one another as in Fig. 3.49d, leading to either very low or very high stability. These cases best express what is known as "mode coupling." Cases

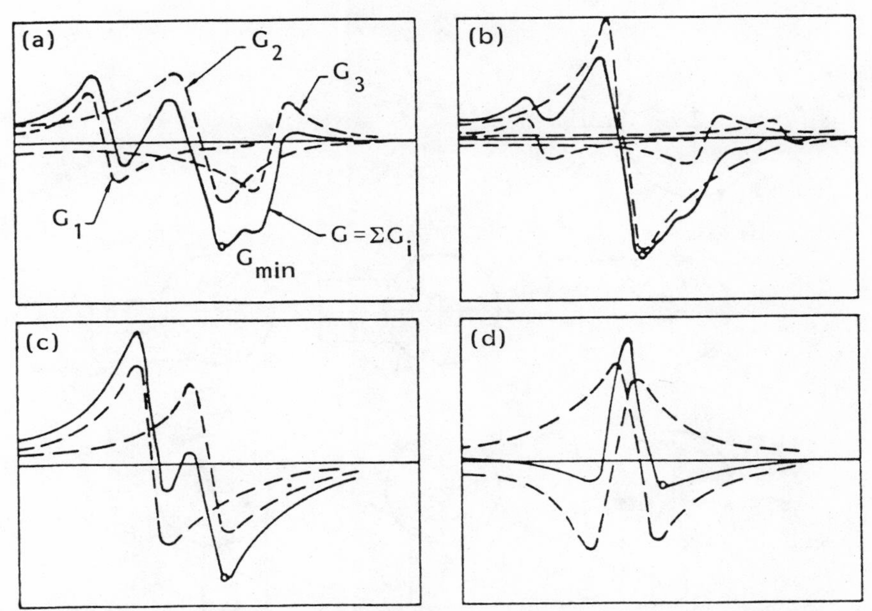

Fig. 3.49 Oriented transfer functions arising from various combinations of modes.[8]

of two dominant modes correspond to structures like a boring bar, ram on a vertical boring mill, overarm on a milling machine, etc.

On some machine tools the directional orientation of the cutting process varies little, as on a lathe (see Fig. 3.50a) where the normal to the cut surface is always horizontal and the direction of the cutting force varies in a small range. On other machines a great variety of orientations occur. In face milling, as shown in Fig. 3.50b, the average directions of the cutting force and of the normal Y vary with the mutual position of the cutter and the workpiece, and with the direction of feed.

From the preceding simple presentations and explanations the following conclusions may be drawn for the structure of a machine tool:

(a) The "basic limit width cut $b_{b,\lim}$" without chatter (if no damping is generated in the cutting process and no interference of wave regeneration is introduced) is indirectly proportional to the "cutting stiffness" (specific for the given workpiece material) and to the maximum dynamic flexibility between tool and workpiece (the magnitude of the minimum of the real part of the oriented transfer function between tool and workpiece). For simple dynamics this means that $b_{b,\lim}$ varies with the stiffness and damping of the system. For actual, more-complex dynamics, the design of a machine tool may require increasing the stiffness of selected parts of the structure, even decreasing the

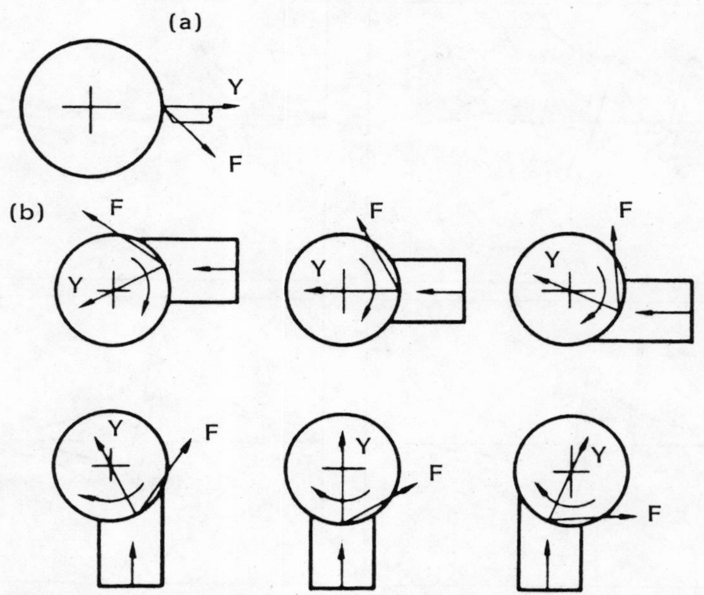

Fig. 3.50 Directional orientations (a) on a lathe and (b) in face milling.[8]

stiffnesses of some other parts, selecting optimum mass distribution, and increasing the damping in selected parts of the structure.

(b) A particular machine tool is used in various configurations, in each of which the cutting process may have various "directional orientations." Limit of stability may substantially vary in these cases.

As mentioned at the beginning of this chapter it is often worthwhile to carry out a detailed analysis of structural dynamics of the machine. One such example was quoted with reference to Figs. 3.42–3.44. Let us now give one more illustration, one which is well related to the business of end milling. This example will also demonstrate one of the methods used for the identification of the structural model. This particular method is suitable in such cases where some of the significant coordinates of the model are not accessible for the measurement of the transfer functions; in other words the experimentally determined mode shapes (eigenvectors) are incomplete. This example will deal with the modeling of the headstock and spindle assembly of a vertical milling machine. Figure 3.51 shows some typical mode shapes of this assembly.

At the bottom of the figure it is shown that the headstock and column provide two lower-frequency (140 and 300 Hz) modes. As shown in the upper part of the figure there are three higher modes (500, 990, 1600 Hz) associated mainly with the spindle and the end milling tool. These three mode shapes can only be measured on the exterior part of the spindle, chuck, and tool and on the headstock but not on the spindle inside the headstock.

A corresponding model is shown in Fig. 3.52. Springs k_1, k_2 represent the stiffness of the column whose motion is measured as x_1. Springs k_3, k_4 connect the column with the headstock on which we measure displacements x_2, x_3. Springs k_5, k_6 represent spindle bearings. Coordinates x_4–x_{11} are associated with the tool, the chuck, and the spindle end. The displacements x_1–x_{11} are accessible for measurement. The spindle inside of the headstock is represented by coordinates x_{13}–x_{25}, which are inaccessible.

Let us now first discuss all such cases where some relevant parts of the mode shapes are missing, after which we will return to our particular case.

The equation of natural vibration of the system

$$[s^2 m_x + k_x]\{x\} = 0$$

leads to the characteristic equation

$$|s^2 m_x + k_x| = 0, \tag{3.40}$$

which is valid separately for each of the measured natural frequencies $s = v_1, v_2, \ldots, v_d$ and it contains the unknown elements. Using the d measured eigenvalues we get d nonlinear equations in the unknown elements $k_{x,i,j}$ and $m_{x,i,j}$ of the stiffness and mass matrix, respectively; actually, since these elements consist of combinations of local masses m_i and stiffnesses k_i, the equations contain them as unknowns. To solve these nonlinear equations an

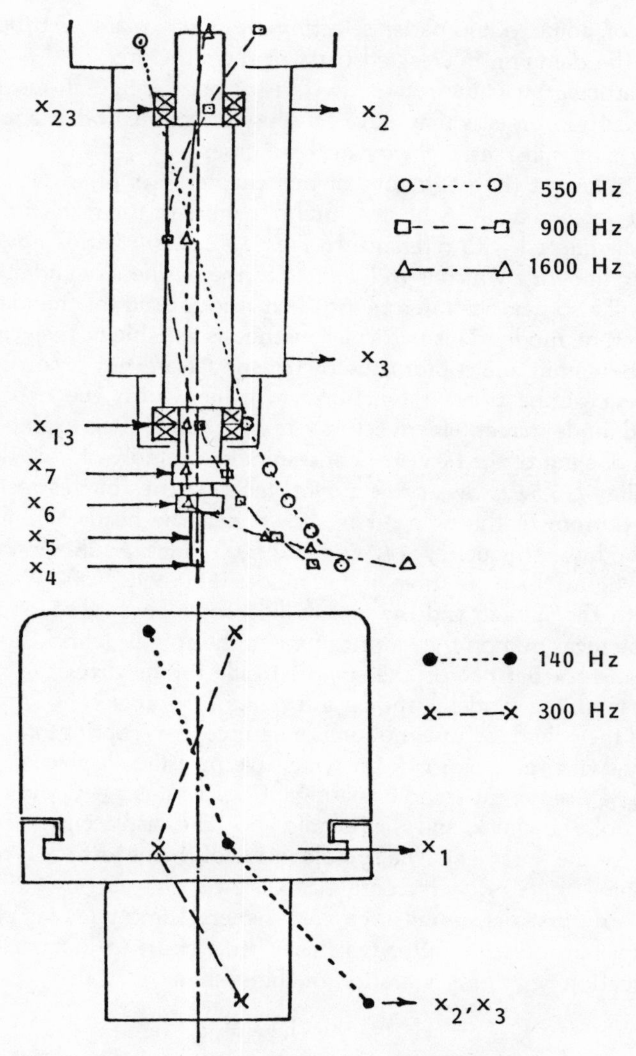

Fig. 3.51 Mode shapes of the milling machine.

optimization routine can be used starting with an initial estimate of the unknowns and minimizing the sum of the squares of residuals R_i:

$$U = \sum_{i=1}^{d} R_1^2 = \left| -v_i^2[m_x] + [k_x] \right|. \tag{3.41}$$

As an alternative method, an optimization routine could be used based on repeatedly solving the eigenvalue problem and comparing its results with the measured data. First an initial estimate of the unknown local springs k_i

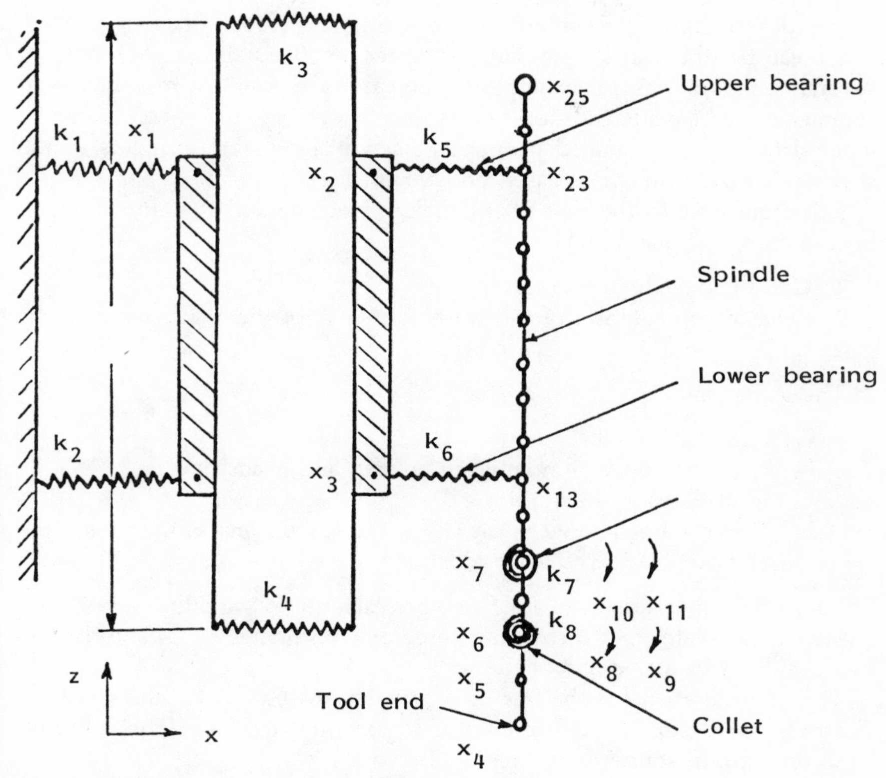

Fig. 3.52 Simplified model of a milling spindle.

and masses m_i is used to compute the eigenvalues ν_i^2, the eigenvectors $\{\phi_i\}$ with elements p_{ij}, and the modal stiffnesses K_i^{rr}, and an objective function is formulated:

$$U = \sum_j^d \phi_i[(\nu_i - \nu_{iex})/\nu_{iex}]^2 + \sum_j^d \beta_i[(K_i^{rr} - K_{iex}^{rr})/K_{iex}^{rr}]^2$$

$$+ \sum_i^l \sum_j^d \gamma_{ij}[(p_{ij} - p_{ijex})/p_{ijex}]^2, \tag{3.42}$$

where the parameters with the subscript "ex" are those established experimentally, α, β, and γ are weighting coefficients; and d is the number of measured coordinates.

The unknowns in $[m_x]$ and/or $[k_x]$ are to be changed so as to minimize U. Other objective functions may be formulated as in refs. 6 and 7.

In order to avoid unrealistic results it is necessary to bound the acceptable solutions by introducing constraints. These may simply require that all $k_i > 0$ and $m_i > 0$, or they may set ranges for the individual stiffnesses and masses.

In order to improve the solutions, larger amounts of input data are needed. These can be obtained by measuring transfer functions and mode shapes in several configurations; for example, in the case of the milling machine, measurements can be made with tools of various lengths and masses. Some of the input data may be obtained by measurements of separate elements of the structure or by calculations based on the drawings of the machine.

Returning now to the case of the milling machine, let us assume the following computational inputs:

1. Good knowledge of all the masses.
2. Reliable computation of the stiffness matrix of the spindle and of the tool.

The stiffnesses are:

k_1, k_2 of the column;
k_3, k_4 of the connection between column and headstock;
k_5, k_6 of the spindle bearings;
k_7, k_8 as rotations between the chuck and spindle and of the tool in the collet are unknowns to be identified.

Inputs from measurements are: five eigenvalues ν_1–ν_5 and the corresponding five incomplete eigenvectors measured on coordinates x_1–x_7 and the five modal stiffnesses at the end of the tool.

The procedure used is the one corresponding to Eqs. (3.41) and (3.42). In order to increase the amount of input data, experiments were done with four tools differing in stiffness and, mainly, in mass.

The procedure just described was related to vibrations in the direction X, horizontal and parallel with the table axis. A similar arrangement applies to vibrations in the direction Y, horizontal and perpendicular to X. The modes in X and Y combine as it was shown in the text referring to Fig. 3.27. It is possible to optimize the design of the spindle and of its mounting and that of clamping the tools so as to achieve an optimum with respect to milling chatter. Those structural characteristics that are eventually obtained and that can again be experimentally verified once the machine is built can then be used in selecting the best cutting conditions, especially the best spindle speed for optimum stability. This will be discussed in the next chapter.

Selecting Optimum Spindle Speed

A change in spindle speed affects the limit of stability, especially in milling. There are two different reasons for this effect. First, it is the effect of the geometric condition for the phase ε of undulations in subsequent cuts, which was mentioned with reference to Fig. 3.5, Eqs. (3.2) and (3.3), and Fig. 3.11. However, in the "Theory of Chatter" section the limit of stability was considered with the assumption that the phase ε was free to adjust itself for maximum regeneration or that spindle speed was such that the phase ε corresponded to the worst case of the "borderline stability." Now, we will disregard this assumption and derive the limit of stability with respect to the value of ε as it varies with spindle speed. Secondly, as the spindle speed is varied the cutting speed varies accordingly and this affects the "damping in the cutting process." This will be discussed in the second part of this section.

Effect of the Phase Between Subsequent Teeth on Stability

In this section we will still disregard any damping generated in the cutting process itself and concentrate on the role phasing only. In the "Theory of Chatter" section it was indicated with reference to Fig. 3.11a that the phase shift ε between undulations produced in subsequent cuts as expressed by Eqs. (3.2) and (3.3) affects stability. It was stated that this effect is strong only if the number of waves between subsequent cuts is small, which can only happen in milling and, especially, in high-speed milling. In turning the effect is so small that it can be neglected. We then did not pursue this and considered only the case of "borderline stability" in which the phase ε is such that it leads to the lowest possible value of the limit width of chip b_{lim}.

The general, rather simple theory related to Fig. 3.11a applies if the phasing between subsequent cuts (between subsequent teeth in milling) is related to one particular oriented transfer function. This involves the directional orientation between the cut and the structure. In milling, however, every tooth has a different direction Z of the normal to the cut surface, see Fig. 3.27, and, correspondingly, every tooth involves a different directional factor and a different OTF. Proper consideration of these aspects was possible in the "Theory of Chatter by Time-Domain Simulation" section, where the method of time-domain simulation was used. One special feature was revealed: in turning it was found that a system with two mutually perpendicular modes was 2.5 times more stable when the two natural frequencies were equal than when they stood in a ratio $R = 0.94$, and this would also be so in a multitool operation in which all teeth would be equally oriented. The reason for the good stability with $R = 1$ was that the two modes vibrated in phase and did not produce an elliptical motion needed for the "mode-coupling self-excitation." It was, however, found that in milling the system with $R = 1$ produced elliptical motion, see Fig. 3.30, and was actually only about half as stable in slotting as a system with $R = 633/660 = 0.96$. This was caused by the different directional orientation of the individual teeth.

The method of time-domain simulation is not suitable for deriving general trends. Therefore, we will first derive the effect of the constraint on phase ε

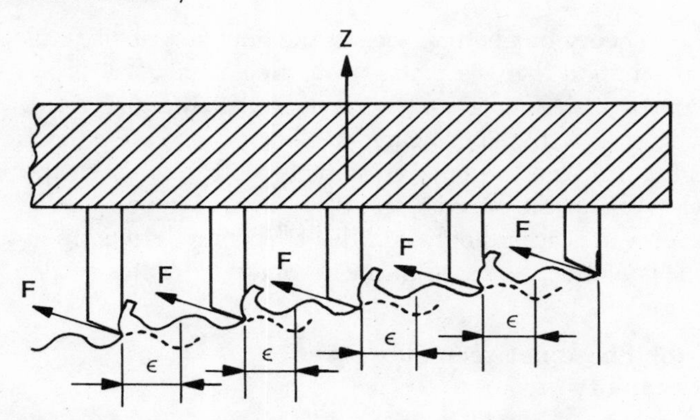

Fig. 3.53 Simplified model of a milling cutter.

under an assumption that will permit a general treatment. We will assume that all the cutting teeth have the same normals to the cut surface. This would only be so if the cutter had an infinite diameter; this is actually the case of broaching. However, the solution will also correspond quite well to face milling with large-diameter cutters when the workpiece width is substantially smaller than the cutter diameter. The diagram in Fig. 3.53 expresses the simplifying assumption being made. All teeth move together and each of them encounters the undulations produced by the preceding tooth with the same phase shift. The normals Z to the cut are parallel for all the teeth and their forces are in phase and parallel. With m teeth cutting simultaneously, their effects are simply added.

Let us now recall the theory as it was presented in relation to Fig. 3.11a and modify it to suit our case. The diagram of Fig. 3.11a is shown in a slightly modified form in Fig. 3.54. The transfer function $G(\omega)$ as presented has the shape of an OTF of a system with two or more degrees of freedom. The limit of stability was derived in Eqs. (3.7)–(3.13). A similar procedure is used here with reference to Fig. 3.54.

The harmonically variable component F of the cutting force is located in the real axis. It is the sum of the forces acting on all the m teeth. This force produces harmonic vibration of the cutter with amplitude Z, which is determined by the OTF of the system. It lags behind the force by a phase shift ϕ. It was shown that this phase shift must be $\phi > \pi/2$ because only the part of the OTF applies in which the real part is negative.

At the limit of stability the vibration has a constant amplitude Z and it produces undulations with the same amplitude. The undulations Z and Z_0 produced by two subsequent teeth have the same magnitude but they are phase shifted by the angle ε, the value of which was expressed in Eq. (3.3). It is

$$\varepsilon/2\pi = f/nz - N, \tag{3.43}$$

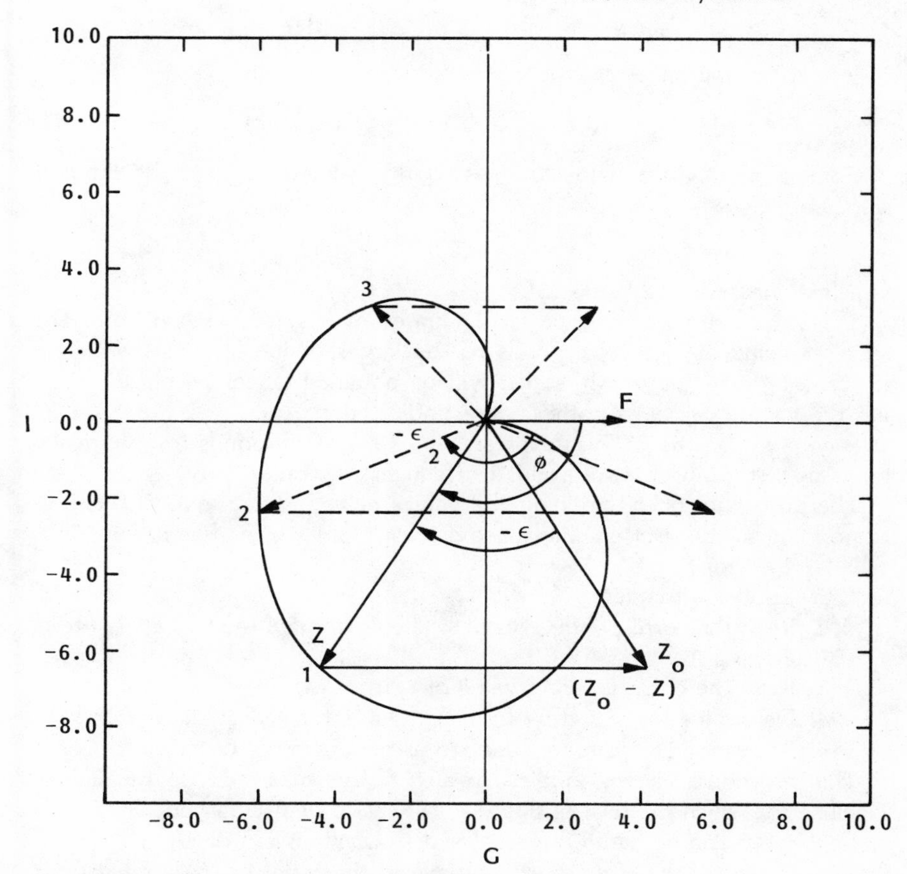

Fig. 3.54 Diagram for deriving limit of stability.

where f is the frequency of the vibration, n is the spindle speed (rev/sec), z is the number of teeth of the cutter, and N is the largest integer for which $\varepsilon/2\pi$ is still positive. Of course, $\varepsilon/2\pi < 1$ and, as it was defined in Fig. 3.7, this phase was measured from Y to Y_0; therefore, it is indicated in Fig. 3.54 as $(-\varepsilon)$ when shown from Z_0 to Z. The geometric condition (3.43) locates the vector Z in the diagram, Fig. 3.52, because, obviously, at the limit of stability

$$\phi = \pi/2 - \varepsilon/2. \tag{3.44}$$

The value of the limit chip width is obtained from the force formula (3.7) as modified here for the simultaneous cutting of m teeth:

$$F = mCb(Z_0 - Z). \tag{3.45}$$

This relationship determines the location of Z_0 in the graph, because with C (cutting force coefficient) and b (chip width) both being real numbers, the vector $(Z_0 - Z)$ must be in phase with F, i.e., parallel with F, or real in the given representation. And, it is seen from this graph that

$$Z_0 - Z = -2\ \mathrm{Re}(Z) = -2F\ \mathrm{Re}[G(\omega)]. \tag{3.46}$$

Combining (3.45) and (3.46) leads to the value of b_{lim} at the limit of stability:

$$b_{\mathrm{lim}} = -1/2mC\ \mathrm{Re}(G), \tag{3.47}$$

which is analogous to Eq. (3.13).

The values of Re(G) and, correspondingly, of b_{lim} will depend on ε. The smallest value of b_{lim} corresponds to the largest value of Re(G), which is denoted by 2 in the graph, and it will be obtained when the phase shift ε has the value $\varepsilon_2 = \varepsilon_m$. Another case is indicated at 3.

Each point on the OTF as expressed by $G(\omega)$ corresponds to a particular frequency; so point 1 corresponds to frequency f_1, point 2 to f_2, etc.

The procedure of determining the values of $b_{\mathrm{lim},i}$ for a given OTF, corresponding to different spindle speeds n_i, can be designated as follows.

Do 1 to 4 for $i = 1$ to k:

1. Choose a frequency f_i,
2. Read the corresponding value of Re(G_i), obtain the value $b_{\mathrm{lim},i}$ from Eq. (3.47),
3. Read the corresponding value of ε_i from the graph of $G(\omega)$,
4. Determine the speed n_{ik} from Eq. (3.43), for $N = 1, 2, \ldots, k, \ldots,$
5. Rearrange the pairs $(b_{\mathrm{lim},i}, n_i)$ in ascending order of n_i.

This procedure can be applied to an OTF as measured on the milling machine for which the relationship (b_{lim}, n) has to be established. Otherwise, steps 2 and 3 can be based on a mathematical expression of $G(\omega)$.

An example of the effect of the geometric constraint of ε on stability, or, in more practical terms, of the effect of variation of spindle speed n, is given in Fig. 3.55a. It applies to the case of a single degree-of-freedom system with parameters as given in the inset and with the corresponding OTF, the real part of which is shown. The vertical coordinate of the diagram is the stability increase ratio $q_s = b_{\mathrm{lim}}/b_{\mathrm{lim,min}}$, where $b_{\mathrm{lim,min}}$ is the chip width that is obtained for the phase shift ε_m which is most favorable for self-excitation. It is the "borderline stability" which was discussed in the "Theory of Chatter" section and which was expressed by Eq. (3.14). The horizontal coordinate is proportional to spindle speed n; it is $v = zn/f = 1/(N + \varepsilon/360°)$; see Eq. (3.43). The diagram has the form of "stability lobes." Periodically, the stability increase ratio becomes $q_s = 1.0$; there is no increase in stability. These points correspond approximately to $\varepsilon = 225°$, i.e., to $v = 1/(N + 0.625)$. Between these points stability is increased. The maxima correspond to $\varepsilon = 0°$, i.e., to such speeds that would give integer numbers of waves between subsequent teeth. At low spindle speeds the number N of waves between subsequent

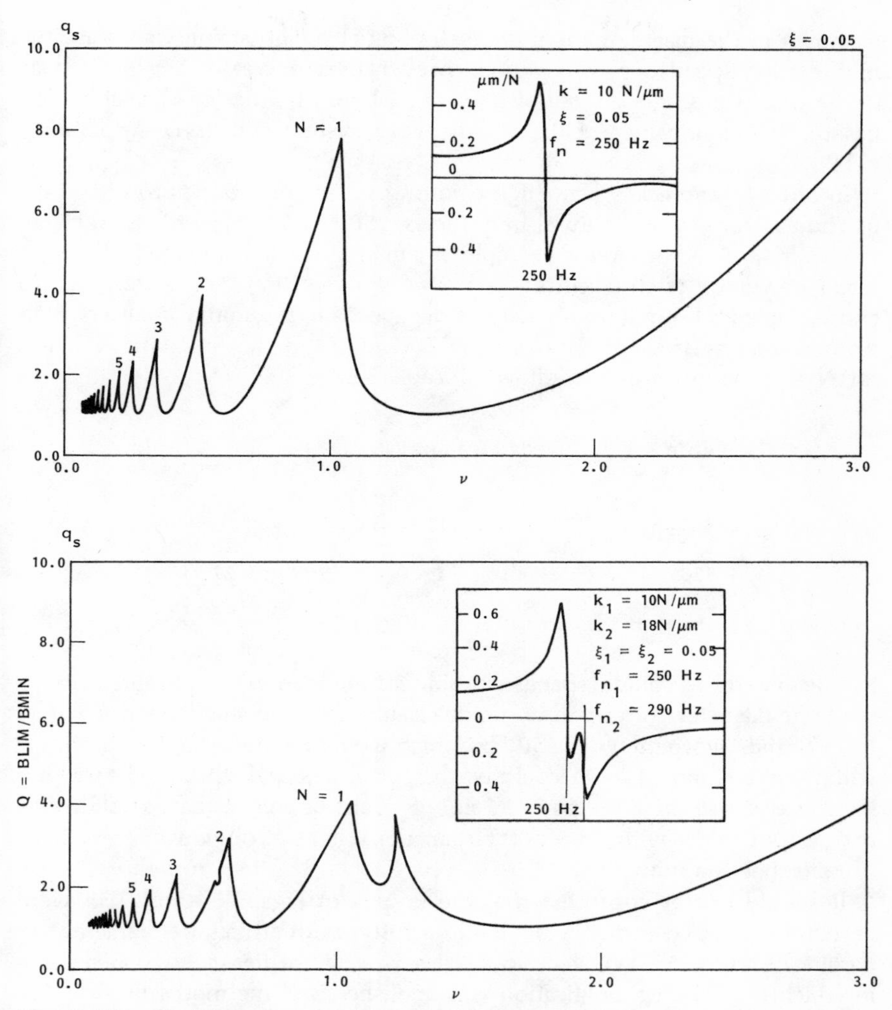

Fig. 3.55 Stability lobes for (a) a single-mode system, simple model, and (b) a two-mode system, simple model.

teeth is high. The lobes are close to one another and stability increases are small. As spindle speed increases the waves become longer and, for two and a fraction waves the ratio $q_s > 4$. For $N = 1$, $q_s = 8$ and for spindle speeds for which there would be less than one wave between subsequent teeth, stability gradually increases indefinitely.

For systems with more than one degree of freedom the fields of high stability are split and narrowed down. The example in Fig. 3.55b shows the stability lobes for a system with two degrees of freedom, the OTF of which is shown in the inset. Again, it shows potentially high stability increases at high cutting speeds, up to $q_s = 4$. However, these increases are lower than in the case of a single degree of freedom. Obviously, the (q_s, v) relationship has to be individually established for every system and its corresponding OTF.

In order to appreciate how high a cutting speed corresponds to $N = 1$ let us choose three typical natural frequencies: 250 Hz, 550 Hz, and 1600 Hz as they were pointed out in the example of a milling machine in Fig. 3.51. Let us assume a typical pitch of cutter teeth $p = 25$ mm (1 in.). The corresponding cutting speed v [m/min (fpm)] and spindle speeds n (rev/min) for cutters with diameters $d_1 = 250$ mm (10 in.), $d_2 = 100$ mm (4 in.), $d_3 = 30$ mm (1.2 in.) for $N = 1$ are given in the following Table 3.1.

Table 3.1 Speeds for Lobe $N = 1$

f	v, m/min (fpm)	n_1	n_2	n_3
250	375 (1230)	<u>500</u>	1250	10,416
550	825 (2700)	1100	<u>2750</u>	<u>22,917</u>
1600	2400 (8350)	3200	8000	<u>66,666</u>

The underlined spindle speeds in Table 3.1 are the most applicable because the large-diameter cutter will be rigidly clamped to the spindle and it would be the headstock mode, $f = 250$ Hz, which would be involved; for the cutter with $d_2 = 100$ mm (4 in.) it could be the spindle mode with $f = 550$ Hz; and for the end mill, $d_3 = 30$ mm (1.2 in.), it could be either the spindle mode or the tool mode with $f = 1600$ Hz. In the last case, obviously, very high spindle speeds are involved.

It should not be forgotten that the graphs of Figs. 3.54 and 3.55 were based on a simplified theory assuming a cutter with an infinite diameter, according to Fig. 3.53. We shall now present results of *time-domain simulations* in which the above simplification was abolished and the model of Fig. 3.27 was used. These results are rather realistic. In Fig. 3.56 the lobes of stability are shown for $v = nz/f$ in the range $0.5 < v < 4.44$ and a system with two mutually perpendicular modes with equal frequencies, $f_x = f_y$. Two sets of curves are included, for (a) half-immersion up-milling and (b) slotting. It is seen that for $N = 1$ an increase of stability is obtained, $q > 12$, for both (a) and (b), and that for spindle speeds much higher than that which gives $N = 1$ (Table 3.1), stability gradually increases. In Fig. 3.57 a similar graph is given for a system with unequal frequencies of the two modes in a ratio $R = f_x/f_y = 0.94$. Stability increases of about $q > 10$ for $N = 1.0$ are obtained. In this case stability of half-immersion up-milling is not much higher than that of slotting as in Fig. 3.56.

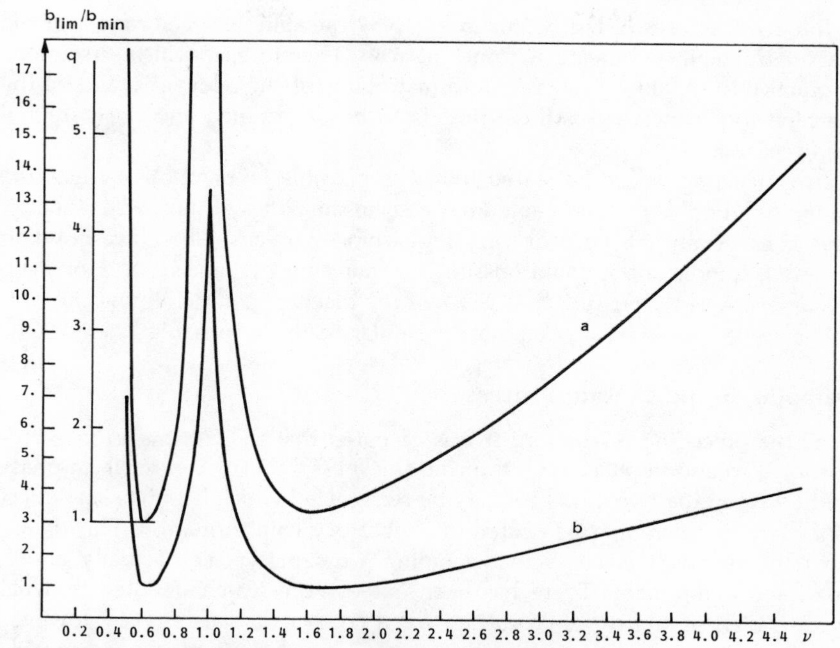

Fig. 3.56 Stability lobes for a two-mode system, $f_x = f_y$.

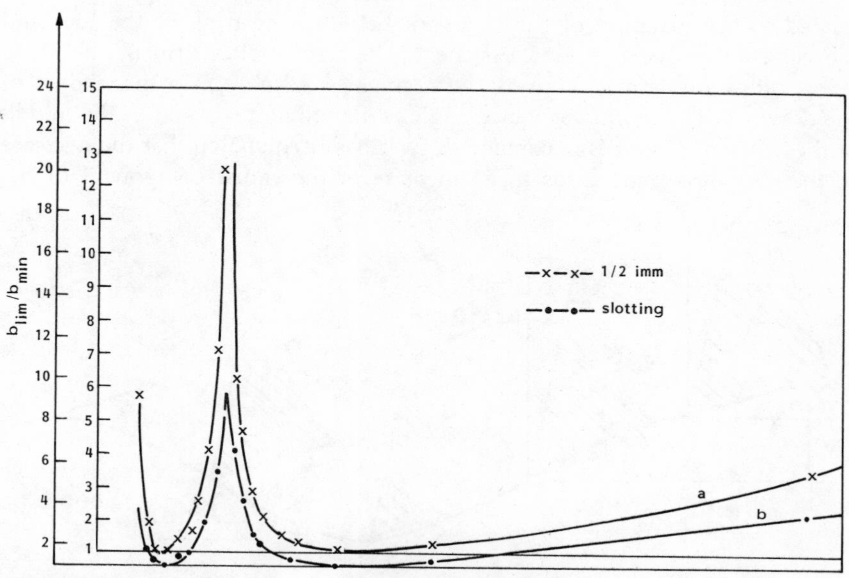

Fig. 3.57 Stability lobes for a two-mode system, $f_x = 0.94f_y$.

In both the cases of Fig. 3.56 and Fig. 3.57 the spindle speed range for the maximum stability increase is rather narrow. Therefore, it will be necessary in practice to obtain rather exact information about the relevant OTF's of the machine tool concerned if this speed is to be determined and the potential gain realized.

In conclusion, it can be stated that it is possible to practically utilize the "stability lobes" and choose spindle speeds in the zones of increased stability. This is especially efficient for very high spindle speeds. The corresponding optimum spindle speed could possibly be found by trial and error. However, the best way is to measure the OTF's of the machine and determine the suitable spindle speeds by a computation similar to those presented here.

Damping in the Cutting Process

In all the preceding discussions it was accepted that the cutting force is proportional to and in phase with the chip thickness variation. In reality a phase shift between the two variables may be recognized which has the same effect as if there was damping generated in a vibratory chip formation. This damping may be positive, i.e., actual damping, or negative, i.e., actually a self-excitation component. There has been extensive research devoted to what was denoted as the dynamic cutting force coefficient; see ref. 12. However, its practical significance can be presented in a rather simple way, omitting most of the theoretical deliberations.

The generation of positive damping in vibration while cutting is generally related to the variation of the clearance between the flank of the tool and the cut surface; see Fig. 3.58. At the top (A) and bottom (B) of the wave being cut the clearance angle is the nominal, "mean" γ_m. In the middle of the downward slope the clearance angle is minimum, $\gamma_{\min}$, and in the middle of the upward slope it is maximum, $\gamma_{\max}$. It has been shown that the decrease of the clearance angle leads to an increase of the thrust component of the

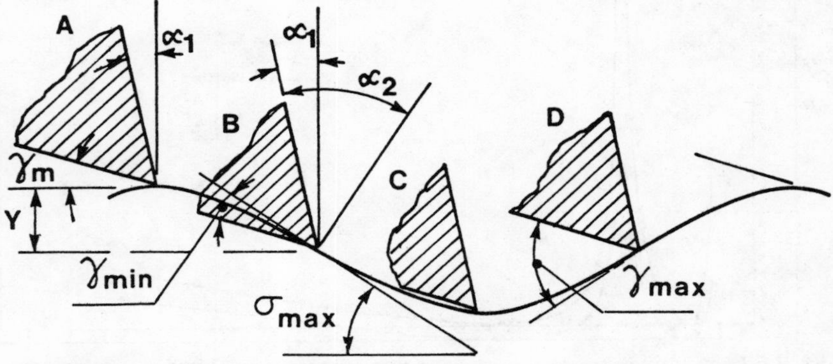

Fig. 3.58 Mechanism of damping in the cutting process.

cutting force. Correspondingly, during the half-cycle in which the velocity is downward (from A to C) it is opposed by a greater force than the force which drives the tool upward (from C to A). This variation of the thrust force is in opposite phase with velocity (90° out of phase with displacement) and it represents positive damping. This damping is larger for short waves because they have steeper slopes. Shorter waves are obtained at low cutting speeds. Correspondingly, it is found that stability increases strongly at very low cutting speeds. This will be documented later. Shorter waves are also obtained at high frequencies. This has a special implication for end milling.

The wavelength w of the undulations produced on the cut surface is

$$w = v/f = \pi dn/f, \tag{3.48}$$

where v is cutting speed, f is frequency of vibration, and n is spindle speed, all in appropriate compatible units.

It has been shown in Fig. 3.27 that the modes associated with a milling spindle are found in three ranges; the headstock mode with $f = 120-250$ Hz, the spindle bending mode, see Fig. 3.59, mode A, with $f = 500 - 600$ Hz, and the end mill mode B with $f = 1600-2000$ Hz. The mode A is associated with

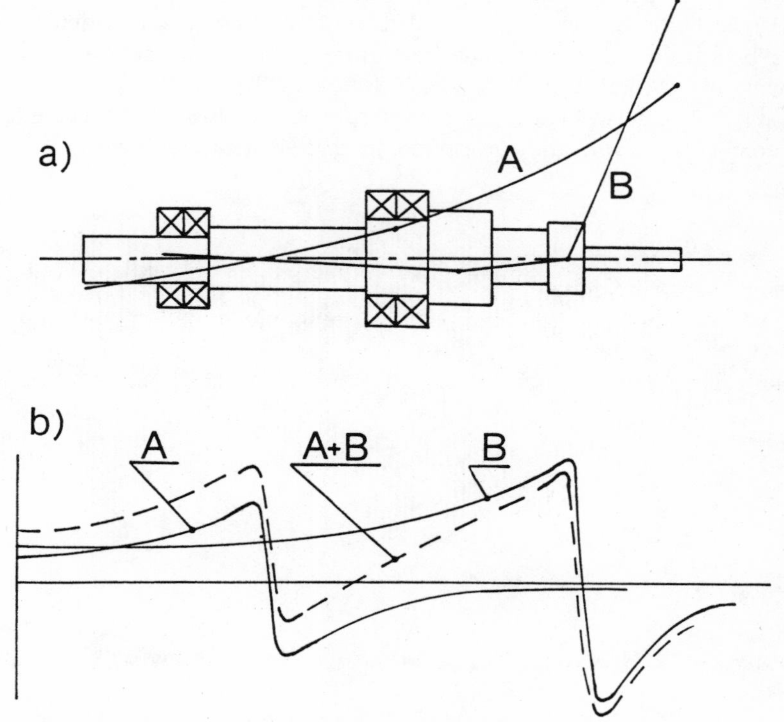

Fig. 3.59 Two basic modes of spindle with end mill.

shell end mills and short end mills, and it is stiffer than the mode B associated with long and slender end mills. The resulting TF has two significant minima with the one at A giving higher stability than the one at B. However, at cutting speeds used for machining steel, e.g., $v = 30$ m/min (98.4 fpm), the high frequency at B would lead to very short waves, $w = 0.25$ mm (0.001 in.), which would produce very high damping in the cut. Correspondingly, chatter occurs at the lower frequency A [$w = 1.0$ mm (0.004 in.)] in spite of the higher stiffness at this point of the TF.

There is another interesting phenomenon; at these lower cutting speeds cutting with long end mills is more stable than with short end mills. This is explained when it is realized that the chatter mode is the spindle mode A at which the long end mill acts as an attenuator of the cutting force variation. This is illustrated in Fig. 3.60, reproduced from ref. 9. Case a shows vibrations obtained in slotting with an end mill diameter 25.4 mm (1 in.), 50 mm (2 in.) long with four teeth. The workpiece material was aluminum alloy but the cutting speed was rather low, $v = 94$ m/min (308 fpm), feed rate 358 mm/min (14 in./min), axial depth of cut $a_a = 22$ mm (0.87 in.). The vibrations are represented by the force signal measured on a table-type dynamometer. The high-frequency (2000-Hz) mode would have given $w = 0.8$ mm (0.03 in.); chatter occurred at $f = 650$ Hz resulting in $w = 2.45$ mm (0.01 in.). Case b is obtained at the same conditions except that the end mill was 100 mm (4 in.) long. Vibration was 20 times smaller.

Table 3.2 correlates cutting speeds v, spindle speeds n, cutter diameters d, and frequencies f with the corresponding chatter mark wavelengths.

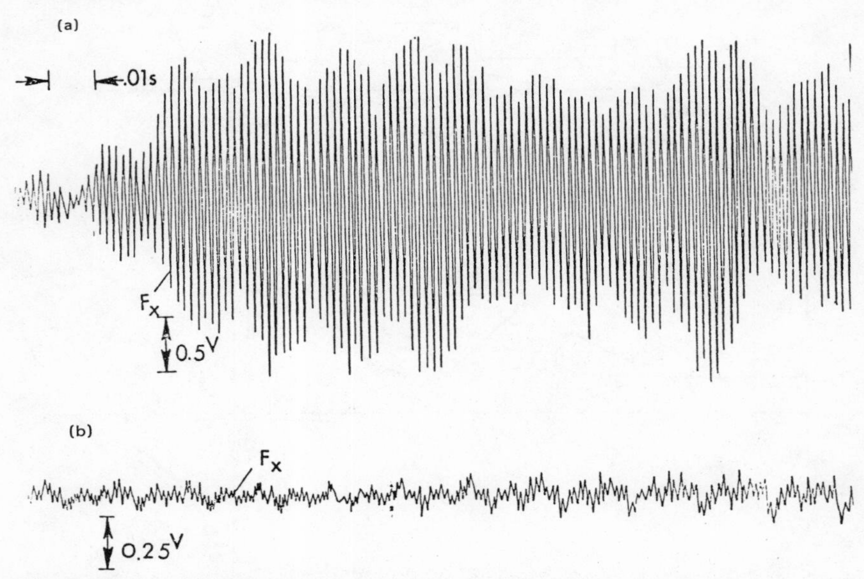

Fig. 3.60 Chatter vibrations with (a) short and (b) long end mills.

Table 3.2 Chatter Mark Wavelengths w [mm (in.)]

Mode	Frequency (Hz)	30 (98)	60 (197)	300 (984)	1000 (3280)	Cutting speed v, m/min (fpm)
		127	254	1270	4244	d = 75 mm (3 in.) } n(rev/min)
		381	762	3820	12732	d = 25 mm (1 in.) } n(rev/min)
Headstock	200	2.5 (0.01)	5 (0.2)	25 (1)	83 (3.3)	
Spindle	200	0.77 (0.03)	1.54 (0.06)	7.7 (0.3)	25 (1)	
End Mill	1800	0.28 (0.01)	0.55 (0.22)	2.8 (0.01)	9.25 (0.37)	

In general, with a very coarse approximation one may exclude chatter with wavelengths $w < 0.3$ mm (0.01 in.) and $w > 50$ mm (2 in.). The strongest tendency for chatter will be for the combination of frequencies and cutting speeds along the diagonal double band of Table 3.2 as indicated by the underlined w values. Another general type of diagram derived in ref. 10 is reproduced as Fig. 3.61. It shows the steep increase of stability (for an arbitrary scale of b_{lim}) at low cutting speeds in machining 1045 steel. This increase of stability is larger, the higher the corresponding natural frequency of the decisive mode of the structure.

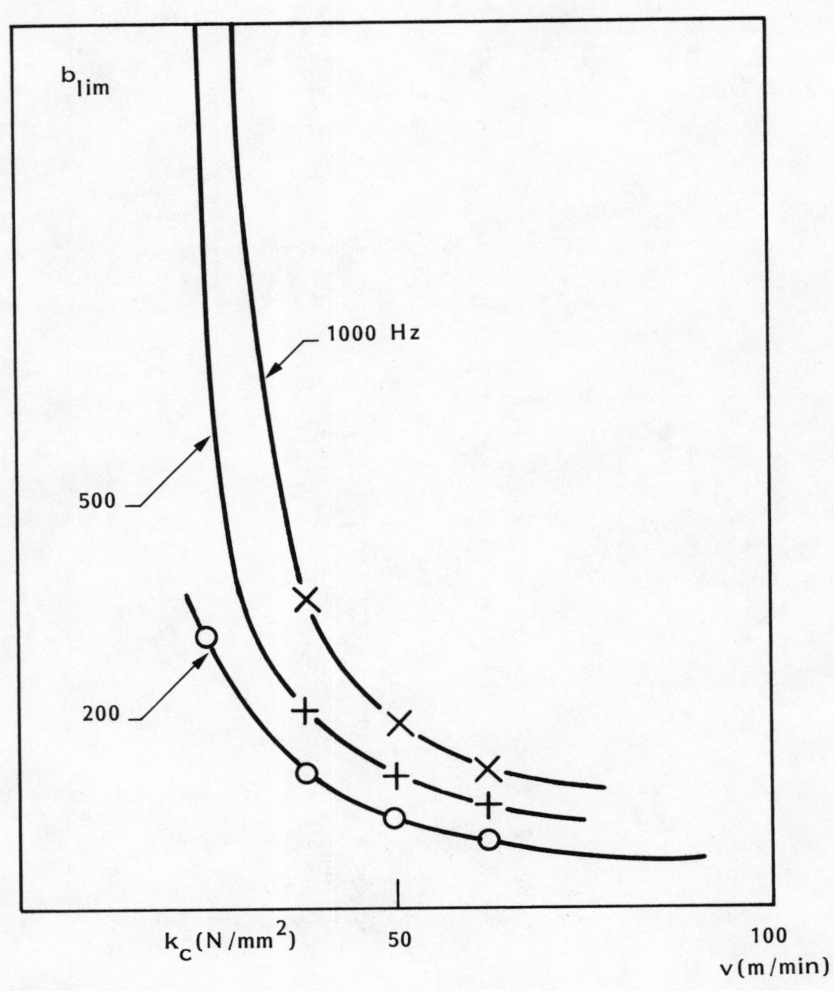

Fig. 3.61 Effect of cutting speed on stability.

It is interesting to state that at higher speeds generation of negative damping in the cut occurs, which leads to a decrease of stability. The mechanism of the negative damping is not yet understood, although an attempt of its explanation was presented in ref. 11. Both the positive and negative damping were well illustrated in special tests described in ref. 10. A nonregenerative cut was taken in the manner of Fig. 3.14. The tool was attached to a special rig illustrated in Fig. 3.62. This rig consists essentially of a mass M on two flat springs representing a simple vibratory system. The damping in this system is provided by an exciter EXC, which is fed from a signal from the vibration pick-up VP, after phase shift ϕ and amplification A, and it can be set to high or low or even negative values. The experiments consist of, first, striking the rig outside of the cut and registering its decaying natural vibration. Subsequently, the rig is struck while the tool T is engaged in the nonregenerative cut. Since the system is single mode, it cannot self-excite in mode coupling either. Any change of the decay rate can only be caused by the damping (positive or negative) generated in the cut. Examples of test results are reproduced in Figs. 3.63 and 3.64. They correspond to turning 1045 steel, 170 BHN, with mean chip thickness $h_m = 0.2$ mm (0.008 in.), tool rake angle 7°, relief angle 7°, and natural frequency of the rig $f = 200$ Hz.

The tests shown in Fig. 3.63 were done with the damping ratio of the rig set at $\xi = 0.012$ and the decay of the vibrations outside of the cut is shown in record a. Records b–e were obtained during cutting with speeds from $v = 15$ m/min (49 fpm) to $v = 45$ m/min (148 fpm). The corresponding mean damping ratios are written on the graphs. It is seen that in all these instances positive damping was generated in the cut; the damping was very large at $v = 15$ m/min (49 fpm) ($\xi = 0.1$) and decreased with the increase of cutting speed.

In Fig. 3.64 the records are shown for cutting with speed $v = 64$ m/min (210 fpm). Case a corresponds to the rig damping set still at the value shown

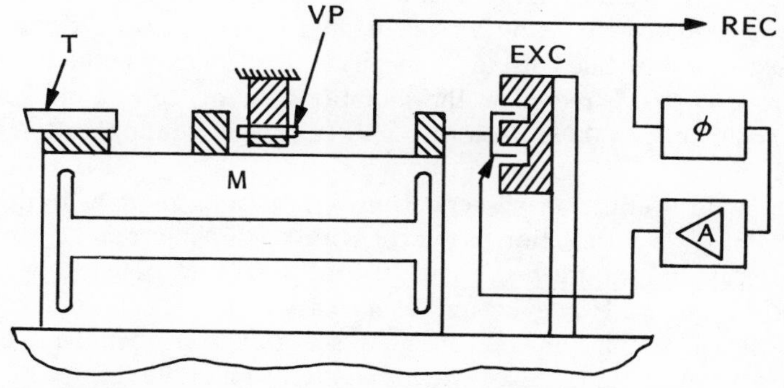

Fig. 3.62 Test rig for damping in the cutting process.

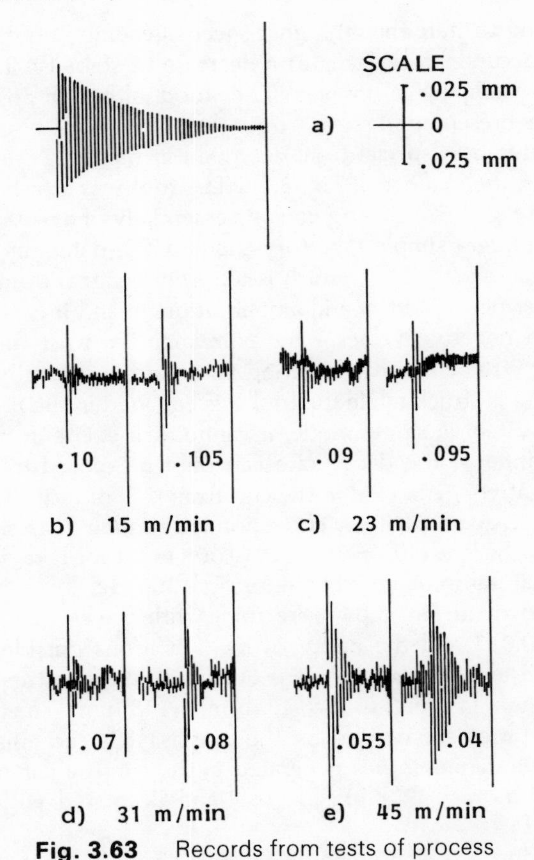

Fig. 3.63 Records from tests of process damping.

in Fig. 3.63a. This time vibration did not decay at all, but increased. This proves that negative damping was generated in the cut. In order to obtain decaying vibration while cutting it was necessary to increase the damping of the rig to $\xi = 0.025$, record b; the vibrations during cutting shown in b1, b2 and again to $\xi = 0.055$, record c with vibrations during cutting shown in c1, c2.

In Fig. 3.65 the effects of higher cutting speeds are shown. With the rig set at $\xi = 0.055$ as in a, vibration in the cut with $v = 90$ mm/min (295 fpm) still did not decay and the rig had to be set to $\xi = 0.09$ to obtain a decay with $\xi = 0.06$ in the cut. However, further increase to $v = 126$ m/min (413 fpm) led to positive damping in the cut again; see graphs at c with no cutting at left and cutting at right. Also, with $v = 180$ m/min (590 fpm), case d with the rig at $\xi = 0.055$, damping in the cut was positive and the decay damping ratio increased to 0.08 and 0.075.

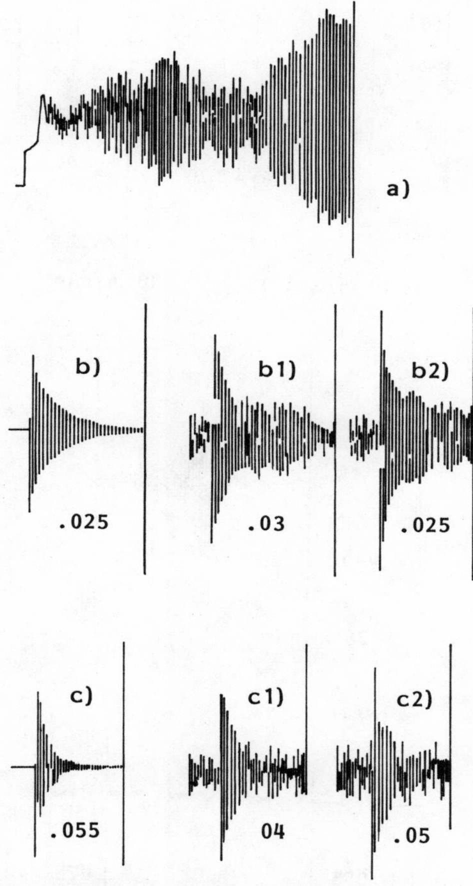

Fig. 3.64 Records from tests of process damping.

It is seen that there is, first, very high damping in the cut at very low cutting speeds [e.g., $v = 15$ m/min (49 fpm)]. Then there is a range, between $v = 60$ and 95 m/min (197 and 312 fpm) in which the cut generates negative damping. Above this range, damping in the cut is positive again. This, at least, was shown here for workpiece material 1045. The damping generated in the cut is usually expressed quantitatively by the value of the "imaginary inner cutting force coefficient." For its definition see ref. 12. It is not necessary to present the full definition here; it is sufficient to accept that it is a relative measure of the damping in the process and it may be included in the computation of the limit of stability once the OTF of the structure is known. The cutting process damping is associated mainly with the thrust component of the cutting force, and it affects stability of cutting much more strongly for

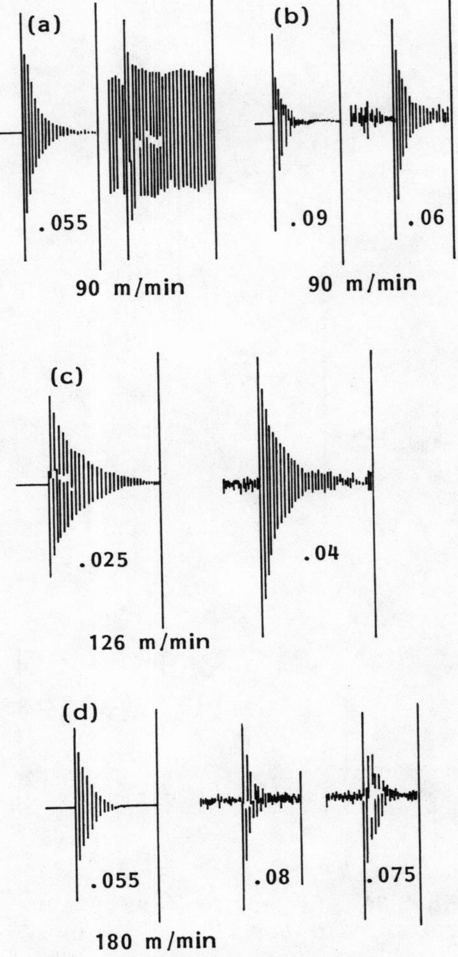

Fig. 3.65 Records from tests of process damping.

vibration modes with directions (defined by the angle α of Fig. 3.9) close to the normal to the cut ($\alpha = 0°$).

Extensive tests of the cutting process damping were described in ref. 10. Results for workpiece materials 4140, BHN 250, and aluminum alloy 7075, BHN 113, are presented in Figs. 3.66 and 3.67. They are represented in the way they affect the limit chip width b_{lim} for a system with the minimum of a DTF, $G_{\text{min}} = -0.2$ μm/N (-2.9×10^{-6} ft/lb) and two alternative directional orientations, $\alpha = 0°$ and $\alpha = 45°$. In both cases the effects are much

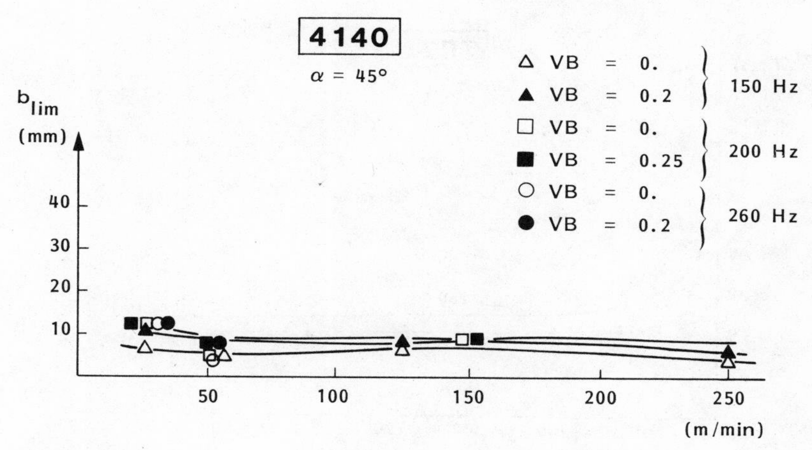

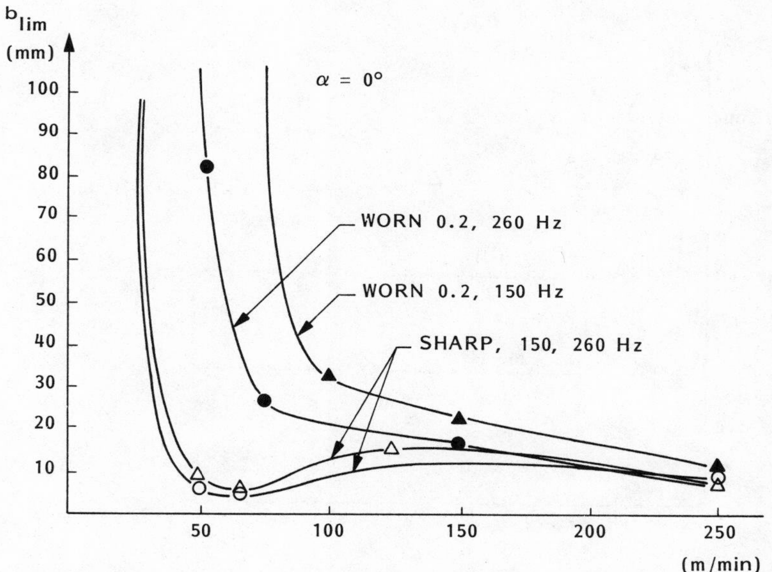

Fig. 3.66 Effect of process damping on stability in cutting 4140 steel.

stronger for $\alpha = 0^\circ$, which is more applicable to end milling and which we are going to briefly discuss. The frequencies for which these results were obtained are rather low, 150–260 Hz, because the test rig was not suitable for higher frequencies. However, data for higher frequencies are not yet available. Also, included is the effect of tool wear. It is obvious that the mechanism of the positive damping as indicated in Fig. 3.58 is much stronger for worn tools.

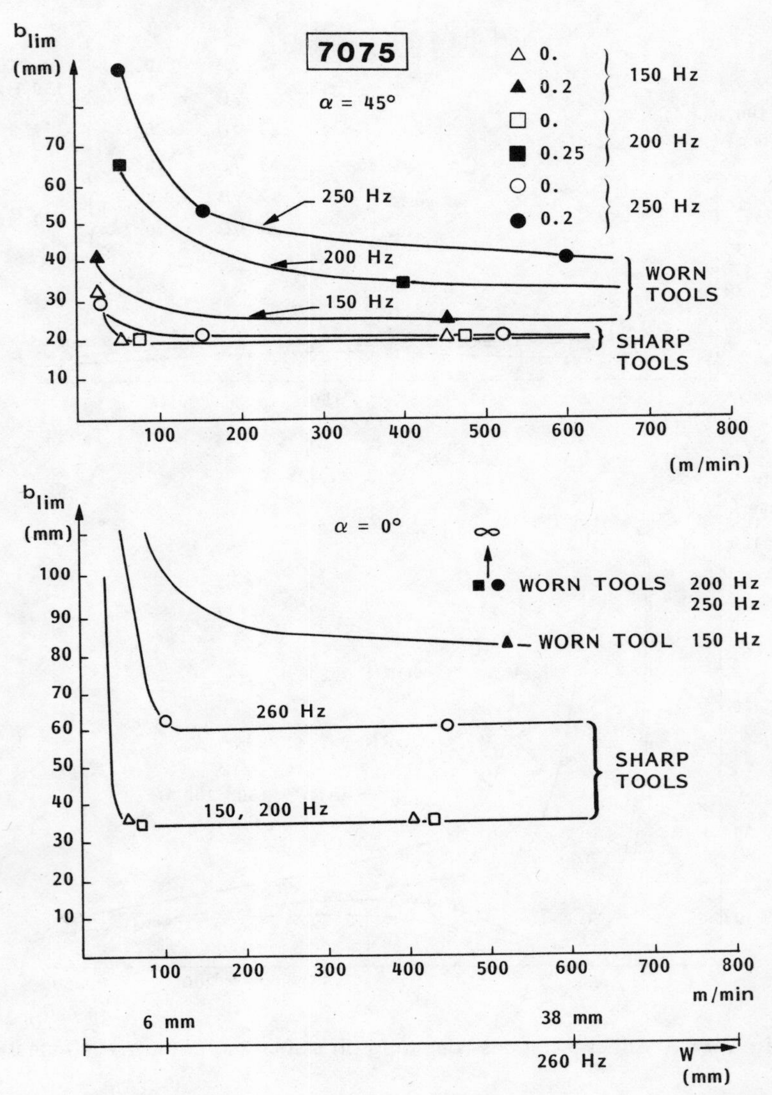

Fig. 3.67 Effect of process damping on stability in cutting 7075 aluminum.

For the steel 4140, Fig. 3.66, and sharp tools, positive damping is strong below $v = 50$ m/min (164 fpm). With worn tools it is still much stronger; it extends to 150 m/min (492 fpm) and increases with frequency. For sharp tools there is a range of very low stability (due to negative process damping) between $v = 50$ and 70 m/min (164 and 230 fpm). Above $v = 150$ m/min (492 fpm) there is very little change and also wear and frequency play a lesser role.

For aluminum 7075, Fig. 3.67, very high damping is found for sharp tools below $v = 100$ m/min (328 fpm) and for worn tools below 200 m/min (656 fpm). For speeds higher than these there is almost no change. However, wear and frequency still have strong effects. It should be interesting to extend these investigations to much higher cutting speeds.

In conclusion, it may be stated that spindle speed variation has a strong effect on stability of milling due to two entirely different effects. The first effect as expressed by the "stability lobes" is strongest at high and very high cutting speeds. The other effect as expressed by the cutting process damping is strongest at very low cutting speeds. For the former effect, the theory is rather well developed but its implementation needs development of practical experimental techniques for determining the OTF's at the end of the tools and of the software and its implementation on suitable computer hardware for determining the optimum spindle speeds. For the latter effect more tests are needed especially for higher frequencies and higher cutting speeds. It seems, however, that this effect will not be very important for high-speed milling.

Special Cutters: Their Effect on Stability

Various special designs of milling cutters have been developed that increase stability against chatter. These are

1. cutters with nonuniform tooth pitch;
2. cutters with alternating helix;
3. cutters with serrated and undulated edges.

In all these cases increase of stability is achieved by disturbing the regeneration of surface waviness, and their effects are most pronounced over particular ranges of cutting speeds depending on the parameters of the cutter design.

The action of disturbing the regeneration of waviness is indicated in Fig. 3.68. In 3.68a a cutter with nonuniform tooth pitch is shown. The teeth 1, 2, 3 vibrate together and each one is cutting the surface produced by the preceding one. Because of the different pitches p_{12} and p_{23}, the phase ε_{12} of vibration of tooth 2 with respect to the waviness produced by tooth 1 and phase ε_{23}, and similarly for teeth 2 and 3, are different. Obviously, the phasing that would be most favorable for regeneration of waviness [recall Fig. 3.54 and Eq. (3.47)] cannot be obtained for both pitches, and this leads

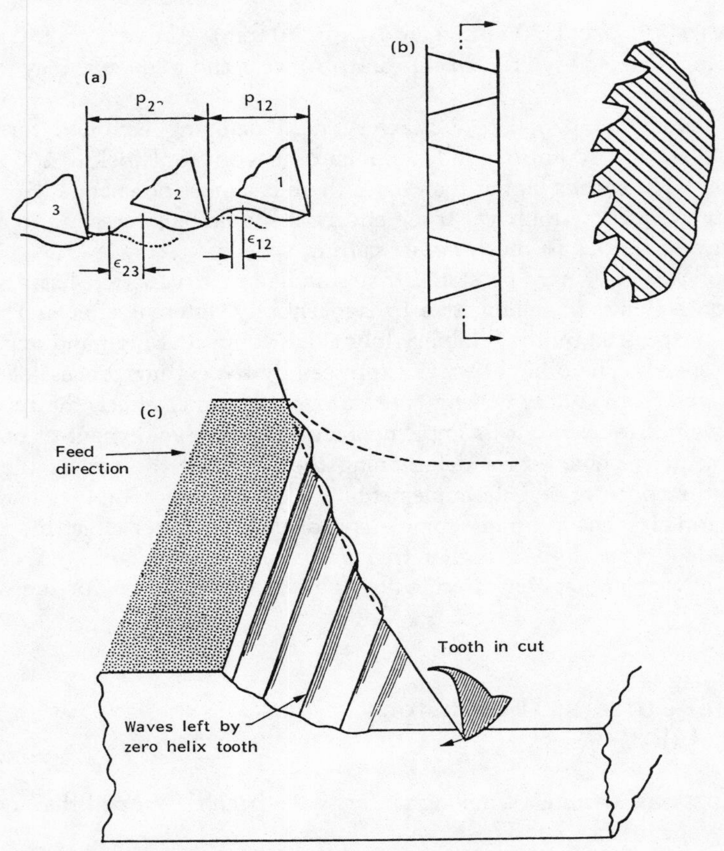

Fig. 3.68 Disturbing regeneration of waviness.

to an increase of stability. In 3.68b a cutter with peripheral teeth with alternating helix angles is shown. As indicated in 3.68c, the waves left behind by a straight tooth when being removed by an inclined tooth produce very little force variation and the regenerating action is very weak.

Similarly as for the regular milling cutters the theoretical analysis of stability of milling with the special cutters can be carried out:

1. *In the frequency domain.* This approach needs simplifying assumptions, the most significant of which is the assumption that all the cutter teeth are parallel, corresponding to a cutter with an infinite diameter. This leads to rather substantial errors, but the results of the analysis are explicit, and they are obtained in a rather general way, with modest computing expenditures. Similarly as before, we shall use this approach first in order to show the general trends.

2. *Using time-domain digital simulation.* This approach gives results that are close to reality, but it involves high computing cost. We shall use it in the second stage to indicate the magnitude of corrections necessary to be applied to the result obtained in the first stage.

First, *cutters with nonuniform tooth pitch* will be discussed. The analysis will be made with reference to the one used for cutters with uniform pitch in the "Selecting Optimum Spindle Speed" section. The graph in Fig. 3.54 relating to Fig. 3.53 and the corresponding equation (3.51) could be replaced by the graph in Fig. 3.69. Three teeth with uniform tooth pitch are assumed to be cutting simultaneously. Their vibrations Z_1, Z_2, Z_3 are all in phase and they lag by ϕ behind the variable component F of the cutting force. The wavinesses produced by the preceding teeth are denoted Z_{01}, Z_{02}, Z_{03} and they are all mutually in phase. At the limit of stability their amplitudes are constant and equal in magnitude to the amplitudes of the vibrations:

$$Z_{01} = Z_{02} = Z_{03} = Z_0, \quad Z_1 = Z_2 = Z_3 = Z, \quad |Z_0| = Z. \tag{3.49}$$

There is a phase shift ε between the wavinesses and the vibrations:

$$\varepsilon/2\pi = n/zf - N, \tag{3.50}$$

where N is the maximum integer for $0 < \varepsilon/2\pi < 1$, n is spindle speed, and z is the number of teeth on the cutter. The latter is distinct from the number m of teeth cutting simultaneously.

The force F is the sum of forces on all the $m = 3$ teeth:

$$F = Db(Z_{01} + Z_1 + Z_{02} - Z_2 + Z_{03} - Z_3) = mCb(Z_0 - Z), \tag{3.51}$$

which is

$$m(Z_0 - Z) = -2mF\,\mathrm{Re}(G). \tag{3.52}$$

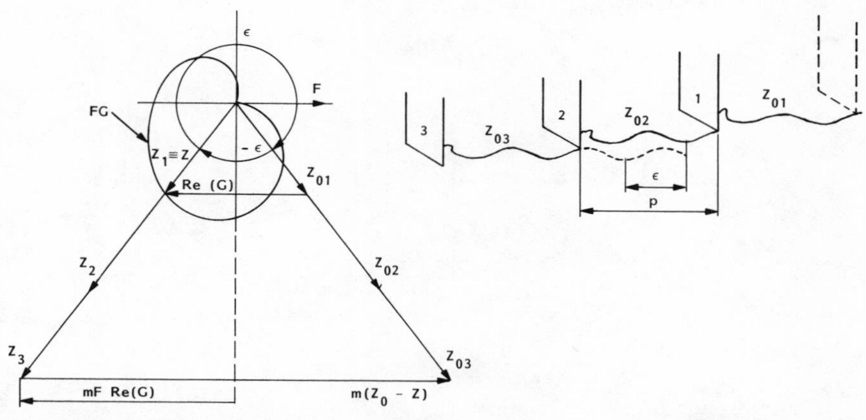

Fig. 3.69 Graph for limit of stability for a cutter with uniform tooth pitch.

Correspondingly, with respect to (3.51), it is

$$F = -2mCbF\,\mathrm{Re}(G)$$

and

$$b_{\lim} = -1/[2mC\,\mathrm{Re}(G)] \tag{3.53}$$

Equation (3.53) is identical to Eq. (3.50) and even if it may seem that it was obtained in a more roundabout way than (3.50), it will help to derive the conditions for the cutter with nonuniform tooth pitch.

Similarly as before, the vector $m(Z_0 - Z)$ is parallel with F meaning that the sum of the chip thickness variations is in phase with the force, corresponding to C and b being real numbers.

Let us now assume a cutter where the pitches vary linearly. Each subsequent pitch is greater by Δp than the preceding one (see Fig. 3.70), and, referring to Eq. (3.50), the phase angles ε_i will also be increasing by $\Delta\varepsilon$:

$$\Delta\varepsilon = \Delta p\varepsilon/p. \tag{3.54}$$

Vibrations Z_1, Z_2, Z_3 are equal and in phase because all the teeth vibrate simultaneously but they meet the undulations Z_{01}, Z_{02}, Z_{03} with different phase angles ε_1, ε_2, ε_3. The magnitudes $|Z_0|$ of the amplitudes of the undulations are still all equal to the magnitudes of the amplitudes $|Z_i| = |Z|$ but each of the undulations must be represented by a vector Z_{0i} of a different direction. The first vector Z_{01} is located at the angle ε, which corresponds to the pitch p and the vectors Z_{02} and Z_{03} are inclined by $\Delta\varepsilon$ and $2\Delta\varepsilon$ with respect to Z_{01}. The sum ΣZ_{0i} is denoted B_0 while the vector mZ is denoted B. However, the total sum of chip thickness variations V is again in phase with the force, which is proportional to it:

$$F = CbV = Cb\left(\sum Z_{0i} - mZ\right), \tag{3.55}$$

where we may denote

$$B_0 = \sum Z_{0i}, \quad B = mZ, \quad V = B_0 - B. \tag{3.56}$$

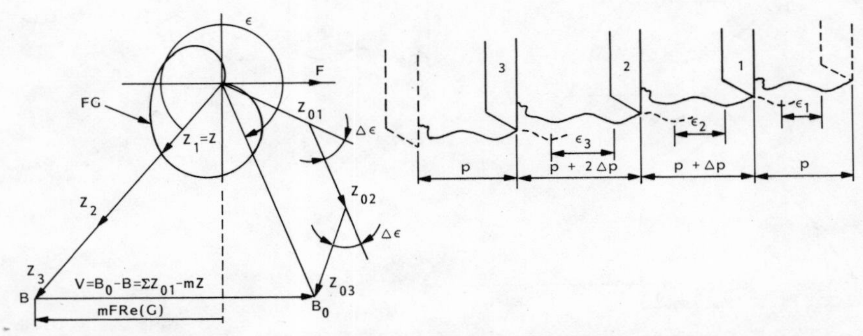

Fig. 3.70 Graph for limit of stability for a cutter with nonuniform tooth pitch.

The vector B_0 is now shorter than in Fig. 3.69,

$$|B_0| = |Z_{01} + Z_{02} + Z_{03}| < 3|Z_0| \tag{3.57}$$

and also the vector V is shorter than $[-2mF\,\mathrm{Re}(G)]$,

$$V = q[-mF\,\mathrm{Re}(G)] \tag{3.58}$$

where

$$q < 2 \tag{3.59}$$

Combining Eqs. (3.55) and (3.58), we have

$$b_{\mathrm{lim}} = 1/[qmC\,\mathrm{Re}(G)] \tag{3.60}$$

With respect to Eq. (3.59) the value of b_{lim} obtained from (3.60) is larger than that obtained from (3.53), for the same f and ε.

In both instances, of the uniform pitch as described by Fig. 3.69 and Eq. (3.53) and of the linearly varying pitch as described by Fig. 3.70 and Eq. (3.60), the values of b_{lim} vary with spindle speed n. The procedure of obtaining b_{lim} is analogous to that described in the "Selecting Optimum Spindle Speed" section following Eq. (3.51):

Do 1 to 7 for $j = 1$ to k: (3.61)

1. Choose frequency f_j. This locates the vector Z on the transfer function $G(\omega)$.
2. Choose the shortest pitch p and pitch increment Δp.
3. Choose the number m of teeth cutting simultaneously.
4. Construct the vector

$$B_0 = \sum_{1}^{m} Z_{0i},$$

where $|Z_{0i}| = |Z|$.

5. Determine $V = B_0 - mY$ as shown in Fig. 3.70, and $q = V/(mZ)$.
6. Determine b_{lim} from Eq. (3.60).
7. Determine ε from Fig. 3.70 and obtain the spindle speed n using the basic pitch p:

$$\varepsilon/2\pi = fp/(\pi dn) - N,$$

for $N = 1, 2$, etc.

8. Rearrange b_{lim} versus n in ascending order of n.

Examples of cutters with nonuniform pitch are shown in Fig. 3.71. In both cases the cutter has four teeth. The amount of pitch nonuniformity is different. In case a, it is $p = 82.5°$, $\Delta p = 5°$; and in case b, it is $p = 63°$, $\Delta p = 18°$. It is obvious that the first cutter will provide sufficient variation of the phase angle ε between subsequent teeth for shorter waves while the other one will

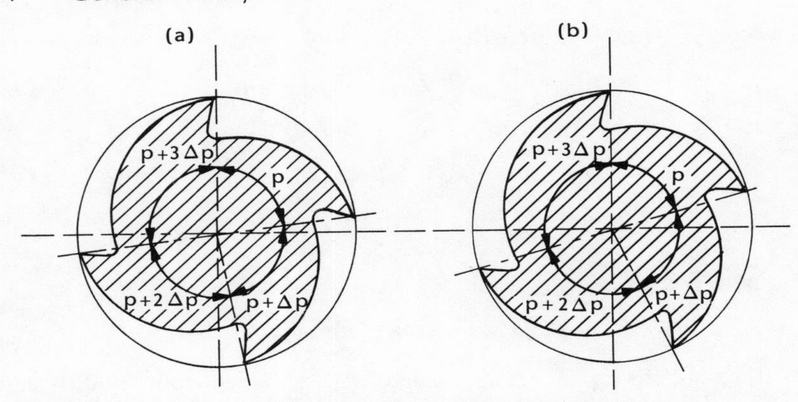

Fig. 3.71 Examples of two cutters with nonuniform tooth pitch.

be more efficient for longer waves. This is shown in the diagrams in Figs. 3.72 and 3.73. In both of them the top line corresponds to "stability lobes" of a cutter with four teeth and uniform tooth pitch, and the bottom line corresponds to the cutter with linearly varying pitch. On the horizontal scale is the number

$$\nu = 1/(N + \varepsilon/2\pi) = nz/f \tag{3.62}$$

where n (rev/sec) is the spindle speed, z is the number of teeth, and f is frequency of vibration; $N + \varepsilon/2\pi$ is the number of waves between teeth for the regular cutter. On the vertical scale is the ratio $q = b_{\text{lim}}/b_{\text{lim,min}}$, where $b_{\text{lim,min}}$ is the value of the "borderline stability." However, the graphs mainly offer a comparison of stability between the cutter with nonuniform

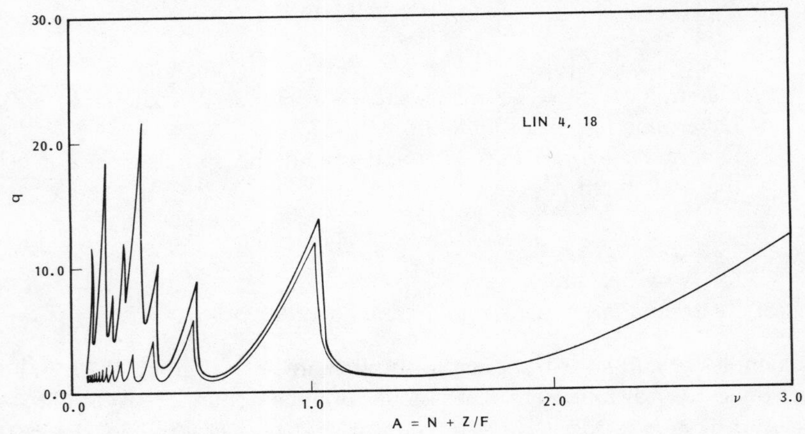

Fig. 3.72 Stability obtained with the cutter with nonuniform tooth pitch of Fig. 3.71a.

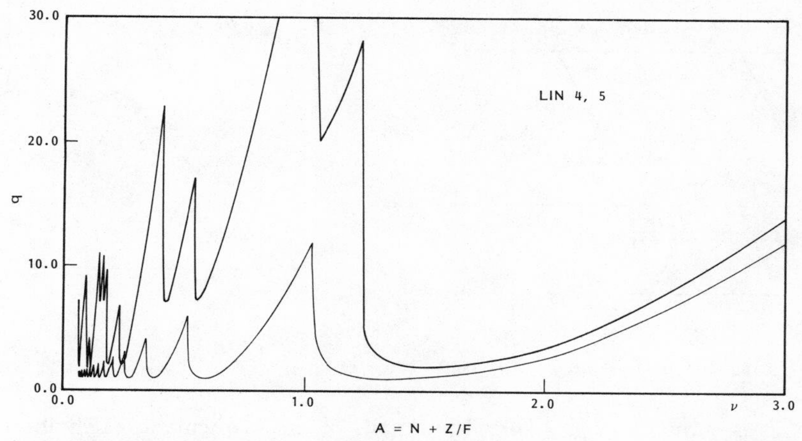

Fig. 3.73 Stability obtained with the cutter with nonuniform tooth pitch of Fig. 3.71b.

pitch and the one with the uniform pitch. It may be seen from Fig. 3.72 that the cutter with a nonuniformity $\Delta p/p_{av} = 1/18$ is most efficient for lower spindle speeds where the number N of waves between teeth is between three and five and, again for $N = 7$. For N about four, $q = 22$. In another way, assuming frequency of vibration $f = 600$ Hz, $m = 4$, the best spindle speed is in the neighborhood of $n = 37.5$ rev/sec $= 2250$ rev/min. For $f = 1600$, it would be $n = 6000$ rev/min.

For the cutter with the coarser nonuniformity of the pitch, see Fig. 3.73, the best increase of stability is over $q = 30$. A stability increase of $q > 20$ is obtained in the range of about $v = 0.5$ to $v = 1.25$ which, for $f = 600$ Hz, would correspond to $n = 75$ to 187.5 rev/sec $= 4500$ to $11{,}250$ rev/min and, for $f = 1600$ Hz to $n = 12{,}000$ to $30{,}000$ rev/min. This is then the range of ultra-high-speed milling. Results similar to these are obtained also for cutters with other types of pitch variation like the sinusoidal or the white-noise variations.

However, let us not forget that these results were obtained by a simplified analysis based on the assumption illustrated in Fig. 3.53.

It is interesting to apply the same type of analysis to two other forms of special cutters, those with alternating helix and those with serrated or sinusoidal edges.

The cutter with alternating helix, see Fig. 3.74a, can be represented as a sum of narrow cutters with the width of Δb and with many pitches p_i differing by Δp. The diagram for the limit of stability of Fig. 3.70 will now be modified in such a way that the vector B_0 consists of a large number of vectors Z_{0i} corresponding to each section of the cutting edge. In the extreme the edge is taken as straight and the vector B_0 becomes the chord of a circular

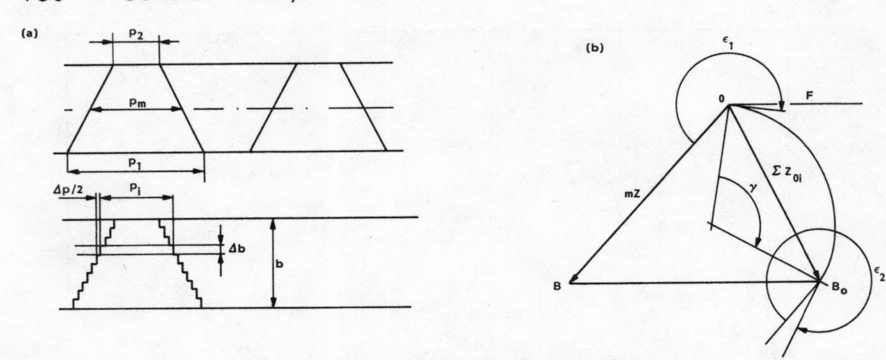

Fig. 3.74 Diagram of a cutter with alternating helix.

arc, as is shown in Fig. 3.74b. The details of the mathematical definition of the vector B_0 for this case are given in ref. 13. The angle $\gamma = \varepsilon_2 - \varepsilon_1$. The limit of stability can now be determined by the same computational procedure as the one denoted (3.61).

The result of the computation of an example of a cutter with four teeth and a ratio of $p_2/p_1 = 2$ is given in Fig. 3.75. The horizontal coordinate is:

$$\nu = nz/f = 1/(N + \varepsilon/2\pi)$$

where N and ε would be related to the mean pitch p_m. When compared with Figs. 3.72 and 3.73 it is seen that this cutter is very efficient in increasing stability by $q > 10$ over almost the whole spindle speed range, up to the speed of about $n = 1.25f/m$. Thus, for a system with natural frequency $f =$ 600 Hz this top speed would be $n = 187.5$ rev/s $= 11{,}250$ rev/min and for a

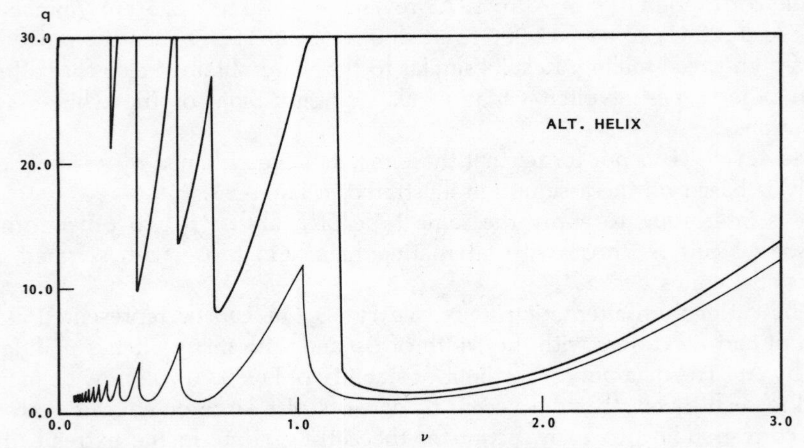

Fig. 3.75 Stability obtained with a cutter with alternating helix.

system with $f = 1600$ Hz (a flexible end mill) this speed would be $n = 30{,}000$ rev/min. Of course, this limit speed could be increased by choosing the parameters of the cutter differently.

This, again, is the result of an analysis under the simplifying assumption of Fig. 3.53.

Cutters with sinusoidal or serrated edges are depicted in Fig. 3.76. In Fig. 3.76a the cutter with serrated edges is shown in a plan view and in Fig. 3.76b with its circumference unfolded. It is shown that the serrations

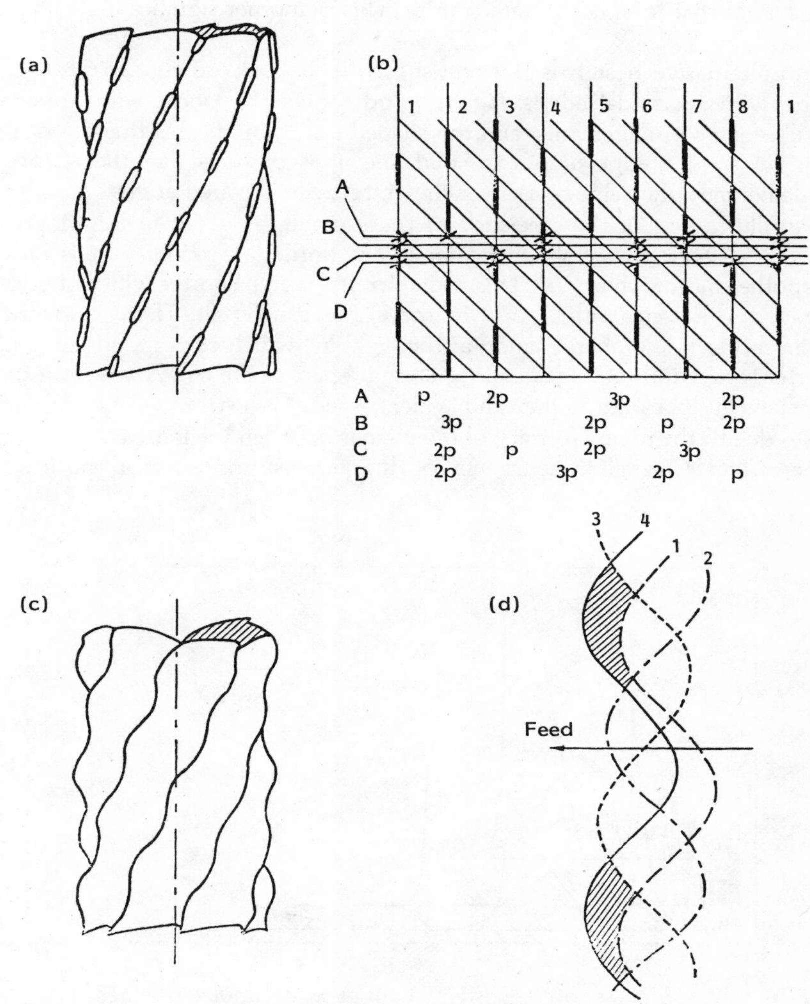

Fig. 3.76 Cutters with serrated and with undulated edges.

are created on a helix. The thick lines represent portions of the cutting edge with the thin lines in between representing the missing parts of the edge. Four sections—A, B, C, D—on four horizontal levels are indicated and dashes show where they intersect an active edge. It is seen that in any given section the edges follow each other with a single pitch p or with double or triple pitch. The corresponding sequences are written below the diagram for each of the four sections. From these illustrations it is obvious that these cutters act as cutters with a very coarse pitch nonuniformity. In addition, the total cumulative width of cutting edges b_{cum} engaged in the cut is only about one-half of that for a nonserrated cutter. This is another significant stabilizing factor.

An alternative design is the one shown in Figs. 3.76c and 3.76d. It is a cutter with sinusoidal edges. Figure 3.76d shows how these edges overlap in subsequent engagements, and the shaded portion indicates the size of the chip cut by the edge drawn in solid line. It is obvious that the action is fundamentally the same as that of the cutter with serrated edges.

An illustration of the efficiency of these cutters is given in the graph in Fig. 3.77. Over the whole speed range the borderline stability b_b is raised to another basic stability b_1. This is due to the overall shorter width of edges engaged in the cut with respect to the b_{lim} width of cut. There is another, similar bottom-line phenomenon at the level b_2, which corresponds to a still shorter total width of edges cutting during a part of the cut. Apart from this, the stability lobes are higher and wider, especially just below $\nu = 1.0$, and this is due to the nonuniformity of the pitches between the individual sections of the edges. Overall, the efficiency of this cutter seems less than those with

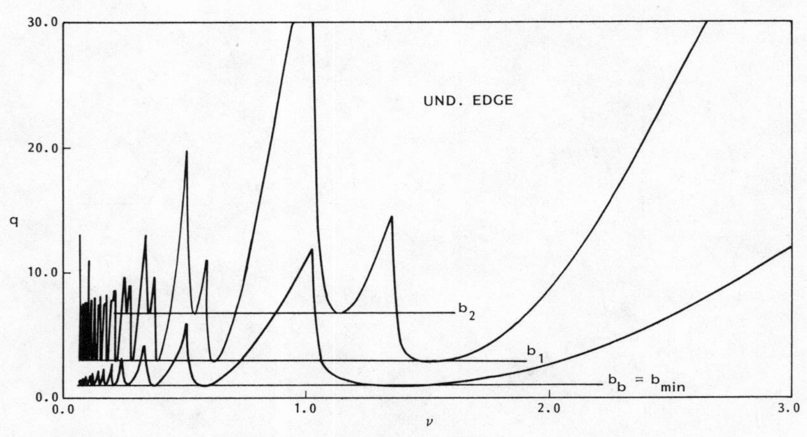

Fig. 3.77 Stability obtained with a cutter with undulated edges.

alternating helix or with nonuniform tooth pitch. However, the latter type is fully inefficient over certain speed ranges while here we have at least the b_1 level extended over the whole speed range. If it is also considered that the cutters with serrated or undulated edges are easier to manufacture than the two other types, and that they produce well-separable chips, it is understandable why they are rather popular, especially for milling steel. On the other hand, they do not give good surface finish and, therefore, are not suitable for finishing cuts.

All the preceding discussions were based on a strongly simplified theoretical approach related to the diagram in Fig. 3.53. In order to assess how valid those results are, one of the cases reported in Fig. 3.73 was also analyzed by time-domain simulation based on the model of Fig. 3.27. This latter approach is much more demanding on computing time and this is why all the previous analyses were carried out by the much simpler procedure denoted by Eq. (3.57). The results of the time-domain simulation approach are presented in Fig. 3.78. They are shown by solid lines and denoted a_n, b_n. In the same graph the limits of stability are reproduced from Fig. 3.56, and they are shown by broken lines. In both cases a corresponds to half-immersion up-milling and b to slotting. The curves a and b correspond to a cutter with regular tooth pitch with four teeth, and the curves a_n, b_n to a cutter with four teeth

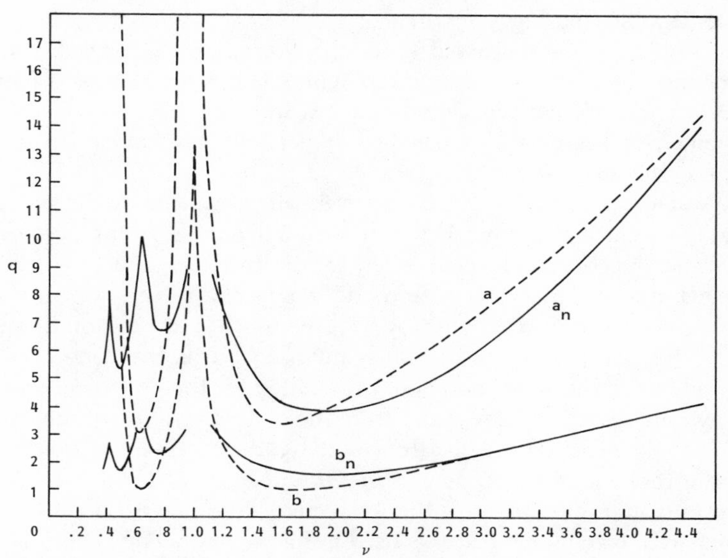

Fig. 3.78 Stability of the case as in Fig. 3.73 obtained by time domain simulation.

and nonuniform tooth pitch, with $\Delta p = 0.2p_m$. In both instances the results were obtained by time-domain simulation. The comparison shows that the regular cutter gives a higher stability increase in a rather narrow spindle speed range around $\nu = 1.0$. The cutter with nonuniform tooth pitch gives only about $q = 2$ increase of stability, but it is spread over a spindle speed range $0.55 < \nu < 1.4$.

In conclusion it may be stated that there are several designs of special cutters to choose from. They represent significant means of stability improvement if designed for a proper match to the conditions of use, including high- and ultra-high-speed milling. The corresponding computational methods have been developed. In applications, input data about the dynamics of the spindle and tool are also needed. It is necessary to expend further efforts to package and streamline these methods for good practical use.

B. EFFECT OF END MILL DEFLECTIONS ON ACCURACY

In this section we will assume that cutting is stable and no chatter vibrations occur. We will, however, consider deflections, both static and dynamic, as they are caused by the cutting force, and the effect of these deflections on the accuracy of the surfaces, especially in applications to pocket milling which involves flexible thin webs.

The problem to be discussed is first illustrated in a simplified form with reference to Fig. 3.79, in which cutters with straight teeth are considered. In Figs. 3.79a and 3.79b a two-fluted cutter is shown in an up-milling cut with the "immersion" ratio $a_r/d = 0.5$ (half-immersion); a_r is radial depth of cut and d is cutter diameter.

The surface generated in this cut is essentially produced by the cutter teeth as they pass through point A, which is found on the radius perpendicular to the feed direction. However, when the tooth is in point A as shown in Fig. 3.79a, the chip thickness is zero and the cutting force is very small, and only a very small deflection arises. When the tooth is in position B as shown in Fig. 3.79b, the cutting force is maximum and so is the deflection of the cutter; however, this deflection is not imprinted on the generated surface. In this way only very small deviations from the desired position of the machined surface would be obtained. If a four-fluted cutter is used as shown in Fig. 3.79c the tooth in B produces a large cutting force and the corresponding deflection is imprinted in A on the generated surface, which is made deeper than required. If the same cutter is used in the down-milling mode, as shown in Fig. 3.79d, again a large deflection is imprinted on the generated surface. This time, however, the surface is made higher (thicker) than required.

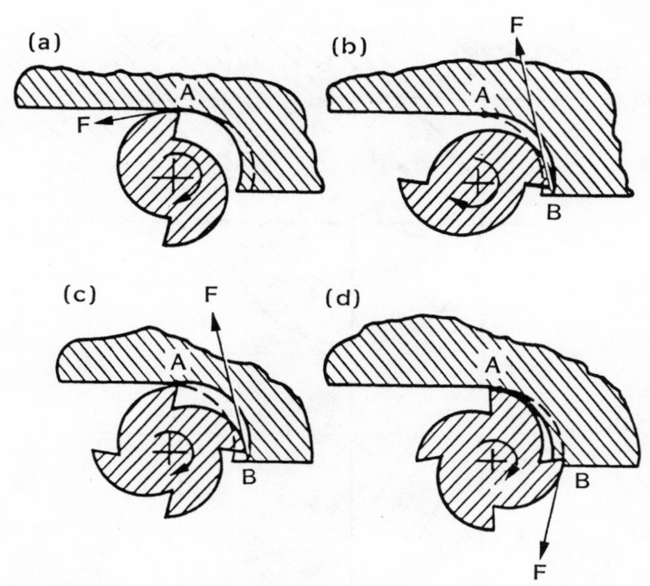

Fig. 3.79 Deformations of end mills and their effect on accuracy of milled surface.

In practice, end mills with helical teeth are used and the relationship among forces, deflections, and the form of the machined surface is more complicated. The variation of the force depends on the combination of both the radial and axial depths of cut. It will be discussed in more detail subsequently. The deflections may have a significant dynamic component, especially in high-speed milling. The envelope of these deflections, which is the generated surface, is now produced by the helical tooth passing through the radial plane perpendicular to the direction of feed. The point of the cutting edge which passes through this plane moves upward on the line of contact; see Fig. 3.80. Every tooth now extends over an arc of spread ψ. This is indicated in the lower part of the picture, which represents the unrolled periphery of the cutter. One cutting edge is first shown in position 1. It is shown as a straight line under the helix angle β and its leading point is just entering the line *A–A* of the contact of the cutter periphery with the milled surface at the bottom end of the axial depth of cut a_a. The spread of the edge in the direction of the cutting velocity v is ψr, where r is the cutter radius. Subsequently, this cutting edge moves through positions 2 and 3 in which the point of the edge which generates the milled surface moves to the mid-height and the upper end of the line *A–A*, respectively. The deflection of the cutter varies during this

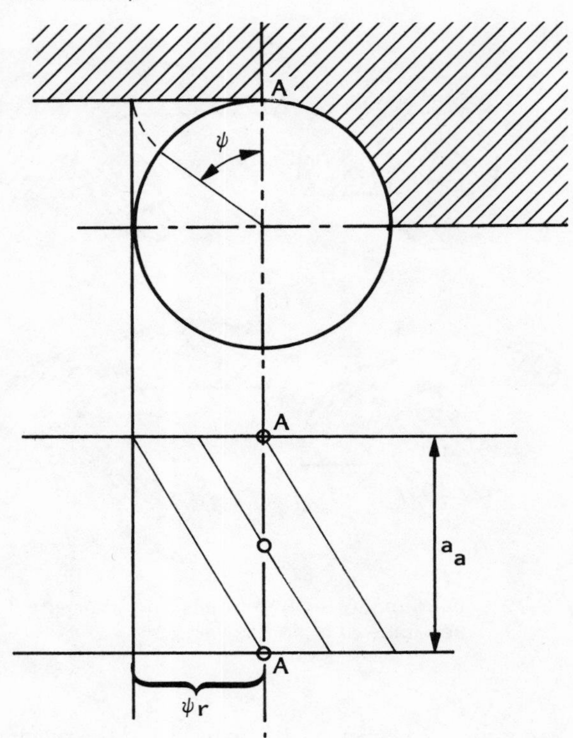

Fig. 3.80 Axial movement of the imprint of cutter deflection on the machined surface.

period as the forces vary on this tooth and also on some preceding tooth which is also engaged in cutting. Correspondingly, the profile of the generated surface along *A–A* is not necessarily parallel with the cutter axis or not even straight. It may be concave or convex.

An illustration is presented in Fig. 3.81, taken from ref. 14. It shows the deviations of surfaces produced in, from top to bottom, up-milling with helix angle 15°, up-milling with zero helix, and down-milling with helix angle 30°. A two-fluted cutter was used and the radial depth increased continuously during the cut from zero to $a_r = d$. In the top case an overcut was produced which increased toward the bottom of the cut and in the bottom case an undercut, almost parallel with the cutter axis, was produced. In the middle case the errors were very small, for reasons explained in Fig. 3.79. Various other cases will be discussed in a latter section.

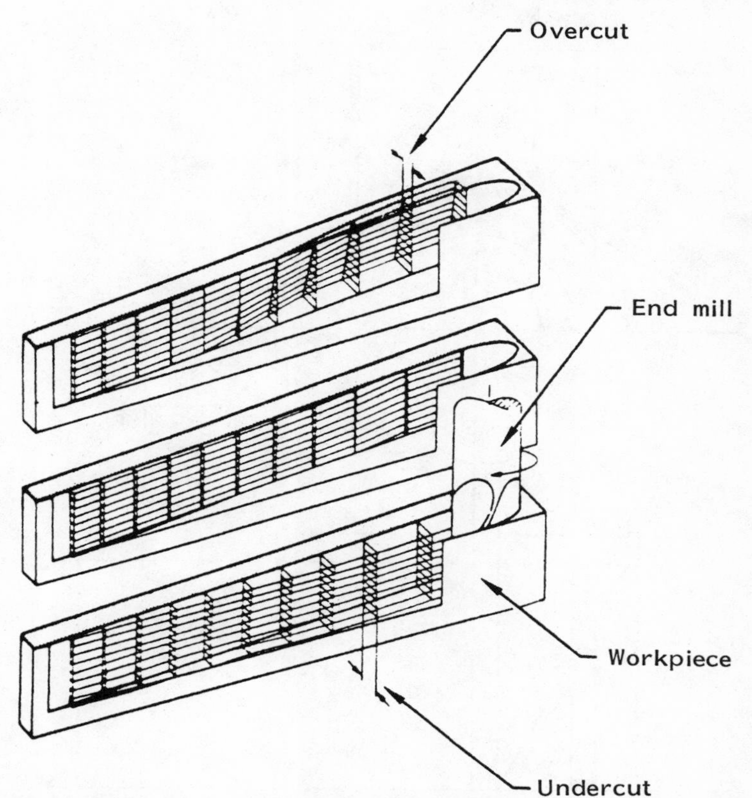

Fig. 3.81 Types of errors caused by cutting force deformations in end milling.[14]

Cutting-Force Variation

Let us consider Fig. 3.82, in which a cutter with helical teeth is used for milling with feed in the direction X. The axial depth of cut is a_a, the radial depth of cut is a_r and it extends from the angle ϕ_s of the start of cut to the angle ϕ_e of the end of cut for every point of a cutting edge. These angles are measured from the point O, which is in a radial plane perpendicular to the feed direction. The top view is a section at the height z by a plane Z–Z perpendicular to the cutter axis; let us consider it as a slice with thickness dz. The pitch angle between subsequent teeth is ϕ_p and the "spread" of a tooth is ψ; it is

$$\psi = a_a \tan (\beta/r). \tag{3.63}$$

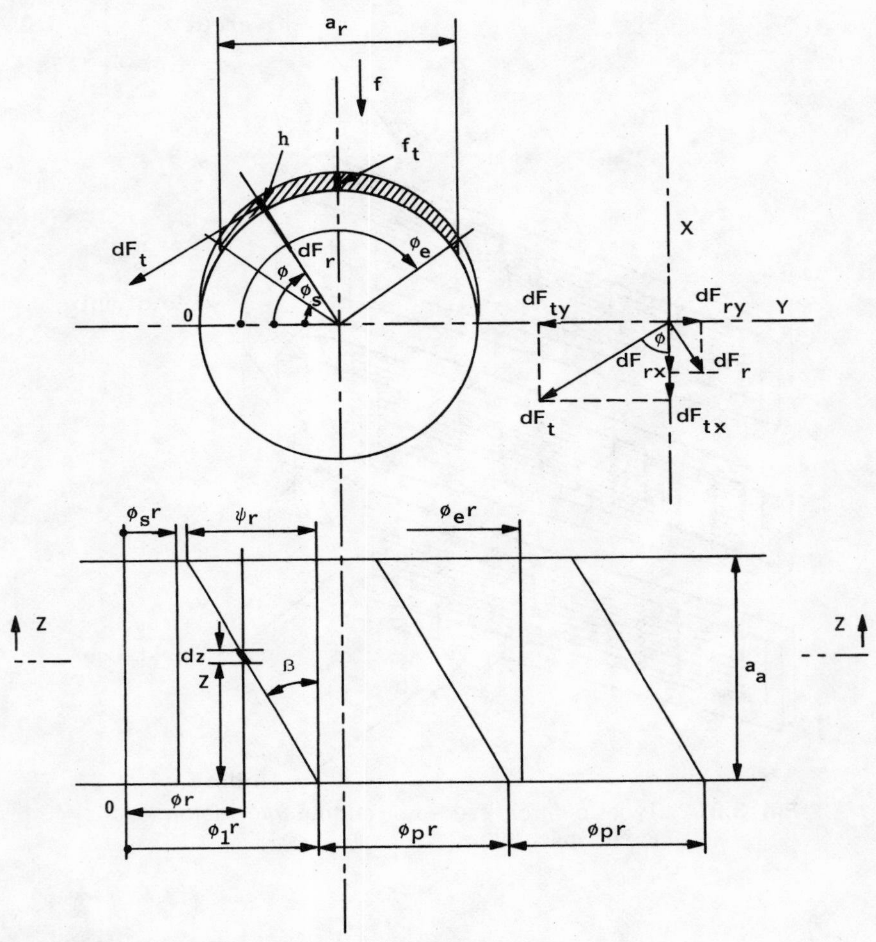

Fig. 3.82 Deriving milling force computation.

The chip thickness h at the angle ϕ at which the incremental length dz of an edge is found is

$$h = f_t \sin \phi, \tag{3.64}$$

where f_t is the feed per tooth. The tangential incremental force produced by this part of the cutting edge is

$$dF_t = K_s \, dz \, h, \tag{3.65}$$

where K_s is the "specific force" and $(dz \, h)$ is the incremental chip area. The corresponding radial force is

$$dF_r = c \, dF_t, \tag{3.66}$$

where the ratio c is commonly about 0.3. These forces will be decomposed into the X and Y axes:

$$dF_{tx} = dF_t \cos \phi = C \sin \phi \cos \phi, \tag{3.67}$$

where

$$C = K_s f_t \, dz \tag{3.68}$$

and

$$dF_{ty} = dF_t \sin \phi = C \sin^2 \phi, \tag{3.69}$$

$$dF_{rx} = C\, dF_t \sin \phi = C\, dF_{ty}, \tag{3.70}$$

$$dF_{ry} = -C\, dF_t \cos \phi = -C\, dF_{tx}. \tag{3.71}$$

In order to obtain the forces F_x and F_y corresponding to a particular angular position of the cutter, e.g., to the angle ϕ_1 associated with the leading point of tooth 1, it is necessary to integrate:

$$\begin{aligned} F_x &= \int_\phi (dF_{tx} + dF_{rx}), \\ F_y &= \int_\phi (dF_{ty} + dF_{ry}) \end{aligned} \tag{3.72}$$

over the intervals $(\phi_1, \phi_1 - \psi)$ for tooth 1, $(\phi_1 + \phi_p, \phi_1 + \phi_p - \psi)$ for tooth 2, etc. However, it is necessary, in Eqs. (3.67)–(3.71) to consider the constraint

$$\phi_s < \phi < \phi_e \tag{3.73}$$

outside of which the dF_t and dF_x forces are zero.

The practical approach to determining the cutting force and its variation for various milling situations as defined by particular values of a_r, a_a, r, β, ϕ_s, ϕ_e, ϕ_p is to use a computer program to carry out Eqs. (3.67)–(3.73) while proceeding in increments dz along the teeth of the cutter in a particular position defined by ϕ_1 and incrementing ϕ_1 in small steps from ϕ_s to $\phi_e + \psi$.

We will give examples of the various milling operations. First, however, let us indicate some combinations of the parameters of operations in which there is no variation of the milling force, that is, at least, if the milling cutter has no run-out.

One such case is slotting with a four-fluted end mill. The drawing in Fig. 3.83 represents a section through a four-fluted cutter in a slotting operation; we will consider a slice with axial thickness dz. At whichever axial level this slice is taken there will be two edges in the cut between $\phi_s = 0°$ and $\phi_e = 180°$; one at the angle ϕ_1 and the other one at the angle $\phi_2 = \phi_1 + 90°$. The magnitudes of the forces on the two edges will be:

$$dF_{t1} = K_s\, dz\, h_1, \qquad dF_{t2} = K_s\, dz\, h_2,$$

$$dF_{r1} = 0.3\, dF_{t1}, \qquad dF_{r2} = 0.3\, dF_{t2},$$

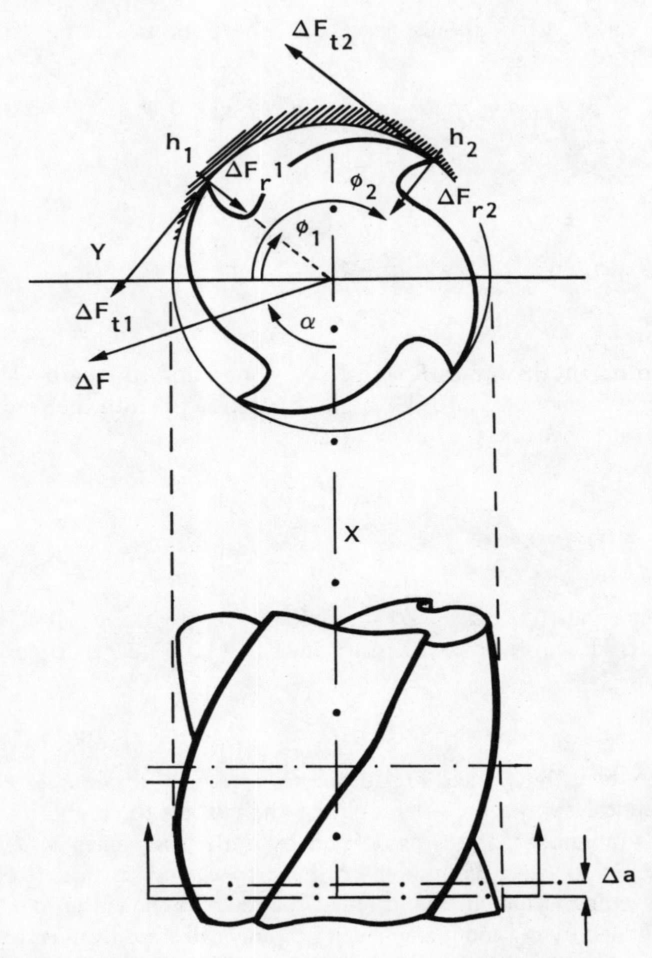

Fig. 3.83 Diagram for forces in slotting with a four-fluted cutter.

and the forces will have directions as shown in the drawing; it is $h_1 = f_t \sin \phi_1$, $h_2 = f_t \sin \phi_2$.

The forces can be decomposed according to Eqs. (3.67) and (3.69)–(3.71). Denoting $C = K_s\, dz\, f_t$, it is

$$\begin{aligned} dF_x &= dF_{t1x} + dF_{t2x} + dF_{r1x} + dF_{r2x} \\ &= C(\sin \phi_1 \cos \phi_1 + \sin \phi_2 \cos \phi_2 + 0.3 \sin^2 \phi_1 + 0.3 \sin^2 \phi_2). \end{aligned}$$

For $\phi_2 = \phi_1 + 90°$, it is

$$\sin \phi_2 = \cos \phi_1, \qquad \cos \phi_2 = -\sin \phi_1.$$

Hence

$$\begin{aligned} dF_x &= C(\sin \phi_1 \cos \phi_1 - \sin \phi_1 \cos \phi_1 + 0.3 \sin^2 \phi_1 + 0.3 \cos^2 \phi_1) \\ &= 0.3C \\ dF_y &= dF_{t1y} + dF_{t2y} + dF_{r1y} + dF_{r2y} \\ &= C(\sin^2 \phi_1 + \sin^2 \phi_2 - 0.3 \sin \phi_1 \cos \phi_1 - 0.3 \sin \phi_2 \cos \phi_2) \\ &= C(\sin^2 \phi_1 + \cos^2 \phi_1 - 0.3 \sin \phi_1 \cos \phi_1 + 0.3 \sin \phi_1 \cos \phi_1) \\ &= C. \end{aligned} \tag{3.74}$$

Result: both the dF_x and dF_y components remain constant. Consequently, the resulting force on the slice is constant and has a constant direction α:

$$dF = C(1 + 0.09)^{1/2} = 1.044C$$

$$\alpha = \tan^{-1}(F_y/F_x) = \tan^{-1} 3.333 = 73.3°. \tag{3.75}$$

All the dF_x and dF_y on all the slices are equal and constant, and the resulting force dF on each slice has the same value and direction as they are given by Eq. (3.75). The total force is constant and it has a constant direction:

$$F = 1.044nC$$

$$\alpha = 73.3°,$$

where n is the number of slices.

In conclusion, a slotting operation, with $a_r = 2r$, with a four-fluted cutter yields constant cutting force irrespective of the axial depth of cut a_a.

Another such situation in which the resulting force on the cutter does not vary arises if the axial depth of cut a_a is such that the spread ψ of a tooth equals the tooth pitch ϕ_p; the force is then constant irrespective of the radial depth of cut or of the initial and final engagement angles ϕ_s and ϕ_e. This particular axial depth of cut is shown in Fig. 3.84 as $(a_a)_a$ and it is

$$(a_a)_a = \phi_p r/\tan \beta = 2\pi r/z/\tan \beta, \tag{3.76}$$

where z is the number of teeth of the cutter.

An example is depicted of a cut with such a radial depth of cut a_r that the cut spreads from line A to line B, over an angle of engagement ϕ_b. In Fig. 3.84a the unfolded circumference of the cut is shown. The lines inclined by the helix angle β spread over an angle ψ which is exactly equal to the pitch angle. Two teeth are shown engaged simultaneously in the cut between A and B and their cutting portions are indicated by a thick line. A small section of the cutting edge is denoted Δ and it corresponds to an angle ϕ which

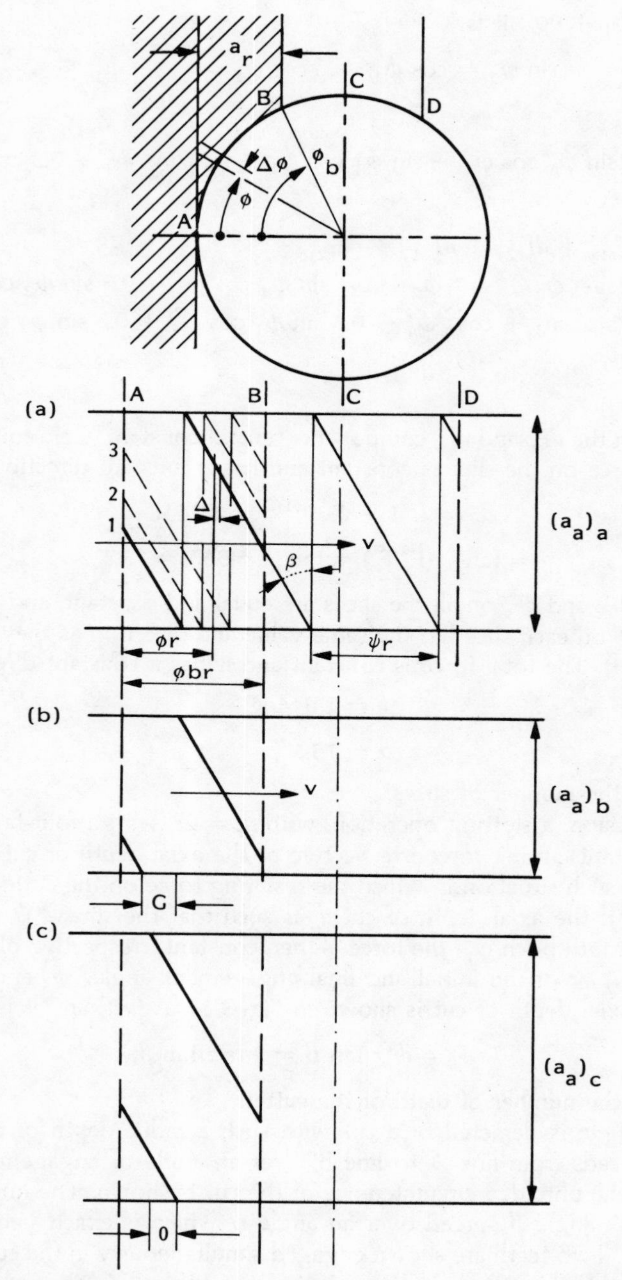

Fig. 3.84 Effect of axial depth of cut on force variation.

can be incremented by $\Delta\phi$ from O to ϕ_b. A particular chip thickness $h = f_t \sin \phi$ corresponds to each section Δ as well as the corresponding force increments dF_t and dF_n. As the cutter rotates the teeth move on the periphery of the cutter with cutting speed v. This is shown in the graph in Fig. 3.84a where successive positions 1, 2, 3 are drawn of the cutting sections of the two edges with the two latter positions shown in broken lines. It is seen that in Fig. 3.84b there is always only one section Δ that corresponds to each value ϕ as it is incremented by $\Delta\phi$ from O to ϕ_b. The same statement would also apply to a cut engaging between lines A and D in which case sections of four cutting edges are cutting simultaneously and, for that matter, it would apply to any cut whatever its radial depth. This then means that the total force is always integrated over the same total length of cutting edges and it remains constant as the cutter rotates.

In Fig. 3.84b in which the axial depth of cut a_a is smaller there is a gap G within the cutting engagement between A and B. No part of an edge acts over this gap and as the cutter rotates this gap moves and represents a variable "missing" part of the force (as compared with Fig. 3.84a). In Fig. 3.84c where the axial depth a_a is greater than in Fig. 3.84a there is an overlap O and it moves and causes a variable component of the cutting force.

In conclusion, for the value of axial depth a_a given by Eq. (3.76) and for integer multiples of it there is no variation of the total cutting force or of its direction. This is, of course, exactly true only if there is no radial run-out of the cutter. Otherwise there is a component periodic once per cutter revolution and its amplitude increases with the run-out. For zero run-out, if the radial depth of cut a_r varies, the force varies accordingly, but this variation is usually very slow compared to the variations with the tooth frequency that occur when the axial depth of cut is not equal to or is not an integer multiple of the $(a_a)_a$. This periodic variation of the cutting force is obviously largest if the axial depth of cut has a value in the middle between the integers of $(a_a)_a$.

Let us now look at results of computations and plots of the cutting force for various cases. These cases differ by radial and axial depths of cut as shown in Fig. 3.85. They all involve a cutter with four teeth, radius $r = 15$ mm (0.6 in.), and helix angle $\beta = 30°$. They are denoted as C_{mn}. The subscript m indicates the radial depth of cut $a_r = 7.5$ (0.3), 15 (0.6), 22.5 (0.9) mm (in.) of up-milling for $m = 1, 2, 3$, respectively, and $a_r = 15$ mm (0.6 in.) of down-milling for $m = 4$. The subscript n indicates $a_a = 10$ (0.4), 25 (0.98), 50 (1.96) mm (in.) for $n = 1, 2, 3$, respectively.

The value of the critical depth of cut of Eq. (3.76) would be $(a_a)_a = 40.8$ mm (1.6 in.). It is seen that for $a_a = 10$ mm (0.4 in.) ($n = 1$) there is a large gap G between subsequent teeth and for $a_a = 25$ mm (0.98 in.) ($n = 2$) there is still a gap G but smaller and for $a_a = 50$ mm (1.96 in.) ($n = 3$) there is an overlap O.

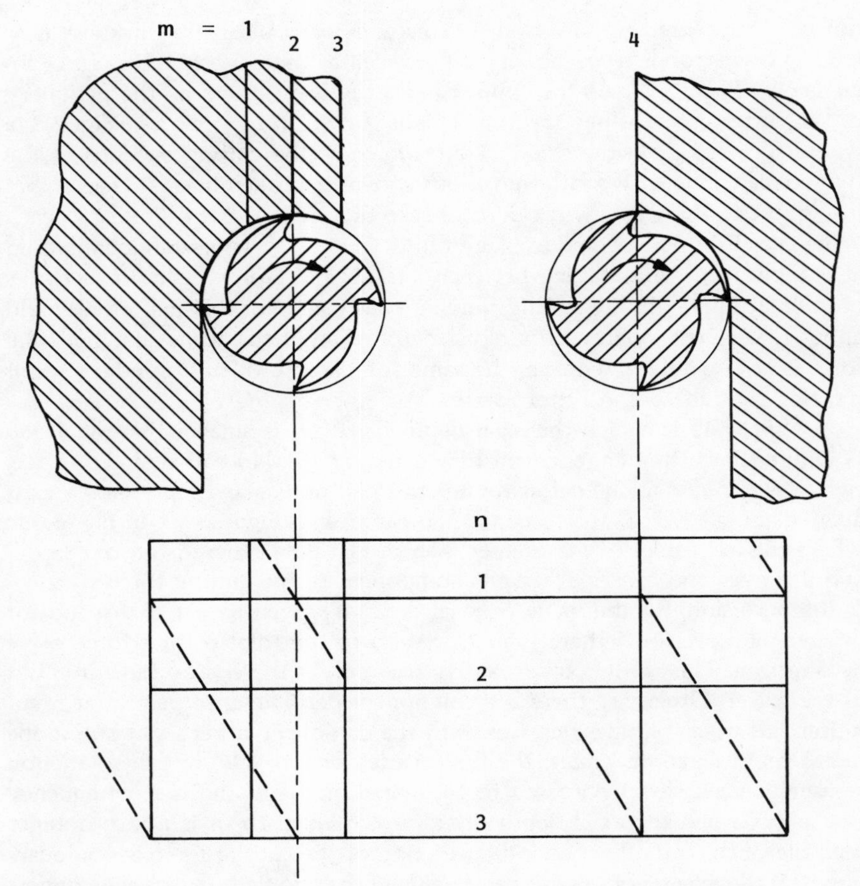

Fig. 3.85 Specification of 12 cases for milling-force computations.

The variations of the cutting forces for all the 12 cases are plotted in Fig. 3.86. The four graphs—a, b, c, d—correspond to the various radial engagements as denoted by the first subscript $m = 1, 2, 3, 4$, respectively, and in each of them the force is plotted for the three axial depths of cut.

In all cases a periodic force is seen with the frequency of the fundamental harmonic component equal to the tooth frequency f_t:

$$f_t = nz = 4n \tag{3.77}$$

where n (rev/s) is the rotational speed of the cutter. Depending on the cutting speed used this may vary, e.g., for aluminum from $n = 3600$ rev/min $=$ 60 rev/s to $n = 36{,}000$ rev/min $= 600$ rev/s (ultra-high-speed milling) and, correspondingly from $f_t = 240$ Hz to $f_t = 2400$ Hz, or more. The peak-to-peak values of the force variations are almost the same [about 2000 N (450 lb)]

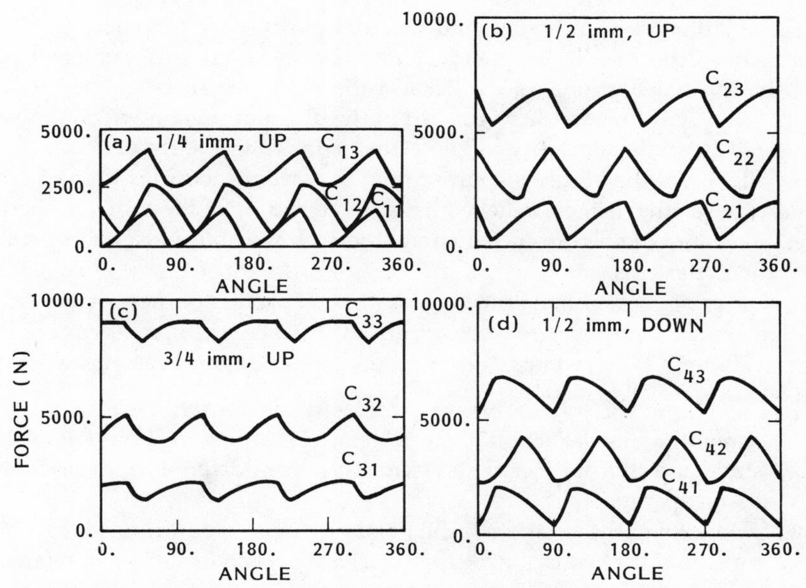

Fig. 3.86 Milling force for the cases of Fig. 3.85.

for all the cases except $m = 3$ with the largest angle of engagement of 120° where they are about half of those for the other three cases.

According to the analyses presented in the preceding the force variations would be zero for slotting, $a_r = 2r$, and for any a_r if $a_a = 40.8$ mm (1.6 in.), but these cases are not included in the above plots.

It is to be noted that, for the milling of aluminum, the cutting-force variation may have strong dynamic effects. Their fundamental frequencies are high enough to present the danger of resonance with one of the natural modes of the structure of the machine of the spindle, or of the end mill itself. In order to avoid these resonances it is necessary to know the natural frequencies of the system for all cutters with corresponding holders and to choose spindle speeds accordingly.

Realizing that there is always a certain radial run-out of the cutter it is also necessary to avoid resonances between the structural natural frequencies and the spindle rotational frequency n (rev/s).

Cutter Deflections and Workpiece Accuracy

The cutting forces produce deflections on the cutter, and, sometimes, as in the case of pocketing with thin webs between pockets, also on the workpiece (deformations of the webs). The deformations on the cutter consist of those of the spindle, the tool holder, and the tool itself. They may be significant, especially in the case of long and slender end mills. As was explained previously, these deflections cause errors of the machined surface. These errors are of two types that are distinguished in the directions along which they are measured. First, there are errors due to the variation of the radial depth of cut especially when the cutter is approaching an inner corner. This type of error is measured from the line parallel with the feed direction of the cutter. Secondly, deviations are obtained from the straightness of lines on the surface, parallel with the cutter axis.

It is possible to determine the magnitudes and profiles of these errors either by experiment or by computations. In the latter case the computer program runs in discrete time instants and performs three tasks in every instant:

1. Computing the force. This was discussed previously.
2. Computing the deflection of the system. This is possible if the flexibilities (static and dynamic) of the system have been determined. This is best done experimentally.
3. Determining the position of the point on one of the cutting edges which is generating the machined surface. This has been explained in reference to Fig. 3.69.

Computer programs of this type have been used to establish basic knowledge about the main features of the errors obtained with different end mills in the various milling situations; see refs. 14, 15, and 16. An example of a typical situation is reproduced from ref. 14 in Fig. 3.87, and it shows the profile of an unfolded surface of a corner radius. The radius starts at the line denoted 0.0. It is seen that just before the cutter starts to form the radius a large negative error (a dip) is generated at the bottom of the milled surface. In this case, which involved an end mill 16 mm (0.63 in.) in diameter and 90 mm (3.5 in)

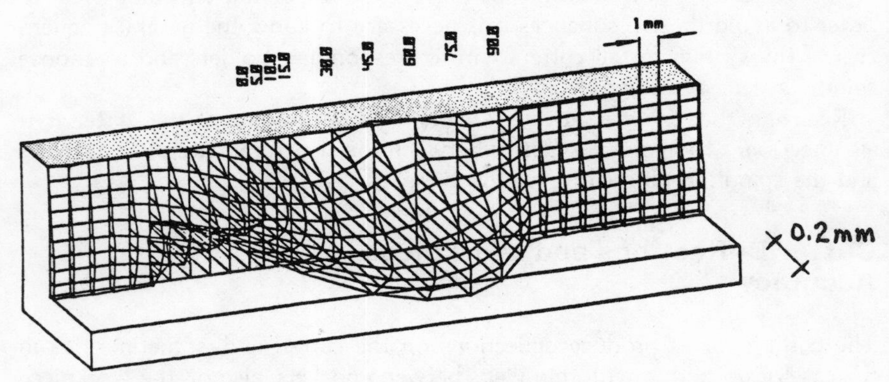

Fig. 3.87 Geometric error of an end-milled corner.

long, with two teeth at helix angle 30°, rotating at 1000 rev/min and moving with feed rate 400 mm/min, the dip is about 0.4 mm (0.02 in.) deep. This may cause a dangerous weakening of the structure. In the middle of the radius there is, on the contrary, considerable thickening at the bottom of the wall.

Studies like this are useful for determining strategies like "cornering routines" which eliminate excessive errors.

References

1. "Technology of Machine Tools," Vol. 3: *Machine Tool Mechanics*, Lawrence Livermore National Laboratory Rep. UCRL—52960-3; distribution by SME.

2. Tlusty, J., "Criteria for Static and Dynamic Stiffness of Structures," Chapter 8.5 of ref. 1.

3. Tlusty, J., and F. Ismail, "Basic Non-Linearity in Machining Chatter," *CIRP Annals*, Vol. 30, 1982, pp. 229–304

4. Tlusty, J., and F. Ismail, "Special Aspects of Chatter in Milling," *Trans ASME, J. Mech. Design*, Paper 81-DET-18. Presented 21 September 1981.

5. Tlusty, J., F. Ismail, A. Hoffmanner, and S. Rao, "Theoretical Background for Machining Tests of Machining Centers and of Turning Centers," *NAMRC* 10 May 1982, SME.

6. Tlusty, J., and T. Moriwaki, "Experimental and Computational Identification of Dynamic Structural Models," *CIRP Annals*, Vol. 25/2, 1976.

7. Tlusty, J., and F. Ismail, "Dynamic Structural Identification Tasks and Methods," *CIRP Annals*, Vol. 29, 1980, pp. 251–255.

8. Koenigsberger, F. and J. Tlusty, *Structures of Machine Tools*, Pergamon Press, Oxford, 1970.

9. Elbestawi, M. A. A., "A Study of an Adaptive Control Constraint System for Milling," Ph.D. Thesis, McMaster University, 1980.

10. Tlusty, J. and O. Heczko, "Improving Tests of Damping in the Cutting Process," 8th NAMR Conference, May 1980, University of Missouri, Rolla, SME, pp. 372–376.

11. Tlusty, J., and B. S. Rao, "Verification and Analysis of Some Dynamic Cutting Force Coefficient Data," 6th NAMR Conference, April 1978, University of Florida, SME, pp. 420–426.

12. Tlusty, J., "Analysis of the State of Research in Cutting Dynamics," *CIRP Annals*, Vol. 2, 1978, pp. 403–412.

13. Zaton, W., F. Ismail, and J. Tlusty, "Effect of Special Milling Cutters on Chatter," *11th NAMRC*, Madison, Wisc., May 1983, SME.

14. Lectures of the Aachener Werkzeugmaschinenkolloquium, WZL Aachen, W. Germany, May 1978.

15. Kline, W. A., R. E. De Vor, and W. J. Zdeblick, "A Mechanistic Model for the Force System in End Milling with Application to Machining Airframe Structures," 8th NAMRC, May 1980, SME.

16. Kline, W. A., R. E. De Vor, and I. A. Shareef, "The Prediction of Surface Accuracy in End Milling," 82-Prod-10, ASME, September 1982.

CHAPTER 4

Cutting Fluids in Industry

C. F. Barth
TRW, Inc.

Introduction

The selection and application of cutting fluids in industry have not always been accomplished in an optimal manner. This chapter will strive to present, on a brief overview basis, useful background information regarding the evolution of modern fluids, their compositions, and simple methodologies to match products to applications. These discussions will be presented from the perspective of the user with the intent to provide a matrix of information to aid potential users in the formulation of effective in-house fluid programs.

Background

In this section, the evolution of cutting fluid technology will be highlighted from early uses of plain water through complex modern formulations available to the manufacturing industry. This development closely followed advances in the metal-removal industry. Early machining operations were conducted at relatively low speeds with carbon steel tools. Improvements in technology, such as tools of high-speed steel or more powerful and faster machines, tended to stimulate developments in other elements of the machining process. Hence, a synergism evolved in which overall progress was made by a series of incremental developments in each of the necessary technological elements. This discussion will focus on the evolution of cutting fluid technology and the characteristics of the resulting cutting fluid industry.

Cutting fluid formulation technology for the manufacturing industry had its more serious beginnings soon after Fredrick Taylor reported improvements in tool life when water was applied to the cut. Early experimenters surmised that water reduced thermal damage to carbon tool steels in use at the time. As a result, the term coolant has since become firmly entrenched as the descriptor of all fluids applied to the tool–chip interface. This term has largely persisted to this day regardless of the functions provided by contemporary cutting fluid formulations.

It was quickly recognized by the manufacturing community that plain water was not a completely satisfactory solution to the question of tool life improvement. The rapid rusting of both ferrous workpieces and machine tool components was of major concern. In addition, observations of used cutters indicated that water had been effective in reducing tool damage through thermal effects but had little effect on reducing mechanical wear. Attempts to use oils instead of water to provide the perceived need for lubricity resolved the rust problem but resulted in smoke, messy shops, rancidity, and inadequate heat abstraction at all but the slowest machining speeds of the day.

The above findings suggested to many experimenters of the time that perhaps combinations of water and oils would be effective by simultaneously providing cooling and lubrication. Since oil and water do not mix, efforts to produce emulsions stable in the shop environment spawned the cutting fluid industry. A fiercely competitive industry emerged following discoveries that additives to an oil–water emulsion provided further performance improvements. In addition to better emulsifiers, the more notable additives were those containing sulfur, chlorine, phosphorus, and fatty acids. The variety of such compounds available to fluid chemists permitted formulation of an essentially infinite series of product compositions, each claimed by its manufacturer to offer advantages over another product for general and specific applications. More recently, synthetic (oil-free solutions) and semisynthetic products have appeared on the market to compete with the older oil–water emulsions and straight oil-base products.

This diversity of products and manufacturers has created a difficult situation for those charged with fluid selection within a production facility. Aggressive marketing practices reminiscent of the old "snake oil" pitchmen did little to aid the selection process. Production engineers became exceedingly skeptical of sales claims for products. Logically, a "prove it to me" attitude developed which resulted in a new set of problems. It soon became an impossible proposition to conduct in-shop tests for products suggested by every salesman without disruption of production or to obtain meaningful data at acceptable costs. Similarly, fluid manufacturers also could not afford the costs associated with laboratory testing of their products under every conceivable combination of cutting parameters and workpiece compositions. This latter problem was further exacerbated by practical difficulties in relating laboratory test data to production line performance.

Despite these difficulties, fluid manufacturers have begun to provide useful technical data relative to recommended fluid applications, expected tool life for various cutter materials, cutting parameters, mixing and disposal requirements, and economic information. The availability of these data has been reasonably useful to production managers, but the variety of products remains a challenge in selecting optimal products.

Product Compositions

This section describes the four basic product formulations available to manufacturing and briefly outlines their function during the cutting event. All currently available cutting fluids can be placed into four generic compositional regimes: neat oils, emulsions, semisynthetics, and full synthetics. Each of these types has specialized properties and applications and is further rated by fluid manufacturers as a light, medium, or heavy-duty product. The following subsections describe the major attributes for each product type, beginning with a discussion of the basic function of the additives used in three of the first four generic types of fluids.

Additives

Compounds containing sulfur, chlorine, or phosphorus are almost universally used as additives to oils, emulsions, and semisynthetic cutting fluids. These additives perform several basic functions intended to reduce cutting forces and tool wear.

Sulfur and chlorine are believed to react with tool materials and workpiece during the high-temperature phase of the cutting event. Generally, sulfur is released from its combined state at temperatures in the 1000–1200°F (538–649°C) range while chlorine becomes similarly active in the 600–900°F (316–482°C) region. The reaction products form low-shear-strength films which tend to isolate the tool from the work and chip. This film formation is contrasted with full hydrodynamic lubrication and boundary film lubrication schematically in Fig. 4.1. This is also known as extreme-pressure (EP) lubrication, and compounds which react with one or both of the opposing surfaces are referred to as EP additives. Sulfur-bearing additives are classed according to their capability to react with workpiece materials and tools. Additives rated as "active" readily form such films at relatively low temperatures, while the more strongly combined forms require higher temperatures to release the active agents and allow film formation. Tailoring the proportions of active to combined sulfur for a given application is necessary to prevent excessive

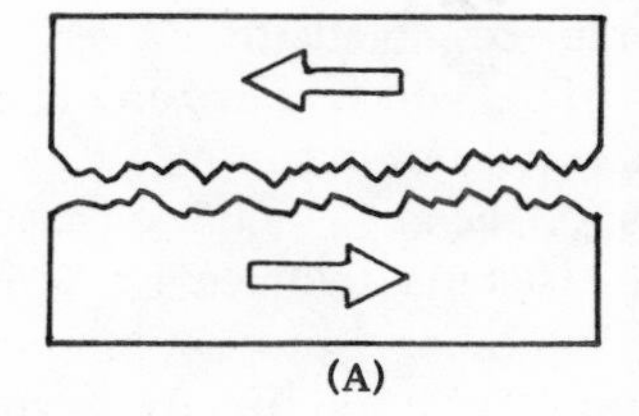

Elastohydrodynamic Lubrication

Continuous lube film isolates opposing surfaces from contact.

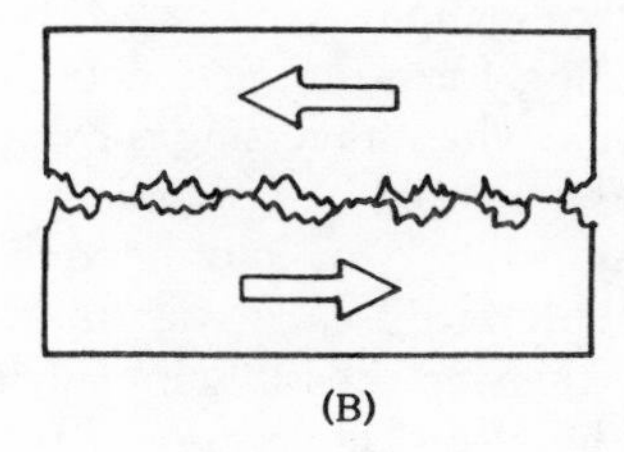

Boundary Layer Lubrication

Lubricant trapped between localized asperity contacts helps support loads.

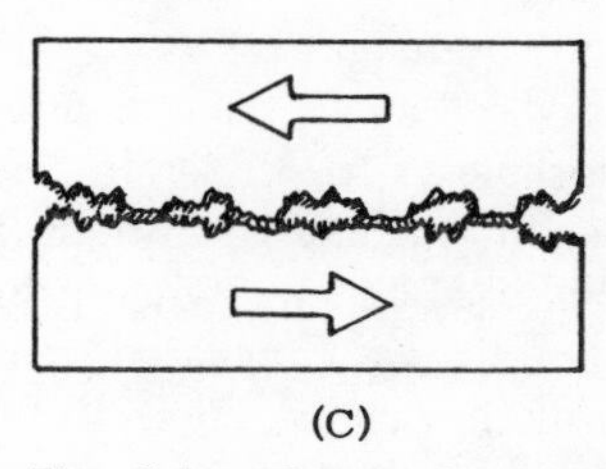

Boundary Lubrication with EP Additives

Heat generated by asperity contacts permits low-shear strength films to form, effectively isolating the two surfaces.

Fig. 4.1 Comparison of common lubrication mechanisms. B and C are more representative of metal-cutting operations.

film-formation tendencies in areas away from the cut. In extreme cases, corrosive damage to the machine tool or workpiece can occur.

Chlorine behaves in a similar manner except that its film-forming tendencies are more operative at somewhat lower temperatures than sulfur. Hence, a sulfochlorinated additive package can function over a greater temperature range than either product by itself. Both of the above additives are used in concentrations reaching 5–7% for heavy-duty cutting fluids, while chlorinated oils and waxes used in drawing and tapping may reach 25–50%.

Phosphorus compounds, usually esters, are also believed to act in the same manner as sulfur and chlorine, but at slightly higher temperatures. Phosphorus additives are used, however, in much lesser concentrations to minimize risks for possible embrittlement of the tools or workpieces. In addition, these additives have much lower solubility in the cutting fluid concentrates and tend to raise the overall fluid viscosity rapidly at concentrations much over 1% when added to neat oils or emulsion concentrates. A second point is that phosphorus compounds represent food sources to bacteria colonies; hence, substantial concentrations should be avoided.

Another aspect characteristic of all three of the above additives was demonstrated by M. C. Shaw. Some interesting experiments were conducted where a volatile chlorinated solvent was painted on a clean steel bar and allowed to dry. The solvent was applied in narrow bands separated by untreated steel. Great care was taken to avoid contact or contamination of the bar with anything but the solvent. Monitoring of the cutting force during a continuous turning cut showed significant drops in the force when traversing the regions that had been exposed to the solvent. Since there could be no cooling effect and there was inadequate time at the cutting temperature for chlorine to diffuse through the chip thickness into the tool–chip interface area, it was suggested that chlorine acted to reduce the flow stress of the steel, possibly by enhanced dislocation nucleation at the free surface of the chip. This proposed mechanism may aid in explaining the effectiveness of sulfur, chlorine, and phosphorus in grinding, where temperatures can approach 1200–1800°F (649–982°C) but chip formation times are very short. Under these conditions, the kinetics required for film formation may not be realized.

The above hypothesis was explored at TRW following data reported by MIT, which showed anodic polarization of workpieces to accelerate film formation on the chip surfaces enhanced low-speed broaching operations. Similar polarization tests at TRW indicated no additional benefits when grinding high-temperature alloys at 6000 sfm (1830 smm). It was concluded that chip formation rates exceeded the rates of film formation even under electrically enhanced conditions. Since chlorine and sulfur additions remain effective under these grinding conditions, the secondary mechanism involving flow stress reduction appeared to be supported, which would probably not be affected by the applied potential.

The function of lard oils and soaps is believed to be by enhanced lubrication. The heat generated in the cutting zone causes changes in the surface tension of fluids containing these additives and results in migration toward the tool–chip interface.

Other additives are also employed to enhance stability of emulsions, to mask odors, to control bacterial growth, for rust control, and for antifoam properties. Dyestuffs are almost universally used to produce distinctive colors for cosmetic purposes or product identification in the marketplace.

All of the preceding additives are considered to be highly proprietary by each fluid manufacturer relative to the specific compounds used to carry the active ingredients in the product. Additives to synthetic products are normally free of the preceding compounds. While considered even more proprietary, many synthetics contain glycol esters, nitrogen compounds, and complex borates. A host of other inorganic and organic compounds are also employed in the formulation of synthetic fluids. It is beyond both the scope and propriety of this brief chapter to discuss synthetic product formulations in any greater detail.

One common caution regarding use of any cutting fluid containing an active additive package relates to the need for separation of the freshly cut chips from the fluid stream as soon as practicable. The logic for this recommendation resides in the fact that the large surface area of freshly cut chips provides a ready medium for reaction with the additive package in the fluid. Although reaction rates are slow, the large surface areas when coupled with extended times can favor significant additive depletion. This recommendation applies equally to chip making or grinding operations, and it is important to minimize swarf buildup in the fluid system or on the machine tool.

Neat Oils

Products prepared from an oil base are utilized when high degrees of lubrication and/or minimal cooling effects are required. Lubrication-sensitive processes include broaching, tapping, some milling operations, and heavy-duty grinding.

Operations such as tapping or broaching are typically carried out at relatively low speeds yet involve high chip loads and heavy cutting forces, particularly for high-strength or heat-resistant alloys. Milling involves rapid intermittent contact between the cutter and the work. Proper management of this process can include use of a highly lubricating, moderate cooling capacity oil-base cutting fluid. Excessively high cooling capacities can enhance thermal fatigue mechanisms and result in chipping of the edges.

In grinding, there is a substantial amount of frictionally generated heat which can be reduced by proper lubrication. A large fraction of the grains on the surface of a grinding wheel are either dull or do not engage the work at a depth sufficient to develop a chip. In either case, the frictional heat, normally carried away in the chip, must be minimized by adequate lubrication. This is important for precision grinding of both heat-sensitive and heat-resisting materials, the former to prevent damage and the latter to reduce grinding forces.

Cutting oils are usually formulated from base stocks of mineral oils having viscosities ranging between 50 and 200 SUS (Saybolt universal seconds) at 100°F (38°C). The most common starting material is a 50–100 SUS mineral seal oil. Products designed for light-duty applications may contain little or no additives beyond cosmetic dyes or scenting agents. Those designed for more-severe applications contain substantial amounts of additives, reaching 30–50% for heavy-duty products for deep-drawing operations.

Table 4.1 General Cutting Oil Compositions

Additive	Light Duty	Medium Duty	Heavy Duty
Active sulfur	0	0–1%	0–1%
Combined sulfur	0–1%	0–3%	0–7%
Chlorine	0–1%	0–3%	0–7%
Fats	0–2%	1–5%	2–20%
Phosphorus	0%	0–0.05%	0.05–0.1%

Additives to neat oils normally involve sulfurized or chlorinated compounds dissolved in an oil carrier. Fats or lard oils and phosphate esters are also added for enhanced performance levels. General compositional ranges for cutting oils are summarized in Table 4.1. Note that the compositions of the more fortified products permit additives to be present either singly or in various combinations. It should also be emphasized that the ranges shown in Table 4.1 represent general levels of additives, and products of some fluid suppliers that can exceed the indicated levels or the relative ratings of heavy, medium, or light duty may not agree completely. The data are shown for broad comparative purposes only leaving room for considerable overlap. Fluid manufacturers can adjust viscosities of their products to suit intended applications. A heavy-bodied cutting oil does not necessarily imply a high level of additives.

Heavy oils tend to cling better to tools like taps and broaches while the lighter oils are more desirable in grinding due to their tendencies to keep the wheels cleaner and wash swarf from the immediate grinding area. As a general rule, the lowest possible viscosity should be employed to permit ease of delivery to the cutting zone, reduce dragout, and maintain good access to the cutting zone. Negative aspects for oils include smoke, airborne mists, fire hazard, and escalating replacement costs.

Emulsions

Emulsions are stabilized mixtures of oil and water designed to combine the strong heat abstraction properties of water with the lubricating qualities of oils. Agitation will readily disperse oils into water to form a milky suspension of tiny oil droplets that readily separates on standing. Surfactants, or emulsifiers, are added to provide emulsion stability. Such compounds are molecular species having a linear structure in which one end is hydrophylic (water-loving) and the other lipophylic (oil-loving). When present in an oil–water mixture, the surfactants are aligned on the surface of the oil droplets and effectively form a transition at the oil–water interface. Another important aspect of this configuration is that the polar molecules produce an electrostatic repulsive charge between oil droplets. This repulsive charge interferes with the normal coalescence process by which oil droplets increase in size and lose the ability to remain suspended.

Plain oil–water emulsions are rarely used in industry. As a minimum, the lowest-cost products contain rust inhibitors and biocides. Additives for increased performance are dissolved in the oil phase much in the same manner and proportion as previously described for neat oils. These additives provide essentially the same functions but with the dramatically improved heat abstraction properties afforded by the water phase. Compounded soluble oils are widely used in industry for their excellent performance, relatively low cost, and lack of drawbacks inherent with neat oils such as costs, fire hazards, smoke, and high dragout rates.

Cutting fluid emulsions are not, however, without their problems. These include loss of water through evaporation, contamination by tramp oils from leakage in the machine tool, and bacterial attack. The heat at the cutting zone and dispersion of the fluid stream upon striking the work zone lead to evaporation losses of the water carrier. Regular checks are necessary to maintain emulsion concentrations within prescribed limits. These checks can range from simple refractometer measurements to more-complex chemical tests involving titration, pH determinations, as well as bacterial counts, additive depletion, and tramp oil levels.

Tramp oils can be a major problem with emulsions in that the surfactants cannot distinguish between fortified oils present in the original product and the hydraulic or lubricating oils entering the sump through machine leakage. Eventually, as the oil concentration increases, the density of surfactant molecules decreases on the oil droplet surfaces and the stability of the small droplet size is lost. Coalescence occurs and the emulsion splits into oil and water fractions. Attempts to maintain an unstable emulsion through mechanical agitation in the machine sump will be eventually unsuccessful as the mixture becomes increasingly gelatinous.

Emulsions are also susceptible to bacterial attack resulting in several types of emulsion failures. Improper system maintenance, through poor system cleanliness, operator-related contamination sources, or inadequate biocide levels lead to rapid growth of aerobic or anaerobic bacteria. As a general rule, bacteria counts below approximately 6.25×10^5–6.25×10^6 in.$^{-2}$ (10^5–10^6 cm^{-2}) are considered acceptable and counts above 6.25×10^7 in.$^{-2}$ (10^7 cm^{-2}) are undesirable. Anaerobic bacteria will propagate rapidly when biocide levels are inadequate and sump aeration levels are low. This frequently occurs over prolonged shutdown periods, such as over a weekend. This bacterial action produces the characteristic "rotten egg" odor when systems are reactivated. Prolonged aeration can usually eliminate these odors and reduce anaerobic bacterial counts; but once active, repopulation of the sump will occur rapidly necessitating replacement of the emulsion. Additions of biocides are effective, but it must be recognized that operators are also biological creatures and high levels of biocides will be harmful, particularly at sites of small skin abrasions or cuts. Apart from the offensive odors, bacterial colonies will feed on the surfactants and other organic additives, also decreasing emulsion stability.

Aerobic bacteria, while not responsible for producing offensive odors, will eventually adversely affect product stability. Hence, an effective program including concentration control, adequate aeration, proper biocide levels, and removal of tramp oil before it is emulsified can greatly extend product life. This is true whether central systems or individual sumps are used for fluid delivery to the cutting zone.

Filtration to remove chips and fines from the fluid is also necessary to prolong sump life as well as to reduce tool wear and part damage from recirculating particles. Proper filtration is essential to ensure that harmful particles over approximately 0.0006 in. (15 microns) are removed with essentially 100% reliability. In addition to harmful effects by abrasive action, buildup of particles can clog the fluid-delivery system. This can be a particular problem when hollow cutting tools are involved having narrow fluid-delivery passages.

Filter system design and operating methods must be carefully considered to not only provide the 100% reliability factor for particulates larger than the upper limit but to ensure that gradual media clogging does not increase its net effectiveness to remove particle sizes bordering on the emulsion droplet diameters. Filtration at the 2×10^{-5}–4×10^{-5} in. (0.5–1 micron) level will begin to strip additives from the emulsion.

There are also considerable data that indicate that water purity, particularly hardness, exerts a significant influence on emulsion stability and rusting resistance. Sluhan has presented data showing realization of these benefits when using deionized (DI) water instead of tap water for fluid mixing and makeup. This can be important for large central systems in which dissolved salt levels can rise dramatically as makeup fluids are regularly added to compensate for evaporation, dragout, and additive consumption.

Synthetics and Semisynthetics

Synthetic cutting fluids are characterized by being true chemical solutions containing no conventional petroleum oils. Semisynthetics are included in this section because they are hybrid combinations of emulsions and full synthetic fluids.

Although synthetics have been available for some time, recent advances in product formulation have promoted their increased acceptance in industry. Advantages claimed by synthetic products are immunity to bacterial attack, immisicibility with tramp oils, and simplified reprocessing requirements. Early formulations were found to produce undesirable residues which were either gummy or crystalline (gritty). Some formulations developed gelatinous mixtures of fluid and tramp oils which badly clogged machine sumps. Care must be exercised in selection of synthetic products to include residue and oil compatibility tests in an evaluation program.

Products available at the present time have demonstrated major improvements over earlier synthetics and most of the popular emulsion-type fluids. Proper matching of product to application has demonstrated tool life increases ranging between 30% and 1000% over heavy-duty emulsions. One of the critical factors in successful utilization of synthetic products, beyond proper product selection, is the maintenance of fluid concentrations between relatively narrow limits. Experimentation has shown that synthetics are less tolerant than emulsions to concentration variations. Optimal results are obtained only through close control of product concentration.

Since synthetic products do not depend on the classic sulfur–chlorine–fat additive packages, product formulations have been extremely varied. Successful formulations are regarded as highly proprietary by their manufacturers and little background information is available.

Semisynthetics are combinations of conventional emulsions and chemical synthetic cutting fluids. Advantages of this hybrid product are improved stability over straight emulsions and the ability to combine additive packages from both emulsions and full synthetics.

Application Methodologies

There are substantial financial and productivity improvement opportunities to be achieved through implementation of a well-planned cutting fluid control system. Returns on investment can easily exceed factors of three or four in virtually any shop, regardless of size or levels of automation involved. The objective of this section is not to provide detailed recommendations for specific applications, which the preceding discussions have indicated to be exceptionally difficult, but to outline a generic methodology for integrating fluid selection and application techniques into the manufacturing planning procedure. A cutting fluid must be viewed as a vital and integral element of the manufacturing process much the same as the more commonly addressed factors, such as machine tool selection, tooling design, and process routing development.

Establishment of an effective cutting fluid control system can be achieved by following a five-step procedure, beginning with an analysis of present practices and culminating in a plan for supplying the machine tools with the proper concentrations and volumes of the optimal cutting fluid using a system that is integrated into the overall manufacturing effort. The succeeding discussion defines these steps and outlines approaches to implement a successful cutting fluid control system.

Analysis of Present System

In this first step, the objective is to determine costs associated with operation of the present fluid delivery system along with any material or equipment limitations inherent in the manufacturing operations being conducted. Cost data are necessary to establish a baseline as well as to provide a structure for cost comparisons with any proposed system modifications. The most effective method would be to construct a model of the operation such that all direct and indirect costs are accurately represented. Generally, the direct costs are readily defined, including fluid purchase, mixing, and disposal of the spent product. However, many of the indirect costs are frequently overlooked or incompletely addressed. Examples of indirect expenses include those associated with fluid storage and transport, present methods of separating chips and fluids, machine downtime for tool change and fluid replacement, tramp oil separation and spent fluid disposal, and direct tool-related costs, which include purchasing and reconditioning requirements. Another cost center to be considered concerns the general shop environment. Costs associated with housekeeping can be significant in cases where neat oils are being used as cutting fluids. The presence of mists in the shop air will also be a major concern relative to operator health and safety. Another factor can be operator absences due to possible dermatitis or related causes arising from poorly maintained machine sumps.

In addition to a comprehensive cost analysis, it is necessary to consider limitations imposed by the materials being processed or any inherently due to the nature of the machine tools present in the shop. For example, the use of chlorinated cutting fluids may be prohibited for certain critical parts or some residues may interfere with post-processing inspections unless an additional cleaning step is included in the process routing. Some older machine tools may not have been designed for operation with water-soluble products, and it may not be economically practical to convert them for such use. This latter problem should not be dismissed out of hand. It may be possible to use the product concentrate as the machine lubricant such that leakage or back-contamination problems can be bypassed.

The preceding factors then provide a rigorous system of boundary conditions for evaluating options for any proposed modification to present practice. One method of introducing a high degree of rigor to the analysis would be to utilize the IDEF series of modeling procedures developed by the Air Force ICAM program. This procedure is applicable to both large and small manufacturing operations and documentation for the modeling is in the public domain. Details are available through the Air Force ICAM Office, Wright-Patterson AFB, Dayton, Ohio.

Screening Tests

The objective of the screening tests is to identify candidate products for serious consideration. There are a number of factors that can be sequentially analyzed to reduce the large number of commercially available products to a reasonable level for more comprehensive evaluation.

Within each of the four classes of products on the market—oils, emulsions, semisynthetics, and full synthetics—are ratings of light, medium, and heavy duty based on the nature of their additive packages. As a first step, information can be requested from fluid suppliers to produce a matrix in which several products can be identified for each of the 12 categories defined by the 4 × 3 class/rating matrix. Consideration of process requirements and machine limitations can then be used to eliminate a substantial number of the original categories. For example, machine capabilities may preclude the use of heavy-duty products because high rate cutting cannot be sustained or, conversely, the rates normally used in the factory may be such that light-duty products are unsuitable. Further reductions may be achieved through constraints imposed by the workpiece materials being processed.

A further reduction in the number of candidate products to be evaluated can be made by constructing a three-dimensional application matrix such as illustrated in Fig. 4.2. Cutting fluid requirements can be defined by the combined effects of three compound factors: material, type of machining operation,

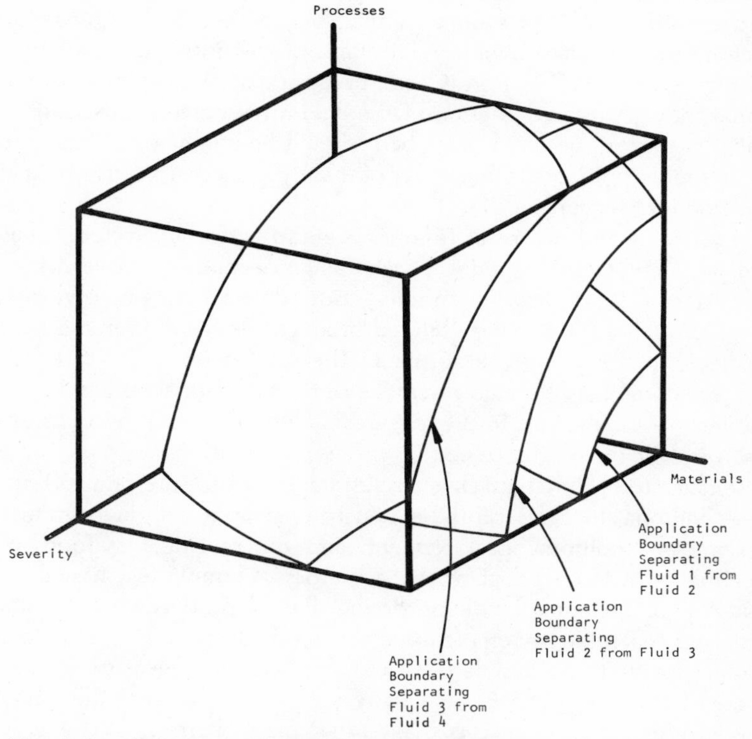

Fig. 4.2 Schematic illustration of application matrix showing how cutting fluids could be defined for various material/process/severity combinations.

and a relative severity index. The last factor—the severity index—represents the degree of aggressiveness for a given machining operation. For example, turning AISI 1010 steel at 100 sfm (31 smm) represents a less-demanding operation than the same cut performed at 600 sfm (183 smm). Similarly, on the material axis, materials can be arranged according to their relative machinability. Machining operations can be arranged along the third axis according to cutting speeds. In this manner, a specific machining operation can be located on the Operation/Material plane and then extended out along the severity axis to locate the relative position of the process in the matrix defined by the three parameters. The surfaces bounding recommended application regions can then be constructed based on data provided by the fluid suppliers. Insertion of representative machining data can then illustrate the types of fluids that will satisfy shop requirements. While this approach may seem to be a formidable effort, it represents a powerful tool for analyzing cutting fluid requirements.

Once these initial reductions have been made, rust, residue, and resistance to bacterial attack tests can be conducted. The rust test is simple and economical to perform. A 25-ml sample of the fluid, diluted to the recommended concentration, is poured over several grams of cast iron chips on filter paper in a Petri-type dish. Cast iron is used because it is the common material of machine tools. After evaporation of the fluid, the degree of rust spotting on the filter paper can be observed. There should be minimal rust stains for the more desirable products. Obviously, oil-base products need not be included in the rust test sequence.

The presence and nature of residues are also important factors for water-mixed products. Highly gummy or crystalline residues could be detrimental to the machine tool following an idle period. Residue tests involve allowing a 25-ml sample of the properly diluted product to evaporate in a clean Petri-type dish and observing the nature of the residue. In many cases, the remaining material may bear no resemblance to the original undiluted product. These tests are important to prevent deposition of harmful residues on various surfaces of the machine tool.

Resistance to bacterial attack is also an important feature of a cutting fluid. While additional biocides can be added to beef up a marginal product, this represents an additional cost element, and the biocide may not be fully compatible with the product or the operators. A simple test would involve inoculation of a standard volume of the diluted product with a culture of aerobic and anaerobic bacteria removed from one of the more mature machine sumps. The time to noticeable product failure can be a measure of the effectiveness of the biocide package. This test may not represent the ultimate in scientific rigor, but it is inexpensive and simple to perform and will at least provide a qualitative evaluation of the product.

Compatibility with workpiece materials can be tested by immersion of test coupons halfway into the properly diluted product for 24 h and observing the result. It is necessary to prepare coupons in a consistent manner to minimize risks that variations in surface finish could influence the test results. The degree of staining or attack would be assessed relative to part acceptability, remembering that most parts are in contact with the cutting fluid for much less time than the 24-h duration of the stain test procedure.

A final evaluation would be to consider product mixing, recycling, and disposal requirements. These data would be available from suppliers of the material. Capital equipment and manpower requirements for these operations could be estimated for those products passing the initial screening tests.

Cutting Tests

The three-dimensional application matrix presented in Fig. 4.2 can be consulted to identify one or two representative cutting conditions for actual product evaluations. The objectives of the cutting tests are to verify the initial screening test indications relative to machine and workpiece compatibility as well as to develop tool performance and sump life data for comparison to present practice baselines. Insertion of these data into the economic model will provide cost-savings data relative to the shop floor component of the overall cost picture.

Disposal and Recycling

This portion of the analysis will focus on the indirect aspects of the cutting fluid control system. The objective is to deliver to the machine tool an adequate volume of clean, properly formulated cutting fluid in a consistent manner at the minimum total cost.

One of the vital elements of an effective system relates to the delivery process at the cutting zone. Little benefit can be realized if the product does not reach the cutting zone in sufficient quantities to be effective. The effect of flow rates is illustrated in Figs. 4.3 and 4.4 for a specific case and for a variety of fluids, respectively. The change in flow rate from 4 to 3 gpm (1.5 to 11.4 liters/min) reduced the tool life some 27% in a turning operation involving AISI 4140 steel. That such a change is hardly noticeable to the casual observer underscores the importance of properly equipping the machine tool to reliably deliver the fluid at all times. The data of Fig. 4.4 show this effect to be realized for a wide spectrum of fluid types, ranging from light- to heavy-duty products.

There are many options for defining the configuration of the remainder of the fluid supply system, ranging from portable batch-type reprocessing units

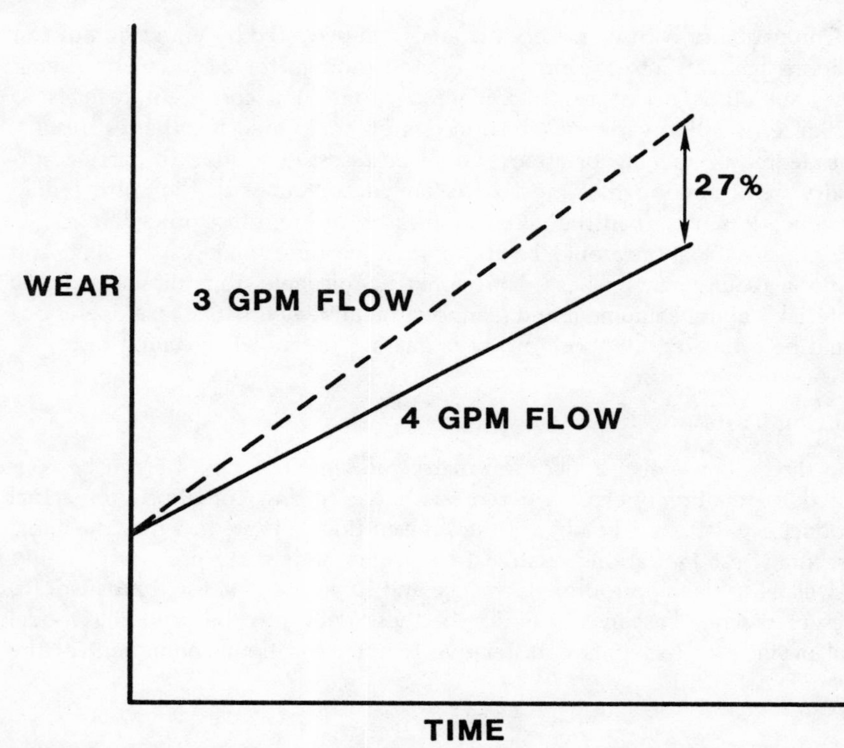

Fig. 4.3 Graph showing increases in tool wear observed during turning AISI 4140 steel for a 25% decrease in cutting fluid flow.

to service individual machine sumps through automated central systems. The individual shop requirements will dictate the range of system sophistication that can be considered in the economic analysis. However, the goal should be to design a system in which the minimum volume of product that must be periodically discharged is at an acceptable minimum level. Disposal costs will continue to escalate, and it is merely a question of time when economics will dictate a zero discharge policy for virtually all manufacturing operations. Thus, batch or central reprocessing systems should be designed with this objective in mind. The products now on the market, particularly the synthetics, are such that many can be reprocessed essentially indefinitely.

Implementation of the System

The preceding four steps of the analysis will develop the system requirements and design information necessary for planning the implementation process. The objective of this activity is to develop a schedule for orderly implementation of the system and to establish a tracking program to ensure the pro-

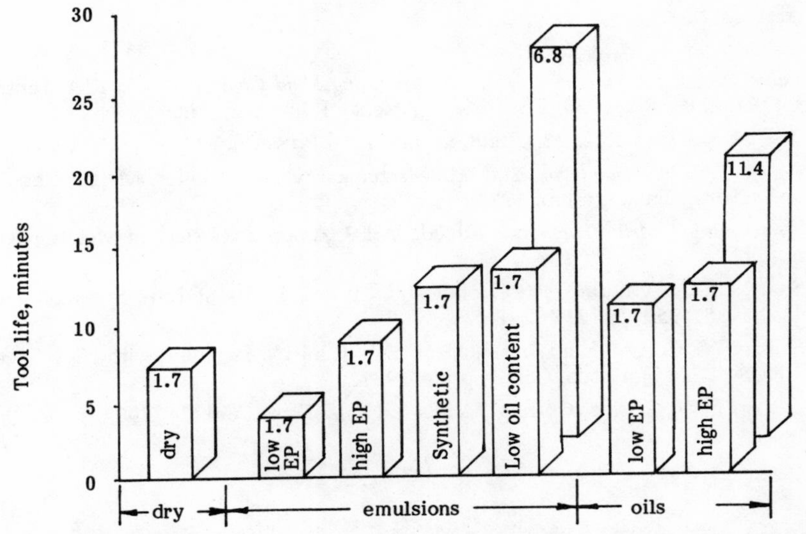

Fig. 4.4 Illustration of cutting fluid types and flow rates on tool wear for machining Ti-6Al-4V workpieces. Numbers on bars indicate flow rate in 1 min.

jected performance is indeed achieved. The tracking process should include both economic and technical performance validations.

Summary

This discussion has attempted to provide, on an overview basis, information relative to the importance of an effective cutting fluid program as an integral part of the manufacturing process.

Included in the overview were historical perspectives, fluid types, functions of the additives present in fluids, and methods to establish an effective cutting fluid program. Perhaps the most important aspect to be retained from this overview would be that cutting fluid management in the factory should be regarded as a vital element of the manufacturing process, and not relegated to the "necessary evil" classes of operations such as building maintenance or office supply services.

The historical perspective was provided to trace the genesis of present product compositions, and the discussion of the role additives play in the cutting process was intended to provide insight into the technology of product formulation. Finally, it is hoped that the five-step procedure for establishing a cutting fluid control system will be of some value for those charged with this responsibility.

References

Leiberman, G. A., *Studies to Establish a Cutting Fluid Control System*, RIA Contract No. DAAAD8-80-C-0033, Rock Island Arsenal, Rock Island, Ill.

Machinery's Handbook, 22nd Edition, Industrial Press, Inc., New York, 1984.

Machining Data Handbook, 2nd ed., Machinability Data Center, Metcut Research Associates, Cincinnati, Ohio, 1972.

"Machining," *Metals Handbook*, 8th ed., Vol. 3, American Society of Metals, Metals Park, Ohio.

Shaw, M. C. "On the Action of Metal Cutting Fluids at Low Speeds," *Wear*, Vol. 2, 1958/1959, pp. 217–227.

Sluhan, W., Cutting Fluid Presentation, SME Clinic, "Machining with High Speeds and Feeds," San Francisco, Calif., Spring 1981.

Taylor, F. W., "Art of Cutting Metals," *ASME Trans.*, Vol. 28, 1907, p. 31.

Part Two

Turning

CHAPTER 5

Turning

C. F. Barth
TRW, Inc.

The basic objective of contemporary turning operations is to reliably generate desired geometries at minimum cost consistent with required quality levels and delivery schedules. Achievement of this seemingly straightforward objective can present challenges to those responsible for establishing and maintaining efficient production. The broad applicability of turning results in wide variations in production requirements, materials, tolerances, lot sizes, and shop facilities that in turn preclude simplistic solutions. In addition, each variable is subject to some degree of natural uncertainty, such as hardness variations in the workpiece, which acts to further compound the problem.

Fortunately, given the long history of turning, a substantial body of technology has evolved to aid the engineer. These aids can range from reliance on past experience or recommendations from the many available handbooks to use of highly sophisticated computer-based algorithms and models. The manner in which these aids may be utilized depends, in part, on the resources and equipment available to the manufacturing engineer. These latter resources may include ready access to computer facilities, shop floor data acquisition and communication systems, various commercial or proprietary software packages, and automated machine tools or integrated manufacturing cells. All the preceding technology is available to the engineer for cost reduction and productivity improvements on the manufacturing floor.

This article will touch on state-of-the-art turning technologies in the ensuing discussion, but will place major emphasis on one element common to all turning operations, the management of gradual tool wear and occasional catastrophic in-the-cut failures. Wear and failures are unavoidable consequences of machining operations. A methodology will be introduced whereby worn tools can be "read" to provide a variety of information relative to the overall process efficiency. While the methodology is essentially universally applicable, it should be particularly useful for highly automated operations, production involving high-value parts, or instances where unscheduled downtime carries a heavy cost penalty.

Background

The primary direct factors influencing production turning rates and process economics may be conveniently grouped into four categories. This grouping is useful for purposes of discussion and to establish a perspective relative to turning process optimization strategies. The four categories are materials, geometries, cutting conditions, and machine tools. Various factors comprising each category will be identified and discussed relative to their interactions and overall influences on the turning process. A number of other factors certainly exist, but will not be treated in this article. Examples of these latter factors include the nature and cost of factory overhead, individual corporate strategies, and external financial considerations.

The ensuing discussion will examine each of the four categories and the factors comprising them at several levels of detail—the objective being to provide an overall perspective while citing specific examples to illustrate how the information might be applied on a practical basis in actual applications. This discussion will remain at the qualitative level to provide the broadest possible applicability to the wide variety of potential turning operations.

Category 1—Materials

This category concerns the net effect of workpiece and tool materials on process responses such as tool life, part quality, and machining forces. The relative machinability of workpiece materials remains one of the more fundamental considerations in turning operations. The basic concept of selecting the most readily machinable alloy for production of a given part is relatively obvious and need not be dwelt upon beyond the fact that considerable data have been published concerning relative machinability indices. Sources include Metcut Associates, American Society of Mechanical Engineers, and American Society of Metals. Such data provide at least an appreciation of the difficulty (or ease) by which a particular alloy may be machined relative to a common low-carbon steel. These data are useful when lattitude exists in material selection. In other cases, part design and performance requirements or customer specifications may dictate use of a particular alloy at a prescribed mechanical property level. Here, the machinability data can indicate the magnitude of the machinability problem to be faced.

Beyond reference to machinability data, metallurgical knowledge can be applied to improve machinability, even when the material selection options are limited. Improved machinability can frequently be achieved through microstructural modifications while meeting mechanical property requirements. Care must be exercised to be assured that possible changes in some of the less common material properties such as fracture toughness (K_{IC}), resistance to stress corrosion cracking (SCC), low- and high-cycle fatigue (LCF, HCF), and electrical conductivity do not become problems. Another consideration is to rigorously determine whether cost savings afforded by the improved machinability are not compromised by possibly added costs to produce a particular microstructure.

Table 5.1 summarizes metallurgical conditions that generally provide optimal turning performance for selected common workpiece material types. These suggestions are necessarily general in nature and other options exist that may be more appropriate for specific process requirements. Several examples will be presented to illustrate some options.

At a given strength level, a medium-carbon, low-alloy steel will machine more freely with a spheroidized microstructure than either a normalized hot-rolled or quenched and tempered condition. If chip breaking becomes a problem during continuous cuts, the martensitic structure will respond more readily to chip breaking attempts but at some expense to tool life. A trade-off

Table 5.1 Machinability Guidelines

	Wrought	Cast
Steels		
Low carbon	Cold rolled	Homogenized, annealed
Medium carbon	Spheroidized[a]	Spheroidized
High carbon	Spheroidized[a]	Spheroidized
Aluminum Alloys		
Low Alloy	As-rolled	Normalized
High Alloy	Solution treated	Solution treated[b]
Copper Alloys		
Copper	As-rolled[a]	As-cast
Brass	As-rolled	Normalized
Bronze	Annealed	Normalized
Beryllium–copper	Annealed/solution treated	Solution treated
Heat Resisting		
Ferritic stainless	As-rolled	Homogenized
Austenitic stainless	Annealed[a]	Homogenized
IN 600s	As-rolled[a]	Homogenized
IN 700s	Solution treated	Solution treated

[a] Tends to be very "stringy"; cold rolling may be beneficial.
[b] Aluminum–silicon alloys such as 390 should be cast to minimize primary silicon particle size.

analysis will be necessary to determine net machine uptime improvements, comparing the benefits of reduced chip nesting and disposal problems against the impact of more-frequent tool changes.

Another aspect involves surface finish and integrity. This aspect is more critical for parts that are finished by the turning operation than for surfaces that will be processed further, such as by hardening and grinding. A "frosty" surface is preferable than a highly specular, burnished surface. The frosty surface is more cleanly cut with little or no subsurface damage, while the burnished surface contains high residual stresses and can have actual surface damage. Microcracks and tears can be smeared over, thus concealing potential damage. Ideal turned surfaces should reveal only the geometric texture imparted by the passage of the cutting tool. This geometric texture is illustrated in Fig. 5.1. The geometrical cusp height is only approximately 100 μin. (2.54 μm) for a feedrate of 0.005 in. per revolution (IPR) (0.127 mm) and a 1/32-in. (0.8 mm) tool nose radius. Tearing effects beyond the geo-

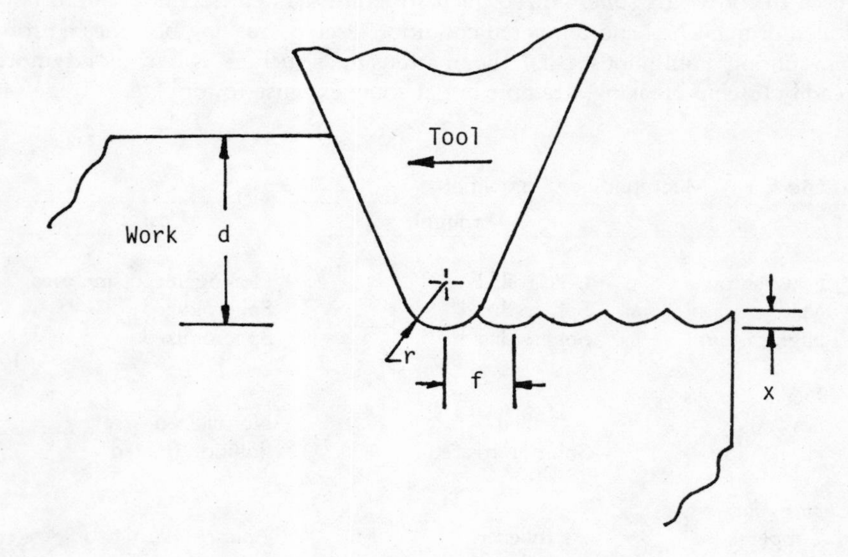

r = tool nose radius
f = feedrate per revolution of part
d = radial depth of cut
x = cusp height

from geometry: $x = r - \frac{1}{2}(4r^2 - f^2)^{\frac{1}{2}}$

Fig. 5.1 Representation of the geometric surface roughness effect in turning.

metrical effects are to be avoided. There are tooling and cutting parameter options to minimize this type of damage, but material conditions can also be used to advantage to minimize tearing damage potentials. Increases in the strength of the work material can provide less damage for critical as-turned surfaces. Strength can be imparted by use of cold-worked stock or changes in heat treatments. For example, in the latter case, age-hardenable alloys can be turned in the solution-treated and aged condition or steels can be quenched and tempered to a somewhat higher hardness level.

Another common method of reducing steel turning costs is to substitute a lower-alloy material that has been cold worked to the strength level provided by the more costly steel alloy. Tool life is increased and cutting forces are reduced when the same cutting parameters are utilized. If the stock envelope and the machine capacity will permit, more aggressive cuts are possible to increase throughput without exceeding previous machine loads and tool change schedules.

Tool material selection can be a relatively straightforward proposition, providing a systems approach is followed. The cost of the cutting edge can be shown to be an almost negligible fraction of total manufacturing costs in virtually all well-engineered processes. Hence, the cost of an individual tool or insert should not be a major consideration in tool material selection. The primary factors are a lack of reactivity with the work material and the limitations (or advantages) of the machine tool to be used in part manufacture.

Ideally, the cutting tool should be inert with respect to the constituents in the workpiece material at the temperatures existing in the cutting zone. Inertness can be considered to include a lack of both gross chemical reactivity and any interdiffusion of one or more atomic species. Three of the factors affecting chemical reactions and atomic interdiffusion of two materials in contact are time, temperature, and unit pressures. Other factors specific for turning which impact the first three include thermal diffusivity, heat capacity, flow stress, and melting points. The basic objective is to select a tool material having the greatest hardness, highest melting point, and the least reactivity with the workpiece material. Such inertness rarely occurs in the real manufacturing environment. An appreciation of these phenomena and an understanding of the conditions under which they occur can be employed to advantage in maximizing the effectiveness of tool material selection.

High-speed steel (HSS) has been used successfully to turn ferrous materials for decades as long as cutting speeds are kept below approximately 100 surface feet per minute (sfm) (30.5 smm). Interface temperatures and hence reactions increase rapidly above this speed threshold, leading to rapid tool degradation. Higher speeds are possible with cemented carbides owing to increased hardness levels, higher melting points of the various carbides, and a higher threshold of reactivity with ferrous materials. Coatings on carbide substrates reduce friction and reactivity giving two related benefits. The first is that with less friction higher speeds are possible before the same surface temperatures are reached, and secondly, high surface temperatures are permissible due to the reduced reactivity. Oxide ceramics offer even greater potential relative to reactivity and interdiffusion. Most metallic workpiece materials exhibit minimal interactions with ceramic tools. Their relatively low toughness, however, dictates careful process setup to reduce fracture risks.

Table 5.2 shows recommended optimum speed ranges for various tool material types in ferrous applications. Similar examples could be developed for other workpiece alloys. The point here is to match the cutting tool material to the speed capabilities of the shop equipment. There are few advantages in using coated carbide inserts at low cutting speeds when HSS tools will perform very well at lower cost. In fact, carbide tools are highly susceptible to edge chipping and BUE (built-up edge) problems when used at speeds more suitable for HSS cutters. Exceptions of this general rule can be devised, but it is usually advisable to follow the recommendations of Table 5.2

Another important general characteristic of cutting tool materials is their relative sensitivity to process parameter variations. Figure 5.2 illustrates the relative performance of three tool materials when used within their respective optimum speed ranges. It is apparent from Fig. 5.2 that carbides and ceramics readily outperform HSS tools, but that they are increasingly less tolerant to

Table 5.2 General Turning Speed Ranges for Ferrous Materials

Tool Material	Ferrous Alloys Cutting Speed Range, sfm (smm)	
	Roughing	Finishing
High-Speed Steels		
M-2, M-3	0–75 (0–29)	50–75 (15.2–29)
M42, T-5	0–100 (0–30.5)	75–120 (29–36.6)
T-15	0–125 (0–38.1)	100–130 (30.5–39.6)
Carbides		
Straight tungsten carbide	100–250 (30.5–76.2)	200–300 (61–91.4)
"Alloy" carbides	100–300 (30.5–91.4)	200–350 (61–106.7)
Coated carbides	300–500 (91.4–152.4)	400–800 (121.9–243.8)
Ceramics		
Straight ceramics	800–1000 (243.8–304.8)	800–1500 (243.4–457.2)
Cermets	800–1000 (243.8–304.8)	800–1500 (243.4–457.2)
Diamond	N.R.[a]	N.R.[a]
Cubic Boron Nitride	600–1000 (182.9–304.8)	800–2000 (243.4–609.6)

[a] N.R. = Not Recommended

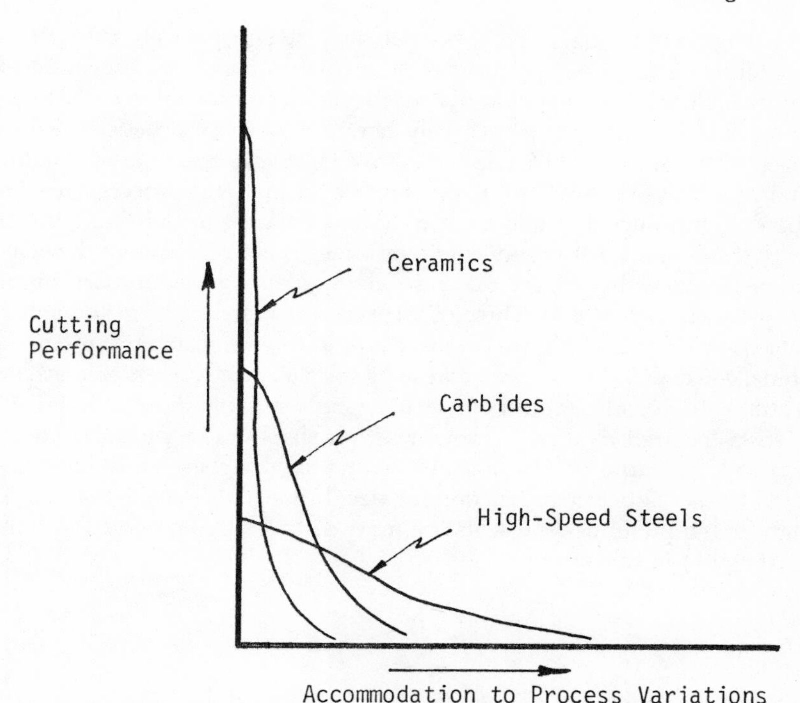

Fig. 5.2 Illustration of the relative performance of three classes of tool materials and the sensitivity or tolerance of these materials to process parameters.

process variations and off-optimum conditions. Disastrous failures can occur if the machine and tool packages have not been designed to utilize the more-advanced cutting tools. Dynamic stiffness, adequate power, and solid cutter support are critical factors in the successful use of carbide and particularly ceramic tooling. Unfortunately, the large variety of proprietary carbides, coated carbides, and ceramics precludes detailed discussion of specific products in this paper. Knowledgeable applications engineers from the various tool suppliers and manufacturers will provide data and recommendations for specific requirements.

Category 2—Geometry

The geometry of the part to be produced is perhaps the most fundamental factor influencing turning process design and economics. The term geometry generally refers to characteristics used in parts classification schemes. A somewhat different context is used in this article in which geometric effects are considered relative to their effects on sustainable cutting rates and options for turning process design. A blocky part may be economically cut from bar, even for large lot sizes, while complex, flanged parts may be more efficiently produced from material that has some degree of prior preform processing. If it is assumed that a totally new part is to be produced and that a reasonable variety of machine tools are available in the shop, a top-down systematic analysis can be performed to define the optimum process. Questions to be examined include exploration of options ranging from turning the part directly from bulk bar to producing a near-net or net preform which requires little or no machining. Figure 5.3 illustrates a rather simple approach for producing a hypothetical part. Three cost curves are added to estimate minimum total cost to produce the part. One curve is the per part cost of the raw material consumed. The second represents costs associated with premachining operations to impart various degrees of shape, while the third indicates turning costs incurred to finish various preform shapes. The curves have been smoothed for purposes of illustration and should include all indirect costs related to the particular production lot size. It is readily seen that turning of rough forged preforms represents the most cost-effective process to produce the part used in this example.

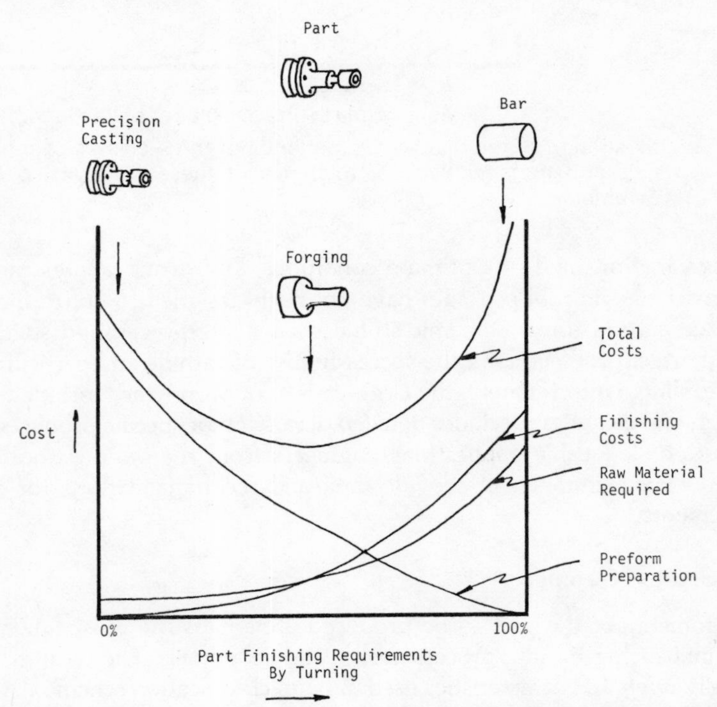

Fig. 5.3 Cost analysis curves to select optimum turning process from several initial shape conditions.

There are far more rigorous methods to perform this analysis. One of the most complete methods has been developed by the U.S. Air Force ICAM Office (Integrated Computer-Aided Manufacturing) at Wright-Patterson Air Force Base, Dayton, Ohio. One of the many ICAM software packages, IDSS-2 (Integrated Decision Support System), is available to industry for process development. Another portion of the ICAM system utilizes comprehensive decision trees to ensure inclusion of all factors in the analysis. A section through the decision network is illustrated in Fig. 5.4, which shows considerations to be evaluated regarding material form, tooling, and machine-tool selection. Extension of the network to the right, for example, will permit evaluation of specific bar sizes relative to cost and availability for the process under development. Given input data regarding process options and costs, the software will aid the engineer or process planner to define the optimum process sequence within existing shop constraints. The ICAM software will also permit rigorous evaluation of new process technologies prior to capital investment commitments. The option exists, of course, to perform the analysis manually or to employ existing resident software packages. The essential

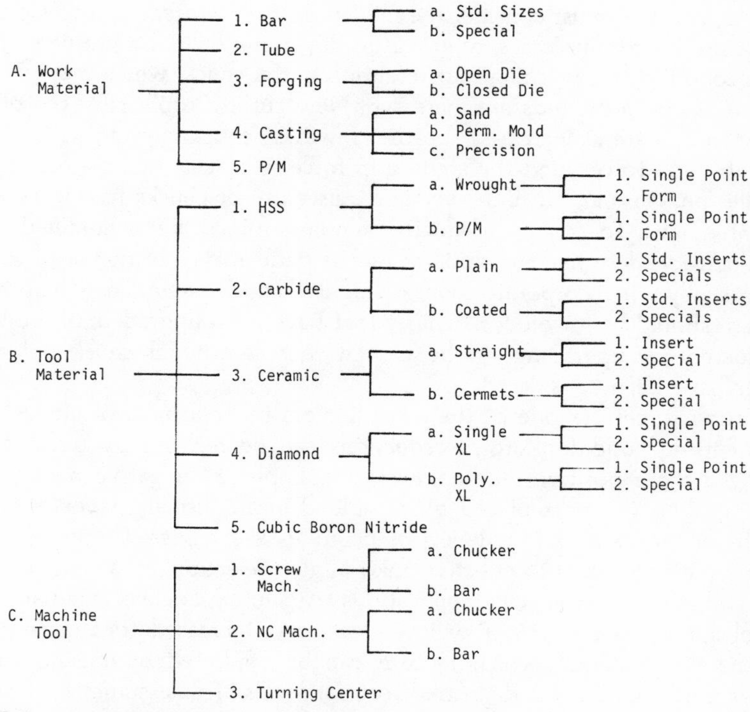

Fig. 5.4 Partial decision tree for turning process selection.

element is to perform a quantitative analysis on a per price basis for the lot size required before a new part geometry is to be released to production.

Category 3—Cutting Conditions

The focus in this section will be on cutting conditions in terms of their relationship with throughput and part quality. It would seem to be axiomatic that turning processes would be operated at maximum rates to achieve the greatest throughput possible. This is, of course, not the case because turning, like any other manufacturing operation, generally involves trade-offs between opposing factors to determine the most advantageous operation conditions.

Part-related factors limiting metal removal rates were discussed in the preceding section and included complexity, continuity of cut lengths, and part strength. Permissible metal-removal rates also depend to some extent on the volume and distribution of the stock envelope. A flanged rod to be cut from bar, for example, may sustain relatively heavy cuts, while a near-net preform shape will require only finishing-type cuts. Other limits involve considerations of cutting forces and part imbalance damages due to high cutting or centrifugal force effects. Attempts to raise metal-removal rates through spindle rpm increases must consider the influence of the preceding forces on part deflection. In extreme cases, retention in the chunk will be a problem. There is a second factor governing part retention in the chuck, which must be considered as cutting speeds are increased. New cutting tools offer the opportunity to operate at increased speeds. However, chuck jaws are also subject to centrifugal forces, and the static grip force will decrease as a function of spindle rpm. Turning at high speeds requires use of chucks having counterweighted jaws to ensure adequate clamping forces are maintained. New machine tools designed to capitalize on the performance of the latest cutting tools usually will incorporate counterweighted chuck assemblies. The primary concern should be for older machines that have been upgraded, or are being considered for upgrading, to be certain that centrifugal chuck unloading effects have been considered.

Cutting forces are one of the many factors influencing tool life. At constant cutting conditions, force reductions can be realized by using a tool having a more positive rake angle. For example, 5° negative rake carbide insert tooling can be replaced by a molded insert, usually a coated grade, which can provide various levels of positive rake angles. Commonly used tools have a 5° or 10° effective rake angle. Selection of a particular design will depend on specific applications. Again, experience or advice from a tool applications engineer will be necessary. Increased tool performance is generally obtained, which in turn can be exploited by raising cutting speeds until tool life has decreased to prior levels. This assumes the original tool life was acceptable, and the part/machine system will sustain higher cutting speeds. The objective is to increase part throughput, not to "save" inserts. Costs of individual cutting edges generally represent small fractions of total part manufacturing costs. Exercise of a reasonably complete cost model will determine optional tool-related cost versus downtime trade-offs relative to total cost and productivity levels.

Cutting fluids and their proper application are also important, but frequently unappreciated, elements of cutting conditions. Turning operations involve longer tool/work contact times in comparison to milling, requiring somewhat different properties in the two applications. Turning basically requires higher heat abstraction capabilities than does milling, and fluids must be matched to process requirements for maximum benefits. The reader is directed to a more complete treatment of cutting-fluid application technology found elsewhere in this volume. Suffice it to say, however, that proper fluid selection can provide surprising benefits, and fluid costs per gallon represent the least significant factor in the overall cost/benefit relation.

An excellent but relatively simple technique to assess cutting conditions is to install one of the commercially available power monitoring systems on the machine. The resulting power consumption data would provide information relative to four important aspects of cutting conditions. These aspects are summarized in Table 5.3.

Interfacing power monitor output signals to a microprocessor-based system affords the means to close the process control loop. The sophistication level of the system can be adjusted to suit process needs, ranging from a simple broken tool detector to an adaptive process optimization controller.

Category 4—Machine Tools

Some aspects of machine tools were discussed within the context of the preceding sections and need not be repeated here. This section will focus on characterizing specific aspects of various turning machine classes relative to their suitability for various applications. Table 5.4 compares characteristics of

Table 5.3 Machine Power Data Applications

1. Net cutting power	Use to evaluate cutting efficiency. Evaluation of new tools, cutting fluids.
2. Tare power	Machine condition/maintenance.
3. Power distribution	The efficiency of NC programming can be quantitatively measured by comparing actual cut time to total cycle time.
4. Broken tools	The absence of a power peak in an NC cycle can be used to detect worn or broken tooling. Excessive power peaks can indicate a previous operation was incomplete; for example, a missing hole or bore for a threading operation.

Table 5.4 Characteristics of Various Turning Machine Types

Characteristics	Engine Lathes		Screw Machines[a]	Turning Centers	
	Manual	NC	Bar or Chucker	Single Spindle	Multispindle
Setup requirements	Little	Programming	Mechanical	Programming	Programming
Tooling requirements					
Costs	Very low	Low	Moderate	High	Highest
Types	Standard	Standard inserts	Form, standard	Special packages	Special packages
Flexibility	Very high	Very high	Low	High	Moderate
Repeatability	Variable	High	Very high	Very high	High
Reliability	High	High	Very high	High	High
Lot sizes	0–100 pieces	1000 pieces	500 pieces	5–1000 pieces	500–5000 pieces
Costs					
Initial	Very low	Low	Moderate	High	High
Maintenance	Very low	Low	Very low	High	High
General applications	Tool rooms Prototypes Specials	Tool rooms Special lots Job shops	High volume	Mixed lots Automated cells Flexible manufacturing	Same as single spindle but higher lot sizes

[a] Some newer design screw machines now feature NC feed controls, which greatly increases their flexibility.

the four major types of turning machines for general applications. The data for Table 5.4 can be useful in selecting appropriate applications for existing machines or selection of new machines.

Sizing of machine tools to the job requirements is also an important consideration to avoid under- or over-capacity conditions. The impact of inadequate, undersize equipment on overall throughput and machine reliability should be obvious. Over-specification of machine tools will result in long machine life, but can represent an inefficient use of scarce capital resources. Power requirements for an intended machine application can be readily estimated from an equation such as:

$$HP = \frac{R\Lambda}{396{,}000}$$

where R = metal removal rate, in.^{3} min^{-1}
Λ = power required for a given work material

and

$$R = (f)(d)(\pi D) \text{ rpm}$$

and f = feedrate, in. rev^{-1}
d = depth of cut, in.
D = cut diameter, in.
rpm = spindle rpm

Values for Λ will vary with large changes in cutting speed and some tool designs, but average values are shown in Table 5.5 for turning operations within conventional speed ranges.

Figure 5.5 was prepared using the preceding data and the horsepower equation to show power requirements as a function of metal-removal rates for common workpiece materials. These data are useful in estimating machine power requirements for new operations or sizing new equipment acquisitions. The data are represented in linear form for general guidelines, although actual values of Λ for specific applications will exhibit some nonlinearity.

The question of machine stiffness is also an important concern relative to size-holding capabilities. Static stiffness, usually expressed in terms of pounds per inch of deflection, is a measure of the machine's ability to resist cutting

Table 5.5 Typical Values for the Power Factor

Material	Λ (power in.$^{-3}$ min^{-1})
Aluminum	$0.2-0.3 \times 10^6$
Copper and brasses	$0.3-0.5 \times 10^6$
Low-alloy steels	$0.4-0.5 \times 10^6$
High-alloy steels	$0.6-1.0 \times 10^6$
High-temperature alloys	$1.5-2.5 \times 10^6$

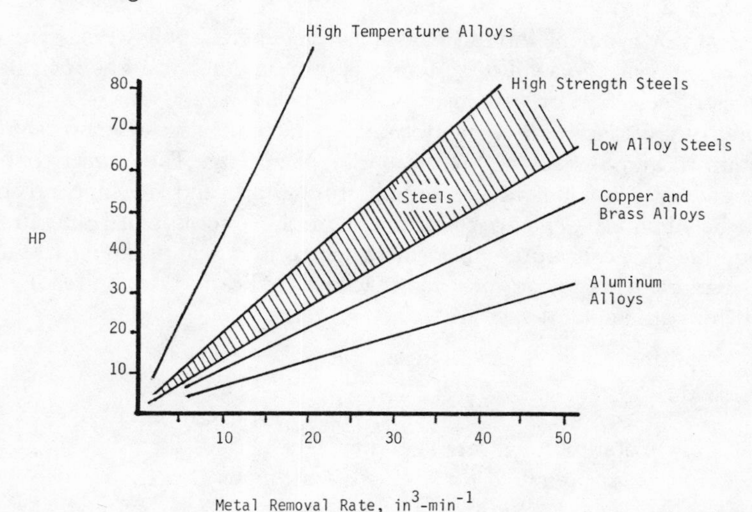

Fig. 5.5 Estimated power required to sustain various metal-removal rates for common workpiece materials.

forces and maintain the desired tool path. Tolerance-holding capabilities are also influenced by the effective dynamic stiffness inherent in the machine design. Sustained, or regenerative, chatter can develop during a cut if turning conditions produce an excitation frequency at or near the fundamental resonance frequency or harmonics of the machine tool. While machine tool builders attempt to design structural components with natural frequencies not likely to be excited during machining, conditions of resonance can still be encountered. Critical damping capabilities are also designed into turning machines. Despite these efforts, regenerative chatter can occur, which is destructive to the tooling and the machine as well as producing generally undesirable surface characteristics on the part. The recourse when chatter is experienced is to redesign the tooling or alter the machining parameters to avoid generation of a critical excitation frequency. This may prove challenging in that, in some cases, resonance conditions are developed in the part rather than the machine. The difficulty arises from the fact that the part is changing in geometry as metal is removed, which in turn continously alters fundamental resonant frequencies. Subtle variations in part programs represent an effective means to detune resonant conditions.

Tool-Wear Management Discussion

Managing tool wear and avoiding catastrophic failures remain a problem for all production turning operations. In a perfect world, all tools used to perform individual cuts would wear at exactly the same rate. Process optimization would be a relatively simple procedure, and tool changing schedules could be readily established. However, small but normal inconsistencies in each element of the turning process combine to produce a relatively broad variation in the performance of successive tools used in a particular cut. The objective is to understand the sources of these inconsistencies and make attempts to minimize their collective impact on the turning operation. Sources of these inconsistencies reside within the workpiece, the cutting tools, the operator, and, to a certain extent, the machine tool.

The workpiece can provide two major inconsistencies, one metallurgical and the other dimensional. Metallurgical variations can arise from compositional inhomogeneities, the size and distribution of second-phase constituents, and hardness variations related to the thermomechanical processing history. Aluminum–silicon alloys and gray irons are classic examples of metallurgically related variability. Primary silicon can form in relatively large sizes during solidification of aluminum–silicon alloys in thick sections or by slow cooling after casting. Tool life can vary by a factor of 2 or more, depending on the particle size of the alpha-silicon phase. Similar levels of tool-life variation occur for gray iron castings. Casting areas containing type-*D* graphite will prove more difficult to machine than more slowly cooled areas having a type-*A* graphitic structure. It is unlikely that absolutely uniform workpieces can be obtained on a sustained, cost-effective basis to eliminate this source of tool-performance variability. An awareness of this factor can be utilized to at least minimize the impact of major metallurgical nonuniformities without serious raw-material cost penalties.

Dimensional variations can lead to high tool entry forces for workpieces larger than nominal or anticipated sizes. Excessive forces can lead to edge breakage or damage to tool holders. Increases in initial offsets will prevent high entry forces on oversize parts but will increase the nonproductive "air-cutting" portion of process cycles.

Variations in cutting tools arise from several potential sources. Possibly the greatest source is related to microsurface damage produced by HSS-tool finishing operations. In-shop regrinding, in particular, can produce serious metallurgical damage through improper or uncontrolled grinding practice. Toolrooms must be carefully operated to minimize surface-damage effects. The risk of such surface damage on new tools can be considered relatively small. Carbide tools are susceptible to the presence of microvoids as a consequence of their method of manufacture. Inserts having low levels of microvoids or inclusions near the cutting edge will be less susceptible to fracture. This is not to say that carbide producers are not diligent in their efforts to eliminate such defects, but mother nature is a tough adversary. Hence, it is not impossible to expect a small, variable amount of microporosity to be present in a portion of carbide inserts. The presence of this randomly occurring microporosity makes a small contribution to uncertainties in tool performance. Carbide tools that are finish ground are also susceptible to surface-damage effects, primarily carbide grain fracture and pullout. Microdamage of this nature does not necessarily represent major deficiencies in tool-manufacturing technology, but these factors contribute to subtle variations in cutting performance.

Operator effects are perhaps the greatest source of variability and possibly the most difficult to control. Tool-changing decisions in most shops are made on the judgment of individual operators. It is vital to establish uniform standards for determining tool-wear endpoints or utilize electronic means of tool-wear measurements. It is essentially impossible to either exercise control of turning operations or evaluate process efficiencies without rigorous and uniform tool-wear standards.

Machine tools have a finite variability inherent in the basic design and, as a consequence, additive manufacturing tolerances affect the many components comprising the machine. All new machines begin their production lives at this point and degrade through wear. Proper maintenance practices are essential to control capability degradation over the machine's useful life. Rigorous maintenance/performance cost analyses are necessary to minimize this source of process inconsistency.

Analytical methods are essential primary methods of process optimization, as indicated in the previous section. The consequences of inconsistencies inherent in the process must also be considered. The purpose of the variability discussion is to highlight the situation and establish some credibility for the value of "reading" tool wear as a mechanism to improve turning process efficiencies.

To the casual observer, tools degrade by various combinations of flank recession, crater development, edge chipping, and fracture. These failure modes are the result of operation of one or more specific tool-wear mechanisms. Not all mechanisms operate at any given speed range, and frequently only one or two dominate under a particular combination of cutting conditions. Figure 5.6 identifies the various wear mechanisms and indicates their general operating ranges relative to cutting speed, or temperature, in turning processes. Characteristics of each wear mechanism are reviewed to establish a basis for interpreting wear scars such that minor process parameter changes aimed at improving tool performance can be effectively accomplished.

At low speeds, adhesive wear is the primary mechanism with some contributions from abrasive effects. Adhesive wear is a microwelding phenomena and is a function of time, temperature, and pressure conditions existing at the tool/chip interface. Workpiece material welds to the cutting tool as a result of this process and produces the BUE (built-up edge) condition. Material flow then occurs over itself rather than the tool surface. A stable BUE condition effectively results in complete isolation of the tool from the chip stream. This

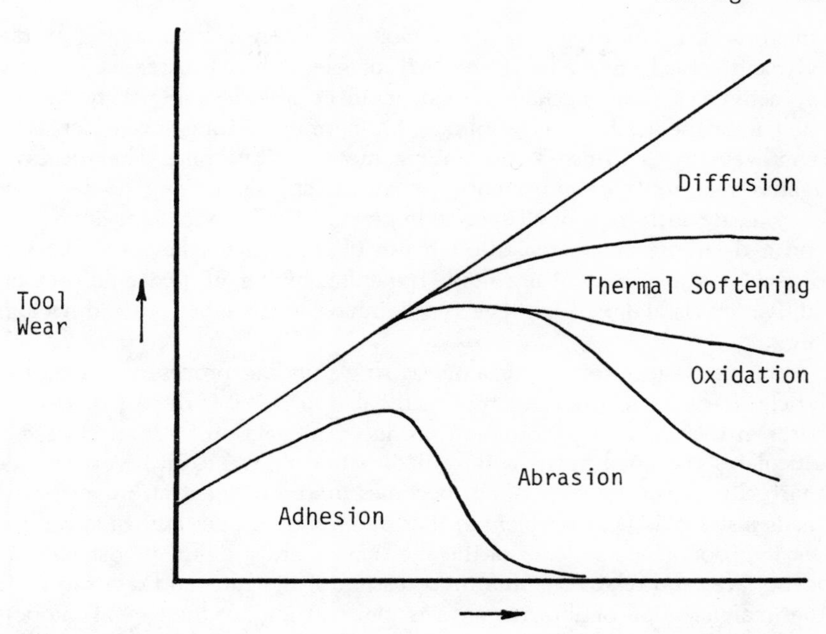

Fig. 5.6 Relative relationships of tool-wear mechanisms with respect to cutting speeds or temperatures.

condition prevents mechanical wear, but can produce thermal damage in both the tool and the work. Unstable BUE conditions are more common in which the adherent work material is periodically torn away, usually taking small amounts of the tool edge along with the dislodged work material. Large-scale occurrences of BUE breakaway produce a ragged, chipped cutting edge, while microchipping is a result of fine-scale localized welding and tearout.

Adhesive-wear characteristics are the presence of metal or chips adhering to the tool with gradual deterioration of the edge by chipping. Damage through chatter-induced chipping not related to resonance conditions of the machine or part also represents a special case of adhesive wear. A dynamic condition can exist in which the BUE develops and breaks away at a more or less stable frequency. When these conditions are experienced, three actions can be taken to improve overall performance. The simplest solution is to increase cutting speeds to reduce the contact interval between incremental chip volumes and the tool, thus reducing the opportunities for welds to develop. The usual action taken is to slow down cutting rates when chipping damage occurs. Unfortunately, this action is counter to maintaining productivity and actually exacerbates the BUE problem. Speed increases, however, may activate a wear mechanism that could be less desirable than the BUE problem, or the machine and tool may not permit a cutting-speed increase. A second remedy is to use a tool having more positive rake. The success of negative rake tools when used at high cutting speeds has engendered a tendency to use such tools at all speeds. In general, BUE is a low-speed phenomenon and is best minimized through use of a positive rake tool. The third possibility is to use a cutting fluid containing higher EP (extreme pressure) additive levels. Higher EP levels will inhibit chip adhesion and welding effects.

Abrasive wear is roughly analogous to a grinding process in which hard particles mechanically remove tool material. Sources of hard particles include phases in the workpiece, corrosion products or scale on the part surface, or particulates entrained in the cutting fluid. Abrasive wear can be experienced at virtually all cutting speeds, but becomes more prevalent at higher speeds. This is just a bulk-flow problem in that more particles per unit time can pass over the tool. Characteristics of this mechanism are smooth, almost polished, tool surfaces and relatively uniform recession of the flank and rake faces. Another manifestation of abrasive wear is the "slotting" of the tool at the point the part surface intersects the cutting edge. Scale and hard surface deposits will enhance tool-slotting effects. Slotting is particularly undesirable in HSS tools that are to be resharpened. Deep grooving would require considerable material removal to restore the cutting-edge geometry. Such actions reduce the total number of resharpening operations for a given tool, indirectly leading to higher tool costs. These costs are experienced even though tool life on the machine, measured in terms of pieces produced per edge, remains acceptable.

Increased tool hardness, improved filtering of the cutting fluids, higher cutting fluid viscosity and lubricity, clean parts, and refinement of second-phase constituents in the work material are effective steps to reduce abrasive-wear effects.

The remaining three wear mechanisms are largely operative at the higher cutting regimes. These mechanisms become particularly important as productivity improvement incentives continue to drive up turning speeds. All three are related to effects of temperatures at the tool/chip interface.

Diffusional wear occurs as atoms from the tool migrate across the tool/chip interface into the chip stream. Characteristics and the intensity of this wear mechanism vary with the relative chemical affinity between tool and workpiece materials. An extreme example would be the rapid transfer of carbon from a diamond tool into the chip stream during attempts to turn ferrous workpieces. Diffusional or chemical wear exhibits a sharp increase in rate with cutting speeds, and wear scars become quite smooth and uniform. There will be a tendency for the flank wear scar to be very flat, as if it were deliberately machined. Early stages involve both crater and flank attack with flank wear eventually predominating as speeds are increased.

In the case of uncoated cemented carbide tools, chemical attack of the cobalt binder phase may be experienced. Recession of the binder will release carbide particles before significant mechanical wear is experienced. Cutting fluids having high concentrations of active sulfur are particularly aggressive to the binder phase and should be avoided. One method is to utilize additives containing chemically combined rather than "free" sulfur. Determination of chemical wear as an active mechanism requires microscopic examination of worn tools. Evidence of binder recession with minimal wear of individual carbide grains will suggest a chemical wear problem.

The best approach for combating suspected chemical wear is to enlist the assistance of a metallurgist or skilled tool engineer. Careful selection of tool materials is necessary to minimize mutual reactivity between the tool and workpiece. Factors to be considered include solubilities of tool constituents in the work material, tendencies toward formation of interstitial or intermetallic compounds, the relative free energies of formation for tool constituents versus possible reaction products with constituents of the work material, and the relative melting points of the tool and work materials.

Thermal softening is primarily a thermomechanical problem. The hardness and flow stress of all tool materials is subject to some temperature dependency. Plastic deformation begins as the cutting forces become a significant fraction of this reduced flow stress. Close examination of a worn tool experiencing thermal softening will reveal displaced material around the vicinity of the contact zone. The crater edge will be elevated above the surrounding area, and bulges in the flank surface will be apparent adjacent to the flank wear scar.

Methods to avoid thermomechanical damage can follow two approaches. One is to use tools having improved high-temperature properties, and the second is to reduce the temperature in the contact zone. Mechanical property data and tool-material recommendations can be obtained from competent service engineers of tool vendors. Temperature reduction is a task amenable to in-shop efforts. A cutting fluid having aggressive heat abstraction and wetting properties, provided at high pressures and flow rates, will be effective in reducing thermal softening effects on virtually all turning applications. Frictional heating effects can also be reduced by the proper cutting fluid as well as using tools having low friction coatings. Nitride and oxide-coated tools are effective in reduction of frictionally induced heating effects and attendent thermal softening problems.

The third high-temperature mechanism—oxidation—is operative on all but oxide ceramic tools. HSS tools will usually fail by chemical wear or thermal softening before oxidation becomes a significant problem. Cemented carbides can be subject to oxidation attack. The carbide phase exhibits good oxidation resistance and excellent hot hardness, but the cobalt binder can be subject to oxidation in addition to thermal softening. Carbide grains become undermined and are lost from the surface. The primary characteristic of oxidation damage to cemented carbide tools is a general lack of wear striae on individual carbide grains. The grains are lost or displaced before wear damage can accumulate to a significant degree. This mechanism is virtually identical to chemical wear of the binder phase in cemented carbides. Steps necessary to reduce oxidative wear are similar to those presented for chemical wear. It should also be recognized that cutting fluids become thoroughly oxygenated as a consequence of the method of delivery and can supply adequate amounts of oxygen to the cutting zone to support oxidation of the tool.

A brief summary of the characteristics of the various tool wear mechanisms is presented in Table 5.6

The primary objective should be to critically examine worn tools and determine the particular wear mechanisms that are operative. This should be done whether or not the operation is perceived to be operating at acceptable levels. Once the specific wear mechanisms are identified, clearly defined actions can be taken to improve tool performance. These actions are generic in nature and can be utilized in virtually all turning applications, regardless of the particular workpiece material being machined.

Table 5.6 Summary of Tool Wear Mechanisms

Mechanism	Major Characteristics	Necessary Actions[a]
Adhesive wear	Built-up edge, chipping	Increase speed moderately Increase positive rake angles Use EP additives in cutting fluid
Abrasive wear	Uniform wear scars, edge grooving	Use harder tools Clean parts Filter cutting fluid Increase cutting fluid viscosity, flow Refine part microstructure
Diffusion or chemical wear	Smooth surfaces	Change tool material Use coated tools
Thermal softening	Plastic flow of tool edge	Improved tool cooling effects Increased hot strength of tool Low-friction coatings
Oxidation	Discoloration around wear zone, microscopic features	Reduce temperatures Improved cutting fluids Coated tools Change tool material

[a] Speed reductions, while effective in some cases, can represent a compromise to productivity and should only be considered as a last resort.

Summary

The influence of materials, part geometry, cutting conditions, and machine-tool design on turning process performance was reviewed to establish a solid background to the problem of tool-wear management on the production floor. The discussion was targeted toward individuals with in-shop responsibilities for maintaining efficient production rather than a theoretical treatise on turning phenomena.

Material properties and acceptable methods for their variation were discussed to illustrate how both deliberate and random changes could affect turning process performance. Quantitative methods were outlined to evaluate options available for setting up existing or new jobs based on part shapes. The impact of cutting conditions and machine design on throughput and chatter were examined. Data were presented to aid in machine selection relative to capabilities and turning power requirements.

Tool wear mechanisms were defined in detail. Included in the discussion were treatments of adhesive, abrasive, chemical, and mechanical deformations, and oxidative wear mechanisms. Guidelines were advanced to identify specific wear mechanisms such that proper corrective actions could be taken. The objective was to integrate an understanding of process factors, wear mechanism recognition, and improvement options to improve tool performance.

The methodologies advanced are intended as a process optimization procedure to be used even for operations generally perceived to be running at acceptable levels.

References

Armarego, E. J. A. and Brown, R. H., *The Machining of Metals*, Englewood-Cliffs, N.J., Prentice-Hall, 1969.

Bhattacharyya, A. and Ham, I., *Design of Cutting Tools*, Dearborn, Michigan, ASTME, 1969.

Field, M., Koster, W. P., Kohls, J. B., Snider, R. E., and Maranchik, J., Jr., *Machining of High Strength Steels with Emphasis on Surface Integrity*, Cincinnati, Ohio, Air Force Machinability Data Center, Metcut Research Associates, Inc., 1970.

Lieberman, G. A., "Establishment of a Cutting Fluid Control System (Phase I)," Government Contract DAAA08-80-C-0033, January 1981.

Machining Data Handbook, Cincinnati, Ohio, Air Force Machinability Data Center, Metcut Research Associates, Inc., 1972.

Metals Handbook, Vol. 3, Machining, Metals Park, Ohio, American Society for Metals, 1967.

Swinehart, H. J., *Cutting Tool Material Selections*, Dearborn, Michigan, 1968, ASTME.

Tool and Manufacturing Engineers Handbook, Dearborn, Michigan, 4th Edition, SME, 1982.

Part Three

Milling

EDITOR'S NOTE

The work presented in this section reflects the results of two major research and development programs funded by the United States Air Force Systems Command with the General Electric Company in the role as systems manager. The authors are the project managers and/or principal investigators from their respective companies and are contributors in select technical areas of the studies. Readers who intend to apply the data presented herein should refer to the cited references, since it would be virtually impossible to include detail to the depth needed in the allocated pages for systems design of a production facility.

It should be further noted that the field of "high-speed machining" is relatively new and in an era of active development. Some of the results needed for application are preliminary, some are insufficient, and some are missing. However, data are now being collected at a remarkable rate with strong industrial and military support. There is enough information available to alter many production processes to incorporate the high-speed techniques for improved productivity. Each year will see a major expansion of available related data, and the editor anticipates that another 10 years will produce enough information to change standard shop operating procedures.

One of the problems of developing this section (or the entire text to a lesser extent) is the reluctance of the private sector to disclose their research results for incorporation in a text for general use within the technical community. Since high-speed machining data are not yet catalogued, industrial firms use the results of their respective studies to gain a competitive advantage among their competitors. One must assume, therefore, that there may be results that are far more extensive than those presented herein. It is also fair to assume that there are other investigators who do not agree with all of the results presented herein. In fact, there are areas within this text in which there are sets of data which have yet to be rationalized, one to another. However, the authors selected are authoritative in their respective areas and have presented the results as they appear to them in 1985.

R. I. KING

CHAPTER 6

General Theory and Its Application in the High-Speed Milling of Aluminum

T. Raj Aggarwal
Principal Engineer, Metcut Research Associates, Inc.
Formerly R&D Associate, Cincinnati Milacron Industries, Inc.

Introduction

For the metalworking industry to remain profitable, it must continually reduce its cost of manufacturing parts. One important segment of this economic procedure has been the costs involved in removing unwanted material from the workpiece. For example, the airplane part shown in Fig. 6.1 originally weighed 6000 lb (2725 kg) and is machined down to 435 lb (197 kg) after many hours on large, expensive machine tools.

Metalcutting is a convenient way of making one or many pieces of almost any shape from an available chunk of raw millstock. Of the many processes of metalcutting, such as turning, boring, milling, and drilling, we will deal with the milling process for aluminum alloys in this chapter.

Research in the high-speed milling of aluminum, as a way of reducing the cost of manufacturing parts, along with methods of selecting the proper speed and machine-slide feedrate, is extensively covered. A graphical method of

Fig. 6.1 Dramatic evidence of the enormity of the metal-removal task, and the importance of increased machining productivity, can be seen in relation to this huge aerospace part. (Courtesy of Cincinnati Milacron)

selecting the proper spindle speed and machine-slide feedrate is proposed. Examples are cited where the use of these graphs will help one understand the benefits of modifying an existing machine tool and of programming the machine tool with advanced speeds and feeds. Finally, the benefits of buying a new machine tool with today's state-of-the-art technology are described.

History of Milling Research and Development

The growth and evolution of industry have been characterized by the use of increasingly complex machine tools and appear certain to continue into the foreseeable future. Man's ability to shape metal according to his needs is undoubtedly one of the cornerstones of our civilization. The industrial revolution gave birth to machine tools which only a century ago were still relatively simple and crude (Fig. 6.2). They were, however, capable of performing operations that could not readily be done by hand, and almost

Fig. 6.2 By today's standards, the milling machines of a century ago were of relatively simple design. This 1884 machine helped launch what is today Cincinnati Milacron. (Courtesy of Cincinnati Milacron)

invariably they made it possible to manufacture products more accurately and economically.

The ever increasing demand for new, more-sophisticated machinery, coupled with the innovative use of machine tools, has resulted in rapid growth of the metalworking industry. Today, machine tools are more complex, computer numerically controlled, and machine a variety of very complex parts in all segments of industry. Figure 6.3 shows a modern five-axis, NC, high-efficiency, gantry-type profile milling machine having bed lengths up to and over 160 ft (50 m).

Basically, all metalcutting operations are the same, in that a cutting edge removes metal chips to produce the required surface. The basic purpose of a

Fig. 6.3 Today's most advanced profile milling machines are typically three-spindle, five-axis computer numerically controlled machines with automatic tool changers. Multiple traveling gantries carrying the spindle heads share a common machine bed. (Courtesy of Cincinnati Milacron)

machine tool is to drive a cutting tool to remove metal so as to produce a required shape. Hence, to permit the machine tool's highest development, it was essential that the process and the dynamics of metal removal be fully understood. Much of the basic knowledge in this field was generated during the 1940s and 1950s. Today, the research laboratory occupies an important place in the machine-tool industry. As a result, the modern machine tool operates at higher speeds without excessive failures and interruption of service. More recent research in chatter, dynamics of structures, servo systems, and computer technology has been very important, not only from the standpoint of achieving higher productivity, but also, more particularly, to make possible the closer limits and refinements of the accuracy and surface finish required on both reciprocating and rotating close-toleranced industrial equipment parts and their mating surfaces.

Basic research in metalcutting has made it possible to understand the mechanism of the metalcutting process, chip formation, the formation and function of the "built-up edge," the importance of tool geometry, tool life, and the impact of high cut speed on productivity and on tool life. This basic research not only led technological advances in establishing more productive and state-of-the-art application skills in machining, but it also led to major technological breakthroughs in cutting tool geometry and its material.

Early investigators working with the data developed in the 1920s and early 1930s by Dr. C. Salomon, a pioneer in metalcutting research,[1] viewed high-speed machining as a high-volume metal removal technology that flourished somewhere beyond the "valley of death" (Fig. 6.4). It was found that higher speeds could, in fact, produce higher metal removal rates, but only up to a certain extent.[2] Then rapid tool wear, excessive chatter, and catastrophic cutter failure appeared to block all efforts to make more chips with yet more speed, i.e., beyond the "valley of death." However, technological developments continued to increase productive cut speed. In 1959, while describing "10 Years Ahead . . . What's in It for Metalworking?" Dr. M. E. Merchant predicted the continued rapid advancement of tool materials and revealed that, since 1900, the permissible cutting speed had doubled about every 10 years (Fig. 6.5) and indicated that this would continue.[3] This trend from 1960 to the present has since been verified and is expected to continue even further into the future.

Work during the 1970s on high-speed machining at Lockheed[4] and elsewhere eventually led researchers, in general, and the industry into a recognition and acceptance of high-speed machining and the potential benefits it might provide. In 1975, in the same general time frame, Professor von Turkovich[5] proposed five specific metalcutting speed ranges as follows:

1. 0–100 sfm (0–30.5 smm): Low-speed conventional industrial machining.
2. 100–2000 sfm (30.5–610 smm): Conventional industrial machining.
3. 2000–6000 sfm (160–1830 smm): High-speed machining.

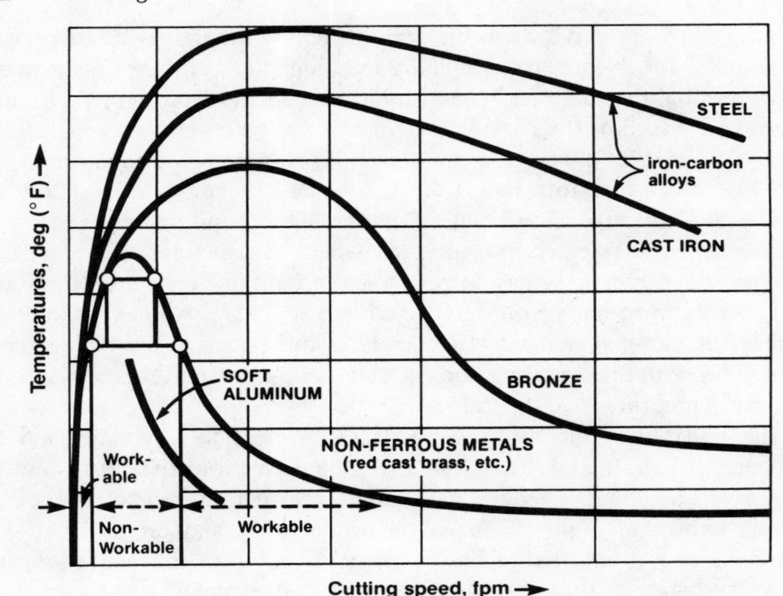

Fig. 6.4 Salomon theorized that cutting temperatures, and therefore tool wear, started to go down once a high, "supercritical" speed, which varied for different metals, was exceeded. (Courtesy of *American Machinist*[2])

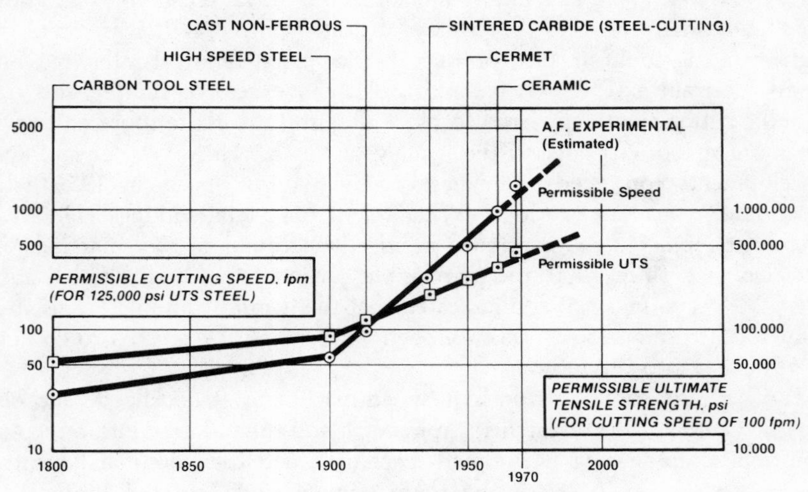

Fig. 6.5 Merchant revealed (1959) that with the rapid improvement in tool materials, the permissible cutting speeds had doubled about every 10 years. He predicted this trend would continue, opening an area of high potential for greater machining efficiency. (Courtesy of Cincinnati Milacron)

4. 6000–60,000 sfm (1830–18,300 smm): Very-high-speed machining.
5. 60,000–500,000 sfm (18,300–152,400 smm): Ultra-high-speed or ballistic machining.

The term "ballistic machining" meant that such ultra high speeds could, under the existing technology, only be generated by firing a bullet-shaped workpiece out of a gun or gunlike device. Early use of this laboratory technique was intended primarily to explore Salomon's theory and was generally considered to, in itself, not have any direct potential in a production environment.

Some of the most recent research in high-speed machining is being carried out under The Defense Advanced Research Project Agency's (DARPA's) Advanced Machining Research Program (AMRP), administered by the U.S. Air Force's Wright Aeronautical Laboratories. This effort, being coordinated through General Electric Company's Research and Development Center in Schenectady, New York, as prime contractor, has already generated important findings that seem to indicate that high-speed machining holds good promise for significant improvements in metalcutting productivity.

In early 1978, the author undertook intensified research work in high-speed end milling of aluminum at Cincinnati Milacron, independent of DARPA sponsorship. The company has had a long involvement with this growing technology and over the years has supplied many of the machine tools essential to its advance.

The material covered in this chapter is largely a result of over 2 years of intensive laboratory testing within the framework of an ongoing, broad-based R&D program focused on high-speed milling. It was decided that our laboratory testing would concentrate on aluminum because it was commanding the greatest attention and because of its extensive use in aerospace applications, as a chassis material for electronic gear, for framing in business and vending machines, and for many other applications, including automotive parts.

Current Trends in Milling of Aluminum

Economic Pressures

The machine-tool industry has always been challenged to invent and manufacture state-of-the-art advanced machine tools that will minimize the real cost of manufacturing quality parts. Here, we will divide the real cost of manufacturing into the following three categories:

1. In-process machining costs.
2. Costs involved with handling parts, such as moving, loading and unloading, inspection, storage, shipping, etc.

3. Fixed costs, or those over which manufacturing personnel may not have any control in a specific plant, such as cost of materials received, buildings, capital equipment, engineering, programming, and other facilities and services.

A major thrust of the basic research in the machine-tool industry has been geared to reducing the in-process costs by shortening the machining time and, secondly, to reducing the handling time. A surge of research in production technology since 1950 has directly addressed the task of shortening the handling time. This has led to the introduction of flexible manufacturing systems, completely controlled and managed by a centralized computer. Automation of the material handling actions on the production floor is now reaching such a level of maturity with pallet shuttles, in-process inspection, robots to continuously load and unload parts, computer controlled movable carts, etc., that the long envisioned unmanned factory has become a virtual reality.

With so much emphasis and technological advances made in reducing the part handling time, in-process machining time has again become a decisive factor in the total cost of manufacturing parts, particularly in the aerospace industry. For example, considerable reduction in the cost of manufacturing the airplane part of Fig. 6.1 could only be achieved by reducing the in-process metalcutting time. This driving need to improve the in-process machining cost was a powerful accelerating force in the metalworking industry for more vigorous research in high-speed machining.

With the increasing use of advanced high-strength aluminum alloys, interest in the high-speed milling of aluminum with small-diameter end mills greatly accelerated during the past decade and resulted in the introduction of numerous high-speed machines.

Results of a Survey

Dovetailed with our decision to undertake extensive laboratory investigations at Cincinnati Milacron, we decided to make a detailed review and study of current attitudes and practices in the end milling of aluminum—in both aerospace and nonaerospace plants. Some of the findings of this study are presented here.

One of the most significant disclosures was that, for convenience, most shops and NC programmers have adopted a conservative set spindle speed and slide feed for each type of cut (roughing, finishing, cornering, ramping), independent of cutter size, cross section of cut, and workpiece material. Operators commonly further reduce these programmed speeds and feeds by as much as 50–60%.

For example, a typical roughing cut in aluminum might be programmed at 1800 rpm and 15–20 ipm (38.1–50.8 cm/min) with a machine load meter reading at about 10–15%. In contrast, many of these existing production machines are easily capable of rough milling aluminum at 3600 rpm and

72 ipm (183 cm/min), with load meter reading at 40–50% of the machine's capacity. One of the newest standard machining centers Fig. 6.6 can easily be programmed at 6000 rpm (with standard optional spindle) and 120 ipm (305 cm/min) with load meter reading at 50–100% capacity. The newest gantry-type profile milling machine can be programmed at 7200 rpm and 150 ipm (381 cm/min) with load meter reading at 70–100% capacity. This machine has removed aluminum metal in excess of 300 in.^{3}/min (4915 cm^3/min) with a 2 in. (5.08 cm) dia., two-flute HSS end mill.

Some of the reasons given, by people surveyed, for using such low speeds and feeds were the tendency to program at the same speed and feed, excessive noise, problems in cleaning chips, probable maintenance problems, and the insecurity of going faster.

The survey showed that, especially in the aerospace industry where the workpiece is of relatively high cost even before machining, the fear of possible work spoilage is a very real deterrent to higher rates of metal removal at least among shop-level personnel.

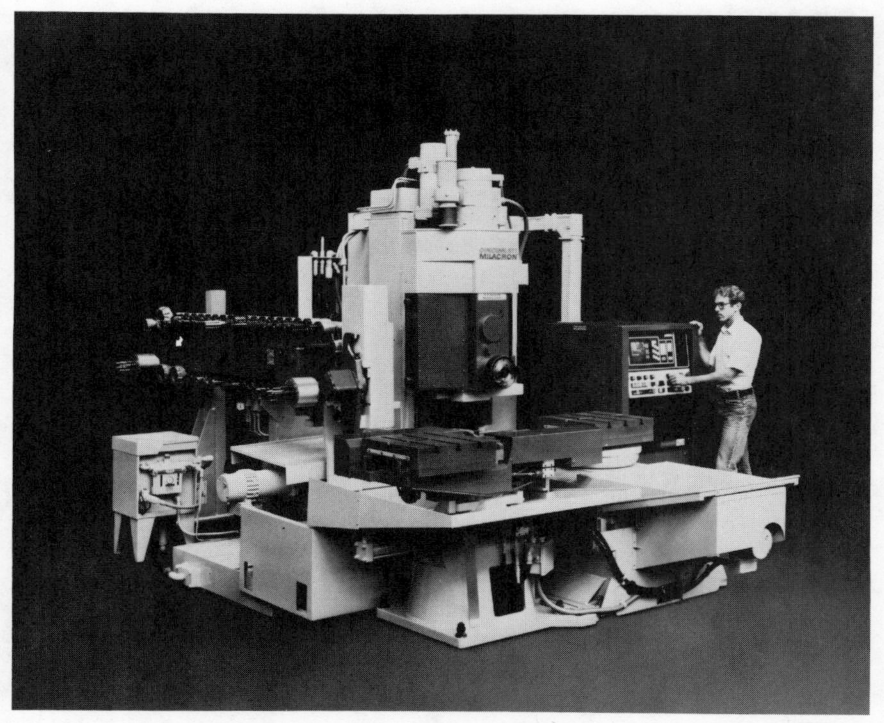

Fig. 6.6 Modern machine tools, like this T-10 machining center, are designed for much higher metal removal rates than were feasible earlier. Industry needs to upgrade its machining practices to take advantage of the technological capabilities of both existing and the newer shop equipment. (Courtesy of Cincinnati Milacron)

In the finishing of thin-web structures for aerospace applications, cutting speeds of 600–700 rpm and feedrates of 5–7 ipm (12.7–17.8 cm/min) were considered by interviewees to be most commonly employed. The primary objective of most finish machining in aerospace is to ensure that the part will pass inspection. And once a part is produced that has met inspection requirements, very little, if any, further program optimization is pursued.

Therefore, from our study, we have concluded that the quickest and easiest way to improve productivity in today's production environment is to further educate and motivate planners, programmers, and machine operators as to the capabilities of existing technology. The newest viable research data must replace the old obsolete data in machinery handbooks and in other published guidebooks which are used by the programmers. This is essential for production operation at 70–100% load capacity of the equipment, rather than the typical 10–15% capacity of today.

Current Reseach in High-Speed Milling of Aluminum

The total on-going research and development program at Cincinnati Milacron in the area of high-speed end milling of aluminum has included extensive laboratory-based research.*

Test Spindles and Methodology

Several spindles were acquired for this program, as outlined in Table 6.1 and shown in Figs. 6.7–6.10. Most of the tests, which involved thousands of cuts,

Table 6.1 Spindles Used in High-Speed Milling Research

Spindle Number	Spindle	Maximum rpm	Maximum hp (kW)
1	Red Head High-Speed Milling Spindle[a]	24,000	10 (7.5)
2	Red Head Internal Grinding Spindle[a]	20,000	20 (14.9)
3	Motorized Active Magnetic-Bearing Spindle	60,000	20 (14.9)
4	High-Performance, Aluminum Cutting Gantry-Type Profiler Spindle[a]	7,200	75 (56)
5	High-Performance, Aluminum Cutting Gantry-Type Profiler Spindle (Modified)[a]	9,000	75 (56)
6	CIM-Xchanger 15HC Machining Center Spindle (Modified)[a]	4,000	30 (22.4)

[a] Products of Cincinnati Milacron.

* Sections of the author's present text are adapted from his earlier SME paper.[6] Permission of the Society of Manufacturing Engineers is hereby acknowledged with appreciation.

Fig. 6.7 Recently designed Milacron Red Head milling spindle, with 24,000 rpm and 10 hp (7.5 kW) capability, is a self-contained, motorized, variable-frequency AC spindle. It has a separate floor-mounted inverter motor control package. The spindle is shown here mounted on a 15HC CIM-Xchanger machining center for laboratory test machining. (Courtesy of Cincinnati Milacron)

Fig. 6.8 Standard Milacron Red Head grinding spindle, shown mounted on a Milacron gantry-type honeycomb mill, is a 20,000 rpm, 20 hp (14.9 kW) unit that under production conditions removed 45 in.^{3}/min (737 cm^3/min) at feedrates of 180 ipm (457 cm/min) for contouring and 360 ipm (914 cm/min) for pocketing, and $\frac{1}{4}$ in. (0.64 cm) depth of cut. (Courtesy of Cincinnati Milacron)

were performed in the Machine Tool R&D Laboratory. An existing standard CIM-Xchanger 15HC Machining Center (Fig. 6.11) was modified to accept various spindles. Further modifications were made to machine and controls to add the capability of 400 ipm (1016 cm/min) programmable feedrates in contouring and to provide 30 hp (22.4 kW) to the standard spindle on this machine. Other, substantial tests were conducted on large profile milling machines in our shop run-off area.

Fig. 6.9 Milacron's laboratory testing at higher spindle speeds included test cuts made with this motorized, active magnetic bearing spindle rated at 60,000 rpm maximum and 20 hp (14.9 kW) maximum. Sensing system, which automatically lowered feedrate to protect limited load capacity of magnetic bearing, prevented use of full 20 hp (14.9 kW) on this particular spindle. (Courtesy of Cincinnati Milacron)

The methodology employed in this testing was reported to the SME.[6] Most of the testing was done on 7075-T6 aluminum alloy, which is widely used in the aerospace industry. Also, it was concluded that the data obtained from this alloy would be applicable, without much loss of accuracy, to other aluminum alloys.

All basic machining tests were run on standard test blocks sawed from plate stock of 7075-T6 aluminum alloy. The test cuts included straight slot-

Fig. 6.10 This Milacron high-performance, gantry-type profiler for machining aluminum aerospace parts carries two 75-hp (56-kW) spindles. Using 2 in. (5.08 cm) dia end mills, this machine is capable of, and has machined at, 7200 rpm and 150 ipm (381 cm), removing 300 in.^{3}/min (4915 cm^3/min) per spindle (Courtesy of Cincinnati Milacron)

ting, contouring with a minimum radius [diameter of cutter plus 0.100 in. (2.54 mm)], pocket milling, and the machining of thin ribs [wall thickness 0.035–0.180 in. (0.9–4.6 mm)]. The cutting tools are widely used standard off-the-shelf, two-flute, high-speed-steel end mills ranging in diameter from $\frac{3}{4}$ to 2 in. (19–51 mm).

Cutting tests were performed with feedrates ranging from 15 to 400 ipm (38 to 1015 cm/min), representing chip loads of 0.002 to 0.035 ipt/rev. Virtually all test machining was done wet, using a Cimcool water-based cutting fluid applied in a pressurized stream of air and coolant. On very-high (60,000 rpm) spindle speeds, the same fluid was applied as a pressurized mist.

Fig. 6.11 Much of the author's laboratory test cuts at elevated speeds and feeds were performed on this standard 15HC CIM-Xchanger machining center, modified when necessary to accept a different machining spindle. (Courtesy of Cincinnati Milacron)

Some of the most significant test results obtained to date are described here.

How Cut Speed Affects Unit Horsepower

Since a major concern of many production shops is that of heavy stock removal, unit horsepower (hp/in.^{3}/min) can be considered as the basic measure of machining efficiency. Thus, the relation of higher cutting speeds (sfm) to unit horsepower is of paramount importance. Various investigators have addressed this question.

We reviewed data extrapolated from McGee and associates,[7] as well as findings of Stelson,[8] which had indicated that unit horsepower is reduced gradually as the cut speed is increased. However, King and McDonald[4] had reported a more dramatic effect of cut speed on machining efficiency. Their data showed that increasing the cut speed by 500% resulted in the in.^{3}/min/hp factor being increased by 300%—a decrease of about 212% in unit horsepower. (It was noted though that their tests possibly involved very light chip loads.) Consequently, we decided to run some additional tests under carefully controlled conditions in our own laboratory to determine the influence of cut speed, as well as other variables, on unit horsepower. The test data we generated are partially represented by Figs. 6.12 and 6.13.

Our data revealed that an increase in cut speed does produce a decrease in unit horsepower, down to a certain value; Fig. 6.12. Throughout our testing, we measured horsepower actually consumed in the cut—subtracting tare horsepower, for any losses through bearings, etc. Holding chip load constant at 0.005 ipt/rev (0.127 mm per tooth/rev), for example, the best value obtainable was 0.2 hp/in.^{3}/min (0.001 kW/cm^3/min) at 5000 sfm (1525 smm). Above this speed we did not see any further reductions in unit horsepower.

Unfortunately, we were not able to obtain a comparable datum point at very high speeds, i.e., 15,000 sfm (4570 smm) using a 60,000 rpm spindle. We were constrained to a very light chip load because of the magnetic bearing's limited load capacity in this particular spindle available for our use in testing. The apparent limit was reached at approximately 11.0 hp (8.2 kW) in the cut. The best test we were able to take with this spindle, using a 1 in. (2.54 cm)

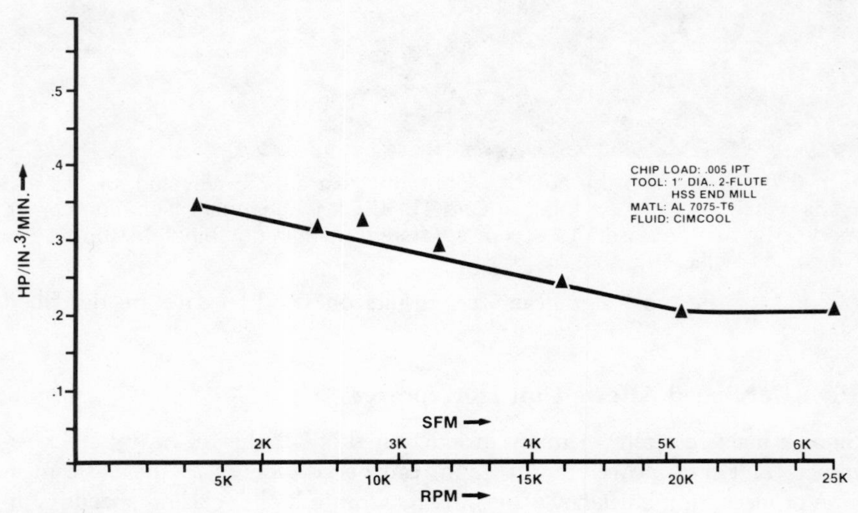

Fig. 6.12 Effect of speed on unit horsepower.

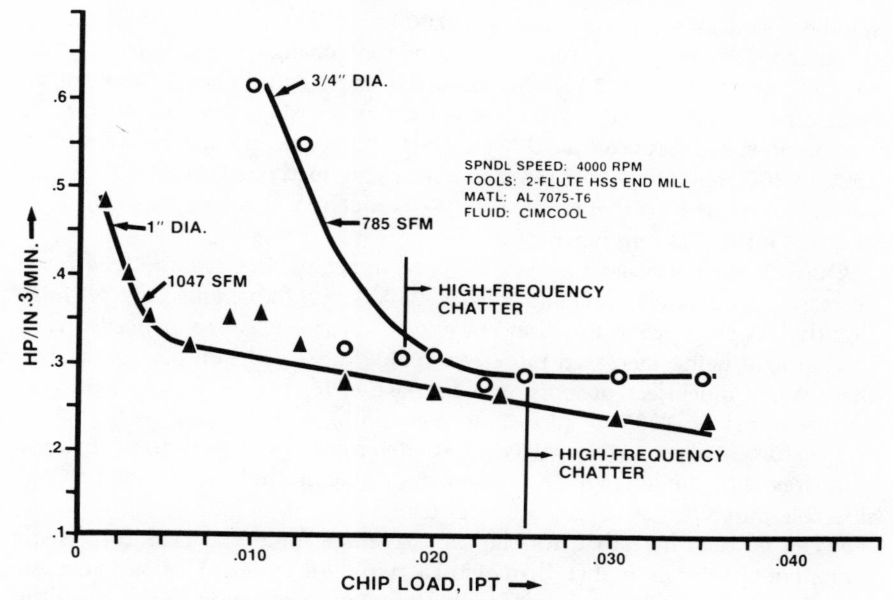

Fig. 6.13 Effect of speed and chip load on unit horsepower.

dia, two-flute solid carbide end mill, was at 1 in. (2.54 cm) radial depth, 0.150 in. (3.8 mm) axial depth, and feed of 200 ipm (508 cm/min), representing a chip load of 0.00167 ipt/rev (0.04242 mm per tooth/rev) and chip removal rate of 30 in.^{3}/min (492 cm^3/min). The resultant unit horsepower was 0.37 hp/in.^{3}/min (0.017 kW/cm^3/min)—considerably higher than what we had experienced with heavier chip loads and cut speed in the order of 5000–6000 sfm (1525–1829 smm).

The conclusion from our data was that increasing cut speed up to 5000 sfm (1525 smm) does improve the metalcutting efficiency. Furthermore, other data from our testing revealed that increasing the chip load and/or the axial depth of cut also improves the metalcutting efficiency. However, as shown in Fig. 6.13, we did reach a ceiling in chip load, beyond which we experienced high-frequency chatter corresponding to the cutter flute's natural frequency of vibration.

Actually there are two limiting factors involved in increasing the chip load on an end mill: first, the point reached beyond which the cutting process becomes unstable, as revealed by our data; and second, the fatigue life of an end mill, which will be covered later in this chapter.

Force versus Speed

We were also interested in what effect increased speeds have on feed forces. King and McDonald[4] had reported reduction in side load to 70% for a spindle speed increase from 400 to 20,000 rpm. They did not disclose the cut speed. McGee et al.[7] found no significant change in force from 3500 to 7000 sfm (1067 to 2134 smm). Schroeder and Hague[9] showed that forces level off at 4000 sfm (1219 smm) and then start rising slowly with further increase in speed. Truncale aand Winiecki[10] showed no change in forces from 3500 to 8000 sfm (1067 to 2438 smm), for a chip load of 0.004 ipt (0.102 mm per tooth). However, there was a 5–10% reduction in forces at a chip load of 0.008 ipt (0.203 mm per tooth).

Our test data showed that as cut speed increases, the average feed force decreases moderately, leveling off at 2500 sfm (762 smm); Fig. 6.14. A slightly less than proportional increase in feed force was experienced owing to chip load being increased more than 0.005 ipt (0.127 mm per tooth), but there was a much less proportional decrease in feed force as chip load was decreased below 0.005 ipt (0.127 mm per tooth).

The forces were further analyzed to determine how peak-to-peak force compares with the average force level. As was expected, we did find a considerable difference as revealed in Fig. 6.15.

It can be seen here that, for the same average force, the peak force with a one-flute cutter is higher than with a two-flute cutter. This becomes an important factor when we consider the fatigue life of the tool shank and/or that of the flute itself.

A further comparison was made between the resultant average force measured from the force transducer and the resultant tangential force computed

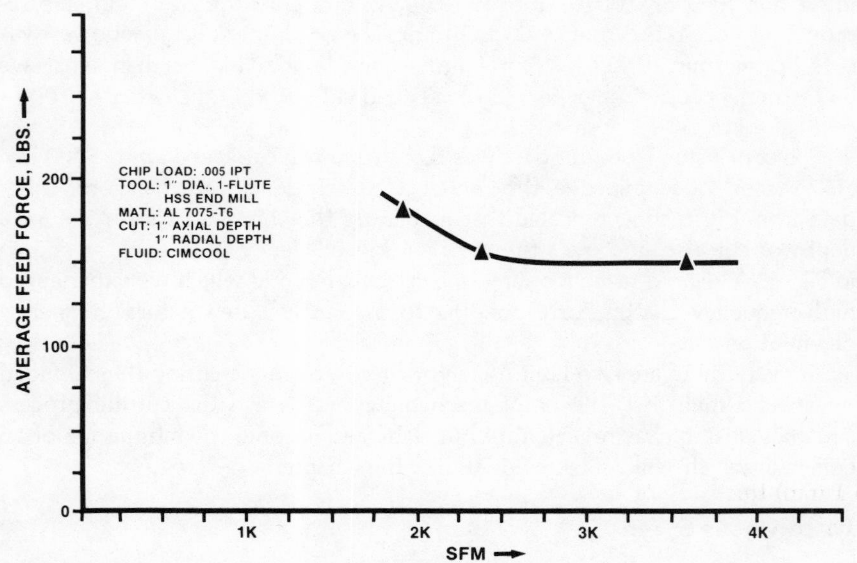

Fig. 6.14 Effect of speed on average feed force.

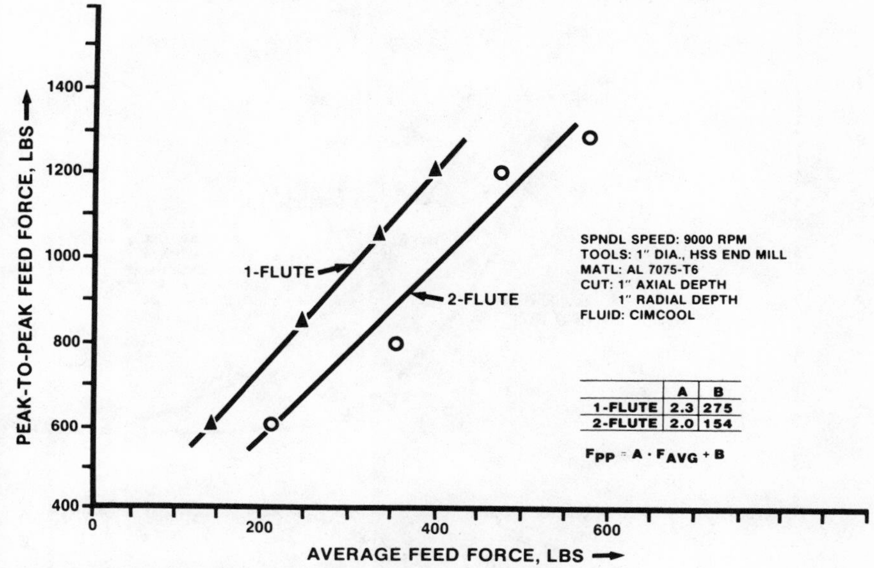

Fig. 6.15 Peak-to-peak force versus average force.

from the horsepower consumed in the cut. These forces were found to compare very well. The data developed are important in ensuring that our calculations would be valid if we use forces computed from the horsepower at the cut for fatigue life.

Surface Finish on Thin Ribs

As reported to SME,[6] a particular problem in the end milling of aluminum is that of holding good surface finish on the sides of thin ribs, which tend to deflect under cutter pressure.

In tests designed to determine whether higher cut speeds can provide a better finish, we produced a series of straight slots, using a $\frac{3}{4}$ in. (19 mm) dia, two-flute HSS end mill. We first milled a slot, in eight passes, to 1 in. (25.4 mm) depth and $\frac{3}{4}$ in. (19 mm) width, leaving a rib on the side in the order of 0.200 in. (5.1 mm) thickness. A finish pass was then taken at full 1 in. (25.4 mm) axial depth to side-mill the rib to a specific thickness, which varied from rib to rib from 0.035 to 0.180 in. (0.9 to 4.6 mm). And, by using different values of feedrate, cut speed, and radial depth of cut, we obtained data with which to evaluate the effects of rib thickness, cut speed, and chip load on surface finish. See Fig. 6.16 for thin ribs and Fig. 6.17 for ribs of 0.200 in. (5.1 mm) thickness or greater.

From these data, it was concluded that surface finish of 75 μin. (1.9 μm) rms or better can be obtained on ribs of 0.125 in. (3.2 mm) or greater thickness. Furthermore, increasing the cut speed while lowering the chip load will

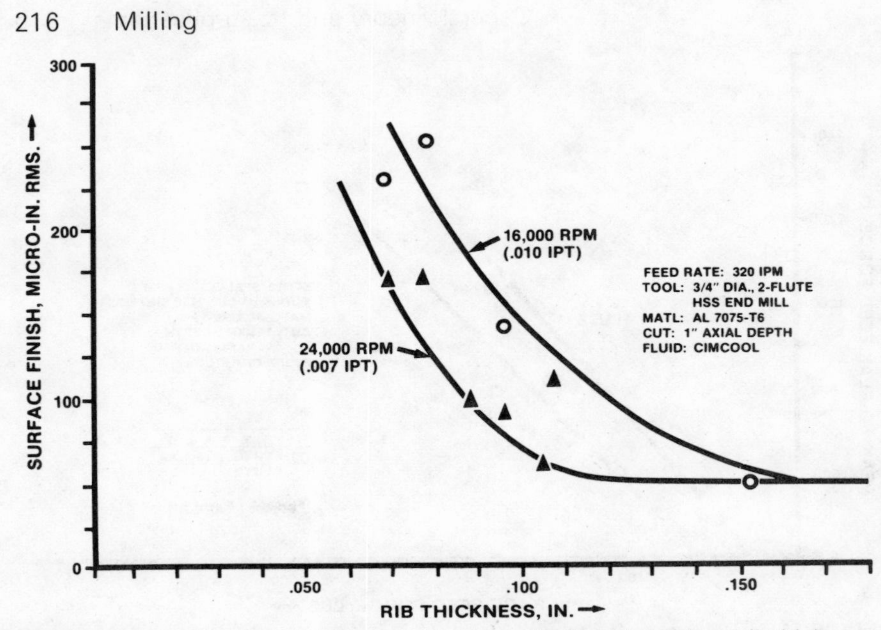

Fig. 6.16 Effect of speed on surface finish relative to chip load and rib thickness.

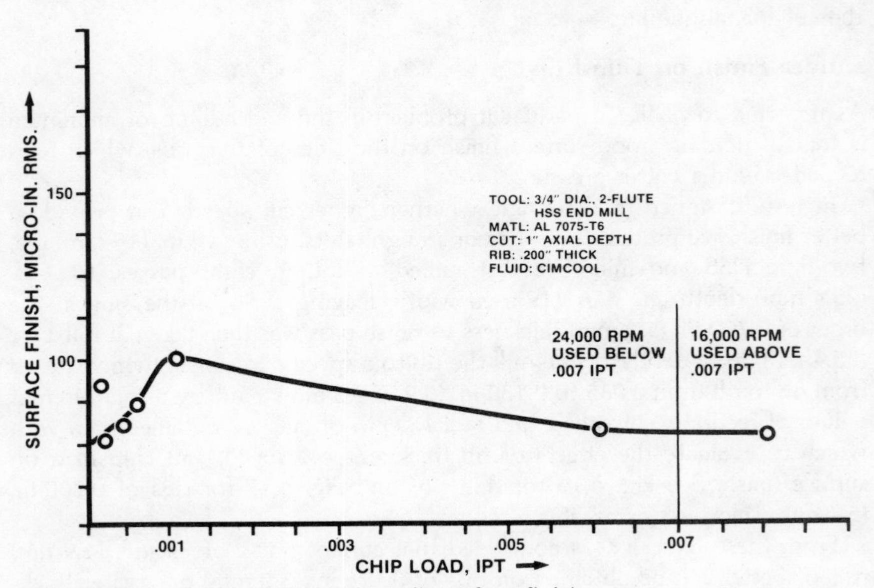

Fig. 6.17 Effect of chip load on rib surface finish.

improve the surface finish on ribs of less than 0.150 in. (3.8 mm) thickness; on ribs of greater thickness there is little benefit. However, see Fig. 6.17, chip load below 0.006 ipt may affect surface finish adversely on ribs of 0.200 in. (5.1 mm) or greater thickness.

Feedrates and Machining Errors

Many people have voiced a concern as to how the machined part will compare with the programmed part when contour milling at high feedrates. This concern has grown out of the fact that the actual tool position trails behind the programmed position and this following error is proportional to the programmed feedrate. Many programmers, and operators, assume that, at high feedrates, the following error in the servo system will cause the cutter to overshoot on corners and produce excessive machining error. This is a major reason why many plants are not using the full capability of present equipment. While following error was a real problem years ago with hydraulic drives, it should not be a problem with late-model machines and controls. Actually, without considering slide inertia, overshoot would be an inverse function of following error.

Two situations must be considered—machining centers and general milling machines and their controls, and profile milling machines and their controls. Controls built for profile milling offer an automatic acceleration/deceleration feature to facilitate cornering and minimize overshoot.

On the 15HC machining center used in our testing, our standard ACRAMATIC CNC control before modification allowed a step command in all axes to 150 ipm (381 cm/min). A series of cuts were taken at feedrates varying up to 150 ipm (381 cm/min). Machining errors were then measured on the workpiece, and some of the data are shown in Table 6.2.

The data indicate that with modern NC controls, if a cut is taken at 150 ipm (381 cm/min), a maximum error of 0.012 in. (0.305 mm) will be experienced while contour milling with a 1 in. (25.4 mm) dia end mill and

Table 6.2 Machined Errors[a] (All Errors Measured at Corners)

	Type of Cut					
ipm (cm/min)	 (square corner path)		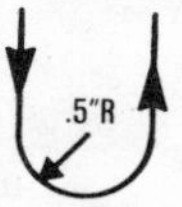 (.5"R path)		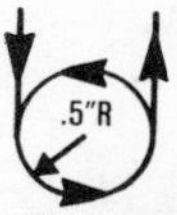 (.5"R path)	
20 (51)	0.000 in.	(0.000 mm)	0.000 in.	(0.000 mm)	0.000 in.	(0.000 mm)
80 (203)	0.002	(0.051)	0.002	(0.051)	0.002	(0.051)
120 (305)	0.004	(0.102)	0.002	(0.051)	0.003	(0.076)
150 (381)	0.012	(0.305)	0.008	(0.203)	0.009	(0.229)

[a] With machining center control. See text.

while making a sharp 90° turn. This order of machining error is, apparently, in general, within acceptable limits. Furthermore, an NC programmer can always use a reduced feedrate, if required by tighter specifications, while making a sharp turn on the programmed cutter path.

Other NC controls that use a different value of servo loop gain than does our machining center control and/or where acceleration and deceleration rates are automatically controlled may provide somewhat different machined errors. Some adjustments may be possible if moving into higher-than-conventional machining rates on a regular basis. And, as machines designed for still higher metal removal rates are introduced, incorporation of new advancements in servo controls and numerical controls permit further control of machined errors.

There are, however, two points that must be emphasized: first, that present-day equipment has plenty of generally unused spindle speed and feedrate capability, and secondly that, when high feedrates are to be used in conjunction with high-speed machining, the NC programmer should know what machine and NC control will be used and what their capacity and possible limitations are, so that he or she can program the cutter path accordingly to ensure that the part passes inspection.

Feedrates and Productivity

Both cut speed and, more importantly, feedrate, which may be expressed in terms of chip load, influence tool life. And a reasonable tool life, commonly considered to be in the order of 60 min, is a vital criterion in evaluating the economies of machining with high speeds.

McGee et al.[7] had published data on the effect of feedrate on tool life for various grades of aluminum alloys. While in the process of obtaining additional tool life data, McGee provided early conclusions on the most economical feedrates, as shown in Table 6.3.

As Kegg and Zdeblick have emphasized, productivity of the milling operation is primarily related to the machine slide feedrate.[11] Because of the spindle horsepower requirement in utilizing both higher spindle speeds and higher feedrates, a series of cutting tests were made using all of the spindles outlined in Table 6.1.

Table 6.3 Most Economical Feedrates[a] (Cobalt HSS End Mills)

Material	Rough Milling (ipt/rev)	Finish Milling (ipt/rev)
Al-A356-T6	0.013	0.005
Al-6061-T6	0.010	0.0025 –0.005
Al-7075-T6	0.008	0.0025 –0.005

[a] Data from McGee et al.[7]

Using each of the spindles, the test part was machined at the maximum rpm with the feedrates increased until some limitation was met. Most frequently, this limitation was insufficient horsepower; whereas, with the 7200 rpm, 75-hp (56-kW) spindle, the limitation was reached when the load on a 1 in. (25.4 mm) dia, one-flute end mill became excessive and broke the cutter. In each case, once the limit was reached, the total run time of the test part was measured by a stopwatch. This was then converted to a production rate in parts per hour.

Figure 6.18 shows the experimental productivity results for all spindles as a function of spindle rpm.[11] The data points represent actual cuts, but the dashed line represents a mathematical projection of what would happen at progressively lower speeds if the loading on the cutter was left at its maximum permissible value. This shows very clearly that if we are limited to available standard spindles of today, more rpm is not necessarily better, and it may actually be worse.

Another way of looking at the same data is to plot the production rate versus spindle horsepower.[11] This is done in Fig. 6.19, which shows a strong correlation between the productivity when milling aluminum parts and the manufacturer's horsepower rating of the spindle.

As a result of this test program, the conclusion is that higher speed spindles *alone* will not necessarily ensure higher productivity in rough milling aluminum parts; in fact, they may reduce it. On the other hand, higher rated spindle

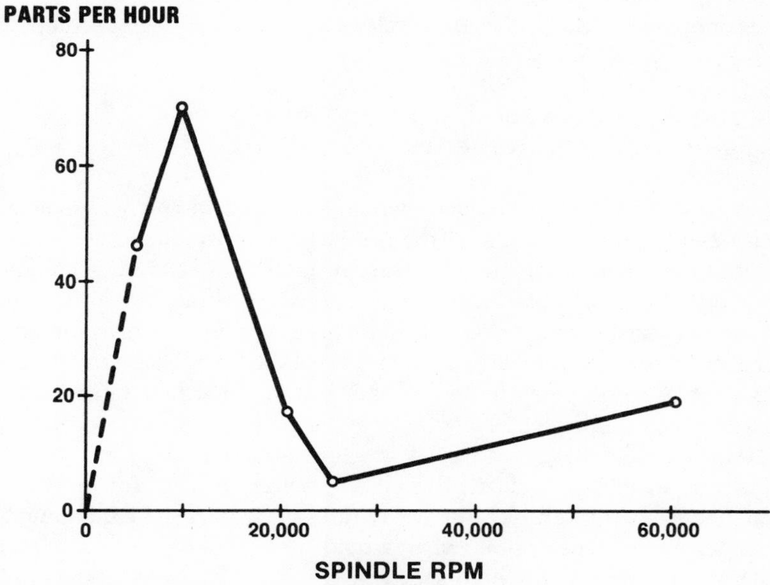

Fig. 6.18 Effect of rpm on productivity.

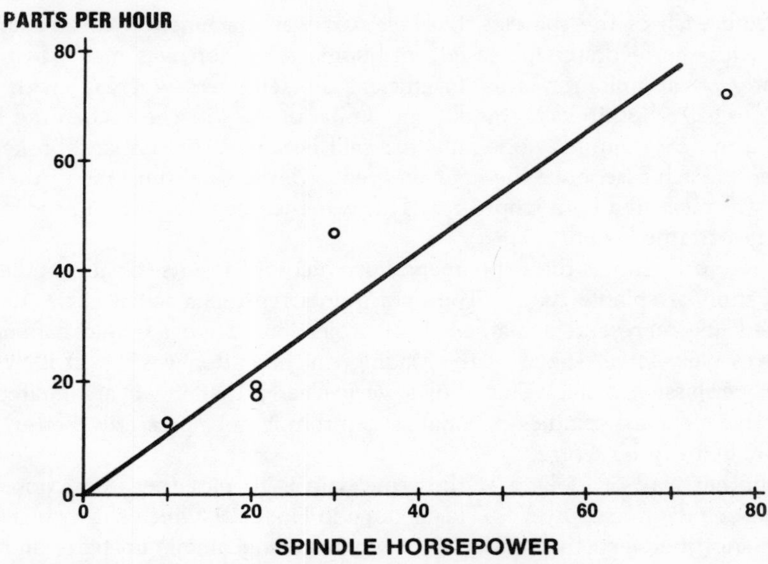

Fig. 6.19 Effect of spindle horsepower on productivity.

horsepower gives significant improvements in production rate. On production cuts where chip load is the limitation of the cut, higher-speed spindles with sufficient horsepower will tend to increase productivity. These conclusions will become more clear in the following paragraphs.

Defining Spindle Speed, Horsepower, and Feedrate Requirements

Associated with the proposed use of high-speed machining with small end mills [$\frac{3}{4}$–2 in. (1.9–5.1 cm) dia] in the production environment, there is need to determine certain basic requirements for machine-tool spindles and controls. The variety of potential production applications complicates the problem. For example, one particular selection of high-speed spindle and machine control may be good for rough milling of deep pockets, whereas the same selection may not be as good when finish machining a part that has thin ribs. In the past, a few very-high-speed spindles providing on the order of 20,000 rpm and 20 hp (15 kW) have been used in the production environment. However, as our analysis will show, this spindle selection may not be the most advantageous when high metal removal rates are a primary requirement.

In order to help determine what would be required of a machine-tool spindle and its control for high-speed machining, let us first look into the basics of metalcutting, i.e., how the metal removal rate is defined.

The metal-removal rate (MRR) is expressed mathematically as follows:

$$\text{MRR} = d_A d_R inN, \tag{6.1}$$

where MRR = metal-removal rate, in.^{3}/min
d_A = axial depth of cut, in.
d_R = radial depth of cut, in.
i = chip load, in./tooth/rev
n = number of teeth or flutes
N = spindle speed, rpm

For all aluminum alloys, the average horsepower at the cut per cubic inch of metal removal per minute (unit horsepower) can be conservatively assumed to be equal to 0.33. Therefore, the horsepower consumed in the cut can be written as follows:

$$\text{HP}_c = 0.33(\text{MRR}) \tag{6.2}$$

$$= 0.33 d_A d_R inN. \tag{6.2a}$$

Next, let us define the cross section of the cut as follows:

$$d_A d_R = K d_c^2 \tag{6.3}$$

and therefore

$$K = d_A d_R / d_c^2, \tag{6.3a}$$

where K = constant
d_c = cutter diameter, in.

By substituting Eq. (6.3) into Eq. (6.2a), we obtain

$$\text{HP}_c = 0.33 K d_c^2 inN. \tag{6.4}$$

For high-speed milling with small end mills, a two-flute end mill will generally be used. Therefore, Eq. (6.4) can be rewritten as

$$\text{HP}_c/N = 0.66 K d_c^2 i. \tag{6.5}$$

The horsepower at the cut per revolution (HP_c/N) and the maximum spindle speed are the most important factors in the selection of high-speed spindles.

Our next task was to define the constant K and the chip load i of Eq. (6.5) for various types of production milling cuts. It was also desired to set goals for metal-removal rate for a range of end mill diameters.

In order to establish realistic practical values of chip load, we considered the following limiting factors:

Tool life
Endurance limit
Surface finish
Rigidity

The chosen chip load should allow a reasonable tool life, normally considered to be 60 min or more. It must not fail in fatigue; forces on the cutter and on the workpiece should be within limits to provide desired surface finish; and, finally, the tool must be rigid enough not to create excessive vibrations and/or chatter marks on the workpiece.

The chip load i stated as inches of tool feed per flute per revolution for 60 min or more of tool life and for speeds to 5000 sfm (1525 smm) were established by McGee et al.[7]; see Table 6.3. Furthermore, to ensure proper endurance limit, an analysis described in Appendix A suggested the endurance limit chip load (i_e), derived from Eq. (6.9A), for various end mills as listed in Table 6.4.

In order to verify these endurance limit chip load values, we performed a series of cutting tests. Initially we took some test cuts on a profile mill in tryout, machining with a 1 in. (2.54 cm) dia, one-flute end mill, 1 in. (2.54 cm) shank dia, 1 in. (2.54 cm) axial depth of cut, 1 in. (2.54 cm) radial depth of cut, $K = 1$, and operating at 7200 rpm, 180 ipm (457.2 cm/min) of feedrate, and

Table 6.4 Endurance Limit Chip Load

Spindle Number	HSS End Mills	ipt/rev., Eq. (6.9A)			
		$K = 1$	$K = 0.5$	$K = 0.25$	$K = 0.125$
1	$\frac{3}{4}$ in. dia. $\frac{3}{4}$ in. shank dia, $l = 1\frac{5}{16}$ in. (two flute)				
	(a) $d_A = 1$	0.010	0.021	0.042	0.084
	(b) $d_A = 0.5$	—	0.016	0.032	0.064
2	1 in. dia, 1 in. shank dia, $l = 1.5$ in.				
	(a) $d_A = 1.5$	0.020	0.040	0.080	0.083
	(b) $d_A = 1.0$	0.015	0.030	0.060	0.125
	(c) $d_A = 0.5$	—	0.012	0.024	0.048
3	$1\frac{1}{2}$ in. dia, $1\frac{1}{4}$ in. shank dia, $l = 3$ in.				
	(a) $d_A = 1.5$	0.0073	0.014	0.029	0.058
	(b) $d_A = 1.0$	—	0.013	0.026	0.052
	(c) $d_A = 0.5$	—	—	0.024	0.048
4	2 in. dia, $1\frac{1}{4}$ in. shank dia, $l = 3$ in.				
	(a) $d_A = 2$	0.0047	0.0093	0.018	0.036
	(b) $d_A = 3$	0.0062	0.0124	0.025	0.050
5	2 in. dia, $1\frac{1}{4}$ in. shank dia, $l = 1\frac{5}{8}$ in.				
	(a) $d_A = 1.5$	—	0.012	0.024	0.048
	(b) $d_A = 1.0$	—	0.088	0.017	0.034

machining 7075-T6 aluminum with a chip load of 0.025 ipt/rev (0.635 mm per tooth/rev). We broke two end mills within three days. Consequently, we took to the laboratory, and, in addition, used our experience of machining with the smaller-diameter end mills in testing for rigidity. As shown in Fig. 6.13, high-frequency chatter, corresponding to the flute's natural frequency, had been experienced at chip loads greater than (1) 0.018 in. (0.457 mm) with a $\frac{3}{4}$ in. (1.9 cm) dia end mill, and (2) 0.024 in. (0.61 mm) with a 1 in. (2.54 cm) dia end mill. Therefore, a few of the higher endurance chip load values in Table 6.4 for smaller tool diameters would be unreachable, because chatter effects would be the limiting factor. We also drew on the experience and thinking of various people in industry, involved in applying advanced metalcutting technology, by an informal sampling survey. The end result was the analysis and values reflected in Table 6.5 to provide some practical guidelines for the end milling of aluminum.

In Table 6.5, for each type of cut, a reasonable value was established for the constant *K*, representing geometry of the cut; see Eq. (6.3a). The feedrate stated as chip load *i* was adapted from the above-stated limiting factors. Then, for each tool diameter covered and for each class of milling cut, we established a metal-removal rate (MRR) that equates with the consensus "goal" of

Table 6.5 Analysis of Various Types of Cuts

Type of Milling Cut	K	i Feed (ipt/rev)	d_c Tool dia (in.)	MRR Maximum (in.^{3}/min.)	HP_c Maximum	$\frac{HP_c}{N}$
Rough	1	0.010	$\frac{3}{4}$	80	27	0.0037
			1	100	34	0.0066
			$1\frac{1}{2}$	150	50	0.001485
			2	200	66	0.0264
Semirough	0.5	0.010	$\frac{3}{4}$	80	27	0.0018
			1	100	34	0.0033
			$1\frac{1}{2}$	150	50	0.0074
			2	200	66	0.0132
Semifinish	0.5	0.005	$\frac{3}{4}$	50	17	0.0009
			1	60	20	0.00165
			$1\frac{1}{2}$	100	34	0.0037
			2	120	40	0.0066
Finish	0.25	0.005	$\frac{3}{4}$	36	12	0.00047
			1	50	17	0.00083
			$1\frac{1}{2}$	60	20	0.00186
			2	75	25	0.0033
Finish with Low Forces	0.125	0.0025	$\frac{3}{4}$	10	3	0.00012
			1	20	7	0.00021
			$1\frac{1}{2}$	20	7	0.00047
			2	30	10	0.00083

various people in industry, universities, and research laboratories. These MRR values range from 3 to 20 times the reported general shop practice.

In each case, the corresponding maximum required horsepower in the cut (HP_c) was derived from the metal-removal rate, using Eq. (6.2). And, lastly, the horsepower per revolution (HP_c/N) was derived using Eq. (6.5).

As reported to SME,[6] data from Table 6.5 and Eq. (6.5) were then used to construct Figs. 6.20–6.24. Each of these graphs represents important spindle and control requirements for a specific class of milling cut provided for in Table 6.5 for the end milling of aluminum. And this approach was possibly the most important contribution made by our program.

On each graph, using log-log scale, an HP_c/N slope line was plotted for each of four cutter diameters representing horsepower at cut versus spindle speed and extending from 1000 rpm to the maximum spindle speed (N_{max}) as derived from Table 6.5:

$$N_{max} = HP_{c(max)}/(HP_c/N).$$

Above these four slopes, another slope of higher value was drawn to include an allowance for typical tare horsepower—representing spindle and motor inefficiencies. Using this higher HP/N slope, an envelope was then completed, covering all the possible cuts for all cutter sizes. Therefore, any spindle which

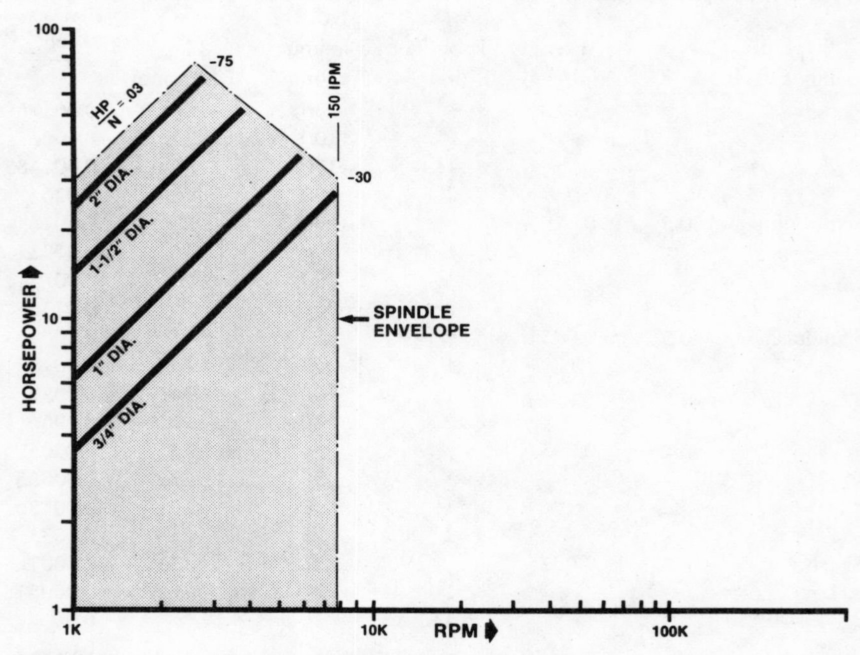

Fig. 6.20 Spindle requirement for rough milling.

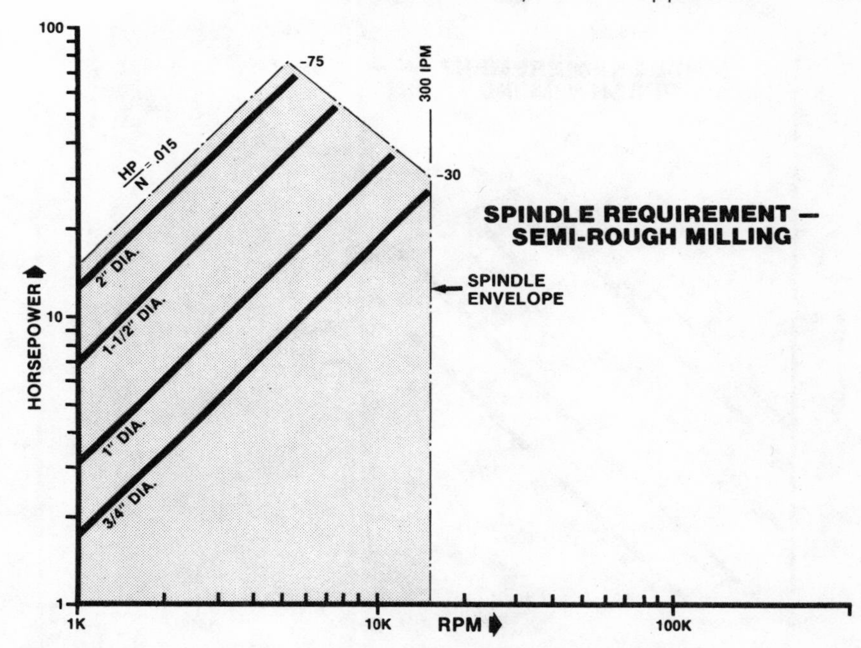

Fig. 6.21 Spindle requirement for semirough milling.

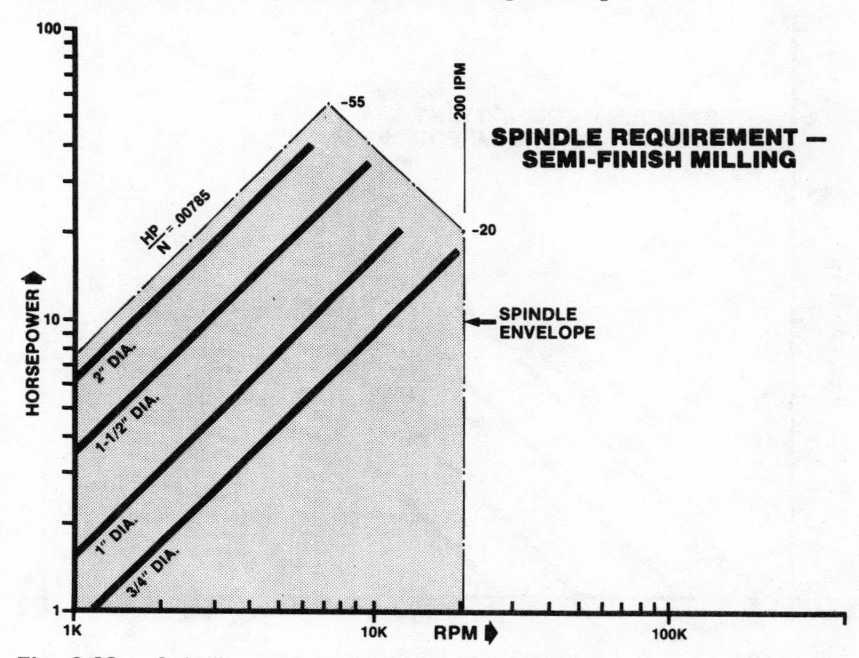

Fig. 6.22 Spindle requirement for semifinish milling.

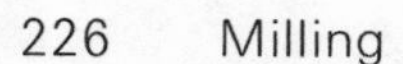

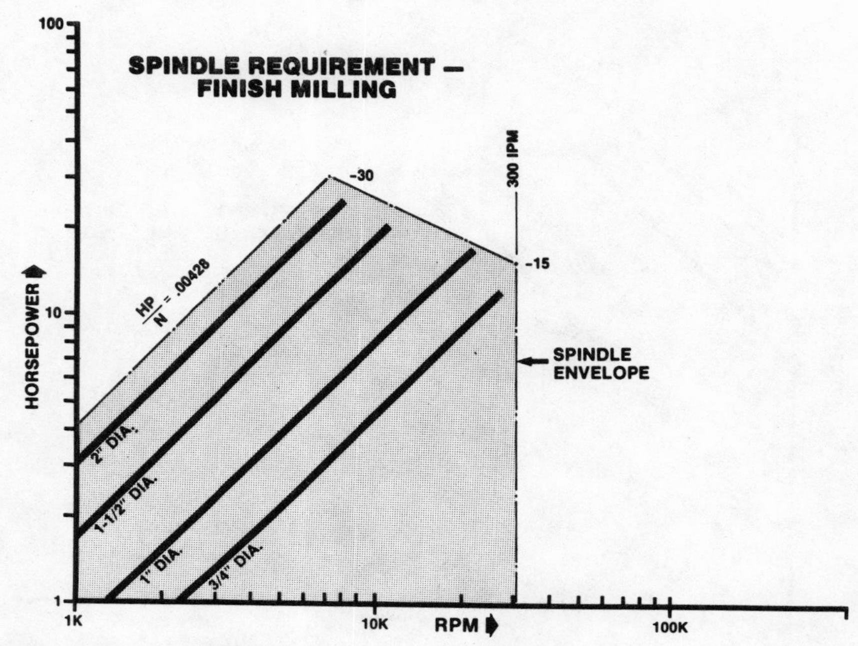

Fig. 6.23 Spindle requirement for finish milling.

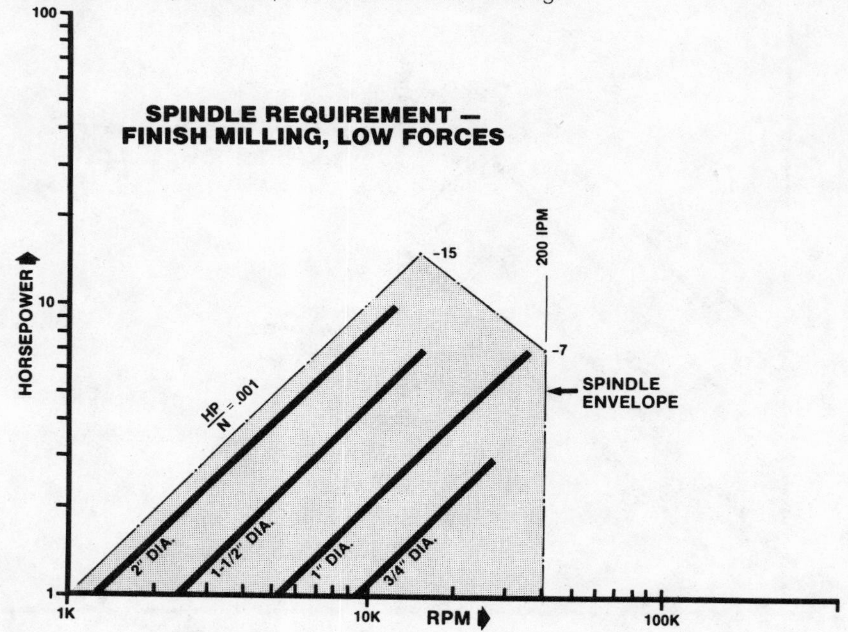

Fig. 6.24 Spindle requirement for finish milling (low forces).

will contain this envelope should be a suitable selection for that specific class of machining.

The maximum feedrate (ipm) that can be used with each type of cut was also specified in the respective graphs (Figs. 6.20–6.24). This was computed by using the following basic equation:

$$\text{ipm}_{\text{max}} = N_{\text{max}} ni. \tag{6.6}$$

We can see the application of these graphs from this example: When rough milling aluminum, see Fig. 6.20. Here we see that a spindle that offers 75 hp (56 kW) at 2500 rpm (HP/N = 0.03) and at least 30 hp (22.4 kW) at the maximum spindle speed of 7500 rpm would be a good choice, using $\frac{3}{4}$–2 in. (1.9–5.1 cm) dia, two-flute end mills. In this case, a maximum feedrate of 150 ipm (381 cm/min) should also be required machine control capability. A typical cut with a 2 in. (5.1 cm) dia, two-flute end mill would be at 2 in. (5.1 cm) axial depth, 2 in. (5.1 cm) radial depth, 2500 rpm, and 50 ipm (127 cm/min), with 0.010 ipt/rev (0.254 mm per tooth/rev) and removing 200 in.^{3}/min (3277 cm^3/min) of metal. Another typical cut with a 1 in. (2.54 cm) dia, two-flute end mill would be at 1 in. (2.54 cm) axial depth, 1 in. (2.54 cm) radial depth, 5500 rpm, and 110 ipm (279 cm/min), with 0.010 ipt/rev (0.254 mm per tooth/rev) and removing 100 in.^{3}/min (1639 cm^3/min) of metal.

Another example is the machining of a part that has thin sections. Here we are to apply high-speed milling to increase production while keeping forces low on the thin sections. As can be seen from the appropriate graph, Fig. 6.24, a high-speed spindle that will deliver 15 hp (11.2 kW) at 15,000 rpm (HP/N = 0.001) and at least 7 hp (5.2 kW) at a maximum spindle speed of 40,000 rpm would be a good choice. Here, a maximum feedrate of 200 ipm (508 cm/min) would be required. A typical cut with a 1 in. (2.54 cm) dia, two-flute end mill would be at 1 in. (2.54 cm) axial depth, 0.125 in. (3.175 mm) radial depth, 35,000 rpm, and 175 ipm (445 cm/min), with 0.0025 ipt/rev (0.0635 mm per tooth/rev) and removing about 20 in.^{3}/min (328 cm^3/min) of metal.

Figures 6.20–6.24 could be used directly to specify a spindle or spindles each dedicated to a specific class of machining. However, a machine tool is generally to be used more broadly. In such a case, a good analysis will be required covering the range of cuts to be assigned to the machine. Using the approach outlined here, a similar graph can be constructed to determine the basic spindle and control requirements. Or a compromise selection can be drawn from Figs. 6.20–6.24, possibly one spindle for rough milling and another for finish milling. The final decision will be based on economics. Figures 6.20–6.24 could also be used to determine the level of present production operation and then determine the attainable benefits from the application of advanced technology.

Examples of Improving Productivity

The graphical approach we have described, together with the equations and tables in this text, can be used to define practical solutions to a variety of manufacturing situations. The solutions of course rest on the application of modern technology in dramatically raising the level of productivity through higher rates of metal removal.

The examples cited here are, in a sense, hypothetical. However, they are based on and closely parallel cases in which we have been involved. The machined part in Example 1 was synthesized from an actual family of parts, and the alternative solutions have been applied experimentally for demonstration.

Example No. 1

Through an analysis of production of many parts in a jobshop, it was determined that those parts similar to the hypothetical production part of Fig. 6.25, and the machine to which they are assigned, constitute the bottleneck for this shop. In order to make significant improvements in the shop's profitability, the production of these parts must be improved severalfold.

Fig. 6.25 Hypothetical machined part, representative of many of the part configurations studied.

These parts are being machined on a non-NC 1960 model, 15-hp (11-kW) vertical milling machine, equipped with hydraulic servo drives. The machine's rated maximum spindle speed is 2240 rpm, and the maximum table feedrate is 30 ipm (76 cm/min). The parts are being machined with a $\frac{3}{4}$ in. (1.9 cm) dia, two-flute HSS end mill at 1000 rpm and 10 ipm (25.4 cm/min) with a maximum of 1 hp (0.75 kW) at cut. How can the high technology improve profitability and productivity of this jobshop?

Solution Through an analysis of the various cuts, we determined that the constant K of the cut cross section [see Eq. (6.3a)] is equal to 1.0 for 15% of the time, and it is equal to 0.5 for 85% of the time. [All rib thicknesses are over 0.150 in. (3.81 mm).] Therefore, the graph shown in Fig. 6.21 was selected to aid our analysis. Three approaches to improvement were studied.

A. *Improve Productivity by Advanced Feedrates:* The surface finish tolerance on parts with side ribs of this thickness (Fig. 6.16) suggested that semirough milling with a chip load of 0.010 ipt/rev (0.254 mm per tooth/rev) would be adequate; see Table 6.5. By applying Eq. (6.6), we determined the maximum feedrate to be as follows:

$$\begin{aligned} \text{ipm}_{\text{max}} &= 1000 \times 2 \times 0.010 \\ &= 20. \end{aligned}$$

After inspecting the machine, it was determined that the above feedrate can be obtained without any problem. Therefore, if the machining feedrates were increased from 10 to 20 ipm (25.4 to 50.8 cm/min), the actual machining time would be reduced to one-half. This, however, was not enough to make a significant improvement in the total productivity for the job shop.

B. *Increase Productivity by Utilizing the Machine Capacity:* Figure 6.26 shows our use of the graph in Fig. 6.21. Here, we have plotted point A to indicate the level of production activity obtainable by increasing the existing machine's feedrate, as described in Solution A. We determine here that the maximum horsepower at cut is on the order of 1.75 (1.31 kW). We also note from this graph that the productivity would definitely be improved by both higher spindle speeds and table feeds. We also know that the rated maximum table feed capacity of the machine is only 30 ipm (76 cm/min). We then used Eq. (6.6) to determine the maximum spindle speed:

$$\text{ipm}_{\text{max}} = 30 = N_{\text{max}} \times 2 \times 0.010.$$

Therefore

$$N_{\text{max}} = 1500 \text{ rpm}.$$

Therefore, we plotted point B in Fig. 6.26 to indicate the maximum productivity level achievable from this machine, horsepower at cut being on the

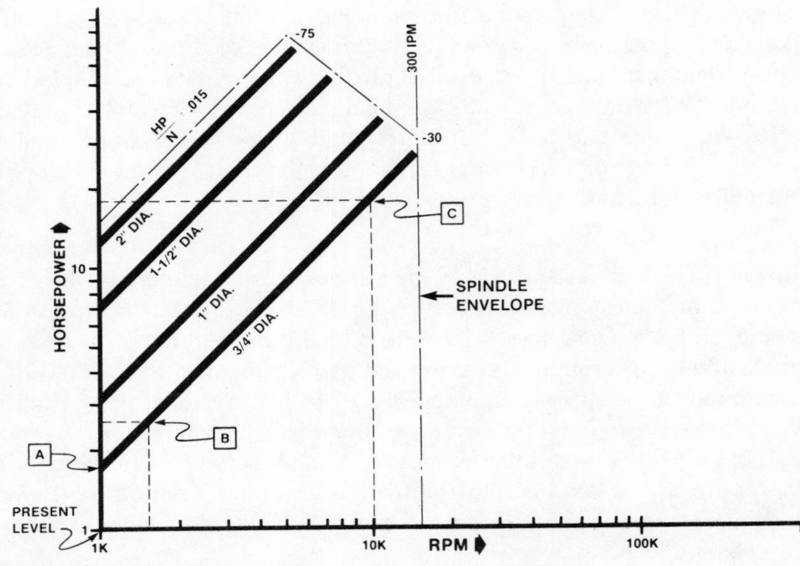

Fig. 6.26 Spindle requirement for semirough milling.

order of 2.5 (1.9 kW). This represents an approximately threefold improvement toward reducing the machining time from the existing level of machining practice.

C. *Improve Productivity by Investing in a New Machine Tool:* We further took note from Fig. 6.26 that a machine tool that will provide 15,000 rpm spindle speed, 30 hp (22 kW), and 300 ipm (762 cm/min) feedrate would provide the maximum attainable productivity with a $\frac{3}{4}$ in. (1.9 cm) dia end mill.

It was further anticipated by the job-shop owner that if a substantial improvement in productivity was made with a $\frac{3}{4}$ in. (1.9 cm) dia end mill, he could get new contracts that would involve machining with $\frac{3}{4}$–2 in. (1.9–5.08 cm) diameter end mills. Therefore, the economic analysis revealed that adequate capital investment in a new state-of-the-art machine tool capable of machining existing and new parts was possible for a good rate of return. Consequently, acquisition of a new NC vertical machining center with 10,000 rpm maximum, 300 ipm (762 cm/min) maximum table feedrate, and 40 hp (30 kW), and with automatic tool changing and chip-handling capability was proposed. This machine would easily accommodate standard HSS end mills to 2 in. (5.08 cm) diameter, in 40-hp (30-kW) cuts, without loss of machine rigidity.

For the parts to be machined with $\frac{3}{4}$ in. (1.9 cm) dia end mills, a new point C is defined in Fig. 6.26, indicating the projected new level of production activity at 10,000 rpm and about 18 hp (13 kW) at cut. This production level represents an improvement of over 18-fold in reducing the machining time from the existing level, which would improve the profitability sufficiently to successfully acquire new business. This projected improvement level does not include the time savings from the automatic tool change operation of the new machine tool.

Example No. 2

It takes about 35 h of continuous semirough milling to machine a typical aircraft spar part, such as that shown in Fig. 6.27, from a 4.25 in. (10.8 cm) thick solid flat plate of 7075-T6 aluminum, 75–150 BHN. At present, all of the machining is performed at 1800 rpm and 20 ipm (50.8 cm/min). It was further noted that the machine assigned to this work is capable of running at 2400 rpm and 30 ipm (76.2 cm/min) maximum with the existing hydraulic servodrive. The maximum rated horsepower of the machine is 20 (15 kW).

After a careful evaluation of all milling cuts, it was discovered that there are three types of cuts being made with the respective percentage of cut time as follows:

1. Pocketing with a 2 in. (5.08 cm) dia, two-flute, HSS end mill, comprising 70% of the total machining time.
2. Surfacing with a 4 in. (10.2 cm) dia, four-tooth, HSS shell end mill, comprising 18% of the total machining time.
3. Profiling the perimeter of the workpiece and taking other cuts with a 2 in. (5.08 cm) dia, two-flute, HSS end mill, comprising the remaining 12% of the total machining time.

The total rough machining is done in four passes with a maximum of 0.6 in. (15.2 mm) depth of cut per pass. This limiting depth of cut value was established to avoid chatter. How can the high-speed technology improve the productivity by reducing the total machining time from the present 35 h to about 15 h or less?

Solution It was realized that the lead role in the situation is filled by the pocket milling, which demands 70% of all the time spent machining the workpiece. Therefore, in order to appreciably reduce the total machining time, we must first reduce the pocket milling time.

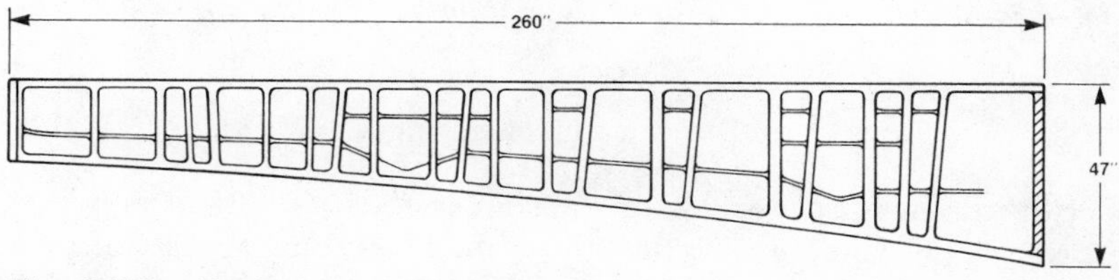

Fig. 6.27 Typical aircraft spar part.

A. *Analysis of Present Pocket Milling:* Presently, the maximum metal-removal rate (MRR) can be computed as follows:

$$\mathrm{MRR}_1 = d_A d_R(\mathrm{ipm})$$

where d_A = axial depth of cut = 0.6 in. (15.2 mm)
d_R = maximum radial depth of cut = 2 in. (5.08 cm)
ipm = feedrate = 20 ipm (50.8 cm/min).

Therefore,

$$\begin{aligned}\mathrm{MRR}_1 &= 0.6 \times 2 \times 20 \\ &= 24 \text{ in.}^3/\text{min } (393 \text{ cm}^3/\text{min}).\end{aligned}$$

Furthermore, from Eq. (6.2)

$$\begin{aligned}\mathrm{HP}_{c-1} &= 0.33 \times 24 \\ &= 8.\end{aligned}$$

The maximum chip load i can also be computed by rearranging Eq. (6.6) as:

$$\begin{aligned}i_1 &= \mathrm{ipm}/(N_{\max} \times n) \\ &= 20/(1800 \times 2) \\ &= 0.0056 \text{ in./tooth/rev } (0.1422 \text{ mm per tooth/rev}).\end{aligned}$$

The constant K for the limiting cut cross-section values as described is computed by using Eq. (6.3a) as follows:

$$\begin{aligned}K &= (d_A d_R)/d_c^2 \\ &= (0.6 \times 2)/2^2 \\ &= 0.3.\end{aligned}$$

With the above-computed values, the present production level for pocket milling is represented by A in Fig. 6.28.

B. *Improving Productivity by Utilizing the Machine's Capacity:* Our research data suggested that in rough milling of aluminum, a chip load of 0.010 in. (0.254 mm) can be taken economically (see Table 6.5). In order to increase the chip load to 0.010 in. (0.254 mm) at 1800 rpm, the maximum feedrate required was computed from Eq. (6.6) as follows:

$$\begin{aligned}\mathrm{ipm}_2 &= 1800 \times 2 \times 0.01 \\ &= 36 \text{ } (91.4 \text{ cm/min}).\end{aligned}$$

But the machine was capable of a maximum slide feedrate of only 30 ipm (76.2 cm/min). Therefore, in order to use the maximum capacity of the machine, we could increase the slide feedrate from 20 to 30 ipm (50.8 to

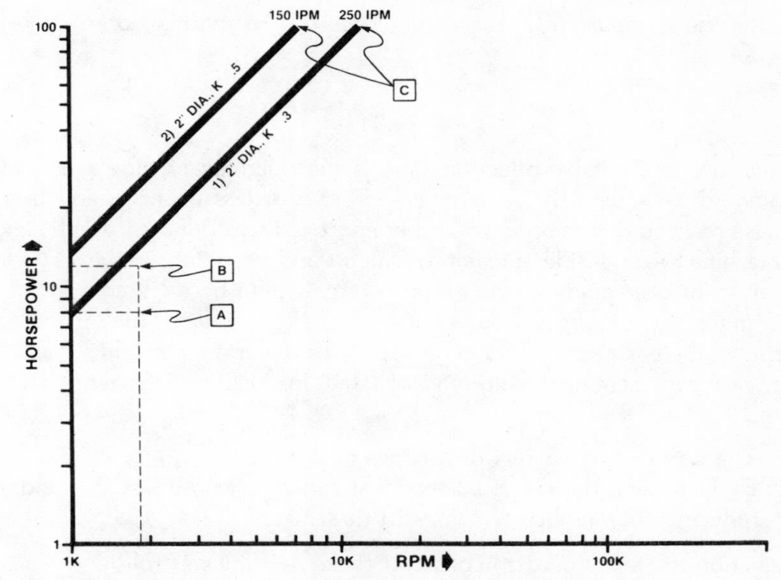

Fig. 6.28 Pocket milling production level for Example No. 2.

76.2 cm/min). The modified horsepower at cut was then computed as follows:

$$\begin{aligned} MRR_2 &= 0.6 \times 2 \times 30 \\ &= 36 \text{ in.}^3/\text{min } (590 \text{ cm}^3/\text{min}). \\ HP_{c-2} &= 0.33 \times 36 \\ &= 12 \text{ (9 kW)}. \end{aligned}$$

This level of modified production activity of milling pockets is represented by B in Fig. 6.28.

Therefore, by modifying the slide feedrate for pocket milling alone from 20 to 30 ipm (50.8 to 76.2 cm/min), an approximate reduction in the total machining time was computed as follows:

$$\begin{aligned} t_s &= \text{Time saving} \\ &= \text{Total machining time} \times \text{percentage time in pocket milling} \\ &\quad \times \text{projected percentage reduction in pocket milling time.} \end{aligned}$$

Based on the increased feedrate, the projected percentage reduction in pocket milling time will be equal to 1 − (20/30) = 33%. Therefore,

$$\begin{aligned} t_s &= 35 \times 0.7 \times 0.33 \\ &= 8.16 \text{ h.} \end{aligned}$$

Now, the modified total machining time (t_m) to machine one workpiece will be

$$t_m = 35 - 8.16$$
$$= 26.84 \text{ h}.$$

This modified total production time is very significant; however, it was not enough to satisfy the requirement of 15 h or less of machining time. It was also realized that in order to further improve the productivity of the existing machine tool, considerable improvements and alterations in the servodrive system, controls, spindle carrier, motors, etc., would be required.

C. *Improving Productivity by Investing in a New Machine Tool:* In order to determine the basic requirements for a state-of-the-art new machine tool, two lines, each representing a chip load of 0.010 in. (0.254 mm), were drawn in Fig. 6.28:

1. The same cut cross-sectional constant, $K = 0.3$.
2. By increasing the cut cross-sectional constant value, $K = 0.5$, and thus reducing the number of passes to three.

Equation (6.5) was used to construct these two lines as follows:

1. $HP_c/N = 0.66Kd_c^2i$
$= 0.66 \times 0.3 \times 2^2 \times 0.010$
$= 0.00792.$
2. $HP_c/N = 0.66 \times 0.5 \times 2^2 \times 0.010$
$= 0.0132.$

We also limited our maximum metal-removal rate to 300 in.^{3}/min (4916 cm^3/min), which from Eq. (6.2) represented a maximum horsepower at cut as

$$HP_c = 0.33 \times 300$$
$$= 100 \text{ (75 kW)}.$$

It was recognized that even with the most efficient and modern chip-removal system, disposing of chips at rates higher than 300 in.^{3}/min (4916 cm^3/min) was not possible. At this rate the chips, from one machine only, would fill 120–150 25-gal (95 liter) drums per hour.

The two lines, 1 and 2, of Fig. 6.28 represent two different solutions as follows:

1. 12,500 maximum rpm, 100-hp (74.6-kW) spindle with a maximum slide feedrate of 250 ipm (635 cm/min).

The new machine must be rigid enough to take at least the same cross section of the cut, i.e., 0.6 axial depth, 2 in. (5.08 cm) radial depth, at 12,500 rpm and 250 ipm (635 cm/min) without chatter and without excessive vibrations and/or noise problems.

2. 7500 maximum rpm, 100-hp (74.6-kW) spindle with a maximum slide feedrate of 150 ipm (381 cm/min.)

Here, we modified the cut cross section to increase the axial depth of cut to 1 in. (2.54 cm), $K = 0.5$, and expected to complete the total machining in, at the most, three passes.

This new machine must also be rigid enough to take at least the cross section of the cut, i.e., 1 in. (2.54 cm) axial depth, 2 in. (5.08 cm) radial depth, at 7500 rpm and 150 ipm (381 cm/min) without chatter and without excessive vibrations and/or noise problems.

With both solutions 1 and 2, the machine-tool spindle must also be capable of adapting and machining with a 4 in. (10.2 cm) dia shell end mill as was required for other types of cuts on the workpiece. (Some spindles that have been used or proposed for high-speed machining will not accommodate a wide range of cutting tools or their standard adaptors.)

With either of the above two alternatives, the projected improvements in the total productivity will be about the same, although selection 1 may be of higher initial capital cost. An automatic tool change operation would further improve the productivity.

An initial estimate of the new production machining time for a workpiece was computed as follows:

(a) New pocket milling time (70% of total milling time) will be

$$= 35 \times 0.7 \times (8/100)$$
$$= 1.96 \text{ h.}$$

(b) With similar productivity improvement in surface cuts with 4 in. (10.2 cm) dia shell end mill (18% of total milling time), the new time will be

$$= 35 \times 0.18 \times (8/100)$$
$$= 0.5 \text{ h.}$$

(c) It was anticipated (assumed) that no improvements in productivity will be attained while performing the profiling cuts and other cuts on the perimeter of the workpiece.

$$\text{Time for profiling cuts (12\%)} = 35 \times 0.12$$
$$= 4.2 \text{ h.}$$

Therefore,

$$\text{Total productive machine time per workpiece} = 1.96 + 0.5 + 4.2$$
$$= 6.66 \text{ h.}$$

This is well within the goal: production time of 15 h or less per workpiece.

Summary Conclusions

1. Many machines now in use are underutilized in terms of machining rates and load capacity when end milling aluminum. More adequate education is needed for NC programmers and equipment users in order to increase the utilization of existing machine tools to 70–100% load capacity. This alone can increase productivity with small end mills by many fold.
2. The metalcutting efficiency, i.e., hp/in.^{3}/min, in aluminum improves with the surface cut speed up to 5000 sfm (1525 smm). Beyond that speed, no further improvements were observed.
3. Cutting forces decrease with an increase in surface cut speed and appear to level off at around 2500 sfm (762 smm). No further reduction in cutting forces were experienced with increase in cut speed.
4. Higher surface cut speed with lower feed per tooth improves the surface finish on ribs of 0.150 (3.81 mm) thickness or less.
5. It is feasible to machine at feedrates up to 300 ipm (762 cm/min). The feedrates for sharp cornering may be limited to 150 ipm (381 cm/min) or less owing to the following-error constraint inherent in servosystems. Considering today's practice of using feedrates on the order of 20–30 ipm (50.8–76.2 cm/min), this is a marked improvement.
6. Higher spindle speeds *alone* will not necessarily ensure higher productivity in rough milling. Higher spindle speeds with higher rated horsepower give significant improvements in the production rate.
7. For production cuts taken with very small diameter cutters, where chip load is the limiting factor, increased spindle speeds will tend to improve productivity.
8. A graphical procedure has been provided, which is one practical approach to determining high-speed spindle and maximum-feedrate requirements for a machine tool.
9. Although the published data indicate that the most economical tool life is possible with feedrates as given in Table 6.3, a detailed tool-life curve with respect to surface cut speed is desired.
10. As industry moves up to machining at higher speeds and feeds, machine-tool technology will be further challenged in providing economical answers to the associated problems such as safety, noise, and chip removal.

The potential benefits of industry's more aggressive use of high-speed machining and high rates of metal removal justify continued research and development aimed at this goal.

A by-product of the more-sophisticated research of the past few years should be that of instilling greater confidence in manufacturing management and shop personnel that, even with today's technology, modern machine tools can be used more productively and with virtually the same assurance as the conventional low speeds and feeds of earlier years.

Appendix: Endurance Limit of End Mills

Widespread concern has been expressed regarding the effects that elevated spindle speeds might have on the fatigue life of end mills, since their fracture while in a cut could present some real problems. The answer can be approached through measuring the effect of increased speeds on feed forces; see Fig. 6.15 and related discussion.

We can establish endurance limit values for end mills—in terms of both horsepower and chip load—and the lower of these two values should be the ruling constraint. The derivation of these values is given here.

Referring to Fig. 6.29, let us define the following for an end mill:

l = flute length
d_{cs} = tool shank diameter
d_c = end mill diameter
d_A = axial depth of cut
P = resultant radial force on end mill

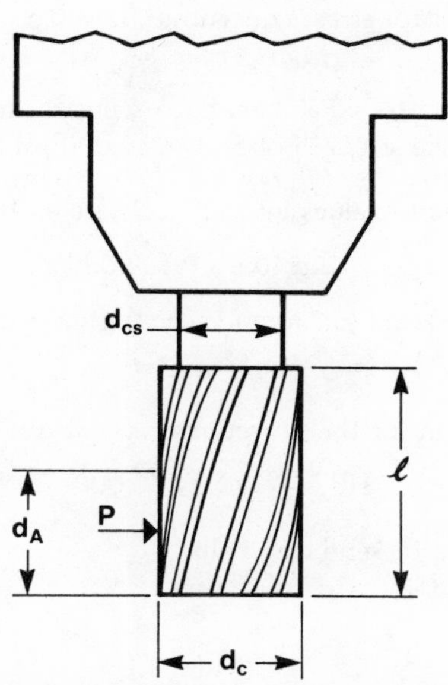

Fig. 6.29 Schematic of a typical old mill.

Furthermore, assuming that the end mill will break from fatigue owing to the maximum bending stress at the weakest section just above the end mill of the flutes, the expression for the bending moment (M) can be written as follows:

$$M = P(l - d_A/2). \tag{6.1A}$$

Also, the bending moment of a round structural member can be expressed as follows:

$$M = S_e I(2/d_{cs}), \tag{6.2A}$$

where S_e = corrected endurance limit
I = cross-sectional moment of inertia

The corrected endurance limit S_e, stated as psi, can be established from the following equation:[12]

$$S_e = K_a K_b K_c K_d K_e K_f S'_e \tag{6.3A}$$

where, for HSS end mills, the following values were selected:

K_a = surface factor = 0.63
K_b = size factor = 0.85
K_c = reliability factor = 0.715 for 99.9% survival
K_d = temperature factor = 1.0
K_e = modifying factor for stress concentration = 0.63
K_f = miscellaneous factor = 0.86
S'_e = 62,000 psi (4.3×10^5 kPa), handbook value for machined steel having an ultimate tensile strength of S_{ut} = 240,000 psi (1.7×10^6 kPa)

Substituting the above values into Eq. (6.3A) we obtain

$$S_e = 12{,}840 \text{ psi } (88{,}530 \text{ kPa}). \tag{6.4A}$$

Equating Eqs. (6.1A) and (6.2A) and rearranging, we obtain

$$P = 2S_e I/d_{cs}(l - d_A/2). \tag{6.5A}$$

The basic expression for the horsepower at cut can be presented as:

$$\text{HP}_c = (2\pi NT)/33{,}000 \tag{6.6A}$$

where T = torque at the end mill, ft-lb
$= P(d_{cs}/2)(1/2)$

and

$$I = [\pi(d_{cs})^4]/64.$$

By substituting these into Eq. (6.6A), we obtain

$$HP_e = [S_e N(d_{cs})^4]/[1.284 \times 10^6 \times (l - d_A/2)], \qquad (6.7A)$$

where HP_e is the endurance limit horsepower at the cut.

Substituting S_e from Eq. (6.4A) into Eq. (6.7A) and rearranging, we obtain

$$HP_e = (d_{cs})^4 N/[100(l - d_A/2)]. \qquad (6.8A)$$

This equation can be used to determine endurance limit horsepower (HP_e) at the cut.

Equating Eq. (6.8A) with Eq. (6.4) of the text, we obtain the endurance limit chip load as follows:

$$i_e = (d_{cs})^4/33Kd_c^2 n(l - d_A/2). \qquad (6.9A)$$

This equation can be used to determine endurance limit chip load (i_e) at the cut, while machining aluminum; see Table 6.4.

References

1. Salomon, C., "Process for the Machining of Metals or Similar Acting Materials When Being Worked by Cutting Tools," German patent No. 523594, April 1931.

2. "High-Speed Machining," Special Report No. 710, *American Machinist*, March 1979.

3. Merchant, M. E., "10 Years Ahead . . . What's in It for Metalworking?" Society of Automotive Engineers, National Aeronautic Meeting, 1959.

4. King, R. I. and J. G. McDonald, (Lockheed), "Production Design Implications of New High-Speed Milling Techniques," *Journal of Engineering for Industry*, ASME, November 1976, p. 1170.

5. Von Turkovich, B. F., "Influence of Very High Cutting Speed on Chip Formation Mechanics," 7th North American Metalworking Research Conference Proceedings, 1979, SME Transactions.

6. Aggarwal, T. R., "Research in Practical Aspects of High-Speed Milling of Aluminum," SME Paper No. MR82-262, Society of Manufacturing Engineers, Detroit, 1982.

7. McGee, J., et al., "Manufacturing Methods for High Speed Machining of Aluminum," Final Technical Report, Vought Corporation Contract No. DAAK-40-76-C-1329, submitted to U.S. Army Missile Research and Development Command, February 1, 1978.

8. Stelson, T. S., "Turning Tests on Aluminum and Titanium," AMRP Report #SRD-81-062, TRW Inc. Contract No. F33615-79-C-5119, submitted to the Air Force Systems Command, 17 August 1981.

9. Schroeder, T. A. and S. Hague, "Conventional to High Speed Turning," G. E. Carboloy Systems Department AMRP Report #SRD-81-062, submitted to the Air Force Systems Command, 17 August 1981.

10. Truncale, J. F. and T. C. Winiecki, "High Speed Machining, Expanded Production Machinability Data for Aerospace Alloys," AMRP Report # SRD-81-086, General Dynamics Convair Div. Contract F33615-80-C-5057, submitted to the Air Force Systems Command, October 20, 1981.

11. Kegg, R. L. and W. J. Zdeblick, "High Speed Milling—Research vs. Today's Practices," Proceedings of 20th Annual Meeting and Technical Conference, Numerical Control Society, Glenview, Ill., 1983.

12. Shigley, J. E., *Mechanical Engineering Design*, McGraw-Hill, New York, pp. 157–191.

13. Tlusty, J. and P. MacNeil, "Dynamics of Cutting Forces in End Milling," *Annals of CIRP*, Vol. 24/1/1975. pp. 21–25.

14. Criger, G. L. (General Dynamics Convair Div.), "High Speed Machining in Production," *SAMPE Quarterly*, April 1981, pp. 12–18.

CHAPTER 7

Machine System Design and Performance

F. J. McGee
Vought Corporation

Introduction

Milling analyses and the influence of new, nonconventional cutting theories have encouraged the development and implementation of high-speed milling for manufacture of both ferrous and nonferrous metal machined parts. Existing data and academic documentation support both economic and technical feasibility of high-speed machining (HSM), specifically milling.

To prove or disprove the supporting data and as a part of its continuing independent and contracted factory modernization efforts, Vought Corporation retrofitted two conventional numerically controlled (NC) milling machine systems (a Sundstrand OM-3 Omnimill five-axis NC machining center and a Bullard VTL three-axis NC machine) with high-speed spindles and associated modifications. The machines were retrofitted; installed in the factory environment; used in existing manufacturing programs; and analyzed with regard to operation, potential, and optimum system specifications. This first venture into high-speed milling provided insight into both the design and performance of high-speed milling machine systems.

High-Speed Milling System Installation and Initial Findings

Installation of Bryant High-Speed Spindle on OM-3 Omnimill Machining Center

A Bryant high-speed spindle was installed on a Sundstrand OM-3 Omnimill that originally had regular 10- to 990-rpm slow-speed and 40- to 4000-rpm fast-speed spindles. To accomplish this, the fast-speed spindle was fed out along its *H*-axis so that it protruded from the swiveled machining head, and from that position, the fast-speed spindle and its housing were removed. Vought designed and fabricated an adapter and housing for the new high-speed spindle to replace the original fast-speed support structure. After the

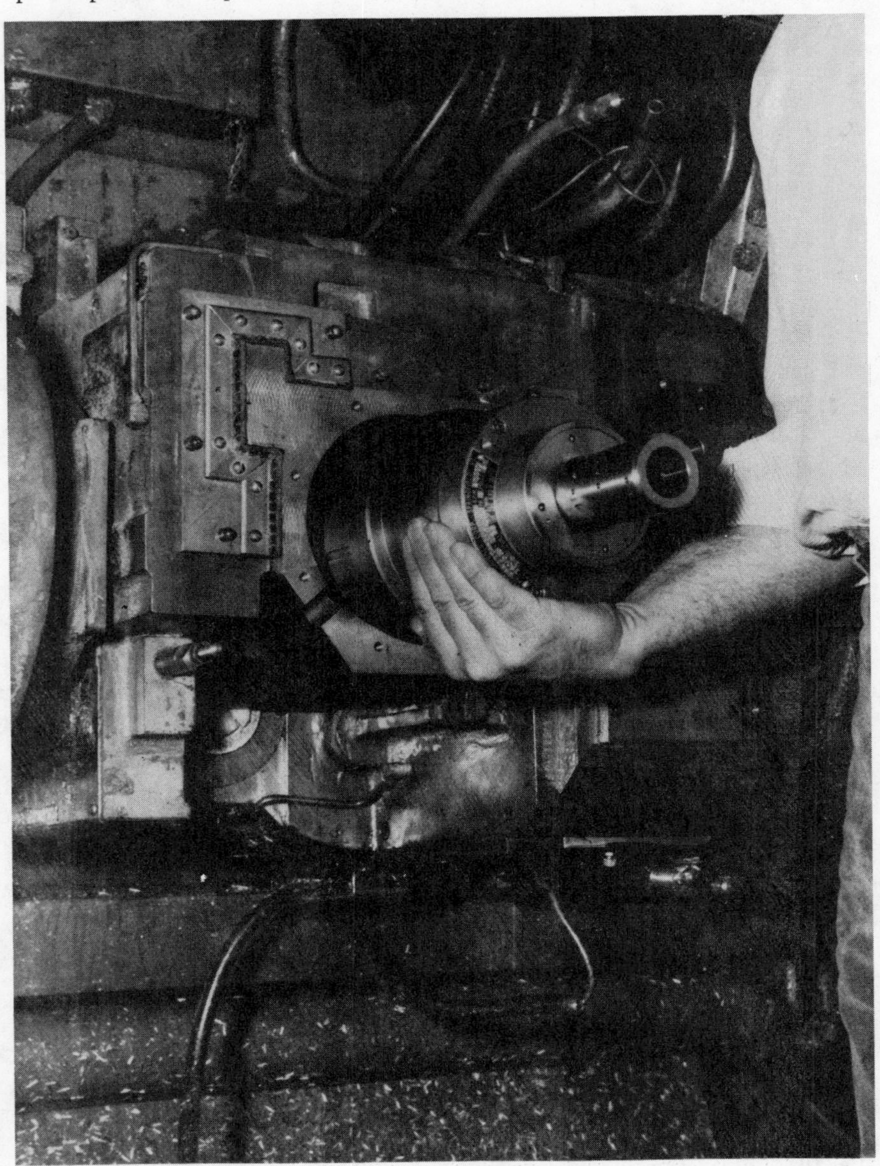

Fig. 7.1 Installing Bryant spindle in Omnimill adapter.

high-speed spindle was positioned in its special housing (as shown in Fig. 7.1), it was clamped in place with a Vought-designed and -fabricated clamp ring. Installation of the spindle was completed when the clamp ring was bolted to the new housing and all electrical, cooling, and lubrication lines were connected.

The Sundstrand fast-speed spindle and the new Bryant high-speed spindle are interchangeable. In approximately 2 h, excluding alignment time, the machining center can be converted from high-speed to conventional milling or from conventional to high-speed milling.

Installation of ECCO High-Speed Spindle on Bullard VTL

An Ekstrom–Carlson Company (ECCO) high-speed spindle was purchased for use on the Bullard VTL. The spindle was bolted to an adapter plate which, in turn, was bolted to a tool-post adapter and clamped in the side tool post of the three-axis NC machine as shown in Fig. 7.2. Alignment for installation was enhanced by a keyway slot provided in the spindle base and duplicated in the adapter plate in two different positions.

To provide needed operational flexibility, tool holders of different lengths were used and an additional locating (keyway) position was machined in the

Fig. 7.2 ECCO spindle mounted on Bullard VTL side tool post.

adapter plate. Feed rates could only be set by trial and error, because table speeds could not be programmed or overridden at the low speeds that had to be used. This setup proved to be useful, but awkward for test purposes.

Technical Findings

Though somewhat crude, these two machine conversions were used to develop manufacturing technology for high-speed milling of aluminum missile components. Additionally, experience gained from the effort led to a better understanding of machine-tool requirements for the HSM process and of the types of parts that could be most effectively fabricated.

The maximum performance parameters demonstrated in milling aluminum on the converted machines with 20-hp (15-kW), 20,000-rpm high-speed spindles and existing machine tools were:

$$\text{Metal-removal rate} = 50 \text{ in.}^3/\text{min } (820 \text{ cm}^3/\text{min})$$

$$\text{Table feed rate} = 500 \text{ ipm } (1270 \text{ cm/min})$$

$$\text{Contouring feed rate} = 100 \text{ ipm } (254 \text{ cm/min})$$

These data indicate that without more power, high-speed milling machines cannot remove metal at a greater volume than conventional-speed milling machines. That indication can be more fully appreciated when two horsepower relationships for machining are considered. By definition, horsepower is the force required to raise 33,000 lb at the rate of 1 ft/min or, in machining operations,

$$\text{HP} = 3.03 \times 10^{-5} F_c V, \qquad (7.1)$$

where HP = horsepower
F_c = cutting force (lb)
V = cutting speed (ft/min)

This relationship shows that once cutting forces have stabilized at some speed [e.g., 2500 ft/min (762 m/min) for aluminum], horsepower requirements will increase directly with cutting speed. Thus, if cutting speeds were further increased threefold, horsepower requirements would also be increased threefold. From this relationship, it is evident that horsepower requirements can be high, and perhaps excessive, for high-speed machining operations.

The second relationship, which can be used to approximate the gross spindle horsepower required to make a given end mill cut in aluminum, is

$$\text{HP} = 0.4 d_A d_R f_m, \qquad (7.2)$$

where HP = horsepower
d_A = axial depth of cut (in.)
d_R = radial depth of cut (in.)
f_m = feed rate (ipm)

If the horsepower and feed rate of a machine are known, this relationship can be used to estimate a maximum cut cross section. For instance, if a machine has a 20-hp spindle and a feed rate of 200 ipm, it can produce a 0.25-in.^{2} cross-section cut in aluminum. That cross-section cut can be increased by decreasing the feed rate, but only up to a point. Feed rate reduction must be limited, since decreasing the feed rate too much can result in chip packing and welding.

From Eq. (7.2), it can also be deduced that standard-cut cross sections (radial cut depth times axial cut depth) cannot be maintained at high feed rates, with 20–30-hp (14.9–22.4-kW) high-speed spindles, because such spindles would be underpowered. However, it is evident from this equation that high-speed milling machines can remove low but comparable cross-sectional areas of material at much higher feed rates than conventional milling machines. Based on these determinations, Vought concluded that high-speed milling was best suited for the machining of castings, forgings, and extrusions, since the higher feed rates could be beneficially used to rapidly remove the relatively small amount of material cut from such parts.

The shape of a part can also affect HSM. First of all, when the cutter must move simultaneously in more than one axial direction to make a prescribed cut, the resulting feed rate may become choked-down or stifled. This would occur if there were not enough reserve force or feedback data available to maintain a constant feed rate along each axis (i.e., each axis would have to share whatever signal or feed force was available, and the resulting vectorial feed rate would decrease accordingly). Second, if feed rates have to be accelerated and decelerated excessively to produce a prescribed cut, permissible feed rates may never be attained. In either case, feed and material-removal rates decrease, and HSM benefits are significantly reduced. Thus, it was concluded that high-speed milling was best suited to making long, straight cuts, although later events were to prove that HSM could also be cost effectively used for "hogging" and pocketing operations.

Economic Findings

Most of the economic gains achieved by installing a high-speed spindle on the Sundstrand Omnimill were lost when the time-saving, automatic tool changer and pallet changer on that machine had to be deactivated to make the conversion. In contrast, a Cincinnati three-axis profiling machine, recently converted into a HSM unit at Vought (see Fig. 7.3) for about two-fifths the cost of a new machine, had neither of the automatic features and has retained the economic gains it derived from HSM. However, this profiling machine could achieve even greater economies and versatility if it were equipped with an automatic tool changer. Other features that could enhance the cost effectiveness and flexibility of the machine-tool system include: five-axis spindle

Fig. 7.3 Converted high-speed milling machine.

movement, interchangeable spindles (low- and high-speed), in-process inspection, interfaces with material-handling systems, and chip-removal systems. Some of these desirable accessories can be procured with new machine-tool systems for an additional cost of about 15% over base price. Considering the initial costs versus the economic gains, versatility, and flexibility to be derived, these devices should be seriously considered when purchasing any new machining system. While it might be difficult to install palletized loading on the existing machine, it would not be difficult to install it on a redesigned machine of the same type. In addition, palletized loading would enable cutters to operate a greater percentage of the time and, thus, be a valuable addition to that machine-tool system. Overall, the converted Cincinnati high-speed milling machine shown in Fig. 7.3 has demonstrated a potential for doing the work of approximately four similar, conventional machines.

High-Speed Milling System Applications

In finding potential applications for high-speed milling in the manufacturing environment, both technical and economic reasons must be considered. Specific applications within the Vought manufacturing operation were made only after consideration of current milling operations and machine tools, the advantages of HSM-capable equipment over current methods, and identification of parts best adapted to HSM. Initial findings pinpointed airframe parts requiring long, straight cuts as those offering the highest potential for improved quality and economy through HSM. The Vought application analysis thus began with an examination of the current machining operations for these identified parts.

Current Milling Methods

Parts identified for potential HSM loading are currently being machined at Vought on conventional tracer and NC mills at cutting speeds of 1800 and 3600 rpm with standard 0.5- to 2-in. (1.27- to 5.08-cm) diameter high-speed-steel end mills. Feed rates vary from 20 to 60 in. (50.8 to 152.4 cm) per minute. The spindles that cut these parts vary in power from 10 to 40 hp (7.46 to 29.8 kW), with 20-hp (7.46-kW) spindles being most common. A 20-hp (7.46-kW) spindle can remove 40 in.^{3} (655 cm^3) of aluminum per minute. However, average metal-removal rates rarely exceed 10 in.^{3}/min (164 cm^3/min). The time required to profile a part varies from approximately 1.5 to 10 h. Cutting fluid used when machining these parts generally consists of a mixture of soluble oil and air. Each machine has its own cutting fluid reservoir. Some of the potential HSM candidate parts are heavily machined, while some of the more promising HSM candidate parts are not.

Potential Milling Method Improvement with HSM

Through increased spindle speeds and feed rates, machining time could be considerably reduced with HSM. Feed rates could be readily increased to 600 ipm (1524 cm/min), or 6–12 times faster than the current method. Such a capability would reduce direct labor costs, provide a faster throughput for parts, and allow net cost savings in the milling operation. It should be emphasized, however, that while feed rates can be increased sixfold by HSM, metal-removal rates do not increase unless a spindle with increased horsepower is provided. This fact is clearly illustrated by Eq. (7.2). When only the feed rate is increased by sixfold without adding additional spindle horsepower, the cut cross section must be decreased approximately sixfold. If that is done, however, the chip load and cutting forces would be significantly reduced. Since high-horsepower HSM spindles are not available, the practice of operating at reduced cut cross sections is generally followed today for HSM.

The low cutting forces generated by HSM make it possible to produce thinner-walled parts than normal. Likewise, these lower cutting forces reduce the number of extra (no-load) cuts required to clean up stiffeners, flanges, and other thin-walled sections left tapered by deflected end mills. Elimination or reduction of these free cuts helps reduce direct labor costs and provides a faster throughput. Additionally, a single high-speed mill has the potential to replace three or four conventional mills. If that potential were pursued, even further economic gains could be realized from reduced machine-tool acquisition and floor space cost savings.

Part Characterization

In examining potential HSM applications, Vought identified an aircraft subcontract that required more than 52 stringer or longeron-type parts in the tail section alone. To produce these parts more economically and faster, Vought converted a conventional milling machine to an HSM machine for the fabrication of all stringers and a number of long, narrow aluminum components required in the program.

The cuts required for these parts are not complex. Most milling operations are straight or involved only simple joggles, pockets, or cutouts. Parts with such cuts were selected for HSM to avoid excessive acceleration and deceleration movements and, thereby, to take fuller advantage of the high-feed-rate capability possible with this method.

Additionally, Vought selected parts for HSM operations that could be manufactured with reasonably stout end mills. This avoided vibration (chatter) problems and conditions conducive to cutter breakage, which are usually more catastrophic at the higher cutting speeds.

Another factor limiting the complexity of cuts made with HSM is machine control. Currently, tolerance effects cannot be controlled very precisely and are generally avoided by slowing down feed rates or simplifying the cut. In contrast, the delicacy of candidate parts does not necessarily limit the complexity of cuts that can be made with HSM. On the contrary, the milling of thin sections can be accomplished better with HSM because the method generates lower side and cutting forces. In summary, while HSM can be used to machine a wide variety of complex shapes and parts, for the reasons given above, HSM processes are presently best suited for making relatively simple cuts and will be largely used for that purpose at Vought.

The HSM operations used to produce the candidate airframe parts include end milling, drilling, and, possibly, face milling. Since cutting forces generally decrease and surface finishes generally improve with increases in cutting speed and because HSM is best suited for removing small amounts of material rapidly, it follows that high-speed machining would be ideal for making finishing as well as rough-finishing cuts on those parts. Vought's machine will be limited to rough-finishing cuts unless the parts can be finished on a three-axis machine. If a five-axis HSM unit were available, the candidate parts could be rough-finished and finished at maximum cutting speeds. End mills would be operated to 20,000 rpm, at 0.010 in.-per-tooth (0.254 mm-per-tooth) feed rates, and a cut cross section of about 0.3 in.^{2} (1.94 cm^2). Drills would be operated to 20,000 rpm with feed rates of up to 100 ipm (254 cm/min).

Tolerances need not be relaxed to use high-speed machining for airframe parts. Missile parts that have been produced with HSM processes have readily passed inspection. Tests have shown that tolerances to ± 0.005 in. (0.127 mm), finishes to 125 root mean square, and good surface integrity can be easily maintained with HSM processes.

Most of the identified candidate parts are fabricated from 7075-T651 aluminum. That material has a Brinell hardness number of 147 and an ultimate tensile strength of approximately 82,000 psi (5.7×10^5 kPa). Its yield strength is about 76,000 psi (5.2×10^5 kPa), and its reduction in area is on the order of 32%. Other aluminum airframe alloys that will be extensively involved in HSM operations include 6061 and 2024.

Milling Machine System Design Requirements and Specifications

While manufacturers of state-of-the-art, high-speed-milling machinery are constantly updating and upgrading their designs for improved performance, available equipment may not fulfill all requirements. For example, the ideal spindle for high-speed milling is one that operates at high speeds with high horsepower. However, current high-speed spindles do not provide this ideal capability combination. There are spindles with relatively low horsepower that operate at ultra-high speeds, or there are spindles with high horsepower that operate at only moderate speeds. The final selection must be a compromise based on the machining requirements and part characteristics of the specific manufacturing tasks to be accomplished.

Equipment also varies in the mobility of the table, column, or gantry in the machining assembly. The range of movement of these components offers both advantages and disadvantages to the machining operation. Final determination of the best mobility characteristics once again depends on the specifications of the milling tasks to be performed.

The subsystem characteristics of available high-speed-milling equipment or components offer a mixture of advantages and disadvantages to planned milling tasks. In the following paragraphs, discussions of each critical subsystem demonstrate the interim analyses and compromises required to develop the machine-tool specification finally released for bid. With the variety of characteristics and combinations, the specification must be prepared in considerable detail to prevent miscommunication with the equipment manufacturers responding to it. The detail required to produce accurate, feasible equipment bids must therefore be identified before release of the machine-tool specification.

Spindle

In Vought's conversion of existing milling equipment to high-speed operation under factory modernization efforts, the specific characteristics desired for the high-speed spindle were based on the product calculated in Eq. (7.2). Horsepower requirements, axial depth of cut, radial depth of cut, and feed rate in inches per minute were identified and input to the calculations.

Based on the mathematical findings, a 20,000-rpm, 100-hp (75-kW) spindle was identified as ideal. However, such a spindle did not exist. A 20,000-rpm, 40-hp (30-kW) spindle was widely acknowledged as the ultimate in state-of-the-art capacity. Vought then contacted a number of equipment manufacturers in an attempt to purchase a 20,000-rpm, 40-hp (30-kW) spindle. None could be located; so in the final evaluation, Vought opted to go with a 20,000-rpm 30-hp (22-kW) Bryant spindle because of its higher feed-rate and cutting-speed capabilities.

The spindle was described by Bryant as being a special high-frequency spindle identified by model or drawing number WRA-178. The spindle has a rated speed range of from 4000 to 20,000 rpm. Spindle braking time is approximately 20 s. That time can be varied by adjusting the dc current flowing through the spindle windings. The spindle motor is a constant torque, linearly variable horsepower, induction motor. Four ball bearings made to ABEC 9 specifications are mounted between the spindle shaft and housing. The diameter of the spindle shaft is approximately 2.5 in. (6.4 cm) at the spindle nose. Cutters attach to the shaft through adapters that have a No. 40 standard milling machine taper (MMT). Vought purchased seven balanced, No. 40 MMT, Weldon-type, end-mill holders to fit shank diameters ranging from 0.5 to 1.5 in. (1.3 to 3.8 cm). With these toolholders, cutters ranging in size from 0.5 to 3 in. (1.3 to 7.6 cm) in diameter can be used with the spindle.

The spindle was purchased with a foot-mounted housing that allowed for either vertical or horizontal positioning. Vought mounted the spindle vertically to minimize costs and to maintain production tooling. In the future, the machine may be converted to a horizontal spindle milling operation for better chip control and disposal and to allow the work area to be more easily encased with a protective shield.

Table/Gantry/Column

The machine Vought converted to an HSM unit had a moving column with a saddle, cross slide, two spindles, and a tracer attachment weighing approximately 6000 lb (2725 kg). The machine also had three axes that were to be modified to move at the following infinitely variable, maximum feed rates:

X-axis: 600 ipm (1524 cm/min)
Y-axis: 400 ipm (1016 cm/min)
Z-axis: 400 ipm (1016 cm/min)

Each axis had a 24- by 24- by 176-in. (61- by 61- by 447-cm) travel area. Vought expected to achieve the above feed rates by changing gear combinations and making other minor modifications. These changes did not materialize, because no source could be found to design and fabricate only one gearbox of the type needed. As an alternative, the machine was disassembled and the *X*-axis rack-and-pinion feed mechanism was redesigned to accommodate a ball-bearing screw-and-nut drive system.

The HSM-modified unit required specific table/gantry/column capabilities to afford proper milling operations and perform the tasks of overcoming weight, opposing cutting forces, and accelerating all these loads at 30 in./s^2 (76.2 cm/s^2) along the three axes. To provide this capability, premium servomotors were purchased. The *X*-axis motor was designed to accelerate the column from 0 to 600 ipm (0 to 1524 cm/min) in 0.33 s or 1.67 in. (4.24 cm) of travel. Both speed and distance of movement along the axes are controlled by the servomotor encoders, tachometers, and the CNC controller. The positional accuracy along each axis was ± 0.001 in. (0.0254 mm) when new. The overall cost for the drive system, including guideways, was about 20% of the retrofit cost. The *X*-axis guideways consisted of two shaft assemblies, four machine-tool bearings, and four bearing housings. The bearing housings that attach to the column were designed and fabricated at Vought.

Cutter

To minimize chatter, maintain tool life, and protect spindle bearings, all cutters must be dynamically balanced before use in HSM operations. Some cutters, especially the smaller ones, are delivered from the manufacturer in balance, while others must be balanced prior to use. Based on that consideration, off-the-shelf, high-speed-steel cutters can satisfactorily machine aluminum at very high cutting speeds, as long as the cutters are ground with excess clearance (e.g., 10° primary and 20° secondary) and with smooth lands to prevent the formation of built-up edges on cutter flanks. Cutting fluid also minimizes the formation of built-up edges on high-speed-steel cutters. Brazed or solid carbide cutters outlast high-speed steel cutters and should be used if economically justifiable. The optimum cutter geometry for HSM aluminum has not been established, but an end-mill geometry of a 25° helix angle and a 5° to 10° radial rake angle is the best currently identified by Vought.

In general, insert milling cutters cannot be used for HSM operations. The high centrifugal forces developed in operation can strip the threads on clamping screws or overcome clamping or wedging forces that bear on the inserts. Owing to the potential damage and danger resulting from the use of such cutters, a rapid detection sensor for tool breakage or cutter imbalance is needed to prevent damage to the spindle or workpiece. Vibration monitors are available, but are not highly recommended by current users. The device used must stop the feed drives since spindle rotation cannot be stopped fast enough (approximately 20 s) to prevent damage.

Workpiece Clamping Device

HSM fixture requirements are not yet firmly established, but because of the lower cutting forces involved, specifications will be less demanding than those for low-speed machining (LSM). In any event, the use of HSM should not increase tooling costs.

NC Programming and Tape Modifications

Since new techniques are continually being developed to enable programmers to design around problems presented by high feed rates at high spindle speeds, NC programming or tape modification should not limit the application of HSM. Instead, the limitations are expected to come from the CNC controller that implements the programmed instructions. For instance, the high-speed mill has the potential to contour at feed rates up to 400 ipm (1016 cm/min). Since no operator can react fast enough to maintain tolerances at those feed rates, such machining operations must be controlled electronically. However, those contouring feed rates cannot be successfully achieved with any existing CNC controller. Most controllers limit contouring feed rates to approximately 100 ipm (254 cm/min), because controller cycle time is not fast enough to process the vast stream of command and feedback signals which are transmitted and received and which tend to overload the processor. The cycle time for a block of data to go through the executive program of many controllers and be acted upon is 10 ms, but that is not fast enough to enable satisfactory contouring at feed rates of 400 ipm (1016 cm/min). The chordal error at that feed rate and cycle time would be 0.066 in. (1.676 mm) and would be noticeable. At a feed rate of 100 ipm (254 cm/min), the chordal error would be 0.017 in. (0.432 mm) and probably would not be noticeable. While the chordal error that can be tolerated is not known at this time, a cycle time of 5 ms may be the maximum allowable for controllers used in HSM operations. That cycle time would almost guarantee that satisfactory contouring cuts could be made at 240 ipm (610 cm/min).

Previously, it was shown that HSM processes were horsepower limited. From the above analysis, it can be concluded that these processes are also cycle-time or control limited and that CNC systems will require an enormous number-crunching capability for HSM applications.

Chip Control

Chips produced in low-speed machining are a major problem. Such chips frequently burn or cut operators, scratch workpiece surfaces, or fall back into the cutting zone where they are recut and accelerate cutter wear. Chips pile up on the machine and surrounding work area and cause machine wear as well as environmental and housekeeping problems.

Since HSM produces more chips at an accelerated rate, chip disposal is a critical problem for high-speed operations. Ideally, HSM is best performed in an enclosed area (e.g., a polycarbonate box around the work area) with a horizontal spindle. With this arrangement, all chips, broken cutters, and cutting fluid are contained within the box, thus protecting personnel. Additionally, spent chips fall onto a conveyor on the floor of the box instead of falling on the workpiece and being recut. In practice, however, it is not always practical to box in the work area, because some parts are too large for such an arrangement and many machines have vertical spindles. For these parts and machines, a shield to safeguard personnel and some type of automatic chip-removal system are needed. Automatic systems such as vacuum cleaners, conveyor belts, or augers are acceptable for chip control. The Vought machine has three augers with a mud-flap deflector mounted behind the spindle which automatically removes approximately one-half of the chips generated. The chip guard is usually adequate, but Vought plans to build a wall around the machine to safeguard personnel passing near it.

Safety Features

In addition to the chip control safety features, HSM machine-tool systems require other protective measures. For example, a spindle turning at 20,000 rpm requires approximately 20 s to brake to a complete stop. If anything should go wrong, it is impossible to prevent some damage. Therefore, efforts have been directed toward stopping the feed and providing interlocks which would head off any trouble. Stopping the feed is not an ideal solution either, because cutters operated at high feed rates could move an inch or two before stopping. Therefore, as demonstrated in Fig. 7.4, considerable effort has been directed toward developing safety interlocks in the machine-tool system.

The high-speed spindle at Vought has interlocks that stop the spindle from running if it gets too hot, is not being lubricated, is overloaded, or is operated at too high a frequency or current. Likewise, interlocks stop the machine if the ways, slides, screws, and bearings are not being lubricated or if the screws are overloaded or overtraveled. The workpiece and spindle are protected by an adaptive control that acts as a protective device. The adaptive control also indicates, to a degree, the amount of cutter wear. It is expected that other interlocks will eventually be added.

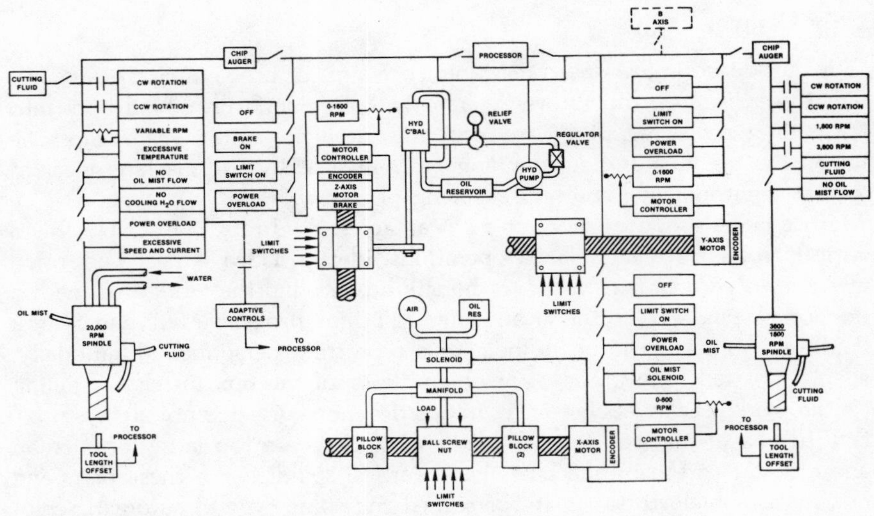

Fig. 7.4 Schematic for prototype machine controls.

Sensors, Instrumentation, and Controls

Sensors on the HSM mill at Vought include thermostats, float switches, relays, digital comparators, pressure switches, a bimetallic thermal overload sensor, and limit switches. The instrumentation consists of a tachometer, thermometer, load meter, elapsed-time meter, horsepower meter, cathode ray tube (CRT), and pressure gages.

The principal controls for HSM operations are listed in Table 7.1. All of the items listed interface in the CNC console. Programmable application logic

Table 7.1 Controls and Sensors Needed for High-Speed Machining Processes

Control/Sensor	Primary Function: Performance	Protection	Technology
CNC Hardware/Software	X		
Spindle Drives	X		
Feed Rate Drive	X		
Encoders	X		
Adaptive Controls		X	
Lube Interlocks		X	
Temperature Interlock		X	
Spindle Overload Interlock		X	
Horsepower Meter		X	X
Limit Switches		X	
Software Interlock		X	
Spindle Tachometer			X

Table 7.2 Axis Information for Vought High-Speed-Milling Machine

Axis	*X*	*Y*	*Z*	*B*
Rapid Traverse (inches per minute – degrees per minute)	600	400	400	3600
Gain (inches per minute per mil – degrees per minute per mil)	1	1	1	1
Gain Break Rate	400	400	200	3600
Home Reference Position	0	0	23	0
Total Axis Travel	158.5	18.5	18.5	45
Plus Directional Travel	158	18	18	+22.5
Minus Directional Travel	−0.5	−0.5	−0.5	−22.5
In-Position Band	0.0005	0.0005	0.0005	0.005
Jog Speed—High	600	400	400	3600
Jog Speed—Medium	200	200	200	1200
Job Speed—Low	10	10	10	3.6
Feedback Resolution	0.00005	0.00005	0.00005	0.001
Programming Resolution	0.0001	0.0001	0.0001	0.001
Axis Calibration Interval (deg)	0.8192	0.8192	0.8192	1.0

Gain Reduction Factor at Gain Break = 1/4
Memory Size = 64K
Machine Home Velocity = 10 inches per minute – degrees per minute
Acceleration/Deceleration Parameters:
Maximum Linear Velocity Ramp = 12 inches per second^2
Maximum Linear Velocity Step = 0.5 inch per second
Maximum Rotary Velocity Ramp = *N/A* degrees per second^2
Maximum Rotary Velocity Step = *N/A* degrees per second

(PAL) software allows the controls to interface in the CNC machine and magnetic input/output rack, and the racks are wired accordingly. Additionally, a post-processor program permits APT programming. The control characteristics for the Vought HSM mill are given in Table 7.2.

Additional Instrumentation and Controls

Additional instrumentation and controls may be needed to successfully implement HSM in a specific factory. Such additions could include a closed-circuit TV system with a video tape recorder to provide maximum safety. Acoustic sensors might be used to provide in-line inspection. Vibration or force sensors could be used to detect broken or unbalanced cutters.

Machine System Implementation

The decision to buy a new machine or to convert an old machine for HSM must be considered. There are several good reasons for buying a new machine, including:

No machine-tool research and development effort is required.
Tighter, more rigid, more versatile, and more accurate machine.
Smaller delta capital investment.

The first step in buying a new machine is to obtain quotes or proposals from various machine tool builders. If an acceptable proposal can be obtained, a detailed specification should be prepared and submitted to a selected machine-tool builder along with the purchase order. While the new machine is being built, a liaison officer from the procuring organization should interface with pertinent personnel to prepare a location and foundation for the machine, provide utilities, expedite deliverable dates, ensure that schedules are maintained, and supervise installation and acceptance tests for the machine.

If an acceptable proposal cannot be obtained, an existing machine must be selected for conversion. None may be found to be ideal; however, the one that most nearly resembles the wanted machine and is most adaptable to conversion should be selected. Prices or proposals for major retrofit items must be obtained to better estimate conversion costs. Once the conversion cost is approved, long-lead-time items should be ordered as soon as possible. The axis drives for the modification can be designed in a parallel effort. While awaiting the arrival of new items, the selected machine should be disassembled, cleaned, painted, and, in general, restored to like-new condition while reassembling.

After the selected machine is overhauled, the conversion design should be finalized. The final design should include a schematic drawing for the controls and sensors, counterbalance design, chip-disposal provisions, plumbing and pneumatic layouts, lubrication system design, and machine layout as shown in Fig. 7.4. Any sensors and controls not previously ordered should be either purchased or designed and fabricated during this interim period so that there will be no unnecessary delays during machine retrofit.

Power for the retrofit, component installation, and system checkout is a pacing item in the HSM conversion. Since power is provided through the CNC system, it is imperative that this system be installed at a reasonably early date. The first task is to write the PAL software that enables all selected controls to interface in the CNC machine and magnetic input/output rack. Preliminary postprocessor parameters should also be written and the program prepared while the data can be readily recalled. The PAL instructions should then be wired into the CNC input/output racks. Once that task is completed, the CNC system can be wired into the high-speed-milling machine so that power is available for the retrofit.

Retrofitting the machine is the next item in the implementation plan. Ideally, the installation of components on the machine should follow a logical sequence so that work flow is not impeded. Enough task overlap should be planned to minimize any unscheduled work stoppages. Delays do occur during the service, alignment, and checkout period, since parts can malfunction and shut down the whole operation until replaced or repaired. Sooner or later, however, such malfunctions will be eliminated, and the high-speed-milling machine will be ready for production implementation.

Introducing HSM into Production

No implementation standard yet exists for introducing a high-speed-milling machine into production. Many scenarios are possible. For example, a faster tape for a candidate part could be developed with allowances for ramping, small radii, and other restrictive cuts. If the tape produces parts successfully, further steps could be taken to speed up the operation. After a few iterations, it should be possible to achieve maximum production rates for that part on the high-speed-milling machine. With experience, the number of iterations for follow-on parts should diminish and HSM cost savings be fully realized.

In summary, successful implementation is as much the culmination of thorough planning as it is the beginning of high-technology production. To ensure success, careful attention must be given to determination of the parts best suited to high-speed machining, selection of the HSM equipment best adapted to those manufacturing tasks, and preparation of detailed specifications and scheduling for that equipment.

References

Aggarwal, T. R., "High Speed Milling of Aluminum," Cincinnati Milacron Publication No. A-313, 1982.

Chaplin, J., J. A. Miller, and R. I. King, "Summary of Recent Lockheed Research Re: High-Speed Machining," Proc. 9th NAMRC Conference, 1981, pp. 311–317.

Flom, D. G., et al., "Advanced Machining Research Program (AMRP)," Annual Technical Report No. SRD-81-062, General Electric Contract No. F33615-79-C-5119, submitted to Air Force Systems Command, 17 August, 1981.

Flom, D. G., et al., "Manufacturing Technology for Advanced Metal Removal Initiatives (AMRI)," Report No. SRD-82-037, General Electric Contract No. F33615-80-C-5057, submitted to Air Force Systems Command, 20 April, 1982

King, R. I., "High Speed Machining Revs Up," *American Machinist*, January 1981.

King, R. I. and J. G. McDonald, "Production Design Implications of New High Speed Milling Techniques," *Journal of Engineering for Industry*, ASME, November 1976.

Jablonowski, J., et al., "High Speed Machining," Special Report No. 710, *American Machinist*, March 1979.

McGee, F. J., "Manufacturing Methods for High Speed Machining of Aluminum," Final Technical Report, Vought Corporation Contract No. DAAK-40-76-C-1329, submitted to U.S. Army Missile Research and Development Command, 1 February, 1978.

McGee, F. J., "Validation of the DARPA Advanced Machining Research Program Operating Regimes of Aluminum, Titanium and Steel Airframe Materials," AMRI Report No. SRD-81-061, Vought Corporation Contract No. F33615-80-C-5057, submitted to the Air Force Systems Command, 20 July, 1981.

Truncale, J. F. and T. C. Winiecki, "High Speed Machining, Expanded Production Machinability Data for Aerospace Alloys," AMRI Report No. SRD-81-086, General Dynamics Convair Div. Contract No. F33615-80-C-5057, submitted to the Air Force Systems Command, 20 October, 1981.

CHAPTER 8

Operational Data

J. F. Truncale
General Dynamics/Convair Division

This chapter will present information that has been acquired through production and experimental programs. The discussion on production experience is specific to our application and is structured like a case history. The experimental data conclude the section and consist of the results of two series of tests.

Production Data

Part Description

The parts currently high-speed machined at General Dynamics Convair are cruise missile airframe components. The cruise missiles are approximately 20 ft (6 m) long and 21 in. (53 cm) in diameter. The airframe is modular and is made up of five aluminum body sections. Four of the body sections are made of machined cylindrical skins and flat circular bulkheads. It is the cylindrical body section skins that are presently machined at high speed.

An interactive process occurs during the early design and manufacturing phases. That is, a structure and method of manufacturing are assumed, and the capability to produce this structure is reviewed. The parts described subsequently are a result of such an interactive process. Some of the alternate concepts for manufacturing the body sections are discussed later.

The raw material for the body section skins is cylindrical forgings made from 2219 aluminum. The forgings are 12–48 in. (30–122 cm) long and about 22 in. (5 cm) in diameter. They have a minimum inside diameter of about 16 in. (41 cm). Prior to high-speed machining (HSM), the cylinders are turned, bored, and welded together to form the rough configuration of a body section (Fig. 8.1).

High-speed end milling is then performed inside the body section. A 20 hp (15 kW), 20,000 rpm spindle is placed inside the cylinder and at a right angle to its surface. Figure 8.2 shows an example of the reinforcing ribs and bosses that are created during machining.

Philosophy of Manufacture

The decision to pursue HSM was based on the selection of a manufacturing method for the cruise missile airframe. Three alternatives were considered: riveted skin and frames, castings, and integral machining. The requirements of leak-tightness, minimum weight, and strength were best met by a machined structure.

Several variations on the machined approach were also tried. Flat plate was machined and then roll-formed. Curved segments were also machined and welded to form a cylinder. Finally, a one-piece cylinder was machined (internally) while mounted on a lathe.

This manual, internal end mill milling was accomplished using an air motor mounted at a right angle on a large boring bar. The cylinder was contoured

Fig. 8.1 Aluminum cylinders are bored prior to internal milling.

Fig. 8.2 An example of the integrally machined details created during internal milling.

by coordinating the movement of the boring bar with manual, incremental moves of the machine's chuck. The next step was to find a machine that could automatically do the same task.

The constraints on the machining system are as follows: A large volume of metal must be removed from these one-piece structural details. The initial inside diameter of the cylinder is small [approximately 16 in. (41 cm)], and part design considerations dictate the use of small radii, which require small-diameter cutters.

HSM has unique advantages that can be applied to solve particular problems. Using HSM, high horsepower can be delivered with small-diameter end mills. The spindle motor itself is very compact. HSM's high rotational speeds can also provide lower cutting forces owing to reduced chiploads. The requirements or constraints of the integral machining method were able to be met by the above HSM characteristics.

It may have been possible to use a right-angled spindle extension to provide 20 hp (15 kW) to an end mill inside the cylinder. The cutter-size limitation, however, would still have prevented the delivery of that horsepower at conventional cutting speeds. HSM, then, was a means for overcoming all of the above limitations. At high rotational speeds a small-diameter end mill

[0.750 in. (19 mm)] can deliver high horsepower and the associated high-metal-removal rates. The relatively small size of high-speed electric spindles also permits them to be placed inside the cylinders. This eliminates the need for a complicated drive extension.

Along with the high-speed spindle requirement came the requirement for correspondingly high feedrates. High spindle speeds *permit* high-metal-removal rates, but high feedrates *cause* the high removal rates. (Metal removal rate = feedrate × axial depth × radial depth.) There were, however, no off-the-shelf machine tools with the required spindle, feedrates, and part-holding capability. Two specially built internal mills were purchased for the task. Figure 8.3 is a picture of one of the machines.

Machine, Spindle, and Tool Description

The internal mill shown in Fig. 8.3 resembles a slant bed lathe. It is a four-axis machine with three linear axes and one rotational axis. The cylindrical parts are held in place by a three-segment chuck and a steadyrest. The end of the part and the steadyrest can just be seen at the left-center of Fig. 8.3. A

Fig. 8.3 Special-purpose high-speed internal mill.

large yoke (the equivalent of a cross slide) travels up and down at 30° off vertical. Sliding through the yoke is the ram (or boring bar equivalent). The spindle is mounted at the left end of the ram. Figure 8.4 shows a close-up view of the part, steadyrest, and spindle. The machine's protective shroud is pulled back to expose the part. The spindle axis is mounted at 30° above horizontal. The spindle moves along its own axis, the *Y* direction, up to 5 in. (13 cm) at up to 100 ipm (254 cm/min). The slant bed or *X* axis can move 24 in. (61 cm) at up to 400 ipm (1016 cm/min). The *Z* axis ram can reach 59 in. (150 cm) into a part at 400 ipm (1016 cm/min). Finally, the rotational *C* axis can turn up to 600 ipm (1525 cm/min) at a 21. in. (53 cm) diameter.

The axes drives are electric, although the slant bed has a hydraulic assist to lift the yoke/ram. The ram itself is hollow and the *Y*-axis drive motor is located at the opposite end from the spindle. The hollow ram also contains the electrical, cooling, lubrication, air, and cutting fluid lines for the spindle.

The spindle is a Bryant 20,000 rpm, 20 hp (15 kW) unit. It is a standard spindle from an operational standpoint, although it has several special features. The power and fluid connections are side mounted to keep the spindle length short. Since the spindle faces slightly upward, compressed air is used

Fig. 8.4 Close-up of a cylinder and the machine's steadyrest and spindle.

to prevent cutting fluid from seeping into the nose of the spindle. The spindle accepts No. 30 taper toolholders that are held in place with a drawbolt. The Bryant units are smooth running, reliable, and tolerant of cutter-path errors that have led to broken tools. Although not required, the spindle is cooled with refrigerated liquid instead of just recirculated. The refrigeration units were part of the original spindle's equipment and have been retained. Other standard features include Grade 9 ball bearings lubricated with an air–oil mist. The instrumentation consists of tachometer, temperature display and interlock, hourmeter, and coolant and lubricant interlocks.

The toolholders are supplied by Bryant and use two vibration-resistant 1/4-28 "Nylock" setscrews to secure an endmill (Fig. 8.5). The upper setscrew is flatbottomed, while the second is round and fits into a matching groove in the end mill. This arrangement has been successful in retaining cutters, although good practice dictates that the setscrews be checked regularly. The end mills for the most part are two-flute solid C-2 carbide with a 26° helix. The carbide has the advantage of being stiffer than high-speed steel. Its flutes

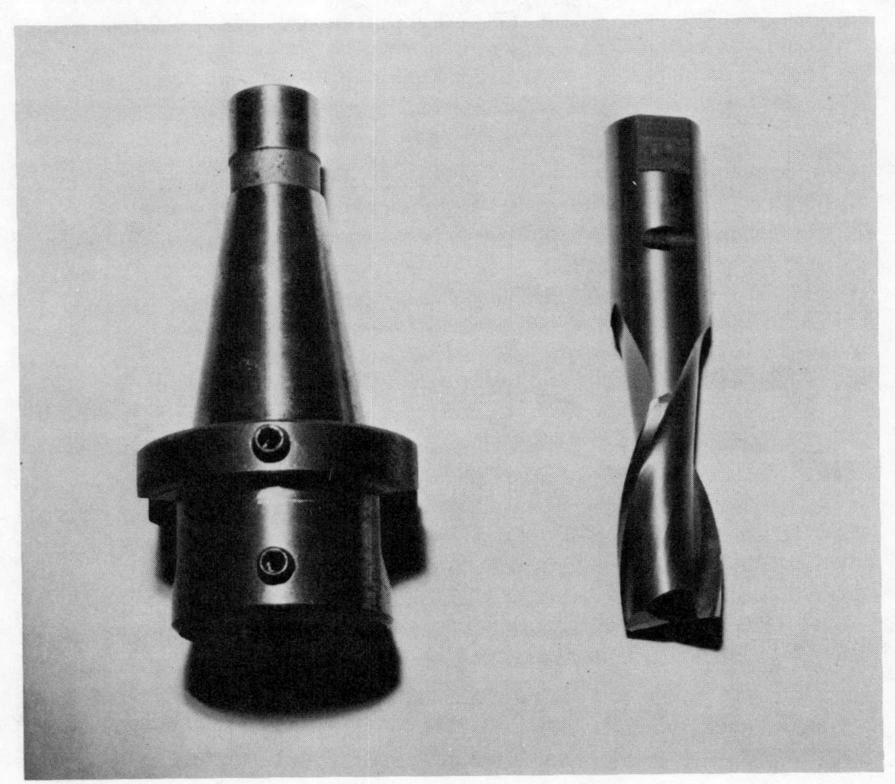

Fig. 8.5 Bryant spindle toolholder and carbide end mill.

are also smoother and help to prevent any aluminum from building up on the tool. Tool life with the carbide is naturally not a problem. T-slot cutters and drills are also used.

Cutting Conditions and NC Programming

As mentioned previously, the cylinders are turned and bored prior to being placed on the internal mill. Then the cylinder is rough machined and finish machined. Part distortion due to the large quantity of material removed (and the attendant relieved stresses) was originally avoided by rough machining in the T-4 condition and finish machining in the T-6 condition. An in-house capability has since been implemented for stretch stabilizing many of the cylinders (i.e., those with constant wall thicknesses). This avoids the necessity of removing the part between roughing and finishing operations and produces an equally good part.

The types of cuts taken inside the cylinder range from fairly long [30 in. (76 cm)] straight cuts to very complex sculptured cuts and pockets. The highest average metal-removal rates are achieved when the amount of contouring is held to a minimum. The long straight runs are typically programmed in the rotational *C* axis and parallel to the integral, machined ring stiffeners.

Roughing cuts, where geometry permits, are 0.750 in. (19 mm) wide by 0.250 in. (6.4 mm) deep at 300 ipm (762 cm/min). They are made with a 0.750 in. (19 mm) diameter solid carbide two-flute end mill. Peak metal removal rates are 56 in.^{3}/min (917 cm^3/min). Average metal-removal rates range from 7 to 31 in.^{3}/min (115 to 508 cm^3/min) for roughing and finishing combined. Total machining times for different parts vary from 0.37 to 6.55 h per part.

The tolerances maintained during the above cuts are 0 ± 0.010 in. (0.25 mm) for 95% of dimensions with some skin thicknesses held to ± 0.005 in. (0.13 mm). Surface finishes are 125–250 μin. (3–6.4 μm).

As mentioned above, the parts are programmed (geometry permitting) to take long sweeping cuts in the rotational *C* axis. This is done to minimize the number of accelerations and decelerations and to maximize the time spent at high feedrates [300–400 ipm (762–1016 cm/min)].

The internal mills do not have an automatic tool-change capability. Tool changes are, therefore, eliminated where possible by requesting design changes to standardize corner radii.

The Valenite power monitor and recorder (Fig. 8.6) are important tools in optimizing the NC program, since our limitation in HSM is spindle power. After programming a new part, a metal proof is run at 50% feedrate while recording the power. NC program sequence numbers are written on the strip chart recording. The recording is later inspected for spikes that indicate power draws of over 50%. The sequence numbers locate the machining sequence where the spike occurred, and the tape is modified before running at 100%.

Fig. 8.6 Valenite power monitor and recorder.

Miscellaneous Features

Parts are loaded onto the machine via an overhead crane and gripper. The parts are clamped in the headstock by a three-segment aluminum-faced chuck. Tooling pins provide positive location of the cylinder, and a steadyrest supports all but the shortest parts.

Single or multiple jets of Cimperial 15 flood coolant are directed at the tool, primarily as an aid in flushing chips away from the machined surface. Another large coolant flood originates from the center of the machine headstock to flush chips out of the cylinder. Misting of the coolant has not been a problem.

The biggest contribution to safety is the nature of the operation. That is, the spindle and tool are inside the body section, which acts as a shroud. There is also a real shroud that can be pulled over the part. This shroud helps contain cutting fluid as the spindle enters the cylinder and protects the operator when openings are machined in the cylinder wall.

The spindle comes standard with temperature limit and lubrication protection. We have added power limit protection by installing a Valenite power monitor and connecting it to the machine's controller.

Experimental Data

In the course of implementing HSM several investigations were carried out to measure cutting forces and specific cutting energies. The first series of tests sought to determine the effects on machinability of aluminum alloy/heat treatment, cutting speed, and cutting fluid application. The second series measured the effects of alternate cutter geometries.

Test Equipment

The machinability tests on this task were carried out on a modified three-axis profile milling machine. The mill is equipped with an adapter that accommodates a 20,000 rpm, 20 hp (15 kW) Bryant spindle. Up to an 800 ipm (2030 cm/min) feedrate capability is available in a single axis with a hydraulically powered, heavy duty slide table. A polycarbonate shield surrounds the hydraulic table and high-speed spindle during all cutting tests. The table is mounted on the existing machine bed and is controlled by a dedicated minicomputer system. The controller provides the capability of programming constant feedrate or variable feedrate cuts. This permits large amounts of force, power, and surface finish data to be gathered during a single test pass. Velocity feedback for the controller is provided via inputs from a linear encoder.

The primary data acquisition sensors included a three-axis force dynamometer platform, power transducers to measure spindle power, and a linear-motion transducer to measure table speed. Appropriate amplifiers, filters, and signal conditioners prepared the analog signals for recording on an FM tape recorder. A time-code generator provided time references, and a microphone provided voice references for locating data for playback. Accelerometers provided vibration data, which were used for troubleshooting and research to find the best setup to reduce vibration and noise in the data.

Initial Problems

Initial tests were conducted using two Kistler dynamometers connected by an aluminum integrally stiffened coupling plate. The plate supported a 7.25 in. × 22 in. (18.4 cm × 56 cm) specimen. A 0.750 in. (19 mm) two-flute cutter was accelerated through the specimen at an axial depth of 0.250 in. (6.4 mm) at chip thicknesses up to 0.008 in. (0.2 mm). Radial depths were one-half or one diameter. Vibration during these cuts was severe enough to loosen a locking nut on the spindle. This vibration was apparently a result of a relatively low resonant frequency for the dynamometer/fixture system (950 Hz) and the high exciting frequency of the cutter spindle system (667 Hz). In order to raise the resonant frequency from approximately 950 Hz, a single dynamometer with a 4 in. × 6 in. (10.2 cm × 15.2 cm) surface was tried. By eliminating the coupling plate and using a smaller test block, the mass of the system was reduced and the resonant frequency was raised. Also, the original exciting frequency of 667 Hz (20,000 rpm with a two-flute cutter) was reduced to 333 Hz by grinding away a portion of one flute. Modification to the 0.750 in. (19 mm) end mill was limited to grinding 0.015 in. (0.38 mm) from one flute. It was also found that full-width cuts produced less vibration than cuts at reduced radial depths. Full-width cuts apparently permit the chip thickness to gradually increase to a maximum rather than subjecting the test block to the impact-type entry of climb cuts at smaller radial depths.

The change from two-flute to essentially one-flute cutters was acceptable for the following reasons: First, the two cutting edges are simultaneously engaged for only a brief period at the beginning and end of each chip (for full-width cuts). Furthermore, this overlapping occurs when the cutting forces are minimal and in no way affects the peak cutting forces that occur about 90° later. The geometry of the active cutting edge was not changed in any way, and the maximum chip thickness was maintained at 0.008 in. (0.2 mm) by reducing the maximum feedrate by one-half.

It is important to realize that the vibration described for the cutter does not occur on the production high-speed-milling machines, where 0.25 in × 0.75 in. (6.4 mm × 19 mm) cuts at 300 ipm (760 cm/min) are routinely achieved. The most probable explanation for this lies in the fact that resonance can be avoided by operating above or below a system's natural frequency. The added mass of production parts and fixtures lowers the resonant frequency of the system, which is then not affected by the spindle/cutter system's higher frequency. This technique cannot be used in the experimental case, however, since operating above the dynamometer system's resonant frequency would result in attenuated outputs.

Test Conditions and Results for First Test Series

The six tested aluminum alloys were: 7075-T6, 2219-T6, 2219-annealed, 6061-T6, 6061-annealed, and 5083 H131. These cover a range of aluminum machinability. The common cutting conditions for all tests were: 20,000 rpm single-flute climb-milling, 0.250 in. (6.4 mm) axial depth of cut, and 0–0.00825 in./tooth (0–0.21 mm/tooth) feedrate. The radial depth of cut was 95% of the cutter diameter [0.750, 1.125, 1.500 in. (19, 28.6, 38.1 mm)]. Cutting speeds [4000–8000 sfm (1220–2440 smm)] were varied through use of the different diameter cutters. Two-hundred and sixteen data sets were recorded and processed. Based on the data, the following observations are given:

1. The variation in the range of *cutting forces* and *specific energies* between the six tested *aluminum alloys* was 26–28%. The variations do not correlate to material hardness or machinability ratings.
2. The larger-diameter cutters (and higher cutting speeds) resulted in decreased average *specific energies* of 8%. Average normal and feed *forces* also decreased by 3% and 11%, respectively.
3. There was an average of only 1–2% difference in the *specific energy* of dry cutting versus cutting with a *cutting fluid.*
4. The effect on the *cutting forces* of a *cutting fluid* was most pronounced at the larger cutter diameters and higher cutting speeds. At these conditions the use of a cutting fluid increased normal forces by an average of 10% and decreased feed forces by 12%.

Tables 8.1, 8.2, and 8.3 summarize the data.

Table 8.1 Material Comparisons—Ratio of Specific Energies and Forces for Other Aluminum Alloys to 7075

Material	Machinability Rating	Specific Energy	Normal Force	Feed Force
7075-T6	B	1	1	1
2219-T6	B	0.85	0.88	0.86
2219-Annealed	C	0.77	0.74	0.74
6061-T6	C	0.83	0.85	0.80
6061-Annealed	D	0.74	0.72	0.73
5083-H131	D	0.82	0.82	0.81

Table 8.2 Coolant Comparisons—Ratio of Dry Cutting Data to Cutting Data with Coolant

Cutter Diameter	Specific Energy	Normal Force	Feed Force
0.75 in. (19 mm)	0.983	0.948	1.027
1.125 and 1.5 in. (28.6 and 38.1 mm)	0.986	0.896	1.125

Table 8.3 Tool (Diameter) Comparisons—Average Specific Energies and Forces (for all Alloys and Coolant Conditions)

Cutter Diameter	Specific Energy[a] (hp/in.^{3}/min)	Normal Force[a] (lb)	Feed Force[a] (lb)
0.75 in. (19 mm)	0.35/0.31	120/202	123/184
1.125 and 1.5 in. (28.6 and 38.1 mm)	0.36/0.28	113/196	114/163

[a] XX/XX
- second value: at 0.008 in. (0.2 mm) chip thickness
- first value: at 0.004 in. (0.1 mm) chip thickness

Objectives of Second Test Series

The second series of tests evaluated alternate tool geometries. Experience has shown that a power limitation has been reached with our existing spindles. One possible solution to this problem is the use of more-efficient cutting tools, i.e., tools that will permit higher metal-removal rates with existing equipment.

Cutting tool manufacturers have claimed for some time that end mills with various sinusoidal patterns ground in the flutes provide multiple benefits. The claimed benefits are reduced power consumption and both reduced and redirected cutting forces. The potential impact of these claims on high-speed machining are several. First, increased productivity could be obtained from very expensive machining centers through higher metal-removal rates. Second, the claimed reduction in side cutting forces (with an increase in axial forces) would permit the machining of even thinner upstanding ribs than is already possible with high speed.

Three- and four-flute hog mills (sometimes called corn cob or roughing end mills) were used to check the validity of manufacturer's claims. Other questions addressed were whether the use of three- and four-flute cutters is possible at high speed since, typically, multiflute cutters tend to pack up in aluminum.

Test Conditions and Results for Second Test Series

Tool diameters were 0.750 in. (19 mm). This diameter minimized balancing problems at 20,000 rpm. Trim-Sol cutting fluid was applied as a mist. Three different cut cross sections were attempted with each cutter. That is, three combinations of axial and radial depths were used and are listed below.

Cut Cross Section	*Axial Depth*	*Radial Depth*
A	0.250 in. (6.4 mm)	0.700 in. (17.8 mm)
B	0.500 in. (12.7 mm)	0.400 in. (10.2 mm)
C	1.0 in. (25.4 mm)	0.200 in. (5.1 mm)

The aluminum alloy used in all tests was 2219-T851 and the dependent variables measured were net spindle power and the three orthogonal cutting forces. Three cuts were attempted with each tool at each test condition.

The use of multiflute end mills prevented the acquisition of any usable force data because of the resonance problems described earlier. Therefore, only power (specific energy) data are presented.

Considering Fig. 8.7, the data are consistent with the exception of the 0.4 in. × 0.5 in. (10.2 mm × 12.7 mm) cut for the two-flute carbide cutter. This curve is based on only one repetition of the cut (in contrast to the three repetitions for the rest) and should be considered less significant. Figure 8.7 compares the three end mills to each other at the same cut cross sections. The curves show clearly that the hog mills are more efficient than the conventional geometry carbide. They also show that of the two hog mills, the

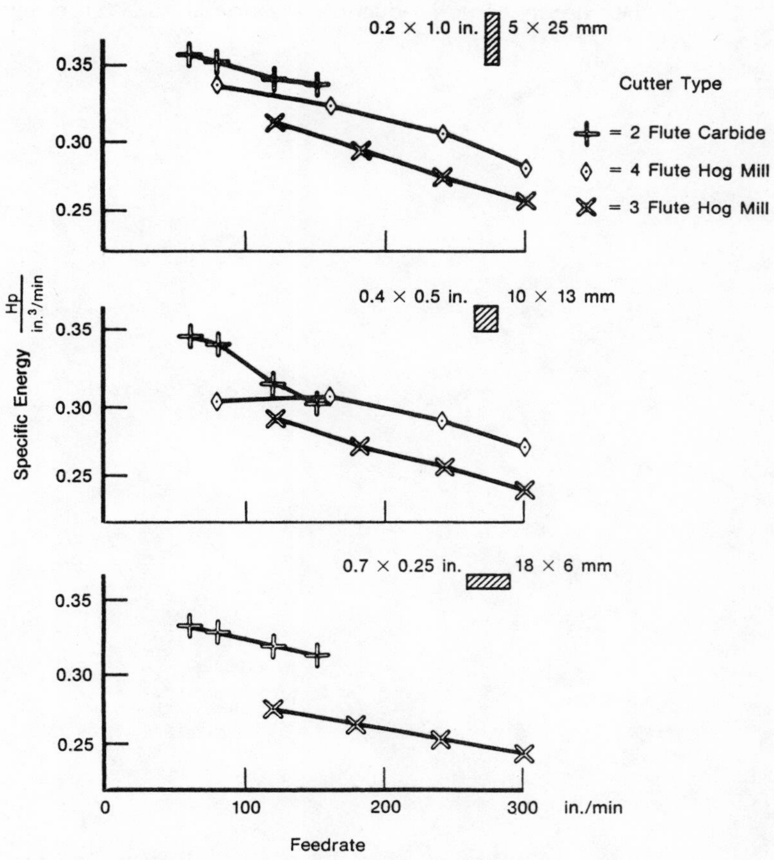

Fig. 8.7 Specific energy versus feedrate at three cut cross sections.

three flute is superior to the four flute. Again, this is consistent with the fact that heavier chip thicknesses are more efficient (the three-flute end mill cuts a heavier chip than the four-flute at the same feed rate). It should be noted that the higher helix of the three-flute end mill could have contributed to the increased efficiency.

The data in Fig. 8.7, then, support the claim by cutter manufacturers for reduced power consumption with hog mills, even though their claims were for conventional rather than high rotational speeds.

Mist lubricant was applied to the end mill during all cutting tests. While the previous tests showed no change in specific energy with or without a cutting fluid, the cutting fluid was shown to be necessary when cutting with hog mills. Initial dry cutting caused the flutes to pack up after a few inches of travel. The problem did not recur after a cutting fluid was applied.

It would seem, then, that the hog mills can solve a power limitation problem, although their use in a true production situation at 20,000 rpm remains to be verified.

Part Four

Drills and Drilling

CHAPTER 9

Introduction to Part Four

S. M. Wu
University of Wisconsin—Madison

Drilling is one of the most fundamental and widely used of machining operations. A single-engine fighter aircraft or a missile could need as many as a quarter of a million holes drilled. Yet drilling, for lack of understanding of the operation, is an art rather than a science. Consequently, drilling costs could be unnecessarily high without being noticed.

The objective of this discussion is twofold:

1. To bring the drilling operation from today's status as an art to the level of a science.
2. To propose some ways of reducing drilling costs, particularly for the aerospace industry.

To bring any engineering work to the level of a science, the first requirement is the development of a mathematical model. It is interesting to note that even though the twist drill has been in use for a hundred years, no comprehensive mathematical models were developed for a drill point geometry until 1977. The 1977 model proposed by Tsai and Wu uses four or five grinding parameters such as point angle, chisel edge angle, clearance angle, and web thickness for conventional as well as improved twist drills. The development of these mathematical models will be explained in Chapter 10.

Based on this model, a seven-axis computer-controlled drill point grinder has been conceived, designed, and manufactured. The grinder is capable of grinding any drill geometry with precision and repeatability. The mechanical structure and the computer hardware and software are given in Chapter 11.

After a drill has been ground, a computer drill analyzer has been developed using the nonlinear-least-squares method to evaluate the performance of the drill grinding operation. The computer drill analyzer is explained in Chapter 12.

Chapters 10–12 lay the groundwork for a better understanding of the twist drill and lead to the feasibility of developing an optimal drill geometry and an optimum drilling condition.

To cut down the cost of drilling and its related operations, particularly for the aerospace industry, nine different areas can be examined:

1. Drill design
2. Drill grinding
3. Drilling operation performance
4. Drilling-related-operation performance
5. Drill guns and tooling design
6. Effect of workpiece material on performance
7. Use of robotics
8. Inspection and quality control,
9. Drilling system performance and economics

In this discussion, only two items will be presented: (1) a new drill with special geometries called the Multifacet Drill (MFD) and (2) a newly developed end effector for a robotic drilling unit.

The MFD was developed in China around 1953 empirically. Tool life has been reported to be doubled or increased in an order of magnitudes. The MFD will be explained and analyzed in Chapter 13. Mathematical models for MFD's have recently been derived by the author and his co-workers, but they will not be presented here.

The development of an intelligent end effector for a robotic drilling unit could greatly facilitate the use of robots for the aerospace industries. The design of the end effector, the sensing devices, and its computer controller are described in Chapter 14.

CHAPTER **10**

A Mathematical Model for Drill Point Geometry

S. M. Wu
University of Wisconsin—Madison

An adequate mathematical representation to describe the drill point configuration is needed for the analysis and design of drill geometry. Based on the mathematical models, a drill point grinding machine can then be built to produce various desired drill geometries accurately with high repeatability.

A generalized mathematical model of a twist drill point geometry in the form of a cone, a hyperboloid, and an ellipsoid has been available since 1977. The flank of the conventional twist drill will be modeled as the surface of a cone (Fig. 10.1), the helical drill as a hyperboloid (Fig. 10.2), and the racon drill as an ellipsoid (Fig. 10.3). The model is comprehensive, simple, easy to use, and capable of incorporating a variety of drill-design features including the spiral chisel edge, the thinned web, and the double cone point.

The Mathematical Model

The derivation of the mathematical model for the drill point is based on the quadratic surface representation of the drill flank configuration. Because of the position of the drill point and the direction of the drill axis with respect to the center and axis of the quadratic surface, coordinate transformations have to be made. This allows the mathematical model to be expressed in a coordinate system that simplifies the mathematical model and facilitates the measurement of the drill point.

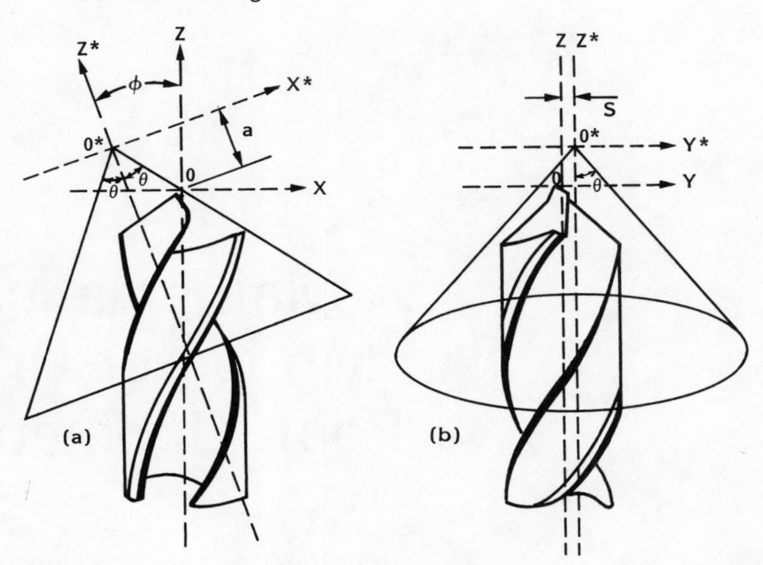

Fig. 10.1 Conic model for a conventional drill.

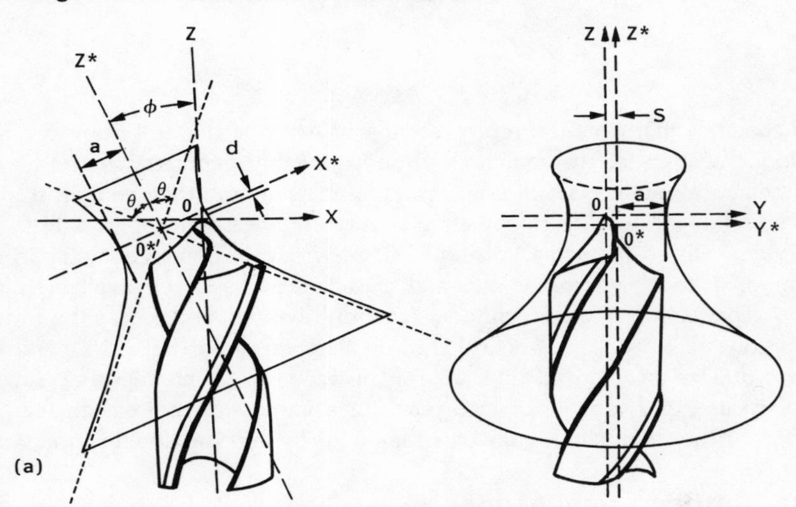

Fig. 10.2 Hyperboloid model for a helical drill.

Quadratic Grinding Surfaces

The drill flank configuration of the conventional twist drill, the helical drill, and the racon drill can be represented by the following quadratic equation:

$$\frac{x^{*2}}{a^2} + \frac{y^{*2}}{a^2} + \delta \frac{z^{*2}}{c^2} = 1, \quad \delta = \pm 1, \tag{10.1}$$

where x^*, y^*, z^* is a coordinate system and a, c are parameters (Fig. 10.4).

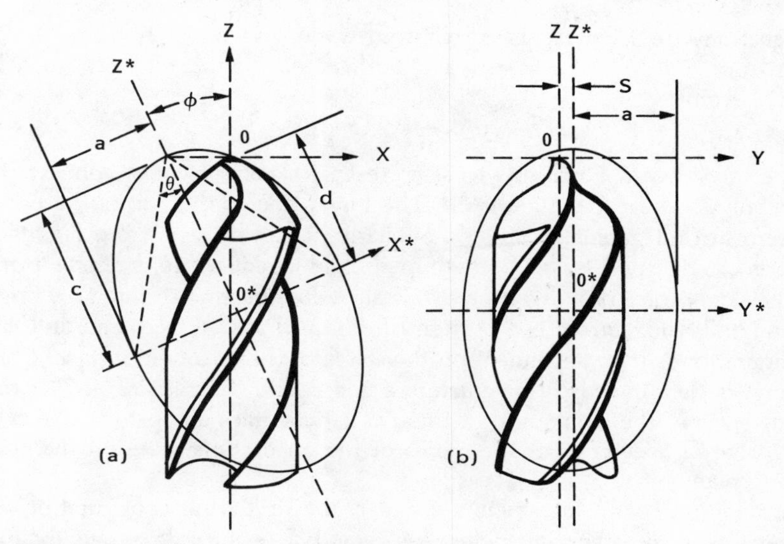

Fig. 10.3 Ellipsoid model for a racon drill.

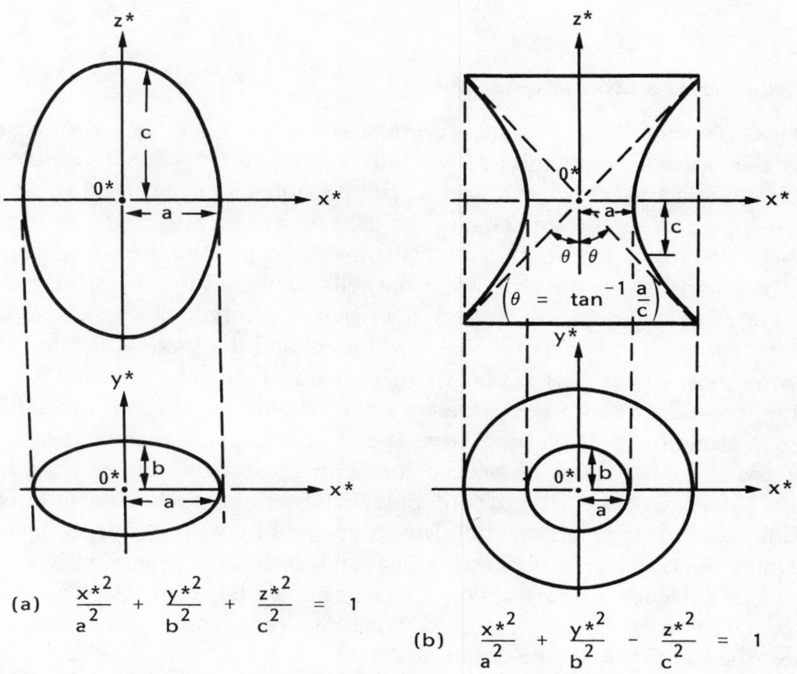

Fig. 10.4 Quadratic surface.

Equation (10.1) represents an ellipsoid when $\delta = +1$ and a hyperboloid when $\delta = -1$. When $\delta = -1$ and both a and c approach zero, then Eq. (10.1) becomes

$$x^{*2} + y^{*2} - (z^* \tan \theta)^2 = 0, \tag{10.2}$$

where $\tan \theta = a/c$. Equation (10.2) represents a pair of cones connected at their points. The cone angle is 2θ. The lower one of the pair is the one of concern in drill geometry studies. Similarly, it can be seen from Fig. 10.4b that, as a and c approach zero, the hyperboloid is reduced to the cone formed by the asymptotes shown by the two dashed lines passing through the origin.

The drill flank surface is a portion of the quadratic surface. The drill flank configuration is then dependent on the shape of the quadratic surface, which, in turn, is determined by parameters a and c, and the sign of δ. The ratio a/c determines the elongation of the quadratic surface along the z direction. The value of a determines the radius of the quadratic surface on the $z = 0$ cutting plane.

In addition to the parameters a and c, the drill flank configuration also depends on three other parameters—d, S, and ϕ—which determine the location of the drill point on the quadratic surface and the direction of the drill axis as explained in the following coordinate transformations. Figure 10.3 shows the relative orientation of the drill point and the quadratic grinding surface.

Coordinate Transformations

In Figs. 10.1 and 10.3, separate coordinate systems have been defined for both the quadratic surface and the drill point. The (x^*, y^*, z^*) coordinate system is selected with the origin of the coordinate system located at the center of the quadratic surface and the x^* axis as the axis of the quadratic surface. This selection simplifies the mathematical expression for the quadratic surface. However, for the analysis of the drill point, it is more convenient to select a coordinate system (x, y, z) such that the origin of the coordinate system is located at the center of the drill point and the z axis coincides with the drill axis. The x axis is chosen to be parallel to the projection of the cutting edge onto the plane perpendicular to the drill axis when the cutting edge is straight. If the cutting edge is not straight, like that of the racon drill, the x axis is chosen parallel to the fictitious straight cutting edge that would have been obtained if the drill point had been ground to have a straight cutting edge. A point on the drill flank is denoted by its (x, y, z) coordinates. Equation (10.1), which describes the quadratic surface, is defined in terms of (x^*, y^*, z^*). Hence, Eq. (10.1) has to be transformed from the (x^*, y^*, z^*) coordinate system to the (x, y, z) coordinate system. The transformation includes the translation of the origin of the coordinate system and the rotation of the directions of the coordinate axes. The transformation is expressed as follows:

$$\begin{bmatrix} x^* \\ y^* \\ z^* \end{bmatrix} = \begin{bmatrix} \cos\phi & 0 & \sin\phi \\ 0 & 1 & 0 \\ -\sin\phi & 0 & \cos\phi \end{bmatrix} \begin{bmatrix} x \\ y \\ z \end{bmatrix} + \begin{bmatrix} \left(a^2 - \delta \dfrac{a^2}{c^2} d^2 - S^2\right)^{1/2} \\ -S \\ d \end{bmatrix}, \tag{10.3}$$

where S and d are parameters that locate point 0 and ϕ is an angle of rotation about the y axis. The rotation takes place only in the x-z plane. Hence, the axis y and y^* remain parallel to each other.

The Mathematical Model of the Drill Point

The shape of the quadratic surface is determined by parameters a and c. The position of the drill point on the quadratic grinding surface is determined by the parameters S and d, while the direction of the drill axis is determined by parameter ϕ. Therefore, from Eqs. (10.1) and (10.3), a mathematical model of the drill point in terms of five parameters—a, c, S, d, and ϕ—can be obtained, expressed in the (x, y, z) coordinate system.

By substituting Eq. (10.3) into Eq. (10.1), we obtain

$$\frac{1}{a^2}\left[(x\cos\phi + z\sin\phi) + \left(a^2 - \delta\frac{a^2}{c^2}d^2 - S^2\right)^{1/2}\right]^2 + \frac{1}{a^2}(y - S)^2 + \frac{\delta}{c^2}(z\cos\phi - x\sin\phi + d)^2 = 1. \tag{10.4}$$

Equation (10.4) represents the model for a racon drill when $\delta = +1$ and for a helical drill when $\delta = -1$.

When the parameters a and c approach zero and $\delta = -1$, Eq. (10.4) reduces to

$$(x\cos\phi + z\sin\phi + \sqrt{d^2\tan^2(\theta - s^2})^2 + (y - S)^2 - (z\cos\phi - x\sin\phi + d)^2 = 0, \tag{10.5}$$

which represents the model for the flank surface of the conventional conical drill with only four parameters—θ, S, d, and ϕ.

Drill Point Geometry

The mathematical model can be applied to the design of drill point grinding by examining the relationship between the grinding parameters and the design parameters. The grinding parameters specify the relative motion of the drill and the grinding wheel surface during the drill point grinding process. The drill flank lies along the grinding surface generated by the relative motion of the drill and the grinding wheel surface.

Generation of Grinding Surfaces

Consider Fig. 10.5 in which the drill flank touches the grinding wheel surface on the straight line *AB*. If line *AB* is revolving about the z^* axis, a conical surface is generated. Hence, in drill point grinding, if a grinding wheel is revolving about the z^* axis with the drill fixed in space, then a conical grinding surface is generated to produce a conical drill flank. In fact, the

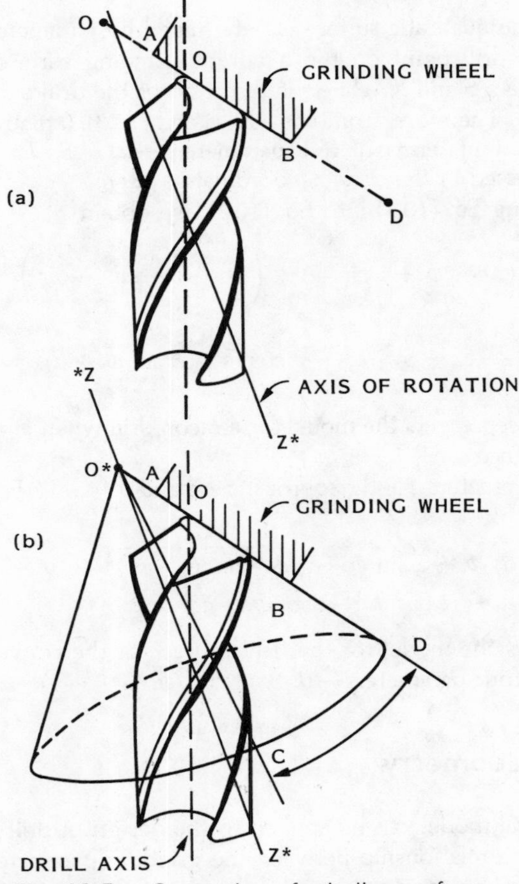

Fig. 10.5 Generation of grinding surface—conical.

grinding wheel does not need to revolve 360° about the z^* axis because the wheel is only actively cutting in the range denoted by arc *CD* shown in Fig. 10.5.

Similarly, when line *AB* is curvilinear, then the hyperboloidal or ellipsoidal grinding surface can be generated by revolving line *AB* about the z^* axis as shown in Figs. 10.6 and 10.7, respectively.

In Figs. 10.5 and 10.7, the drills are considered fixed in space during the drill point grinding process, while the grinding wheel is revolving about the z^* axis. This makes it easier to appreciate the generation of the grinding surfaces and the relative location and orientation of the drill with respect to the grinding surface. But, from a practical grinding process point of view, it is desirable to keep the grinding wheel fixed and revolve the drill as shown in Fig. 10.8.

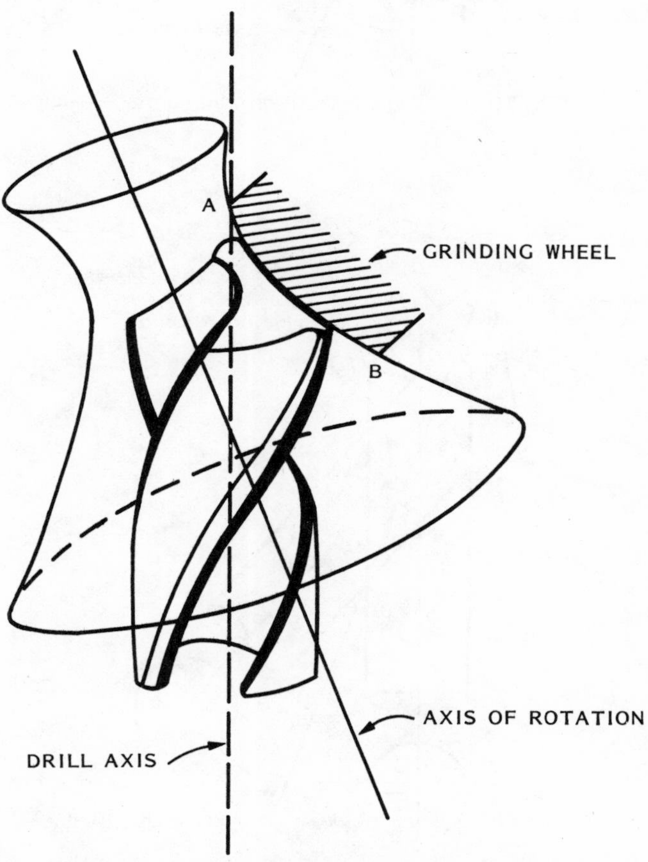

Fig. 10.6 Generation of grinding surface—hyperboloid.

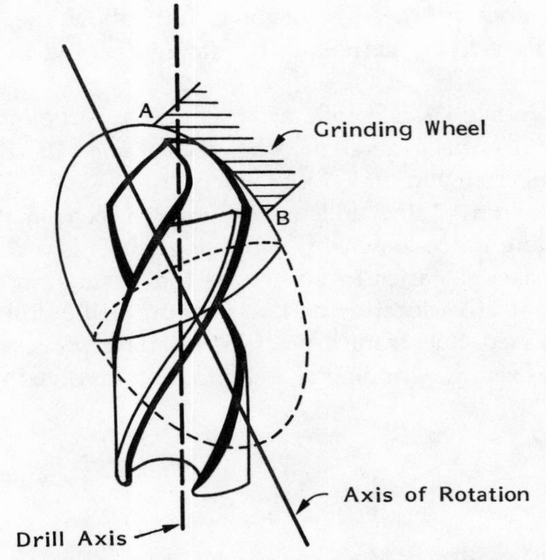

Fig. 10.7 Generation of grinding surface—ellipsoid.

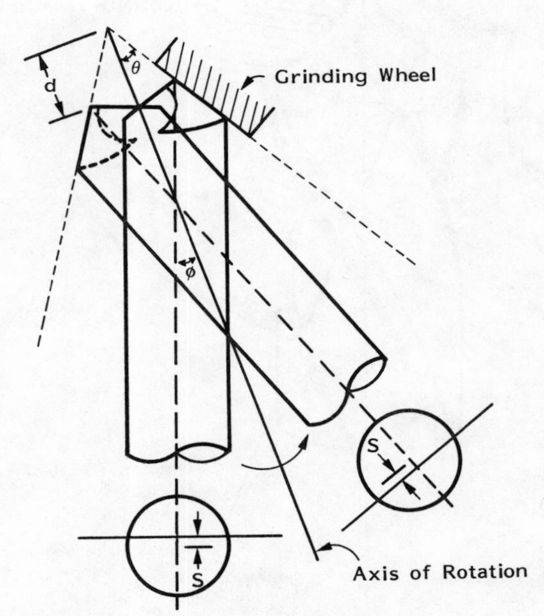

Fig. 10.8 Grinding wheel fixed.

Influence of the Grinding Parameters on Drill Point Geometry

A fundamental question in drill point grinding is the specification and control of the grinding parameters to obtain a desired drill geometry. During regrinding, it may be desirable to alter one of the drill design parameters, such as the point angle, while maintaining the existing values of the other parameters. The mathematical model developed herein enables one to precisely characterize, predict, and control drill point geometry through the specification of the grinding parameters.

The derivation of the relationships between the drill design parameters and the grinding parameters can be obtained by using Eq. (10.4) and the computer analysis of drill point geometry.

For example, the conventional drill design parameters of point angle, chisel edge angle, clearance angle, and web thickness can be derived from the grinding parameters of θ, S, d, and ϕ. A given set of drill flank grinding parameters produces a unique set of design parameters.

The effects of the grinding parameters on the peripheral clearance angle are shown graphically in Fig. 10.9. If all the other parameters are kept constant, an increase in θ or d will decrease the clearance angle. The effect of the parameter ϕ is small. The dashed line in Fig. 10.10 is the combined effect of θ and ϕ with the value of the drill point angle kept constant.

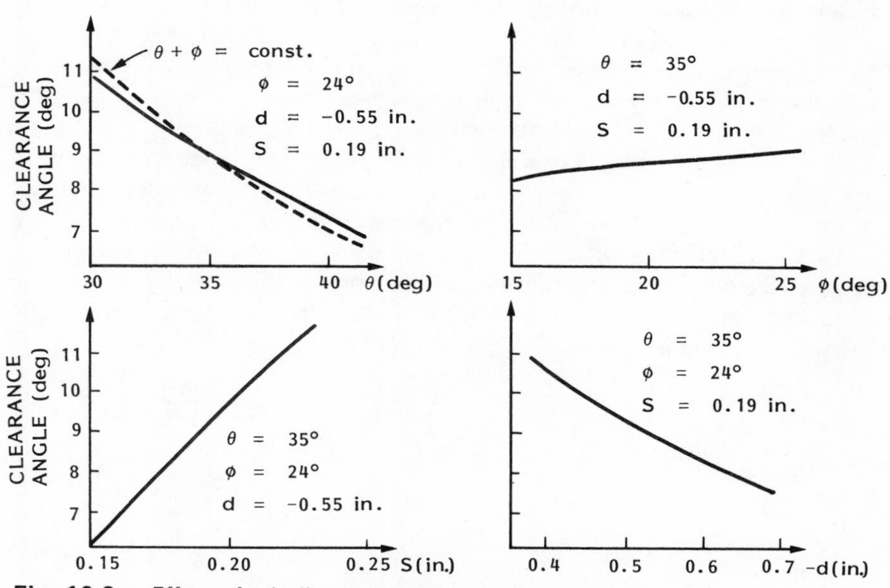

Fig. 10.9 Effect of grinding parameters on the chisel edge angle.

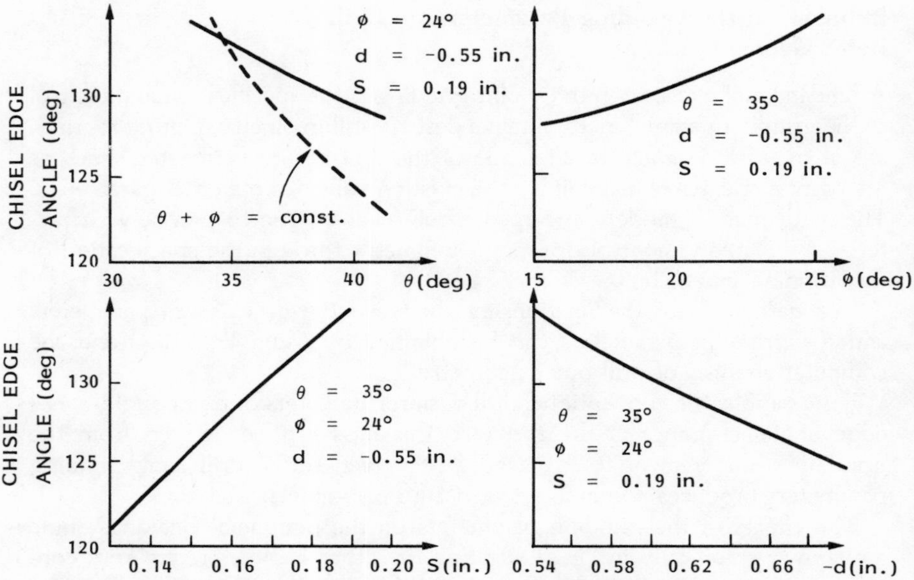

Fig. 10.10 Effect of grinding parameters on the chisel edge angle.

The effect of the grinding parameters on the chisel edge angle is graphically illustrated in Fig. 10.10. The effect on the chisel edge angle is similar to that of the clearance angle; namely, the increase in θ or d tends to decrease both the chisel edge angle and the clearance angle, while the increase in ϕ or S tends to increase both these angles. However, the rates of change of these two angles with the grinding parameters are different.

By using the generalized mathematical model, the drill point geometry of a conventional twist drill, a helical drill, and a racon drill can be quantitatively and precisely described. The mathematical model also facilitates development of the relationship between design parameters and grinding parameters. This will enable prediction of a drill's geometry, and control over the grinding process. By examining the influence of the grinding parameters on the drill angle and the sensitivity of the drill point shape to incremental changes in the grinding parameters, precise control can be achieved.

CHAPTER 11

Microcomputer-Controlled Seven-Axis Drill Point Grinder

S. M. Wu
University of Wisconsin—Madison

The problems of accurate and precise drill point grinding are closely related to the performance of the drills. Over a long period of time, the drill grinding machines available today have evolved based on empirical knowledge and by trial and error. The lack of a precise way of describing the drill point is even evidenced in national and international standards where only the conventional quantities like the chisel edge angle, point angle, clearance angle, and web thickness are used to describe the point geometry. It is not difficult to show that this system is not unique, that is, many different point geometries can be described by an identical set of the above-mentioned parameters. On the other hand, it is common knowledge that even minor variations in point geometry may lead to dramatic changes in drill performance.

At the University of Wisconsin—Madison, in the past decade, considerable efforts have been devoted to the development of ways to precisely characterize the geometry of drill points resulting in a generalized mathematical model, as explained in Chapter 10.

Subsequently, utilizing the exact relationships between the parameters of the model and the geometry of the point, a prototype drill grinding machine has been developed.[1,2]* A microcomputer was used to compute and control the necessary motions involved in the grinding process. But it was not until 1979 that at the General Dynamics' Fort Worth plant the need arose to precisely reproduce the point geometry on solid carbide split point drills for the drilling of graphite composite materials used in the F-16 multipurpose fighter.[3] Based on the first prototype, which demonstrated the flexibility of the approach, a new seven-axis computer-controlled grinder has been developed.

The essential elements in the development of the machine and the hardware configuration are discussed, with special attention paid to mechanical structure, control computer, sensor system, and computer software. The present status and capabilities of the machine are given along with its future potential applications such as the grinding of multifacet drills (MFD), end mills, etc.

Hardware Development

Based on the parametric representation of a comprehensive drill point, a prototype grinding machine capable of grinding drill points associated with quadratic flank surfaces has been developed. Since the hyperboloidal and ellipsoidal drills can be ground by piecewise fitting of cone segments to the desired point shape, the conical drill point model has been taken as the basic underlying model. The machine is constructed so that position and motion of the axes produce a cone of varying shape depending on the required point geometry. A brief description of the major elements of the machine follows.

Mechanical Configuration

The necessary motions have been subdivided into two major groups:

1. The motions associated with the drill bit
2. The motions to be performed by the grinding wheel

Referring to Fig. 11.1, the basic motions the drill bit has to perform are the following:

1. The frame carrying the drill rotates about the vertical axis *HH* providing the ϕ parameter.

* Superscript numbers refer to references listed at the end of Chapter 14.

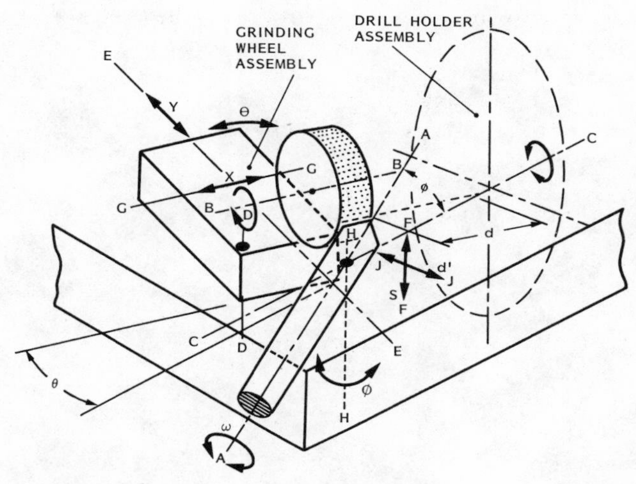

Fig. 11.1 Machine motions and axes.

2. The drill itself can rotate about its axis *AA* exposing one or the other flank surface to be ground to the grinding wheel (ω).
3. The base of the frame has a translatory motion along axis *FF* to set the *S* parameter.
4. In addition, the base of the frame has a translatory motion along the direction *JJ* providing the *d* parameter setting.
5. The entire frame is then oscillated about axis *CC*, i.e., the "axis of rotation," to generate the grinding motion. For the conical model this is the only motion during the grinding process.

The basic motions of the wheel assembly are:

1. The rotation of the grinding wheel about axis *BB*.
2. The wheel spindle and wheel have a translatory motion along axis *EE*, which provides the infeed during rough and finish grinding.
3. In addition the wheel has a translatory motion along the direction *GG*, perpendicular to *EE*, facilitating an exact positioning of the wheel edge in relation to the drill tip for grinding split point drills.
4. Finally the whole wheel assembly rotates about vertical axis *DD* to set the cone angle θ.

Out of the total of nine required motions only the oscillatory motion of the frame and the rotary motion of the wheel are not under computer control but are inherent in the design. The remaining seven axes must be under the control of a suitable computer to provide the necessary flexibility and repeatability of the machine.

Fig. 11.2 Photograph of the seven-drill grinder.

Based on this motion and axis assignment the prototype machine has been developed. Figure 11.2 show the overall view and the photograph of the machine.

Control Computer and Sensor System

In order to have a mechanically simple and yet versatile machine a microcomputer is used for coordinate calculations and control. The computer facilitates the calculation of the needed positions of the seven axes based on the drill geometry specified. This allows for rapid changes from one drill geometry to another in an entirely automatic manner. Furthermore, the computer's capabilities are used to compensate for any manufacturing and assembly deviations from the nominal values. Using a suitable sensing system after "power-on," the computer searches for the absolute zero reference of the machine.

The computer selected is a general purpose eight-bit microcomputer based on the Motorola M6800 chip. It has two parallel ports, each port having 16 I/O lines. This configuration is ideally suitable to generate pulse trains to control the motion of stepping motors used as the actuators for the axes.

Typically, two lines are required to interface the computer with the translator card of the driving motor, one for clockwise and the other for counterclockwise rotation.

A separate optically isolated digital I/O interface board is used to interface the computer with the sensor system used to detect overtravel on each of the axes and to determine the zero reference position. Two limit switches are used on each axis to define the end points of travel in addition to a proximity sensor, having a repeatability of 0.0004 in. (0.01 mm), to determine the zero reference. In the initialization sequence following the "power-on" situation, the computer seeks a high level on one of the limit switches and the proximity sensor to determine the reference zero. During operation where one of the limit switches closes, an emergency shutdown procedure is initiated, designating an abnormal condition.

The command inputs to the computer are handled through the same digital I/O board on a prioritized interrupt basis. The logic and sequence of the commands is checked first before they are accepted or rejected. The requests entered through the operator's console are treated in a similar way. Finally, the computer activates the grinding wheel motor and the motor for the generation of the grinding motion.

A schematic representation of the microprocessor-based controller is given in Figure 11.3. The majority of components are standard off-the-shelf items.

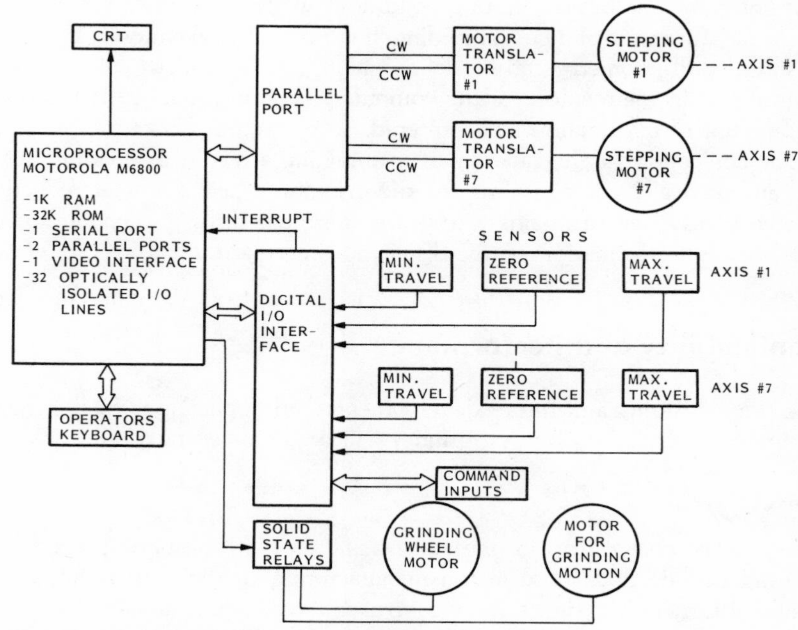

Fig. 11.3 Schematic of the microprocessor controller.

Software

The control software is written on a modular basis and is summarized in Fig. 11.4. The modular approach facilitates easy software maintenance and necessary additions to increase the versatility of the machine in future development stages.

The main module is responsible for all communications on one side and the computer and machine tool on the other. All I/O transactions are treated on an interrupt basis; consequently, one of the basic functions of the main program is to invoke the proper routine to handle the request.

The driver package consists of a set of subroutines handling the acceleration/deceleration functions and keeping track of the positions of each of the axes. It is capable of generating motion in all axes simultaneously. Furthermore, it is responsible for overtravel conditions which may arise as a result of malfunctions.

The arithmetic processor consists of a set of subroutines implementing the basic arithmetic operations and the necessary trigonometric functions. It is also used for the computations of the motion parameters, namely, axes coordinates, amount of travel, and velocity of travel. Finally, a collection of different subroutines is provided for diagnostic and maintenance purposes.

The operation of the machine is extremely simple. The operator has to enter, in an interactive manner, the ϕ, d, S, and θ parameters defining the drill point and the desired cutting conditions in terms of stock removal and infeed rate for rough and finish grinding. If the operator so chooses, a standard or stored drill geometry can be defined along with standard cutting conditions stored in the memory of the computer. This procedure further reduces the amount of information to be entered.

The controller has the capacity of maintaining a record of dressing cycles and automatically compensating for the grinding wheel diameter reduction. Finally, the system maintains a record of different drills ground during the shift, and this information can be displayed on request.

Applicability and Performance

The present status and future developments and applications of the microcomputer-controlled drill point grinder will be discussed next.

Present Status

The primary purpose of the presently developed prototype is the grinding of solid carbide drills used in the manufacturing of the F-16 multipurpose fighter at General Dynamics. Basic control software is included for the generation of conventional conical drills and conical split point drills. The point

	Module	Functions	Description
1	Command/ Communication Program	• Zero reference • Grinding cycle • Dressing cycle • Start/Stop • Display information • Wear and error compensation • Emergency stops	Provides the basic communication between operator and machine on a prioritized interrupt basis. Immediately after power-on seeks the absolute zero reference. Accepts the parameters and cutting conditions for the drills to be ground. On request displays information on current axis position, number of drills ground, wheel wear compensation, etc. Transfers control to specific routines.
2	Driver Package	• Acceleration • Deceleration • Travel at given velocity, given distance • Overtravel detection	Generates the necessary acceleration/deceleration profiles for all the axes. Drives the seven axes simultaneously at the specified velocity the required distance. Checks the state of the overtravel sensors and shuts down the machine if overtravel occurs.
3	Arithmetic Processor (Floating Point)	• Basic arithmetics (+ −/*) • Functions (SIN COS, ARCSIN, SORT, etc.) • Axis coordinate computations • Collision checks	Based on the specified drill geometry computes the coordinates of each axis and the necessary motion parameters. Performs a collision check if more than one axis has to move simultaneously. Computes the safety position for drill change.
4	Diagnostics/Test Routines	• Memory check (RAM/ROM) • Serial and parallel port diagnostics • Sensor system diagnostics • Simple axis test routines	Aids in diagnosis of malfunctions and in routine maintenance. Detects errors in the computer and sensor system. Provides test routines for axis diagnostics regarding direction, speed, and positioning accuracy.

Fig. 11.4 Control software.

splitting, or sometimes referred to as web thinning, operation is performed by keeping the oscillating frame stationary close to its lower position, setting the remaining axes into a position dictated by the desired parameters, and infeeding the wheel in the usual manner. All parameters are maintained within the tolerances prescribed by the national and aerospace standards. However, for most of the angular parameters the machine provides an order of magnitude better tolerance than is prescribed by the mentioned standards. This fact is important especially since even minor changes in point geometry may result in significant improvements in drill performance. On the other hand, high repeatability must be provided in order to reproduce drills with commensurate performance.

Future Developments and Applications

Efforts are being devoted to the incorporation of the oscillating motion about the *CC* axis under computer control to result in an eight-axis machine. This addition will then allow the manufacture and regrinding of multifacet drills, which, based on some preliminary experiments at the University of Wisconsin—Madison, showed a dramatic decrease in thrust and torque as compared to more conventional point geometries.

The eight-axis machine will be ultimately equipped with the necessary control system to generate almost any general grinding surface.

Finally, the applicability of the present prototype for end mill, reamer, and countersink grinding is being investigated. Some basic endmill grinding operations have already been successfully demonstrated. The key problem is the development of suitable mathematical models for the description of the cutting edge geometries and the establishment of the relationships between the model and the motion parameters of the machine.

Conclusions

1. Using the mathematical model as a design tool, a microcomputer-controlled drill grinding machine has been developed.
2. The conical model has been implemented for the definition of the basic motions of the machine.
3. The mathematical drill point model enables the generation of conical, hyperboloidal, and ellipsoidal split point drills.
4. The machine is mechanically simple because all the machine settings are under computer control.
5. The grinder facilitates a tool for reproducing desired drill geometries with a high degree of accuracy and repeatability, hence providing uniform performance characteristics of the ground drills.

6. The developed mathematical models are more precise and lead to an exact and unique description of the point geometry, unlike the conventional way of defining the point, which does not yield unique representations.
7. Since the machine provides motion in seven axes simultaneously, the addition of a new axis being under consideration and the enhancement of the software package will lead to an even more versatile machine capable of grinding multifacet drills, end mills, etc.

CHAPTER 12

Drill Analyzer

S. M. Wu
University of Wisconsin—Madison

After grinding the flank surface, determining the accuracy of the ground surface may be desirable to modify the grinding process. Also, to reproduce a given point configuration, the flank surfaces need to be measured to obtain the required grinding parameters. An apparatus named the "drill analyzer" has been built at the University of Wisconsin—Madison to achieve this. It can automatically measure the flank geometry and yields desired results with the aid of a computer (microprocessor).

Measurement of the Drill Flank Configuration

Concept

To measure the drill flank configuration, a coordinate system should be defined and the measurements should cover every area of the drill flank surface. The number of observations depends on the complexity of the drill flank configuration.

The Coordinate System

The drill flank configuration is measured in a rectangular coordinate system (x, y, z) with the origin located at the drill point center; the Z axis coincides with the drill axis and the X axis is parallel to the cutting edge, as shown in

Fig. 12.1. This coordinate system is the same as the one used for the drill point mathematical model discussed in Chapter 10.

For certain drills, namely, those whose cutting edges are not straight, the direction of the X axis cannot be easily determined. However, by introducing an additional grinding parameter ω, the X_ω axis may be chosen in any direction perpendicular to the drill axis and at an angle ω with the cutting edge. For convenience, before fitting the model, the X_ω axis may be chosen to be parallel with the chisel edge. The drill flank may be measured in this new coordinate system and then transformed into the (x, y, z) coordinate system in which the drill model is expressed. The angle between the X_ω axis and the X axis is ω, which is positive in the direction shown in Fig. 12.1. The (x, y, z) coordinates can be obtained from $(X_\omega, Y_\omega, Z_\omega)$ by rotating an angle

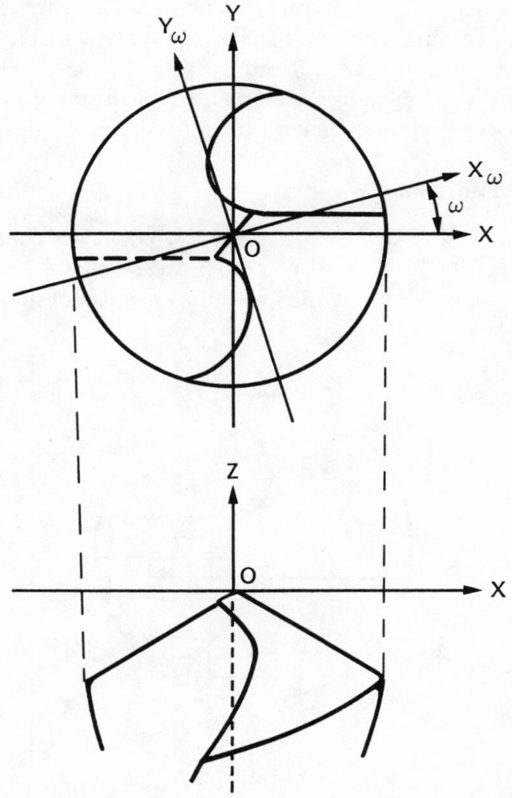

Fig. 12.1 The coordinate system to measure the drill flank configuration.

of ω about the Z axis:

$$\begin{bmatrix} X \\ Y \\ Z \end{bmatrix} = \begin{bmatrix} \cos\omega & -\sin\omega & 0 \\ \sin\omega & \cos\omega & 0 \\ 0 & 0 & 1 \end{bmatrix} \begin{bmatrix} X_\omega \\ Y_\omega \\ Z_\omega \end{bmatrix} \tag{12.1}$$

The value of ω can be estimated simultaneously with the grinding parameters by fitting the model.

Number of Observations

For a conical drill, which has a simple drill flank configuration, the number of observations required for fitting the model need not be large (30 or less). The hyperboloidal and ellipsoidal drills have more complex drill flanks. Hence, more observations (100–200) are required so that surface variations can be precisely measured. The measurements may be made sequentially in the direction of increasing X and increasing Y along the drill flank surface, and distributed over the drill flank surface as shown in Fig. 12.2. The drill flank surface is divided into a grid, and the (x, y, z) coordinates at each node can be measured by a transducer. The number of observations will affect the precision of the parameters evaluated from the data.

The Objective Function

If the drill flank fits the drill model well, the errors between the measured drill flank coordinates and the drill model are small. The best fit is determined by an iterative method, which uses the criterion of minimizing the sum

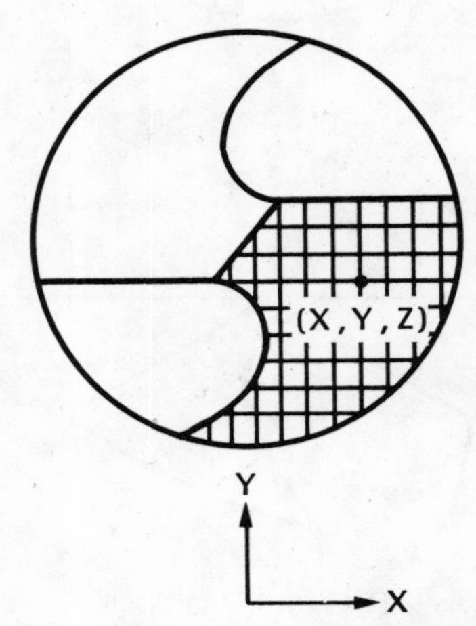

Fig. 12.2 Grid set-up for measuring the drill flank configuration.

of squares of errors. The objective function to be minimized in the nonlinear-least-squares method has an effect on the speed of convergence and the accuracy of the model obtained.

In order to distinguish the drill flank configuration obtained from the model from the directly measured configuration, let $(\hat{x}, \hat{y}, \hat{z})$ denote the coordinates obtained from the model and let (x', y', z') denote the measured coordinates. If a model is good, $(\hat{x}, \hat{y}, \hat{z})$ should be close to the measured values (x', y', z'). Hence, minimizing the error between (x', y', z') and $(\hat{x}, \hat{y}, \hat{z})$ can be used as an objective function in fitting the model, i.e.,

$$\sum_{i=1}^{N} \varepsilon_i^2 = \sum_{i=1}^{N} (z_i' - \hat{z}_i)^2 = \sum_{i=1}^{N} [z_i' - f(x_i - y_i)]^2, \tag{12.2}$$

where subscript i denotes the ith observation and N is the total number of observations.

Initial Guess Values

In fitting the drill point model, an initial guess of the parameter values is required. The nonlinear-least-squares method computes the value of the objective function based on the initial guess and searches for parameter values that minimize the objective function. The new parameter values are then used as the next set of guess values and so on, iteratively, until some minimizing criterion is met. Good initial guess values are needed to ensure that the program will not converge to a local optimum and to keep the total computational time (number of iterations) to a minimum.

Because some of the parameters are correlated, the initial guess value of each parameter is not determined independently. After fitting the drill model, the information data measured from the drill flank are condensed into five parameters:

- a and c: determine the shape of the quadratic grinding surface and have the dimension of length.
- d: determines the location of the drill point center on the quadratic surface and has the dimension of length.
- S: the skew distance larger than half of the web thickness in value; it provides positive clearance angle on the drill flank.
- ϕ: the direction angle, in radians, between the axis of the quadratic grinding surface and the drill axis; it increases the effective clearance angle in the cutting process.

The initial guess values of parameters can be estimated from the geometrical relationship between the drill points and the quadratic surface. For example, in the conical drill, a and c approach zero; $\theta + \phi = p =$ half-point angle, etc.

Apparatus to Measure and Model the Flank Geometry

Using the concept discussed previously, a microprocessor-controlled device was developed to automatically measure the z coordinates of the drill flank surfaces, fit least-squares models to the data obtained, and yield the model grinding parameters of the flank. This section discusses the mechanical and electronic features of this device.

System Components and Mechanical Design

The device is shown in Fig. 12.3. The drill is held vertically (point up) in a chuck, which is mounted on a stage. The stage consists of two slides that

Fig. 12.3 Microprocessor-controlled device to measure and model the flank geometry.

can be moved in perpendicular directions (x and y axes) by precision lead-screws. One turn of the leadscrew moves the stage 0.025 in. (0.635 mm).

An appropriate means of driving the stage is through the stepping motor. These motors are designed to produce incremental movements in response to electrical pulses provided by a computer. They can be operated in an open-loop fashion and still provide sufficient positioning accuracy. Two four-phase, bifilar dc stepping motors (SLO-SYN M061-FD08) with positioning accuracy and 35 oz-in. (2.5 kg-cm) holding torque were chosen. With a 1.8° step size (200 steps/rev), these motors can produce incremental movements of 0.000125 in. (0.0032 mm) in either the x or y directions when coupled to the leadscrews.

Measurements in the z direction (along the drill axis) are made by a linearly variable differential transformer (LVDT), which is an electrical transformer with a separable noncontacting core producing an electrical output proportional to the displacement of the core. It has a long mechanical life (due to frictionless operation), ruggedness, good resolution, repeatability, and input/output isolation. A dc-exited LVDT (Hewlett-Packard 24 DCDT-100) with a nominal displacement range of 0.1 in. (2.54 mm) was employed. This LVDT has a built-in carrier oscillator and demodulator; therefore, no external circuitry is required. The nominal value of the scale factor is 90 V/in. (35.4 V/cm) and that of maximum nonlinearity is 0.5% of the full scale. However, during the calibration of the LVDT, it was observed that the relationship between the actual LVDT core displacement and the value acquired by the computer was not quite linear—probably because of the circuitry interfacing the LVDT to the microcomputer. In order to overcome this problem, the actual displacement was related to A/D convertor output by five line segments instead of one, as shown in Fig. 12.4. This approach involves more programming, but provides higher accuracy and increases the useful range of the LVDT. The core of the LVDT is connected to a stylus that can rest on the flank surface of the drill. In order to make the movement of the stylus-core accurate and easy, two linear bearings are mounted on the sides of the LVDT. Movement of the stylus is carried out by a solenoid that absorbs a soft iron core which is connected to the other side of the LVDT core. The stylus-LVDT-core-iron core is pulled up by a magnetic force when the solenoid is energized, and moves downward under its own weight when the solenoid is deactivated. A two-way pneumatic damper with its piston connected to the iron core prevents the stylus from accelerating excessively while falling downward, which might damage the stylus and/or the drill flank surface. It also prevents oscillation when the solenoid is activated. The damper used is model 95 AIRPOT, which can be adjusted to provide damping coefficients from 0 to 1 lb/in./sec (0 to 0.18 kg/cm/sec). A drawing of the stylus assembly is shown in Fig. 12.5.

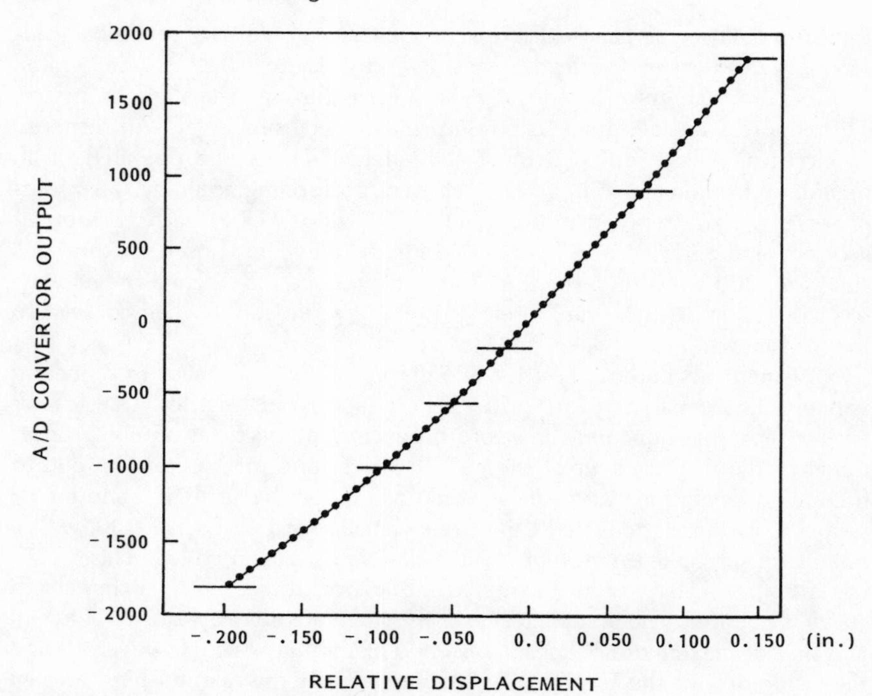

Fig. 12.4 The LVDT calibration curve.

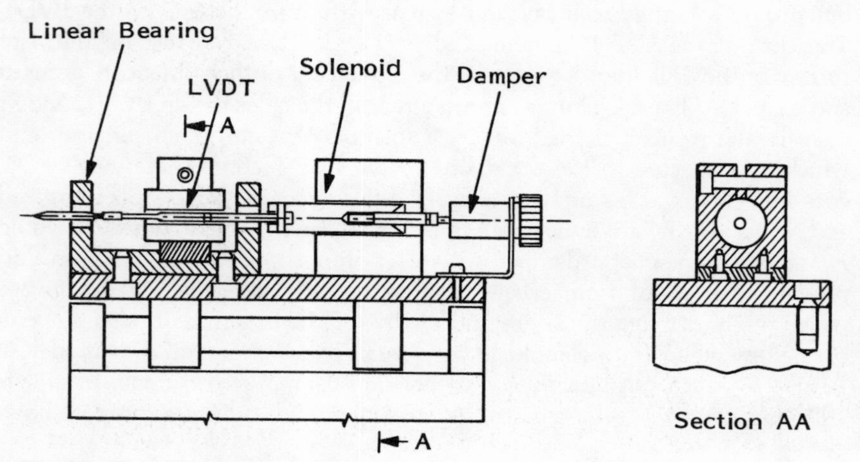

Fig. 12.5 The stylus assembly.

Electronic Design

The Microcomputer The microcomputer used to perform the control and computational tasks is a Midwest Scientific Instruments 6800, based on the Motorola 6800 eight-bit microprocessor with 16K bytes of memory. The computational power and speed of the basic microcomputer are greatly enhanced by the addition of an arithmetic processing unit (AM 95 11). This single chip is capable of calculating a large number of transcendental functions in addition to basic arithmetic operations on both fixed-point and floating-point numbers. It is superior to software packages, since routine calculations can be performed more easily and at a much higher speed. The microcomputer, the video terminal, and the measuring device are shown in Fig. 12.6.

The Interfacing of the Computer with the Other Components Figure 12.7 shows how the computer is interfaced with the various components. The interfacing is accomplished through a standard parallel interface port.

The LVDT is interfaced with the computer through an analog-to-digital conversion board, which turns the output voltage of the LVDT into a binary number. The resolution of the measurements in the z direction is about 0.001 in. (0.025 mm).

Through a solid-state relay (CLARE 223B11A1A) the solenoid is controlled by the computer.

Fig. 12.6 Drill point measuring and modeling system.

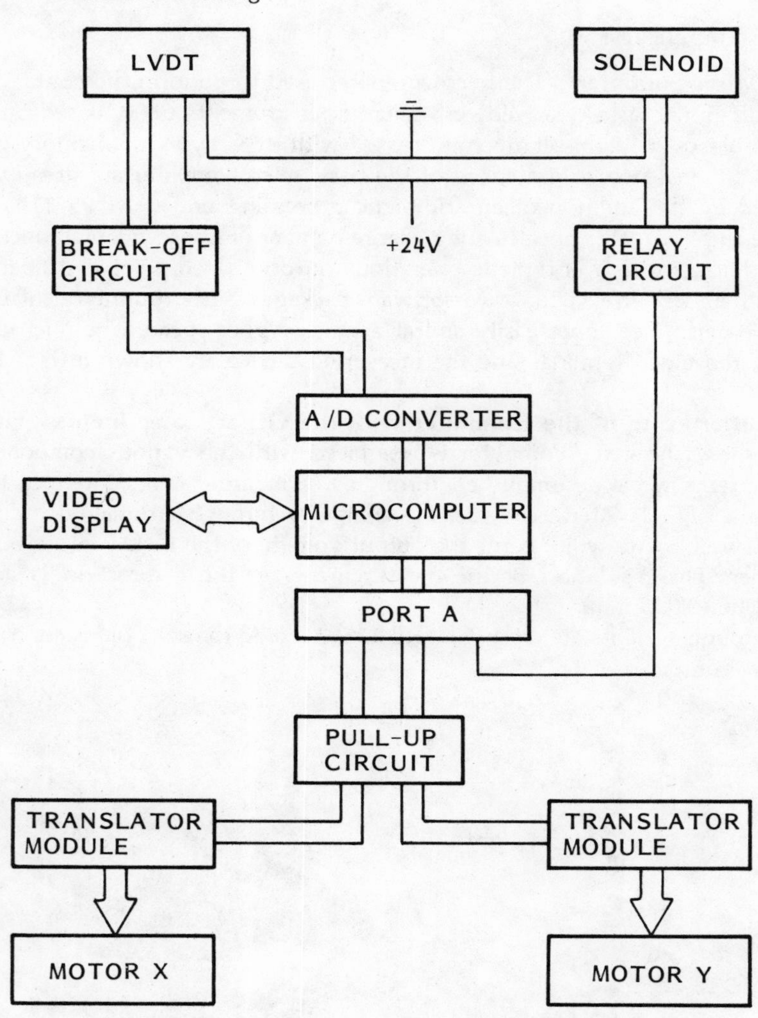

Fig. 12.7 The computer/component interfacing.

CHAPTER 13

Multifacet Drills

S. M. Wu
University of Wisconsin—Madison

Since the introduction of the conventional twist drills about 100 years ago, various drill point geometries, such as the double conic drill, split point drill, racon drill, and helical drills, have been introduced to improve drill performance. Around 1953, a new drill point geometry appeared in Beijing, China, which, in this chapter, is called the Multifacet Drill (MFD). In the subsequent 20 years, more than 20 new types of MFDs have been developed dealing with various workpiece materials and drilling conditions.

The features of a typical MFD are summarized first. Then, the designs of the MFD in terms of its function are analyzed including force reduction, drill life, heat transfer, chip ejection, and hole quality. Several representative MFDs are briefly presented, and the MFD designed for aluminum alloy drilling is analyzed in more detail. Finally, a suggested MFD for drilling composite material is offered.

A Typical MFD

A typical MFD is shown in Fig. 13.1. This MFD has six main facets (Flanks 1, 2, and 3 on both cutting edges). Flank 1 is ground as a conventional twist drill. A cylindrical flank designated as Flank 2 is ground on Flank 1 to form an arc cutting edge in the inner part of the cutting lip. Flank 3 is ground to form another cylindrical surface near the center of the drill to reduce the chisel edge length.

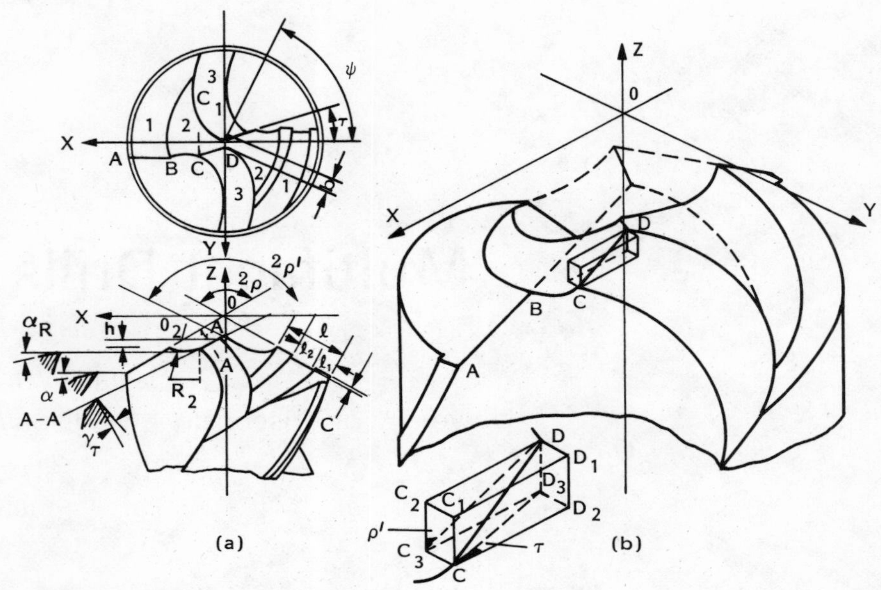

Fig. 13.1 Typical type of MFDs.

The geometry parameters of a typical MFD are explained as follows.

1. Chisel edge length b, tip height h, and point angle of the inner cutting edge $2\rho'$: The chisel edge length b is only $0.03D$ (i.e., 3% of the diameter of the drill) to reduce the thrust force such as the split drill and web thinning. In order to strengthen the reduced chisel edge length b, the tip height h (which is unique to the MFD owing to its double-tipped cutting edge) is designed to be $0.3D$ with an inner point angle $2\rho'$ of 135°. The large value of $2\rho'$ is also required to avoid high-temperature generation by heat transfer in the reduced chisel edge.

2. Cutting edge length l and radius of curved edge R: On a conventional drill, the rake angle decreases from the periphery toward the drill center. At the point about $0.2D$ from the drill axis, the rake angle is close to zero. Thus the straight cutting edge length l is designed to be $0.2-0.3D$. An arc cutting edge is made to increase the rake angle with a radius of curved edge R equal to $0.1D$. The arc cutting edge is effective in dividing chip and drill centering.

3. Inner cutting edge inclination τ, rake angle γ_τ, and the relief angle α, α_R: The inclination angle ranges from 20° to 30° and depends on the length of the inner cutting edge, CD, the chisel edge length, and the rake

angle of the inner cutting edge. The outer relief angle α of the MFD is a slightly greater than the conventional drill because the feed rate for the MFD is usually greater. The range of α is 10–15°. The curved edge relief angle α_R is 2–3° greater than α.

4. Point angle 2ρ: Point angle is related to the rake angle, cutting force, chip ejection, and heat transfer. As the point angle increases, the rake angle and the thrust force increase, while torque will decrease. The chip will be ejected and broken easily, but the temperature of the periphery will increase. The point angle of MFD equals 125° as compared with 118° for conventional drills.

5. Special feature for MFD: If the diameter of the drill is greater than 0.6 in. (15 mm), one or two grooves are ground on one of Flank 1 to divide the chip, making it easier to brake and remove. The total width of the groove is half of the outer edge length, that is, $l_2 = \frac{1}{2}l/Z$. Another half of the outer cutting flank is equally divided, so the cutting edge length l_1 can be determined as $l_1 = \frac{1}{2}l/(Z + 1)$.

The parameters of a typical MFD are shown in Table 13.1 for drills with diameters ranging from 5 to 60 mm.

The Functional Design of an MFD

The design of a MFD is for

1. Reducing cutting forces
2. Strengthening the center of the drill
3. Speeding up heat transfer
4. Improving centering tendency
5. Facilitating chip ejection

Cutting Force

On the conventional twist drill, the rake angle decreases toward the drill center and approaches large negative values. Thus, while the outer edge produces a smooth chip, the inner edge does not. The material under the chisel edge is subjected to deformation by displacement. Therefore, the chisel edge creates a great thrust force.

The MFD is designed to reduce the cutting force by

1. An arc cutting edge in the middle of the cutting lip to increase the rake angle.
2. A reduced chisel edge length b to decrease the thrust force.

Table 13.1 Parameters of a Typical MFD

<table>
<tr><td>Drill diameter (mm)</td><td>5–7</td><td>>7–10</td><td>>10–15</td><td>>15–20</td><td>>20–25</td><td>>25–30</td><td>>30–35</td><td>>35–40</td><td>>40–45</td><td>>45–50</td><td>>50–60</td></tr>
<tr><td>Drill tip height, h (mm)</td><td>0.2</td><td>0.28</td><td>0.36</td><td>0.55</td><td>0.7</td><td>0.85</td><td>1</td><td>1.15</td><td>1.3</td><td>1.45</td><td>1.65</td></tr>
<tr><td>Radius of curved edge, R_2 (mm)</td><td>0.75</td><td>1</td><td>1.5</td><td>1.5</td><td>2</td><td>2.5</td><td>3</td><td>3.5</td><td>4</td><td>4.5</td><td>5</td></tr>
<tr><td>Outer edge length, l (mm)</td><td>1.3</td><td>1.9</td><td>2.7</td><td>5.5</td><td>7</td><td>8.5</td><td>10</td><td>11.5</td><td>13</td><td>14.5</td><td>17</td></tr>
<tr><td>Cutting edge length, l_1 (mm)</td><td>—</td><td>—</td><td>—</td><td>1.4</td><td>1.8</td><td>2.2</td><td>2.5</td><td>2.9</td><td>2.2</td><td>2.5</td><td>2.9</td></tr>
<tr><td>Groove width, l_2 (mm)</td><td>—</td><td>—</td><td>—</td><td>2.7</td><td>3.4</td><td>4.2</td><td>5</td><td>5.8</td><td>3.25</td><td>3.6</td><td>4.25</td></tr>
<tr><td>Chisel edge length, b (mm)</td><td>0.2</td><td>0.3</td><td>0.4</td><td>0.5</td><td>0.6</td><td>0.75</td><td>0.9</td><td>1.05</td><td>1.15</td><td>1.3</td><td>1.45</td></tr>
<tr><td>Groove depth, c (mm)</td><td colspan="3">—</td><td colspan="5">1</td><td colspan="3">1.5</td></tr>
<tr><td>Quantity of grooves, z</td><td colspan="3">—</td><td colspan="5">1</td><td colspan="3">2</td></tr>
<tr><td>Point angle of outer cutting edge, 2ρ (deg)</td><td colspan="11">125</td></tr>
<tr><td>Point angle of inner cutting edge, $2\rho'$ (deg)</td><td colspan="11">135</td></tr>
<tr><td>Chisel inclination ψ, (deg)</td><td colspan="11">65</td></tr>
<tr><td>Rake of inner cutter edge, γ_τ (deg)</td><td colspan="11">−15</td></tr>
<tr><td>Inner cutting edge inclination, τ (deg)</td><td colspan="3">20</td><td colspan="5">25</td><td colspan="3">30</td></tr>
<tr><td>Outer edge relief angle, α (deg)</td><td colspan="3">15</td><td colspan="5">12</td><td colspan="3">10</td></tr>
<tr><td>Curved edge relief angle, α_R (deg)</td><td colspan="3">18</td><td colspan="5">15</td><td colspan="3">12</td></tr>
</table>

3. A relatively large point angle of 120–125°, which can increase the rake angle of the outer part.

Strength of Drill Center

In order to strengthen the reduced chisel edge of an MFD, the length of the chisel edge b is kept to a limit of $0.02D$; the angle $2\rho'$ is enlarged to 135°, as an example; and the top height h is limited to the range of $0.02-0.03D$.

Heat Transfer

For some materials such as cast iron, heat conductivity is low. In order to decrease the temperature accumulation as well as wear at the outer cutting edge, a double point angle is adopted to make the outer point angle z equal to 70°, as an example. Also, the clearance angle of an MFD is greater than conventional drills to reduce frictional forces and heat generation.

Chip Ejection

One of the important features of the MFD is that the chip form and its ejection are always considered in detail. An arc cutting edge is always used to divide the chip for all MFD designs.

When the elongation and toughness of the workpiece material become large, the point angle is designed to be bigger to facilitate the chip ejection. For example, the point angle of the MFD for aluminum alloy is designed to be 140–170°; for stainless steel, 135–150°; and for titanium alloys, 125–140°. Furthermore, there is one or even two grooves ground on one of the flanks of the big drill to break the chip.

Centering Tendency

The reduced chisel edge b plays an important role for drill centering. For copper and aluminum, the hardness of which is low and the elongation large, the design of the MFD in the tip height h is increased to $0.05-0.06D$ and the angle $2\rho'$ is reduced to about 110° to enforce the centering tendency. For stainless-steel and titanium alloys, when the tip height h increases to $0.06-0.08D$, the chisel edge length b increases to $0.04-0.05D$ simultaneously in order to strengthen the center part of the drill. Also, the arc cutting edge with a relative small radius can improve centering.

Special MFDs

A typical MFD does not always have optimum performance under various working conditions. Therefore it should be modified according to individual needs. A few MFDs are presented with due modifications for illustration.

(1) *For cast iron:* A second conical surface with small $2\rho'$ (70°) is added to reduce the work temperature at periphery (see Fig. 13.2).

(2) *For Plexiglas* (see Fig. 13.3): Two distinguished features are

(a) Smoothing outer corners and forming a new curve cutting edge to improve hole surface quality.
(b) Plane 6 is ground to increase the rake angle of the outer cutting edge. Plane 8 is ground to decrease friction between the drill margin and the hole.

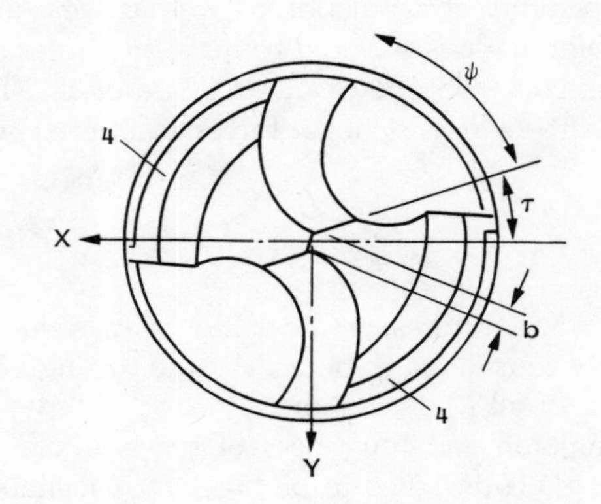

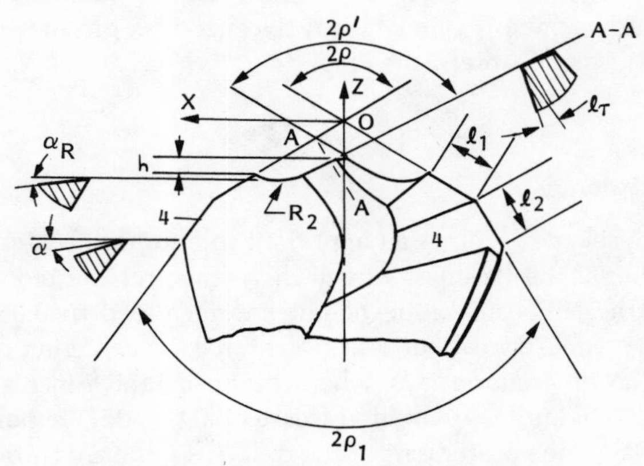

Fig. 13.2 MFD for drilling cast iron.

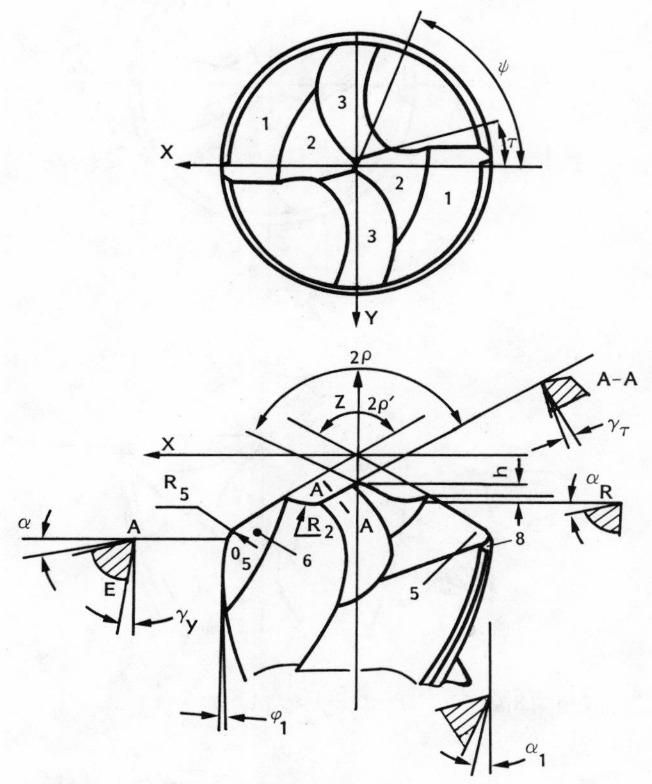

Fig. 13.3 MFD for drilling Plexiglas.

(3) *For thin sheets* (see Fig. 13.4): A sharp tip in the center gives a good centering on the workpiece, and the two sharp corners cut the sheet to form a hole.

(4) *For rubber* (see Fig. 13.5): Two sharp cutting edges are formed on the periphery.

(5) *For finishing work* (see Fig. 13.6): The inclination angle of its cutting edge is positive. In this case chips may not touch the machined hole surface thus forming a better surface finish.

(6) *For enlarging holes* (see Fig. 13.7): The outer cutting edge of this drill must be short enough so that the two tips of the outer cutting edge can touch the surface of the workpiece for centering. The center tip is lower than the two of the outer cutting edges; hence it has no effect on centering.

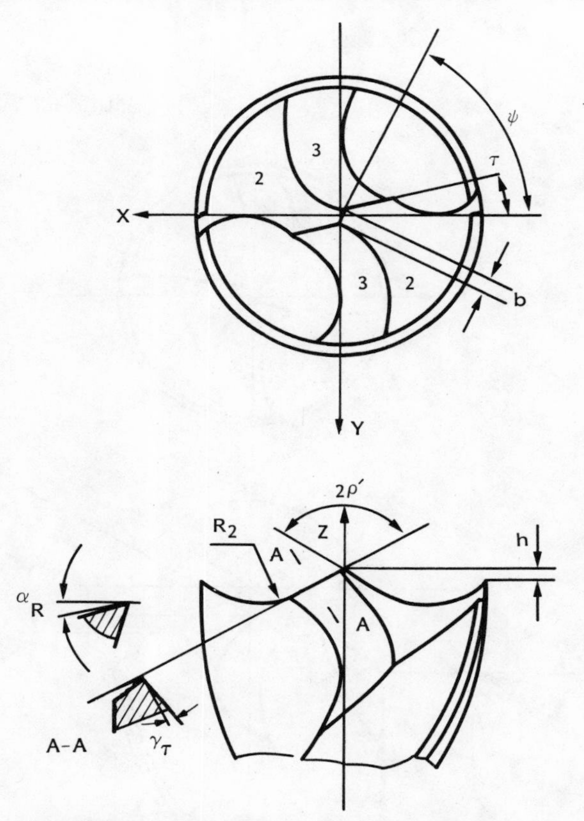

Fig. 13.4 MFD for drilling thin sheets.

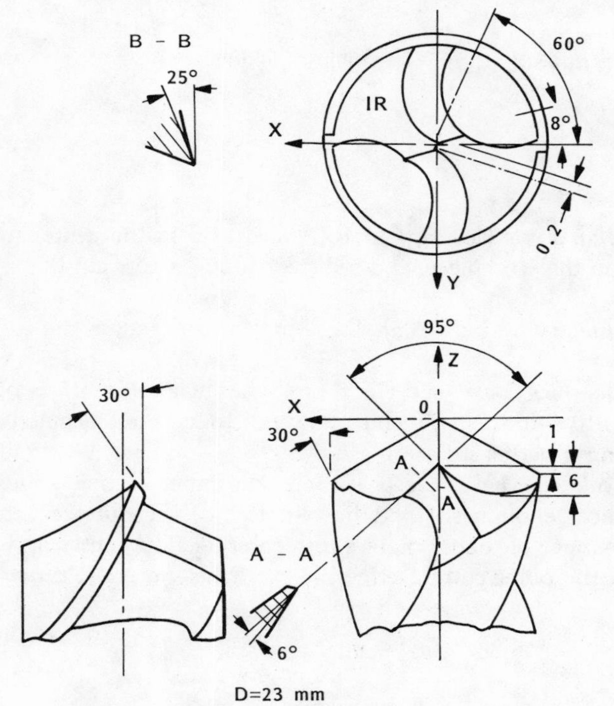

Fig. 13.5 MFD for drilling rubber.

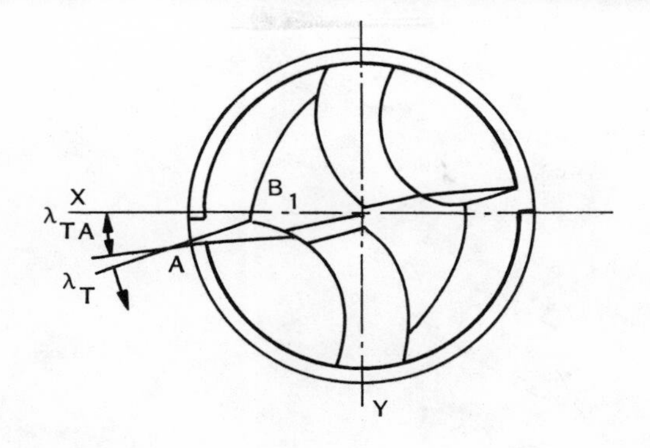

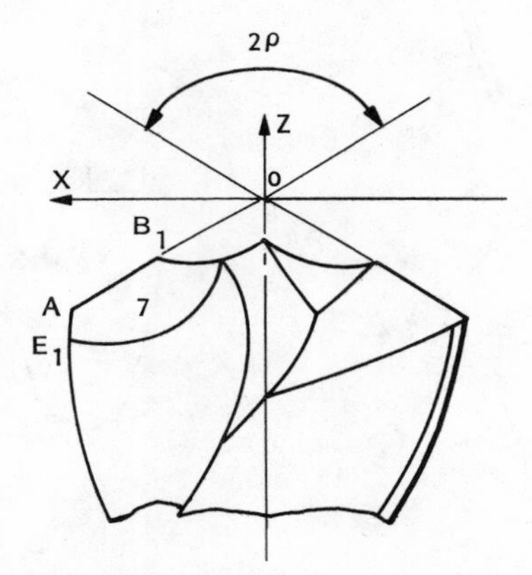

Fig. 13.6 MFD for drilling finishing work.

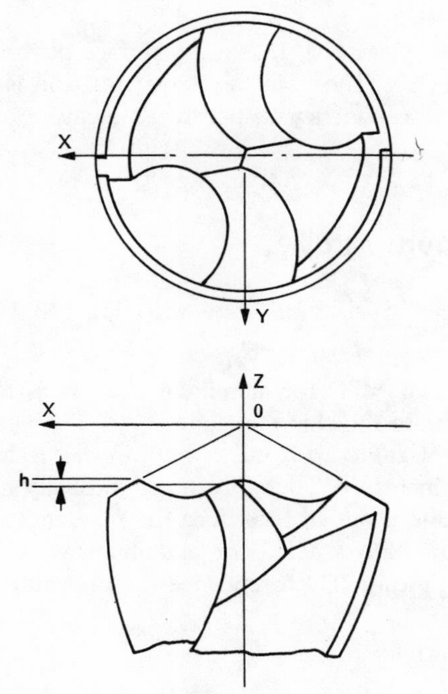

Fig. 13.7 MFD for enlarging holes.

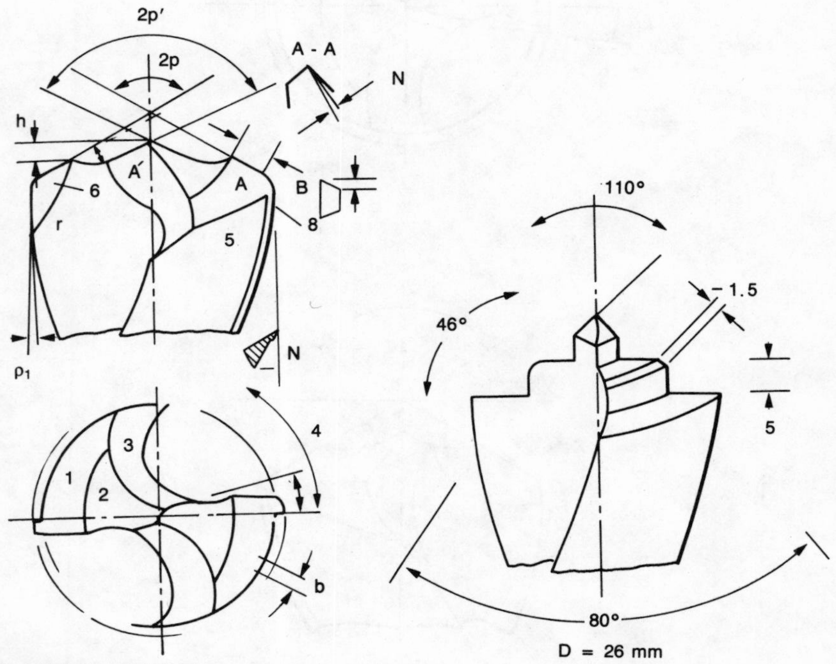

Fig. 13.8 MFD for inclined surface.

(7) *For inclined surface* (see Fig. 13.8): The surface of the workpiece is not perpendicular to drill axis. The point angle of this drill is 180° so that the drill axis will not have a tendency to be drifted away.

MFD for Aluminum Alloy

The MFD for aluminum alloy is shown in Fig. 13.9. Several features should be pointed out:

1. The point angle of MFD for aluminum alloy is as high as 140–170° in order to facilitate the chip ejection.
2. Because the point angle 2ρ is rather high in the design, the rake angle is unnecessarily increased. Therefore, an additionally ground part on the outer cutting edge is added to reduce the rake angle to 8–10°.
3. The center point of the MFD for aluminum alloy is slightly higher than that of a typical MFD to enhance the centering tendency.

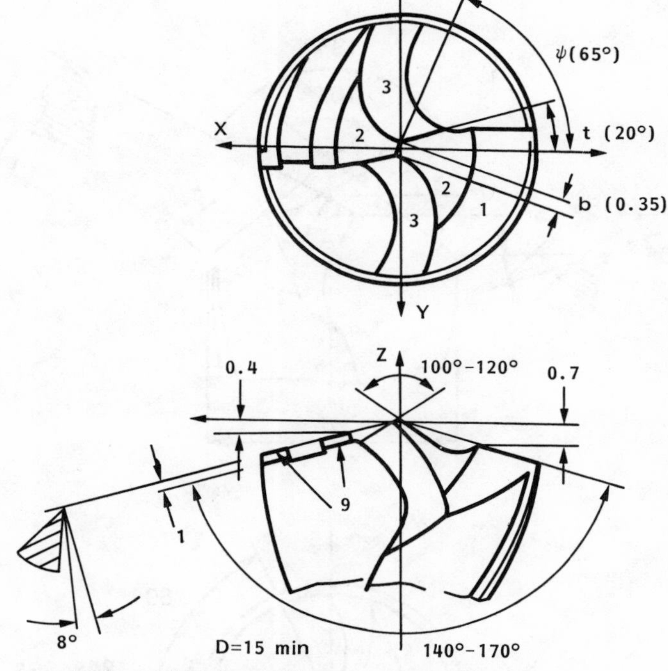

Fig. 13.9 MFD for drilling aluminum alloys.

4. It should be noted that the drill point is not symmetric. The tip height of one cutting edge is 0.02 in. (0.4 mm), but the other is 0.03 in. (0.7 mm). By so doing, ejection efficiency is increased.

A Suggested MFD for Composite Material

There is no MFD for the composite material in existence. We conceive a MFD for composite material as shown in Fig. 13.10, with the following parameters similar to a typical MFD:

$$l = 0.2D \qquad h = 0.3D$$

$$2\rho' = 135^\circ \qquad \tau = 20\text{--}30^\circ$$

$$\alpha_R = 15\text{--}18^\circ \qquad \alpha = 12\text{--}15^\circ$$

$$\gamma_\tau = -10^\circ \qquad b = 0.02D$$

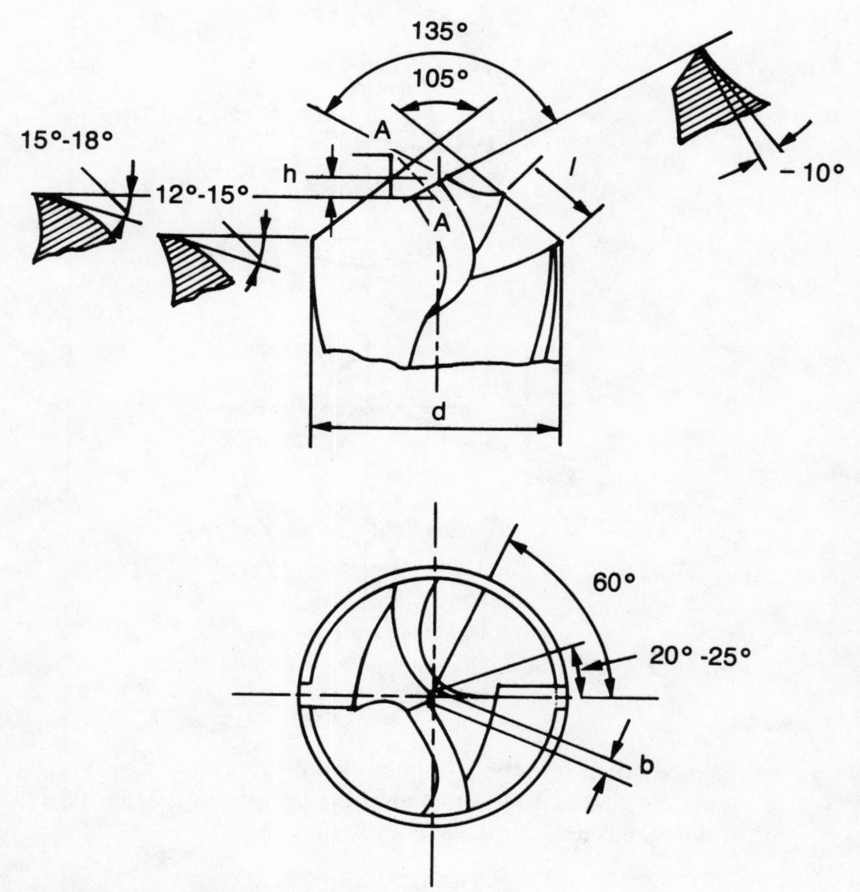

Fig. 13.10 MFD for composite material.

Since the heat conductivity of composite material is very low, the point angle is chosen as 100–110° to reduce the temperature of the periphery.

Since composites are brittle, hard material, the chip is in powder form, which will make the drill wear fast. The most appropriate drilling conditions seem to necessitate a low rotation speed and a fast feed rate to reduce the heat and installation of a vacuum system to remove the chips.

CHAPTER 14

An End Effector for Robotic Drilling

S. M. Wu
University of Wisconsin—Madison

Industrial robots have been used for welding, painting, assembling, and material handling where the required accuracy is usually not very crucial. For drilling processes in the aerospace industry where individual part accuracy and hole quality are important, today's industrial robots have limitations. There are several difficulties in applying the robot to the drilling process. First, no available drilling machine is suitable for use by a robot. Second, the accuracy or rigidity of present industrial robots is not good enough for hole positioning and is not able to cope with the dynamics of the drilling process. Third, the cutting tool, i.e., drill bit, is subject to tool wear and breakage. At present, a commercially available drilling tool was fitted with the robot, but those existing drill tools were designed for use in manual operation and did not yield satisfactory results. Much effort was put into modifying and augmenting these drilling tools to improve their performance with robots. The knowledge gained in the modification of these manual drills has led us to conclude that a special end effector exclusively designed for robot application is necessary to meet all the needs of robot-controlled drilling.

The ideal end effector to be used by robots should meet all the requirements of robotic drilling:

1. Computer controllable and easy interface with the robot controller.
2. Fully automatic quick tool changing to increase efficiency and reduce down time.
3. Precise sensing and control of countersink depth.

4. Light and compact construction for easy handling by the robot.
5. Compensation devices to correct the positioning errors of the robot.
6. An elaborate sensing system to ensure performance, including the ability to detect machine failure, tool wear, and tool breakage.
7. Providing auxiliary equipment such as lubrication, chip removal, etc.

The Mechanical Design of the End Effector for the Robotic Drilling Unit

Figures 14.1 and 14.2 show the design of the end effector for the robotic drilling unit. The unit consists of two subassemblies:

(1) the compliance mechanism and primary motion actuator, which is for the inserting motion and provides small lateral and angle adjustment;

(2) the main machine, which provides all the basic elements for the drilling operation.

The main power for material removal is provided by two air motors that deliver 2 hp (1.5 kW) with 90 psi (621 kPa) air pressure applied to the air inlet. Through a different pinion gear set, the power is transmitted to the spindle with a rated spindle speed of 3600, 8000, or 16,000 rpm. The noise of the air exhaust is reduced by two mufflers. The feed is provided by the pre-

Fig. 14.1 The end effector for the robotic drilling unit.

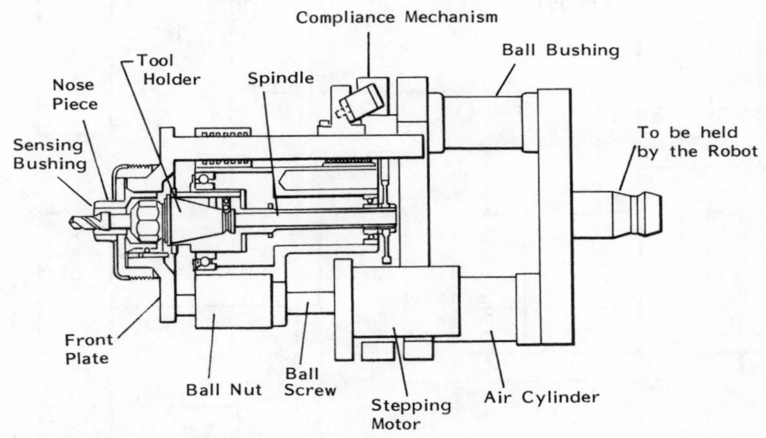

Fig. 14.2 The robotic drilling unit.

loaded ball nut and ball screw, which is driven by a stepping motor. The resolution is 0.0005 in. (0.0127 mm) per step. The carriage, which carries the air motor and spindle, slides on three shafts with four ball bushings. The shafts are also a part of the machine frame joining the front and rear plates The front plate adapts the nose piece (guide bushing). The spindle has a specially designed quick-tool-change mechanism, associated with the quick change mechanism for the nose piece; the unit can release or pick up the nose piece and tool holder simultaneously. The total machine length is 16 in. (41 cm) without taking into account the shank for the adapter to the robot. The main machine before being suspended by compliance mechanisms weighs 18 lb (8.2 kg) and the whole unit weighs 38 lb (17.3 kg).

The Controller

The controller uses a MEK 6800D2 single-board microcomputer, which includes the following devices: MC6800 MPU, MCM6830 ROM with Minibug II monitor, MCM6810 RAM (5 × 128 × 8), two MC6820 Peripheral Interface Adapters (PIA), one MC 6850 Asynchronous Communication Interface Adapter (ACIA), and a MC 6871 B Clock generator.

Two 2708 EPROMS were added with a prestored software program. A few subroutines, such as open-loop drilling, closed-loop drilling, tool release, and tool pick-up, were stored for the robot controller. The CRT monitor provides a visual display to the operation of the current control command and also receives the change command to change the required parameters. The air motor and air cylinder are controlled by two solenoid valves. The stepping

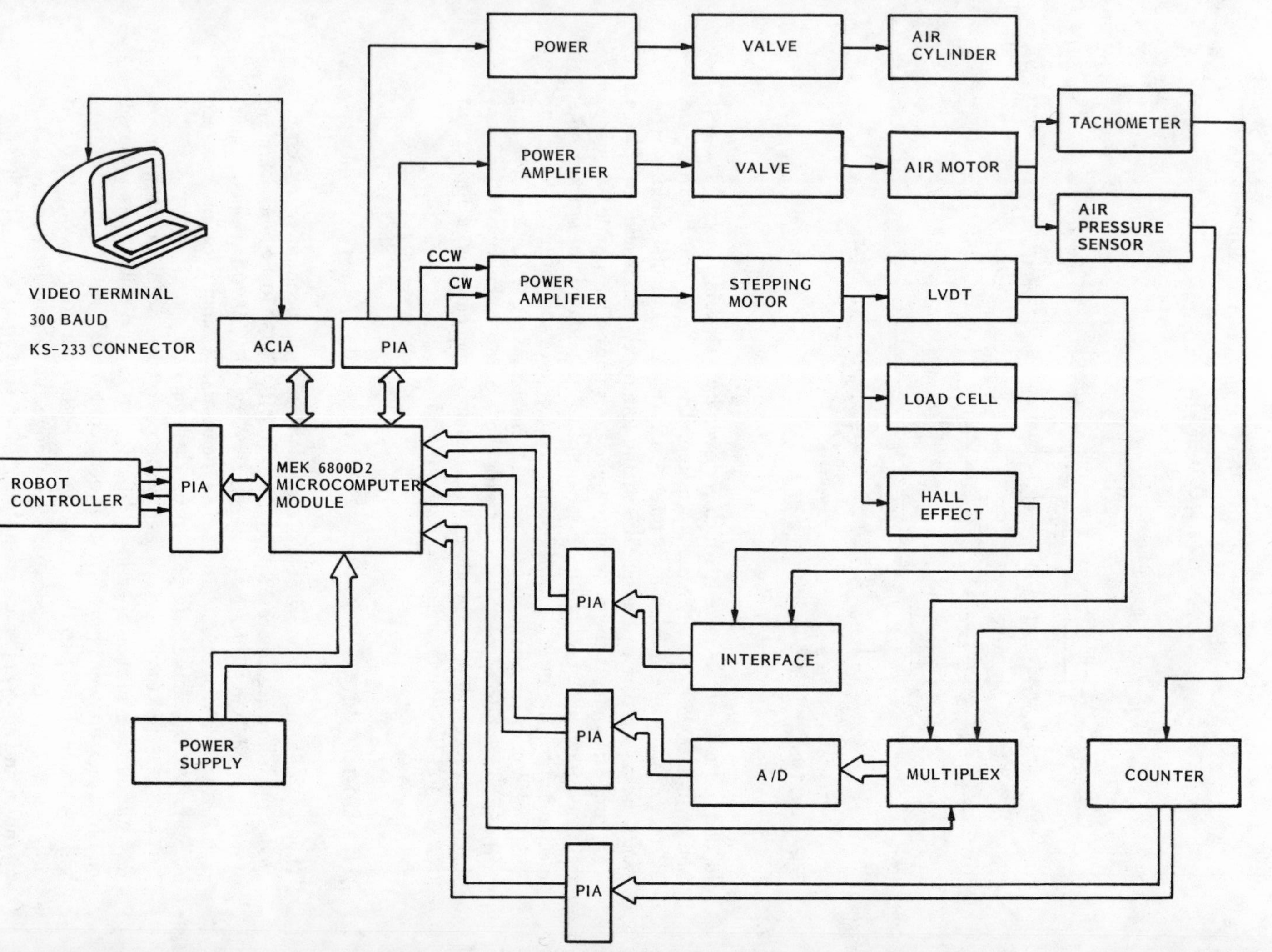

Fig. 14.3 Control system.

motor drive module is included in the controller. The pulse signals are generated directly by the computer. From that, the feed stroke and feed rate can be fully controlled by the computer.

Additional hardware and software expansion are needed in the future for on-line performance monitoring and hole quality prediction. The control system is shown in Fig. 14.3.

Sensor System

In order to get the performance feedback, the sensor system is arranged before the mechanical design. Therefore, the transducers are either built-in or mounted in suitable places. The sensor system includes:

1. Displacement sensor (LVDT) to provide the information of position, displacement, and feed rate.
2. Speed sensor (magnetic pickup) to measure the spindle speed.
3. Thrust force sensor (strain-gaged bridge) designed to measure the thrust force in the drilling process.
4. Pressure transducer to measure the inlet air pressure to ensure performance.
5. Piezoaccelerometer (two pieces mounted in the x and y directions individually) to measure the cross-direction vibration of the drilling unit, which is related to the hole quality and tool condition.

Automatic Tool Changing

The end effector has a quick-tool-change mechanism to change the tool and nose piece for different hole sizes. The drill bits are held in a standardized tool holder, which is not the same but similar to numerical control (NC) machine tooling. A special tool-presetting device is used to accurately fix the drill in the tool holder, especially for length control. The nose piece is used to guide the drilling unit. The dimension of the back cylindrical bushing with the locking groove is standardized for every nose piece, but the dimension of the front template boss varies with the drill size. For drilling aluminum, the nose piece has three large windows to release the chips. For drilling composite material, the nose piece is modified for vacuum chip removal.

For the quick-change mechanism for the drill, the adapter for the tool holder is the nose part of the spindle and includes three mechanisms: (1) main lock mechanism including three steel locking balls, lock cup, connecting rod, push ring, and compression spring (used to lock the tool holder); (2) secondary lock mechanism, including lock pin, spring, and slot on the connecting rod (used to prevent the main lock mechanism from being in a locked condition when tool holder is not in); (3) power transmission mechanism, including a spring, a key, and slots in both the tool holder and the shaft (enables the spindle to turn the tool.)

Countersink Depth Control

There are three control modes to give the desired countersink depth: (1) mechanical stop; (2) open-loop control; and (3) closed-loop control.

The mechanical-stop method uses an adjustable stop to prevent overdrilling.

In open-loop control, the feed of the tool during the drilling cycle is given by a stepping motor and a precision ball screw. Therefore, the depth of drilling a hole can be controlled by an open-loop system. The desired depth is converted to the number of pulses needed for the stepping motor to drive the ball screw, which has a resolution of 0.0005 in. (0.0127 mm) per step.

In closed-loop control, on-line measurement of countersink depth is shown in Fig. 14.4. The distance L_1 between a drill tip and the carriage surface is fixed. The distance L_0 between the end surface of the nose piece and front plate surface is also fixed. The end surface of the nose piece front plate surface is also fixed. The end surface of the nose piece is pressed against the surface of the workpiece. Hence, the countersink depth H can be determined, if we measure the distance X between the carriage and the front plate, by $H = L_1 - L_0 - X$. In real applications we are not really concerned with the ab-

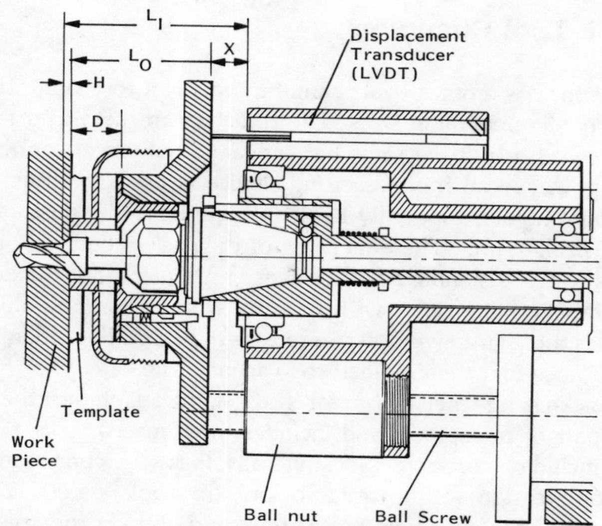

Fig. 14.4 On-line measurement of countersink depth.

solute value of the countersink depth. Instead we are more interested in the LVDT output voltage, which is related to the accurate countersink depth.

The drill is fed by the stepping motor and the countersink depth is checked by the displacement transducer; thus a closed-loop control is formed to obtain high accuracy of countersink drilling.

Compensation for Robot Position Error

The majority of industrial robots are not able to do drilling work in terms of accuracy and rigidity, especially for those interchangeable parts of an aircraft. The developing and building costs for a robot that would have enough accuracy and rigidity to carry on the drilling work could be prohibitive. An alternate approach is through the adequate design of the end effector and tooling system; a template could be and is used to locate the positions for drilling. The nose piece (guide bushing) of the end effector is inserted into the hole of the template to find the exact location for drilling. A remote center compliance system is introduced to the unit to determine both lateral and angular compliance. Between the first and the second plate, there are four laminated elastomeric elements for angular displacement of the drill head with the center of rotation close to the leading edge of the nose piece. The other four shear pads are mounted between the second and the third plates to determine the lateral compliance. Through the tooling of the template and the compliance design of the unit, the rigidity and accuracy requirements are not critical for accurate drilling.

Future Development of Machine Intelligence

One of the main reasons to develop the end effector for the robotic drilling is the need of an intelligent machine that can detect all failures in the drilling process. The failures include tool wear, tool breakage, machine problems, and workpiece failure. The sensor system provides all signals for diagnosis. In normal operation, all the sensor outputs are in a specified range. However, when the sensor output is out of this range, there is clearly something wrong. To distinguish the problem, either a single sensor output or multiple sensor outputs may be used to locate the trouble areas. For example, air motor failure can be checked by the air pressure first, then by the spindle speed. Bearing failure or spindle roundout may be found using vibration signals from accelerometers. Drill wear and breakage can be deduced from higher thrust force and an increased vibration signal. Hence, the spindle speed, thrust force, and vibration are the useful parameters for judging the tool wear and breakage.

The On-Line Prediction of Hole Quality

The errors or imperfections of the drilled hole, such as errors in shape, roundness, geometry, dimensions, and burrs, result from imperfections of the machine tool, the drill bit, and the drilling process. From the dynamic signals in the process, the quality of the hole could be predicted, leading to on-line quality control.

For example, the roundness or dimension error of the hole is essentially due to the accuracy of the spindle bearing and the tool grinding. These two factors also contribute to the machine vibration. Thus, the output of the two accelerometers are a measure of the roundness or dimension error. The burr formation in drilling aluminum and the delamination in drilling composite material are related to the thrust force and the feed. The error in hole shape is related to different vibration patterns or amplitude during drilling. More experimental data are needed to establish the quantitative relationship between the dynamic signals and hole quality.

Acknowledgment

The author wants to thank the Society for the Advancement of Material and Process Engineering (SAMPE) for allowing the use of some materials the author and co-workers have presented at the 14th National SAMPE Technical Conference, October 1982, Atlanta, Georgia.

References

1. Boston, O. W. and Oxford, C. J., Sr., "Power Required to Drill Cast Iron and Steel," *Trans. ASME*, Vol. 52, 1930.

2. Boston, O. W. and Oxford, C. J., Sr., "Torque, Thrust, and Power for Drilling," *Journal SAE*, Vol. 28, No. 3, 1931, pp. 378–383.

3. Boston, O. W. and Oxford, C. J., Sr., "Performance of Cutting Fluids in Drilling Various Metals," *Trans. ASME*, Vol. 55, 1933, pp. 1–29.

4. Chen, L. H. and Wu, S. M., "Multifacet Drills," 14th SAMPE, Material & Processes Advances, Atlanta, Ga., 1982.

5. Croy, M., Wong, T. L., and Wu, S. M., "Analysis of Delamination in Drilling Composite Material," 14th SAMPE, Material and Processes Advances, Atlanta, Ga., 1982.

6. "Development of a Computer Controlled Drill Point Grinder," Report to General Dynamics, Fort Worth Division, 1979.

7. DeVries, M. F., Wu, S. M., and Mitchell, J. W., "Measurement of Drilling Temperature by the Garter Spring Thermocouple Method," *Microtecnic*, No. 6, 1967.

8. DeVries, M. F. and Wu, S. M., "Evaluation of the Effects of Design Variables

on Drill Temperature Responses," *Journal of Engineering for Industry, Trans. ASME,* August 1970, pp. 699–705.

9. DeVries, M. F., Saxena, U. K., and Wu, S. M., "Temperature Distributions in Drilling," *Journal of Engineering for Industry, Trans. ASME,* Series B, Vol. 90, No. 2, May 1968, pp. 231–238.

10. Eman, K., Wu, S. M., and Hawkins, J., "7-Axis CNC Drill Grinder," 14th SAMPE, Material & Processes Advances, Atlanta, Ga., 1982.

11. Fugelso, M. A., "A Microprocessor Controlled Twist Drill Grinder for Automated Drill Point Production," PhD Thesis, University of Wisconsin—Madison, 1978.

12. Fugelso, M. A., and Wu, S. M., "A Microprocessor Controlled Twist Drill Grinder for Automated Drill Production," *Journal of Engineering for Industry, Trans. ASME,* May 1979, Vol. 101, pp. 205–210.

13. Fujii, S., DeVries, M. F., and Wu, S. M., "An Analysis of Drill Geometry for Optimum Drill Design by Computer. Part I—Drill Geometry Analysis," *Journal of Engineering for Industry, Trans. ASME,* Series B, Vol. 92, No. 3, Aug. 1970, pp. 647–656.

14. Fujii, S., DeVries, M. F., and Wu, S. M., "An Analysis of Drill Geometry for Optimum Drill Design by Computer. Part II—Computer-Aided Design." *Journal of Engineering for Industry, Trans. ASME,* Series B, Vol. 92, No. 3, August 1970, pp. 657–666.

15. Fujii, S., DeVries, M. F., and Wu, S. M., "Analysis of the Chisel Edge and the Effect of the d-Theta Relationship on Drill Point Geometry," *Journal of Engineering for Industry, Trans. ASME,* Series B, Vol. 93, No. 4, November 1971, pp. 1093–1105.

16. Fujii, S., DeVries, M. F., and Wu, S. M., "Analysis and Design of a Drill Grinder and Evaluation of Grinding Parameters," *Journal of Engineering for Industry, Trans. ASME,* Series B, Vol. 94, No. 4, November 1972.

17. Galloway, D. F., and Morton, I. S., *Practical Drilling Test,* Research Dept. of the Institution of Production Engineers, Great Britain, 1946.

18. Galloway, D. F., "Advances in Drilling Techniques Arising from Recent Research," *Microtecnic,* Vol. 9, 1955, pp. 135–141.

19. Galloway, D. F., "Some Experiments on the Deflections and Vibrations of Drilling Machines," *Proceedings Institution of Mechanical Engineers,* Vol. 170 No. 6, 1956, p. 207.

20. Galloway, D. F., "Some Experiments on the Influence of Various Factors on Drill Performance," *Trans. ASME,* Vol. 79, 1957, pp. 191–231.

21. Haggerty, W. A., "The Effect of Drill Geometry on Performance," ASTME Technical Paper, No. 254, Vol. 60, Book 1, 1960.

22. Haggerty, W. A., "Effect of Point Geometry and Dimensional Symmetry on Drill Performance," *Int. J. Mach. Tool Des. Res.,* Vol. 1, 1961, pp. 41–58.

23. Hine, C. R., *Machine Tools and Processes for Engineers,* McGraw-Hill, New York, 1971.

24. Horng, S. Y., Wu, S. M., and Van, Y. G., "End Effector of Robotic Drilling," 14th SAMPE, Material & Processes Advances, Atlanta, Ga., 1982.

25. Kronenberg, M., "Drilling Feeds," *Machinery,* Vol. 45, 14 February 1935, p. 661.

26. Law, S. S., DeVries, M. F., and Wu, S. M., "Analysis of Drill Stress by Three Dimensional Photoelasticity," *Journal of Engineering for Industry, Trans. ASME,* Series B, Vol. 94, No. 4, November 1972.

27. Maksimenko, P. G., "Cutting Forces in the Drilling of High-Strength Cast Iron," *Russian Engineering Journal*, Vol. XLVI, No. 10, pp. 80–81.

28. Nashida, S., Ozaki, S., and Motomura, T., "Study of Drilling (I)—Cutting Forces in Drilling," *Journal of the Mechanical Laboratory of Japan*, Vol. 8, No. 1, 1962, pp. 56–58.

29. Naureckas, E. M., and Gabrick, J., "Photoelastic Techniques to Evaluate Cutting Forces During Drilling," *ASTME*, 1969.

30. Nishida, S., Ozaki, S., Nakayama, S., Shiraishi, T., and Nagura, K., "Study on Drilling II—Lip Temperature," *Journal of the Mechanical Laboratory of Japan*, Vol. 8, No. 1, 1962, pp. 59–60.

31. Okochi, M., and M. Okoshi, "Researches on the Cutting Force," *Inst. of Physical and Chemical Res., Science Papers*, 5, Tokyo, 1927, p. 261.

32. Oxford, C. J., Jr., "On the Drilling of Metals I, Basic Mechanics of the Process," *ASME Transactions*, 77, Part 1, February 1955, pp. 103–114.

33. Oxford, C. J., Jr., "Some Recent Research on Twist Drills and Drilling," *Proc. of 23rd. Annual Meeting of ASTME*, Los Angeles, Calif., 1955.

34. Oxford, C. J., Jr., "Some Recent Developments in the Design and Application of Twist Drills in America," *Proceedings of the 8th International MTDR Conference*, Manchester, United Kingdom, 1967.

35. Pal, A. K., Bhattacharyya, A., and Sen, G. C., "Investigation of the Torque in Drilling Ductile Materials," *Int. J. Mach. Tool Des. Res.*, Vol. 4, 1965, pp. 205–221.

36. Reznikov, A. N., et al., "Investigation of Stresses in Twist Drills," *Machines and Tooling*, Vol. 36, No. 9, 1965.

37. Russell, W. R., "Drill Design and Drilling Conditions for Improved Efficiency," ASTME Paper No. 397, Vol. 62, Book 1, 1962, pp. 1–12.

38. Saxena, U. K. and Wu, S. M., "Building Mathematical Models to Predict Transient Drilling Temperature Response," *Trans. ASME*, Vol. 91, 1969, pp. 641–651.

39. Saxena, U. K., DeVries, M. F., and Wu, S. M., "Drill Temperature Distributions by Numerical Solutions," *Journal of Engineering for Industry, Trans. ASME*, Series B, Vol. 93, No. 4, November 1971.

40. Schmidt, A. O., and Roubik, J. R., "Distribution of Heat in Drilling," *Trans. ASME*, Vol. 71, 1949, pp. 245–252.

41. Shaw, M. C., and Oxford, C. J., Jr., "On the Drilling of Metals II, The Torque and Thrust in Drilling," *Trans. ASME*, Vol. 79, 1957, pp. 139–148.

42. Shaw, M. C., "Drilling Fundamentals—A Review of the Theory of Drilling," ASTME Paper No. 396, Vol. 62, Book 1, 1962.

43. Tobias, S. A. and Fishwick, W., "The Vibrations of Radial Drilling Machines under Test and Working Conditions," *Proceedings Institution of Mechanical Engineers*, Vol. 170, 1956, p. 232.

44. Tsai, W. D., "Drill Geometry Models and Dynamics of Drilling," PhD Thesis (1977), University of Wisconsin—Madison.

45. Venkatataman, R., Lamble, J. H., and Koenigsburger, F., "Analysis and Performance Testing of Dynamometer for Use in Drilling and Applied Processes," *International Journal of Machine Tool Design and Research*, Vol. 5, 1976.

46. Wu, S. M., "How to Cut Down Drilling Cost for the Aerospace Industry," SAMPE 14, Material and Process Advances, Atlanta, Ga., 1982.

Part Five

Grinding

CHAPTER 15

Grinding

Robert S. Hahn
Hahn Associates

Introduction

The problems of reducing grinding costs, improving product quality, and operating with personnel having a minimum of grinding expertise confront many plants. Grinding operations often produce parts that do not conform precisely to what is desired. Holding precision, or roundness and concentricity, or taper, or surface finish without burn, in a short cycle time, is often difficult to achieve. These problems arise because of the inability to control certain variables in the grinding process. Therefore, it is important to identify those variables, bring them under control to ensure consistent size, taper, surface finish, surface integrity (freedom from burn), and product quality at minimum cost. With these goals in mind, the following sections are devoted to the various factors that influence the productivity of grinding operations.

Part Processing by Grinding

Multioperation Grinding

Many workpieces often require a number of surfaces to be ground on each individual workpiece. If several of these surfaces can be ground in one operation, production efficiencies can often be achieved. If they cannot all be ground in one operation, then several operations are required. However, production efficiencies can also be achieved if multiple operations can be performed for one staging of the workpiece, thereby reducing the number of additional set-ups and part handling, and ensuring squareness and concentricity. Figure 15.1 illustrates, for example, a CNC grinder with two wheelheads mounted on the same cross-slide, the left-hand head grinding six surfaces in operation 4, and the right-hand head subsequently grinding three surfaces in operations 1, 2, and 3. Multiple grinding operations can sometimes also be performed by using "compound wheels," where several cutting surfaces, or wheels, are provided on the same wheelhead to execute several operations as illustrated in Fig. 15.2. An axial feed (ZFEED) is used to grind operation 1, while a vec-

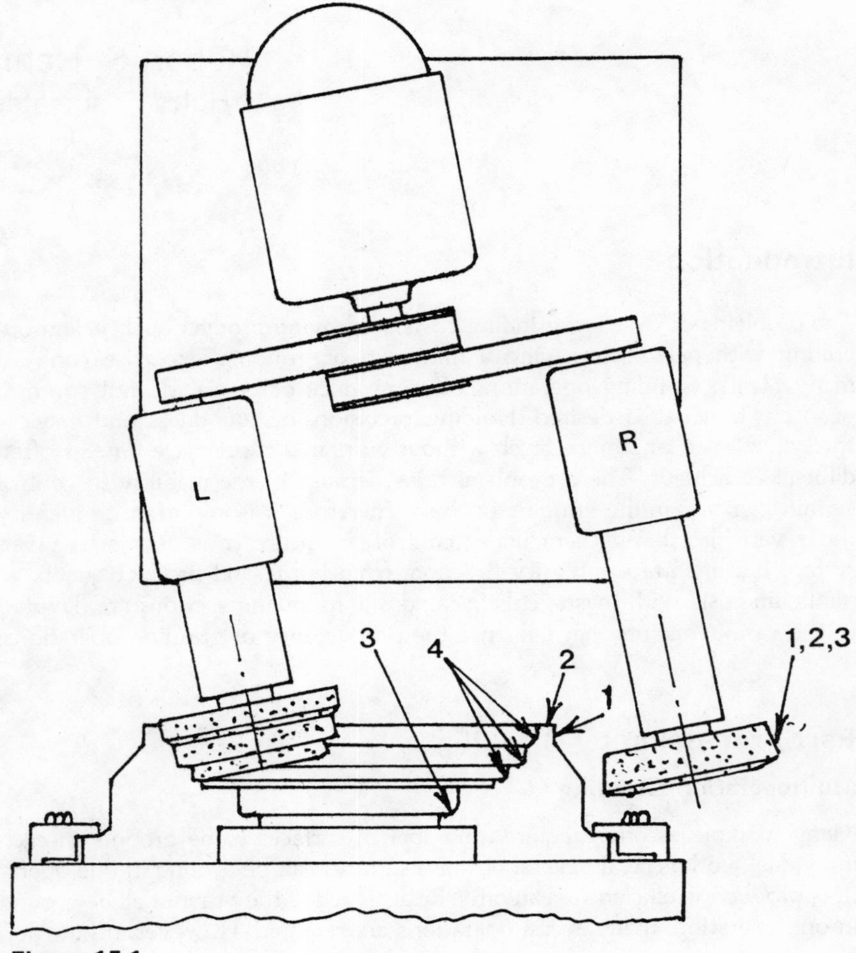

Figure 15.1

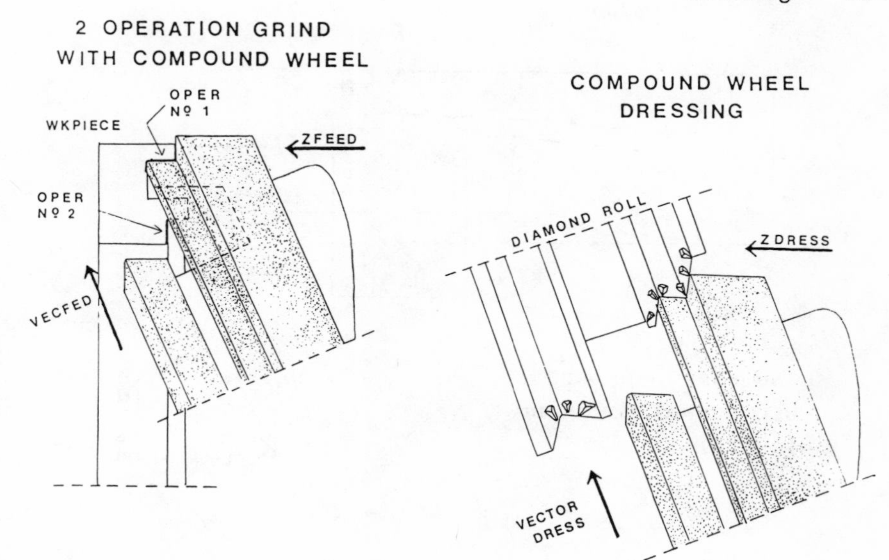

Figure 15.2

tor feed (where computer control provides simultaneous feed rates to both *X* and *Z* axes) subsequently executes operation 2 with the smaller section of the grinding wheel. Also shown is the diamond dress roll for providing compound wheel dressing.

With computer control over both cross-slide and axial motions (*X* and *Z* axes), external or internal grinding of tapered parts can be performed as illustrated in Fig. 15.3b. In this example, the tapered section *A* of the wheel is used to grind the taper *A* where the *X* and *Z* slides under computer control perform a vector reciprocation along the surface *A*. Following this operation the straight section *B* is ground and finally the shallow taper *C* is ground.

Multiplunge Roughing with Reciprocate Finish Grind

Figure 15.3a illustrates a simple straight bore to be ground. In conventional reciprocate grinding the wheel is usually fed in manually to contact the work, develop a spark stream, and then fed axially to traverse the length of the bore. If there is more stock in another section of the bore, excessive forces are developed as the wheel traverses along the bore. Generally, these forces cause either the leading edge or the trailing edge of the wheel, depending on the angular stiffness of the wheel spindle, to break down. If the angular stiffness is high, the leading edge tends to break down. If it is low, the leading

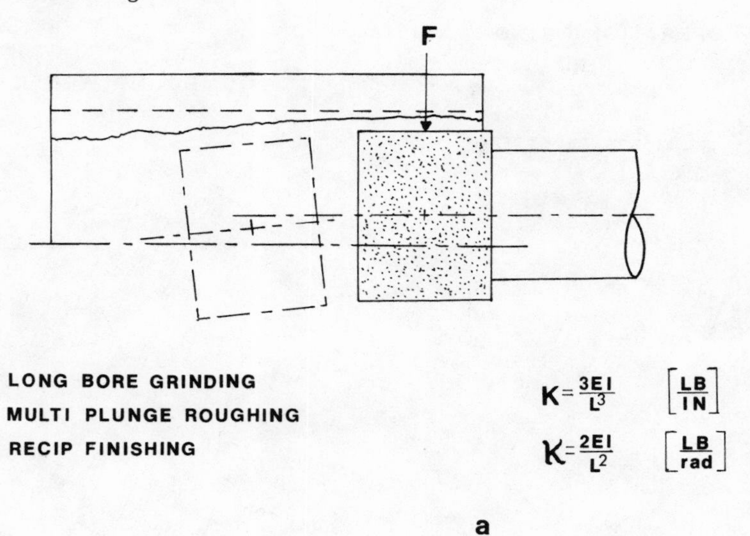

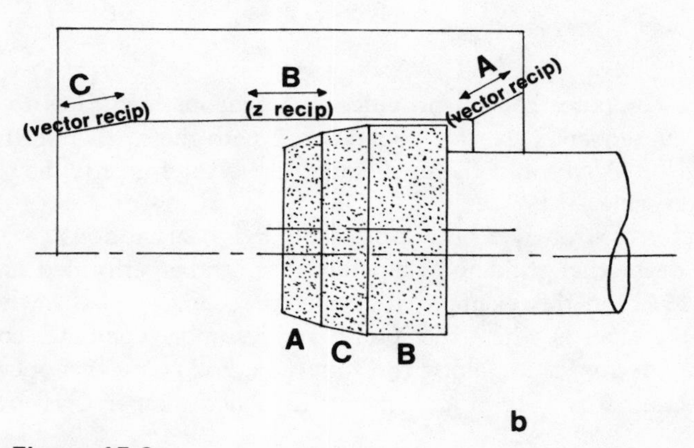

Figure 15.3

edge tends to deflect out of the cut concentrating the grinding force on the trailing edge causing it to break down. In either case the wheel face is not being used effectively. With CNC grinding, these difficulties can be overcome by making a series of adjacently spaced plunge grinds to rough grind the part, correcting tapered stock conditions, and bringing the part to so-called first size, and then to finish grind to size by reciprocating the wheel in the usual manner. This method avoids wheel breakdown during the rough grind, but still uses the wheel aggressively to remove stock.

OD/ID Grinds with Adjacent Shoulders

A number of methods of grinding OD or ID parts with adjacent shoulders are illustrated in Fig. 15.4. Their advantages and disadvantages are discussed below.

Figure 15.4a shows a wheel making a plunge grind in the X direction, where it must feed the entire width of the shoulder before it strikes the OD/ID

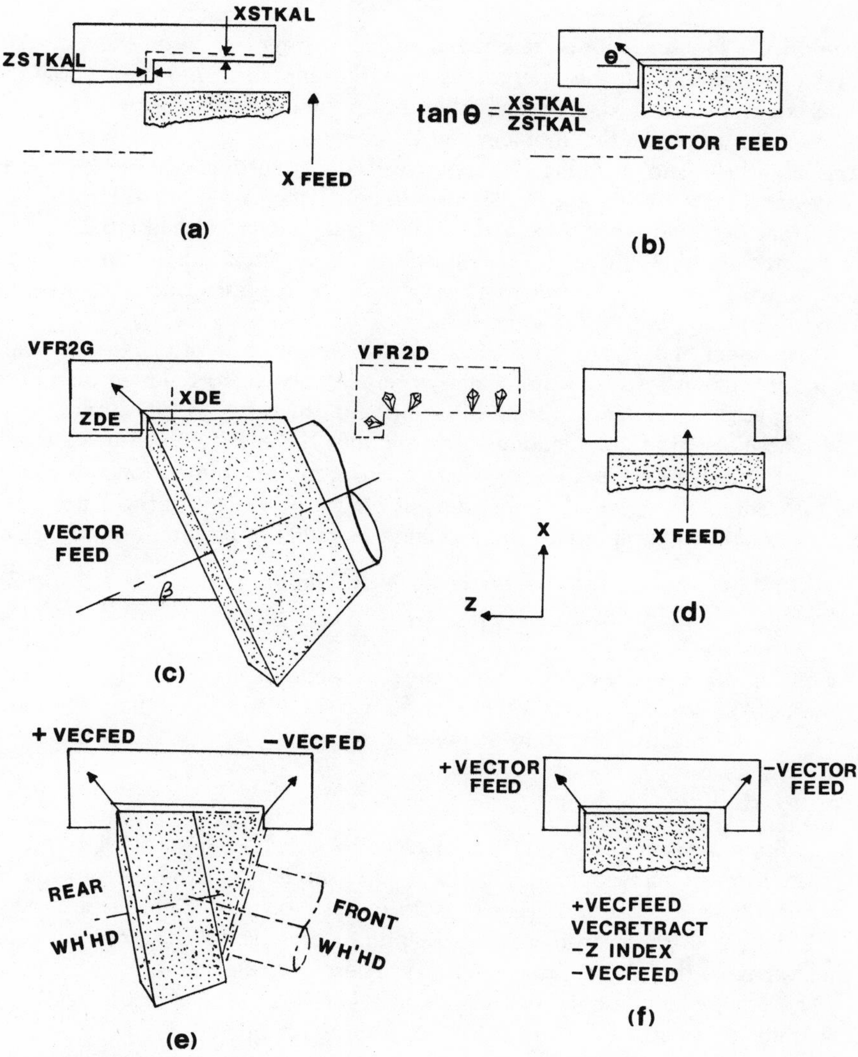

Figure 15.4

stock (XSTKAL). This is a very time-consuming operation, and causes considerable wheelwear on the left-hand edge of the wheel. A somewhat improved situation occurs in Fig. 15.4b where the wheel is rapidly traversed into the corner, and then fed on a vector to grind the shoulder and the OD/ID simultaneously. If the vector angle is chosen such that

$$\tan \theta = \frac{\text{XSTKAL}}{\text{ZSTKAL}}, \tag{15.1}$$

size on the shoulder will be reached at the same time as size on the OD/ID. The disadvantage of this method is that the shoulder may be "burned," cracked, or thermally damaged metallurgically owing to excessive heat.

A cooler grind on the shoulder can be obtained by tilting the wheelhead axis slightly as shown in Fig. 15.4c, where the end cutting face of the wheel now has a curvature in the Z direction, thereby increasing the difference in curvature between wheel face and shoulder [see Eqs. (15.8) and (15.9)].

Figures 15.4d, 15.4e, and 15.4f illustrate similar situations where a "back face" must be ground. Figure 15.4f represents the coolest grind, but requires two wheelheads on the same cross-slide.

A shoulder and adjacent OD/ID can also be ground with only one axis of motion under feed by tipping the work axis relative to the feed direction as illustrated in Fig. 15.5a for an ID grind and 15.5b for an OD grind.

In these cases the shoulder dimension and the OD/ID dimension are coupled through the size-coupling effect, where a change in one dimension will cause an (unwanted) change in the other dimension. Readjustment of the Z position can be made to compensate for a change in radial dimension according to

$$\Delta X = \Delta R \cos \theta + \Delta T \sin \theta, \tag{15.2}$$

$$\Delta Z = \Delta T \cos \theta - \Delta R \sin \theta. \tag{15.3}$$

By setting $\Delta T = 0$, the condition that the shoulder dimension does not change, values of ΔX and ΔZ are obtained that permit readjustment of the cross-slide and table to effect a change in diameter size only.

Form Grinding

In addition to grinding shoulders adjacent to an OD/ID, some parts often require form-grinding operations. These parts sometimes exhibit problems in thermal damage, local variations in surface finish, or loss of profile accuracy. In dealing with these problems it is helpful to understand the grinding process variables that cause these problems. These variables are discussed.

The plunge grinding of a ball track in a ball-bearing race can be used to illustrate the effects found in many form-grinding operations, as shown in Fig. 15.6. There are four local process variables, which vary from point to point along the profile.

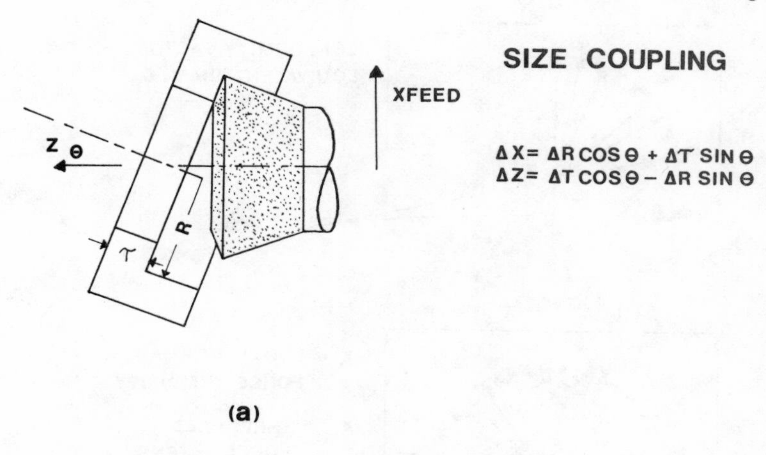

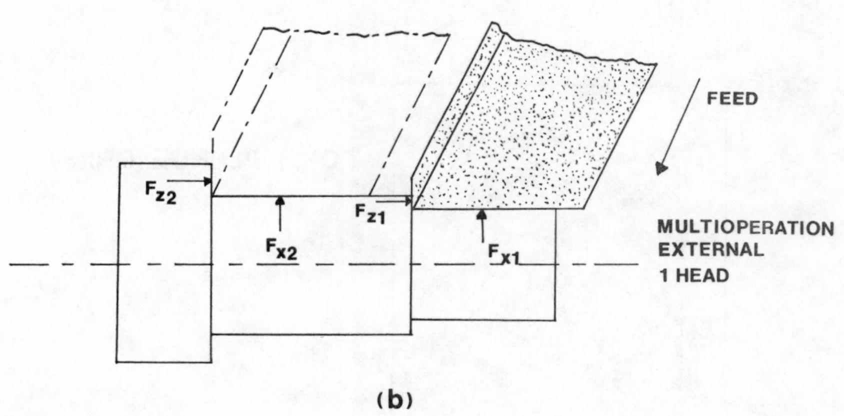

Figure 15.5

Figure 15.6a illustrates the distribution of the equivalent diameter D_e along the profile.* The equivalent diameter is a measure of the difference in curvature between the abrasive wheel and the workpiece in the contact zone, and is described in more detail in relation to Eqs. (15.8) and (15.9).

Figure 15.6b illustrates the distribution of the local normal force intensity (normal force per unit width of cut) along the profile while the dashed line

* A nomenclature list is given at the end of this chapter.

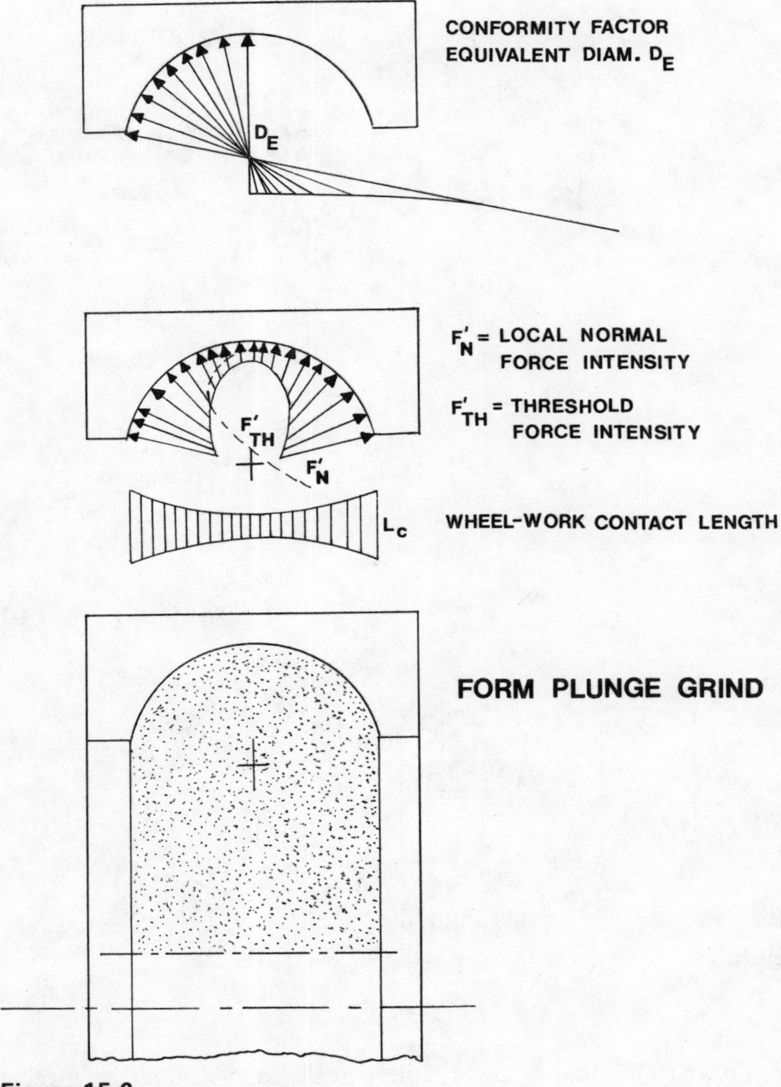

Figure 15.6

represents the distribution of the local "threshold force intensity." These variables are discussed in detail in the next section. The length of the wheelwork contact zone L_c is also shown. The large value of equivalent diameter D_e on the sidewalls, the long length of contact L_c, and the high normal force intensity often combine to exceed the threshold conditions for thermal damage, resulting in burn on the sidewalls (see "Surface Integrity" subsection). Surface finish is generally better at the bottom of the ball track where the induced force intensity is low.

Principles of Grinding

Input–Output Variables

Although many grinding operations perform satisfactorily, occasional undesired results are sometimes produced, namely, thermal damage, size scatter, fluctuating taper, variations in surface finish, etc. These variations in output variables are caused by variations in certain grinding process variables that are not being properly controlled. It is important to distinguish between inputs to the grinding machine, inputs to the grinding process, and outputs from the grinding process as illustrated in Fig. 15.7. The inputs to the grinding machine consist of the various machine settings as well as stock variations, stock runout, and hardness variations. These inputs to the machine cause certain wheelwork interface forces to be generated. These forces, along with the wheel sharpness (defined quantitatively in the next subsection), act as inputs to the grinding process (that which occurs at the wheelwork contact zone). Outputs from the grinding process are listed on the right in Fig. 15.7. In

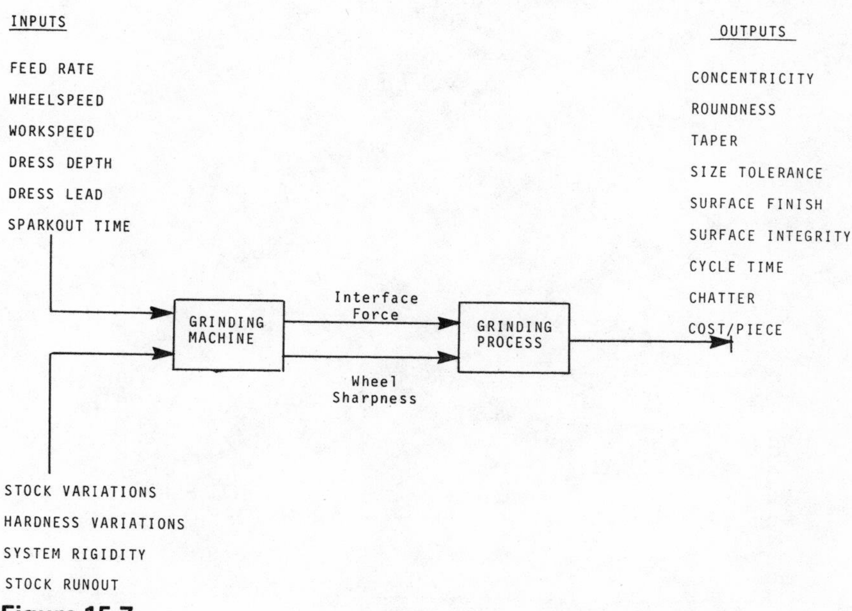

Figure 15.7

order to obtain certain desired outputs it is helpful to understand the relationship of the various outputs to the inputs. These relationships are developed in the following sections.

The Wheelwork Interface

The grinding process takes place at the wheelwork interface. In cylindrical plunge-grinding operations, the interface force intensity (normal force per unit width of contact) is uniformly distributed over the face of the wheel. Accordingly, plunge grinding is the simplest type of grinding, as illustrated in Fig. 15.8. The feedrate $\bar{v}_f$ is applied to the cross-slide of the machine. At the moment the wheel contacts the workpiece, the interface force intensity is zero. As the cross-slide continues to move, the "springs" in the system compress, generating some interface force intensity F'_n. This causes the wheel and work to mutually "machine" each other, the radius of the workpiece decreasing at the rate $\bar{v}_w$, the radius of the wheel decreasing at the rate $\bar{v}_s$, and the deflection x increasing at the rate $\dot{x}$; thus

$$\bar{v}_w + \bar{v}_s + \dot{x} = \bar{v}_f. \tag{15.4}$$

The feedrate of the cross-slide $\bar{v}_f$ only equals the plunge-grinding velocity $\bar{v}_w$ when the wheel wear $\bar{v}_s$ is negligible and $\dot{x} = 0$ (the steady state).

The volumetric rates of stock removal Z'_w and wheel wear Z'_s per unit width of contact,

$$Z'_w = \pi D_w \bar{v}_w \tag{15.5}$$

and

$$Z\bar{v}'_s = \pi D_s \bar{v}_s, \tag{15.6}$$

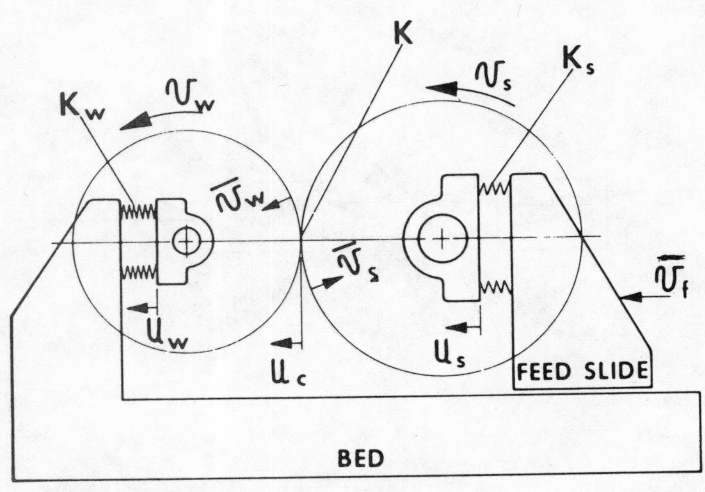

Figure 15.8

are plotted against the normal interface force intensity in Fig. 15.9, resulting in a "wheelwork characteristic chart," which shows how a given wheelwork pair machine each other. The stock removal curve Z'_w (solid dot) has a "rubbing region" at force intensities below F'_{th}, the "threshold force intensity"; a "plowing region" for force intensities between F'_{th} and F'_{pc}, the "plowing–cutting transition"; and a "cutting region" above F'_{pc}.[1] In the cutting region, the abrasive grits remove chips, in the usual way; in the plowing region, they remove material by causing lateral plastic flow and highly extruded ridges to be formed along each side of the scratch, these ridges being removed by subsequent grits. The plowing region is important in obtaining good surface

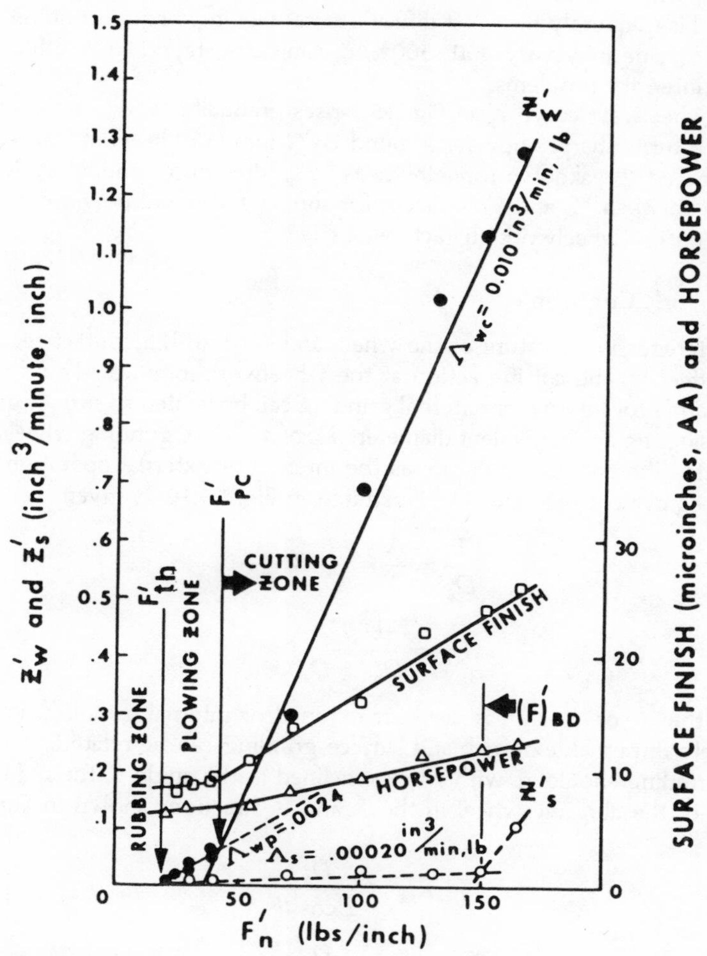

Figure 15.9

finishes. The cutting region is important in rounding up the workpiece and fast stock removal.

The slope of the Z'_w curve is called the "work removal parameter" (WRP), or sometimes Λ_w, and is indicative of the "sharpness" S of the grinding wheel, defined as:

$$S = \frac{\text{WRP}}{V_s} \quad \left(\frac{\text{m}^2}{\text{N}}\right), \tag{15.7}$$

which represents the cross-sectional area of a hypothetical ribbon of material being removed from the workpiece per unit normal force. The sharpness of grinding wheels is one of the most important variables in the grinding process, and is frequently the most difficult to control in practical grinding operations. Its value may vary 400–500%, causing size, taper, surface finish, and surface integrity problems.

The wheelwear curve Z'_s in Fig. 15.9 rises gradually at low force intensity and then turns sharply upward [around 28 N/mm (150 lb/in.) in this case], at the so-called "breakdown force intensity" F'_{bd}. Precision grinding cycles must operate between F'_{th} and F'_{bd}. Curves for surface finish and power can also be shown on the wheelwork characteristic chart.

Wheelwork Conformity

The difference in curvature of the wheel and work in the contact region has some effect on the cutting action at the wheelwork interface. The difference of curvature for internal or external grinding can be related to surface grinding by considering an "equivalent diameter" D_e of a surface grinding wheel having the same difference of curvature as the internal or external operation.

The equivalent diameter D_e, illustrated in Fig. 15.10, is given by

$$\frac{1}{D_e} = \frac{\Lambda}{2} = \frac{1}{D_s} \mp \frac{1}{D_w}, \tag{15.8}$$

$$D_e = \frac{D_w D_s}{D_w \pm D_s}, \tag{15.9}$$

where the + or − sign is used for external or internal grinding. With this parameter, internal, external, and surface grinding can be related.

In grinding shoulders with wheels inclined at the angle β, the radii of curvature of the abrasive wheel in the X and Z directions shown in Fig. 15.11 are

$$R_{sx} = \frac{D_s}{2 \cos \beta}, \tag{15.10}$$

$$R_{sz} = \frac{D_s}{2 \sin \beta}, \tag{15.11}$$

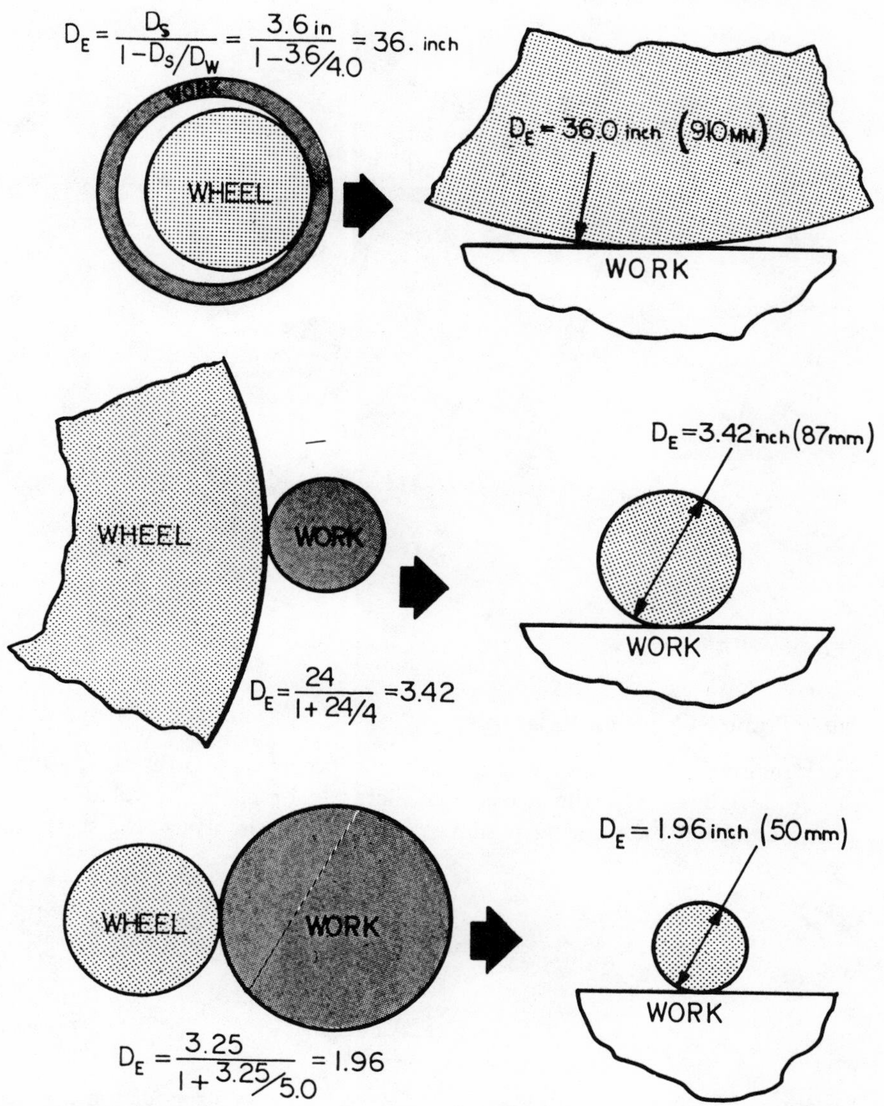

Figure 15.10

so that

$$XD_e = \frac{D_w(D_s/\cos\beta)}{D_w - D_s/\cos\beta}, \tag{15.12}$$

$$ZD_e = \frac{D_s}{\sin\beta}. \tag{15.13}$$

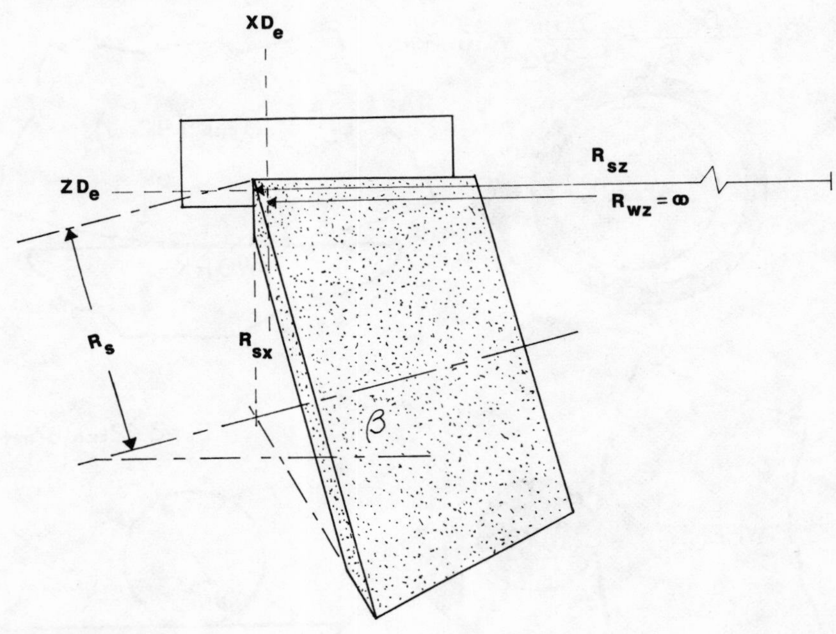

Figure 15.11

Basic Plunge-Grinding Relations

Stock removal, wheelwear, surface finish, and power and force relationships can be developed from the wheelwork characteristic chart illustrated in Fig. 15.9. Neglecting the plowing region in Fig. 15.9 for simplicity, the stock removal relation is*

$$Z'_w = \mathrm{WRP}(F'_n - F'_{th}), \tag{15.14}$$

or

$$\pi D_w \bar{v}_w = \mathrm{WRP}(F'_n - F'_{th}), \tag{15.15}$$

or

$$\bar{v}_w = \frac{\mathrm{WRP}(F'_n - F'_{th})}{\pi D_w}. \tag{15.16}$$

Equation (15.16) gives the plunge-grinding velocity in terms of the normal interface force intensity.

When the wheelwear rate is negligible ($\bar{v}_s < \bar{v}_w$), and a steady state of deflection exists ($\dot{x} = 0$), Eq. (15.4) becomes

$$\bar{v}_w = \bar{v}_f \tag{15.17}$$

---

* Primed quantities signify per unit width.

Equation (15.16) can be solved for F'_n, with $\bar{v}_f$ replacing $\bar{v}_w$; thus

$$F'_n = \frac{\pi D_w \bar{v}_f}{\text{WRP}} + F'_{th} \tag{15.18}$$

This gives the induced force intensity generated by the feedrate $\bar{v}_f$ in the steady state.

The wheelwear curve Z'_s in Fig. 15.9, below F'_{bd}, may be approximated according to Lindsay[2] by

$$Z'_s = \text{WWP}(F'_n)^2 \tag{15.19}$$

where

$$\text{WWP} = 0.068 \times 10^{-9} \frac{l^2\left(1 + \frac{c}{l}\right) N_s D_s}{\left(\frac{D_e}{2.54}\right)^{1.2/\text{vol}} (\text{vol})^{0.85}} \left(\frac{\text{mm}^4}{\text{min}\cdot\text{N}^2}\right). \tag{15.20}$$

With this wheelwear parameter WWP, wheelwear can be estimated for various grinding conditions.

The breakdown force intensity F'_{bd} in Fig. 15.9 may be estimated for Al_2O_3 vitrified wheels[3] by

$$F'_{bd} = 62.3(\text{vol})^{0.55}(D_e)^{0.25} \quad (\text{N/cm}). \tag{15.21}$$

Precision grinding cycles should be designed so that the induced force intensity lies below F'_{bd}.

The wheel depth-of-cut h (advance of wheel per work revolution) is given by

$$h = \frac{\bar{v}_w}{N_w} \quad (\mu\text{m}). \tag{15.22}$$

This relation permits all the results developed for cylindrical grinding to be applied to surface grinding operations.

The "work cutting stiffness" K_{cw} (normal force required to take unit depth-of-cut) is an important quantity governing the rate of rounding up, the sparkout time, and chatter behavior when compared to the "system stiffness" K_m. It is given by

$$K_{cw} = \frac{F_n}{h} = \frac{V_w W}{\text{WRP}} \quad (\text{N/mm}). \tag{15.23}$$

The dimensionless "machining–elasticity number" α, formed by the ratio

$$\frac{K_{cw}}{K_m} = \alpha, \tag{15.24}$$

relates elastic effects in machining or grinding operations to the stiffness of the machine tool.

The power P absorbed in the grinding process is

$$P = F_t V_s \quad (\text{N}\cdot\text{m/sec}); (\text{W}). \tag{15.25}$$

The ratio F_t/F_n varies between 0.3 for a dull wheel and 0.7 for a sharp wheel with an average value of 0.5. Therefore,

$$F_t = 0.5F_n \tag{15.26}$$

and

$$P = \tfrac{1}{2}F_n V_s. \tag{15.27}$$

Using Eq. (15.18), and neglecting the threshold force F_{th},

$$P = \frac{\pi D_w W V_s}{2\text{WRP}} \bar{v}_f \tag{15.28}$$

gives the power required for any feedrate $\bar{v}_f$.

The "specific power" P_s, using Eq. (15.14) and neglecting F'_{th}, is

$$P_s = \frac{P'}{Z'_w} = \frac{V_s}{2\text{WRP}} \quad (\text{J/mm}^3). \tag{15.29}$$

The "time constant" τ_0 of a grinding system governs the time required to build up grinding force or sparkout. It depends upon the system stiffness K_m and on the material being ground, WRP [see also Eq. (15.45)]:

$$\tau_0 = \frac{\pi D_w W}{\text{WRP}\ K_m} \quad (\text{sec}). \tag{15.30}$$

On materials exhibiting a plowing region, the time constant suddenly changes during a sparkout when the plowing region is encountered. The WRP in the plowing region is about one-third the WRP in the cutting region.

The "G ratio"—the ratio of the volume of metal removed to the volume of abrasive consumed—is generally a variable depending on the particular operating force intensity. It is a valid ratio only in the case where the Z'_w vs F'_n and the Z'_s vs F'_n relations are straight lines emanating from the origin in Fig. 15.9. Generally, this is not true.

Machinability Parameters in Grinding

The work removal parameter WRP and the threshold force intensity in Eqs. (15.14) and (15.15), and the specific power, are the principal factors governing machinability in grinding. Some typical values of WRP for various alloys are given in Fig. 15.12, ranging from around 0.06 for T15 to 1.5 ($\text{mm}^3/\text{sec}\cdot\text{N}$) for chrome cast iron. Values of WRP may be estimated for so-called "easy-

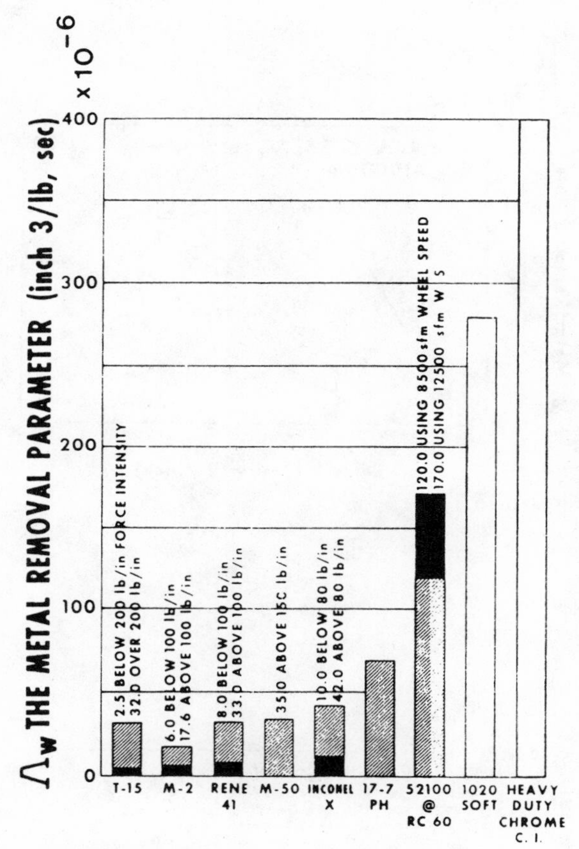

Figure 15.12

to-grind" materials with Lindsay's[2] semiempirical equation:

$$\mathrm{WRP} = 0.00958 \frac{\left(\dfrac{N_w D_w}{N_s D_s}\right)^{0.15} \left(1 + \dfrac{2c}{3l}\right) l^{0.58} N_s D_s}{D_e^{0.14} (\mathrm{vol})^{0.47} d^{0.13} (R_c)^{1.42}} \quad (\mathrm{mm}^3/\mathrm{min}\cdot\mathrm{N}). \qquad (15.31)$$

Values of WRP for "difficult-to-grind" materials are best obtained by measurement. Some typical values for T15 and M50 are shown in Figs. 15.13 and 15.14. These figures also show threshold force intensitites F'_{th}. The threshold force is very important at large D_e, and often causes problems in rounding up, sizing, and obtaining consistent surface finish. The variation of F'_{th} with D_e for several wheel speeds is shown in Fig. 15.15.

The specific power P_s for AISI 52100, M50, and M4 at several wheelspeeds, shown in Fig. 15.16, ranges from 7.7 for 52100 to 320 (HP/in.^{3}/min)

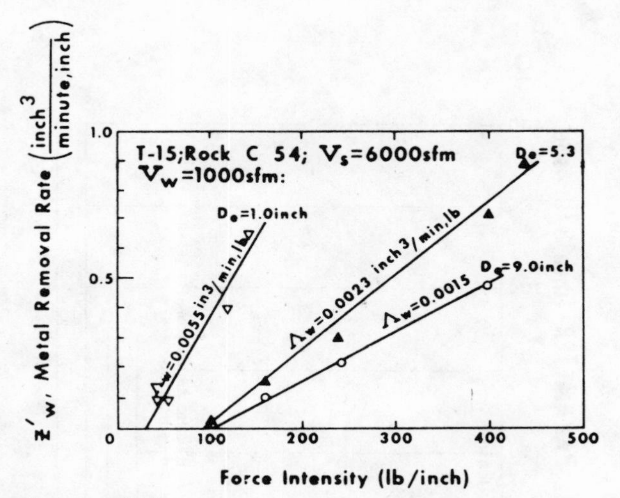

Figure 15.13

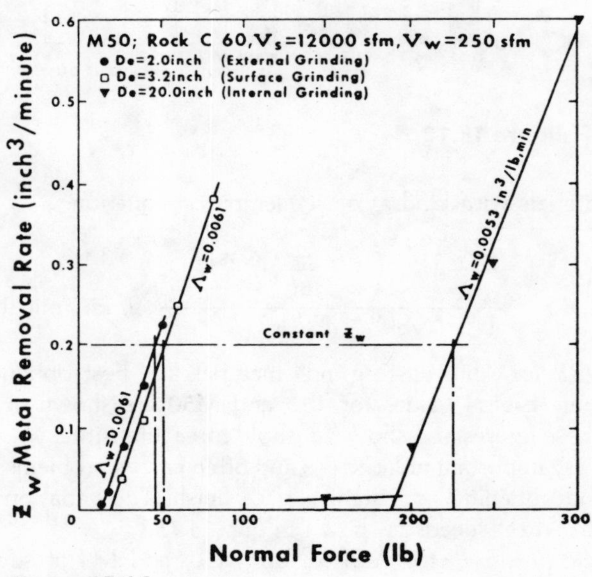

Figure 15.14

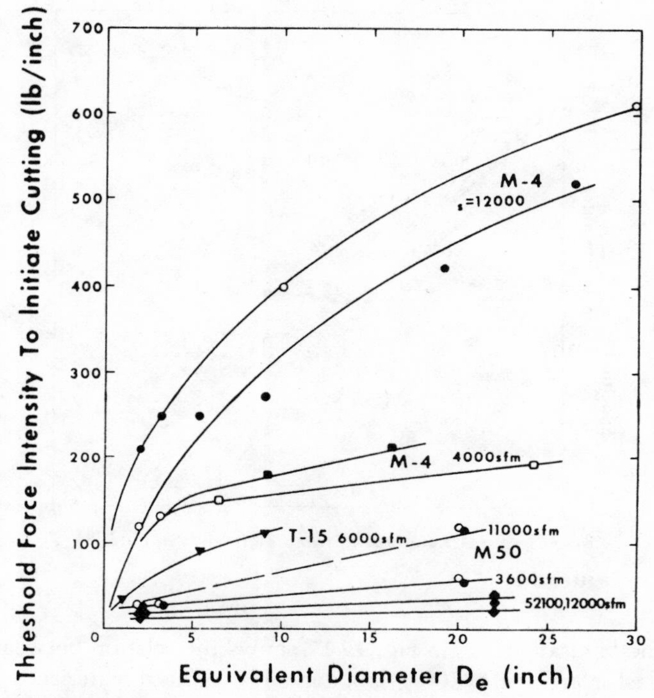

Figure 15.15

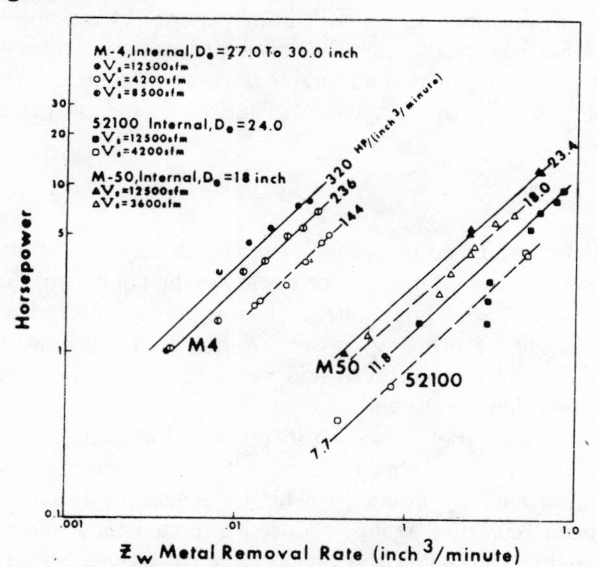

Figure 15.16

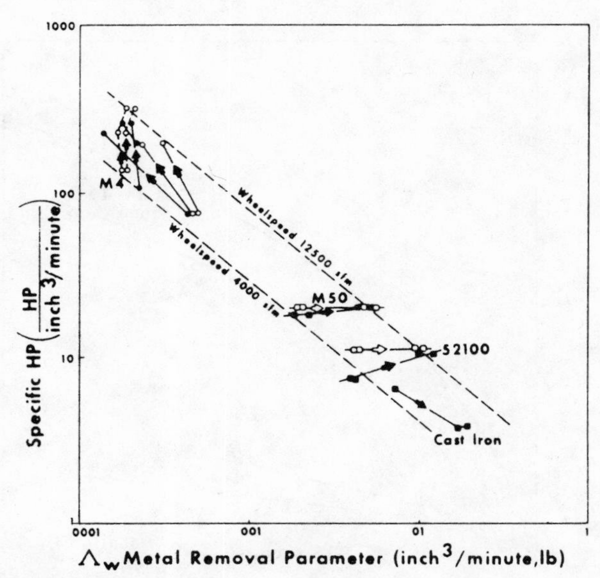

Figure 15.17

for M4. The Lindsay chart[4] in Fig. 15.17 shows the relation between P_s and WRP for cast iron, AISI 52100, M50, and M4. For each material, the arrows indicate the effect of increasing wheelspeed. For cast iron, increasing wheelspeed reduces P_s and increases WRP, both parameters moving in a favorable direction. For M4, increasing wheelspeed increases P_s and reduces WRP, both parameters moving in an unfavorable direction. Accordingly, M4 should be ground at low wheelspeeds, while cast iron should be ground at high wheel-speeds.

Surface Finish

The surface finish produced by grinding wheels depends on the topography of the wheel surface, the interface force intensity, the equivalent diameter, and the wheel grit and grade. The wheel surface topography gradually changes from that produced by dressing to one produced after a long grinding interval. Generally, the surface finish and surface profile (large-scale peak-to-valley roughness) deteriorate with usage.

Figure 15.18 shows the surface finish produced at various stock-removal rates and wheelspeeds in external grinding.[7] It will be seen that at a given Z'_w, the finish is improved by going to a higher wheelspeed. The data in Fig. 15.18 have been replotted against induced interface force intensity in Fig. 15.19, which reduces the five curves to a single curve, and demonstrates that surface finish is, in reality, a function of F'_n. The reduction of surface finish

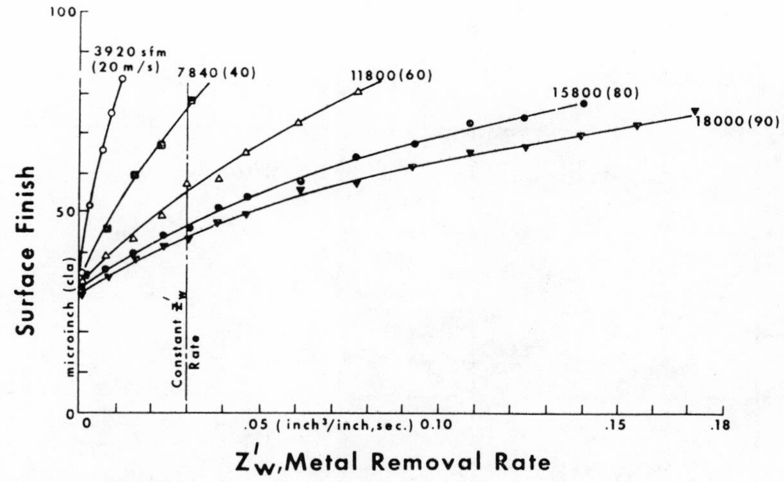

Figure 15.18

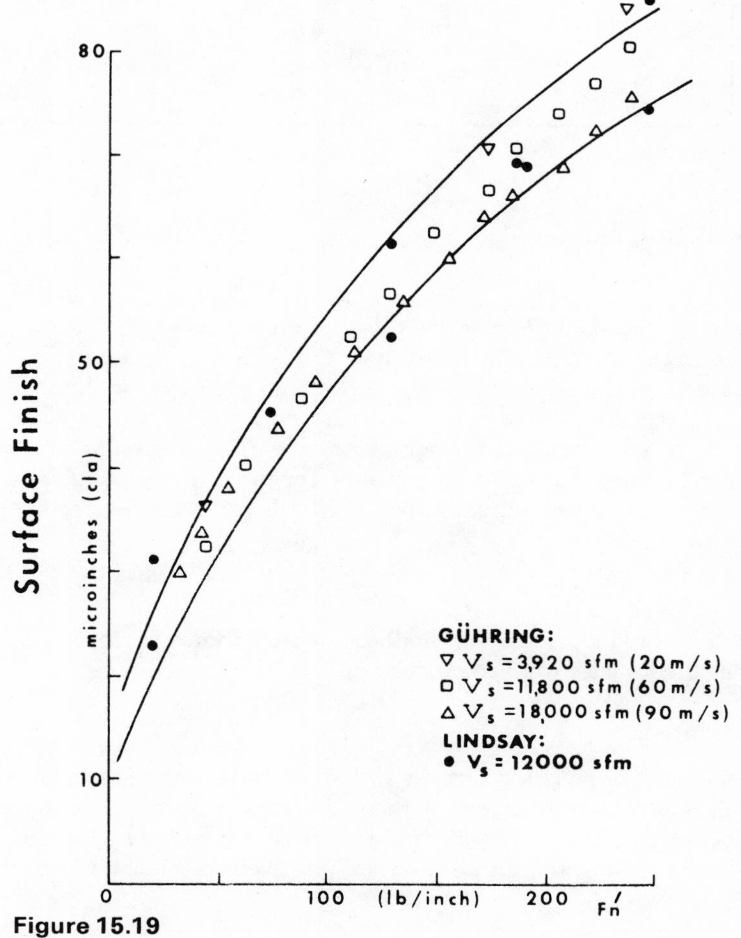

Figure 15.19

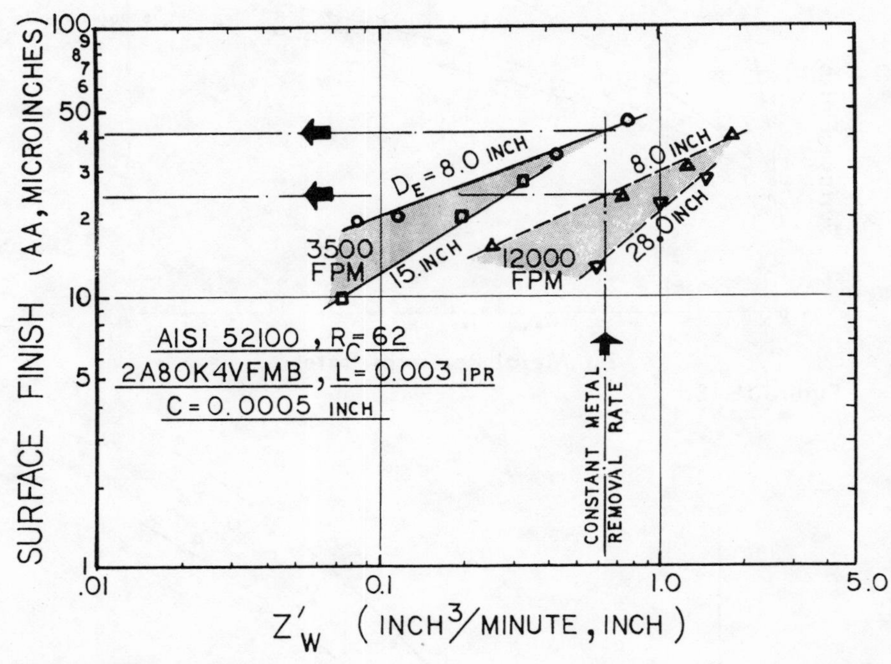

Figure 15.20

at high wheelspeed is due to lower induced force intensities, which result from the increase in WRP with wheelspeed according to Eq. (15.31). Then, with the larger WRP, F_n' is reduced in accord with Eq. (15.18).

Surface finish also depends on D_e. Figure 15.20 shows surface finish plotted against Z_w', again, for several wheelspeeds and D_e. The two bands of data in Fig. 15.20 when replotted against Lindsay's[5] T_{ave} are rectified as shown in Fig. 15.21. From this figure, an equation for the surface finish f for single-point, diamond-dressed, vitreous-bonded wheels has been developed:

$$f = 0.094\left(\frac{dl}{R_c}\right)^{0.5}\left(\frac{c}{l}\right)^{0.3}\frac{(F_n')^{0.33}}{(D_e)^{0.166}(\text{vol})^{0.13}} \quad (\mu\text{m, AA}). \qquad (15.32)$$

The validity of Lindsay's surface finish equation is shown in Fig. 15.22 where 129 measured values were compared with the calculated values.

Surface Integrity

The generation of precision surfaces with good surface finish does not necessarily guarantee a surface free of thermal cracks, residual tensile stress, or untempered martensite. Those detrimental effects are caused by localized high temperatures during the grinding operation. There is generally a threshold

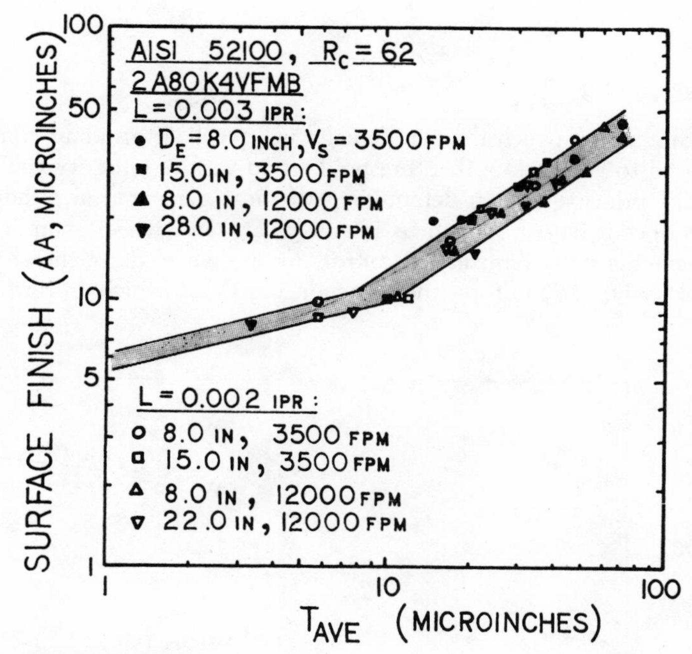

Figure 15.21

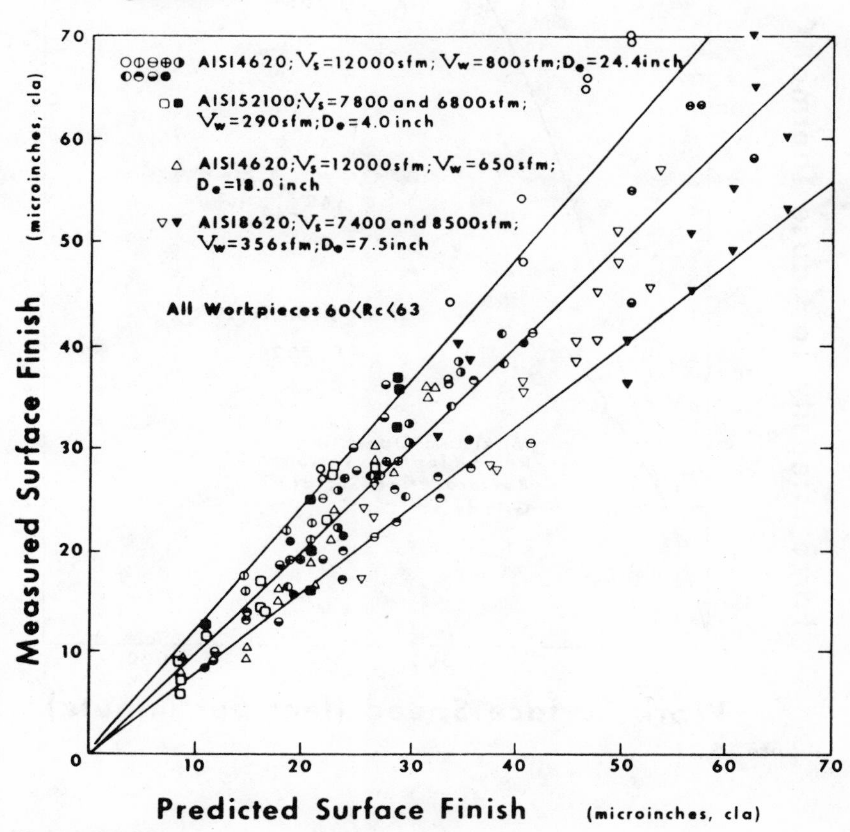

Figure 15.22

temperature above which thermal damage will result. If grinding conditions are adjusted to stay below this threshold, good surface integrity can be obtained. The principal factors determining the heat generated in grinding are the wheelspeed, workspeed, force intensity, D_e, and wheel sharpness, the latter being the most difficult to control. As the wheel dulls, the sharpness S defined by Eq. (15.7) drops, thereby reducing WRP, which, in turn, causes

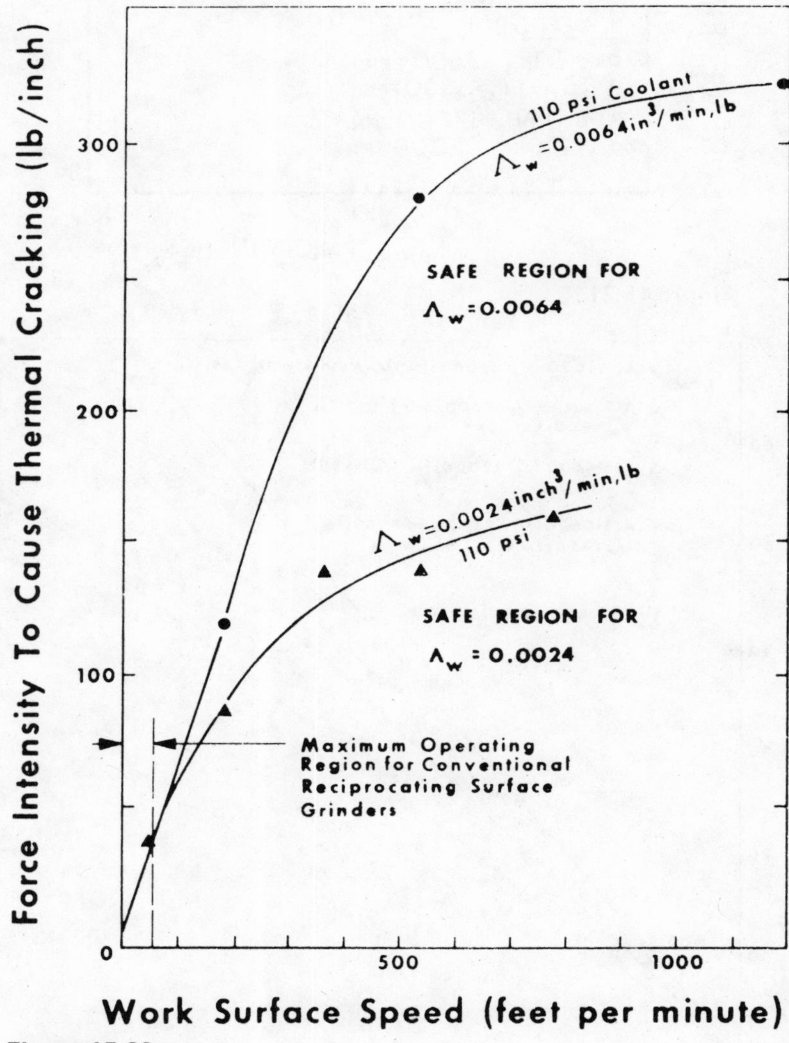

Figure 15.23

a rise in F'_n according to Eq. (15.18). Both of these effects tend to raise grinding temperatures.

The threshold force intensities to cause incipient cracking of AISI 52100 steel after an acid etch are plotted against workspeed in Fig. 15.23[6] for a dull wheel (WRP = 0.1 mm^3/sec·N, the lower curve) and for a sharp wheel (WRP = 0.335 mm^3/sec·N, upper curve). It will be seen that high workspeeds and sharp wheels permit higher force intensities and stock removal rates to be used. When the grinding zone temperature is calculated,[3] the two curves in Fig. 5.23 are reduced to a single function as shown in Fig. 15.24, showing that temperature is the controlling factor.

Good surface integrity can be obtained by operating at high workspeeds, using sharp wheels, and not permitting the induced force intensity to exceed some experimentally determined threshold value. Figure 15.25 illustrates the effect of increasing workspeed and wheel sharpness on residual stress.[6] Point 2 corresponds to a dull wheel (WRP = 0.00288 in.^{3}/min·lb) and a low workspeed (V_w = 50 fpm) resulting in 200,000 lb/in.^{2} tensile stress. Point 3 shows the effect of increasing the workspeed with the dull wheel. Point 7 corresponds to a sharp wheel (WRP = 0.0058 in.^{3}/min·lb) and high workspeed,

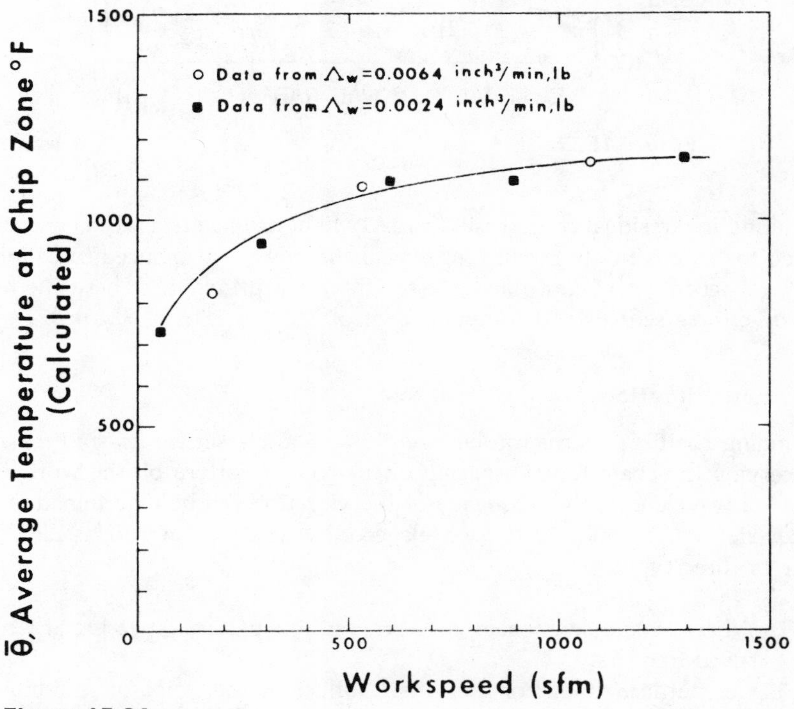

Figure 15.24

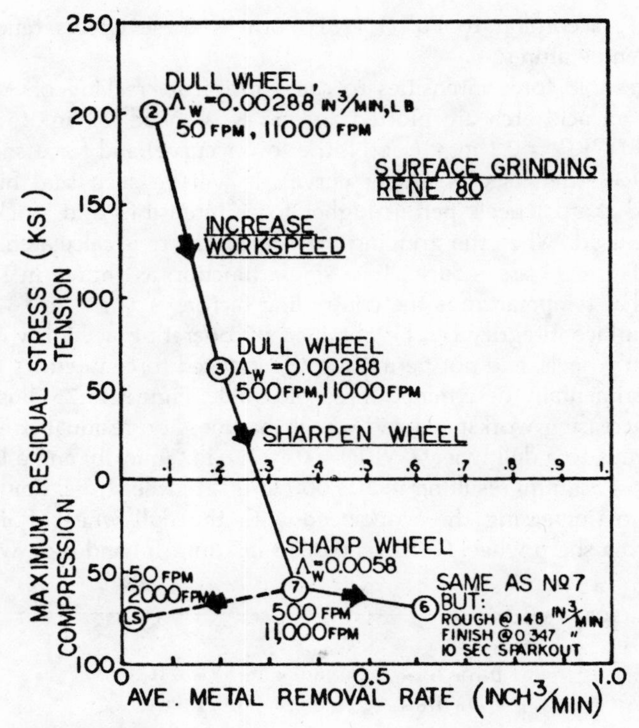

Figure 15.25

resulting in a residual compressive stress. Methods have recently been developed for automatically monitoring wheel sharpness and induced force intensity to accomplish "controlled surface integrity grinding." These methods involve force sensors and computer control.[8]

Chatter, Vibration, Surface Patterns

Grinding chatter patterns can be interpreted and classified into six types. By observing the characteristic spacing of the chatter pattern on the workpiece, i.e., the wavelength, the frequency of the vibration can be determined. If the frequency corresponds to the wheelspeed, the chatter can be classified into one of three types:

1. Wheel or spindle imbalance—a straight-line pattern repeating at wheelspeed frequency.
2. Geometrical runout of the wheel surface at the point of grinding—a straight-line pattern repeating at wheelspeed frequency.

3. Wheel mottle patterns—a nonstraight-line random pattern repeating at wheelspeed frequency.

If the frequency does not correspond to the wheelspeed, it can be classified again into one of three types.

4. General forced vibration due to pulleys, belts, drive motors, hydraulic pumps, etc. In this case, the frequency will change in proportion to the speed of these elements.
5. Wheel regenerative type in which the straight-line chatter frequency is essentially independent of workspeed, drive-motor speed, etc., but corresponds approximately to a natural frequency of the wheelhead, workhead, or machine structure. In this type the wheel gradually wears into a multilobed cylinder.
6. Work regenerative type, again where the straight-line frequency corresponds to a natural frequency but where the workpiece develops a wavy surface similar to a "corduroy road"—tends to occur at high workspeeds.

In the above cases, a straight-line chatter occurs in the workpiece under plunge-grinding conditions. If a spiral chatter pattern occurs, vibration between the dresser diamond and the wheel during the dressing process is probably the cause.

Once the type of chatter has been determined, steps can be taken to eliminate it. Since types 1, 2, and 4 are generally well understood, only types 3, 5, and 6 will be discussed below.

Wheel Regenerative Chatter Wheel regenerative chatter is a type of self-excited vibration and is to be distinguished from forced vibrations. In a self-excited vibration, the periodic driving force is created and controlled by the vibratory motion itself, whereas in a forced vibration the driving force is independent of the vibratory motions. During a grinding cycle, if the grinding wheel is given a small vibratory disturbance, the interface force between wheel and work will fluctuate. This causes a local fluctuation of the instantaneous amount of wheelwear. After the wheel has made one revolution, this local minute wavy wheel surface produces a transient force variation that, in turn, may cause another fluctuation in wheelwear. If the system is stable, these disturbances die out. If the system is unstable, these disturbances build up. Snoeys, Brown and others[9,10] have investigated the stability of grinding. A plunge-grinding system can be represented by the block diagram shown in Fig. 15.26. The input to the system is the feedrate $\bar{v}_f$. This causes a stock removal velocity $\bar{v}_w$, a wheelwear velocity $\bar{v}_s$, a deflection velocity $\bar{v}_m$, and a wheelwork interface force F_n. This block diagram can be reduced to

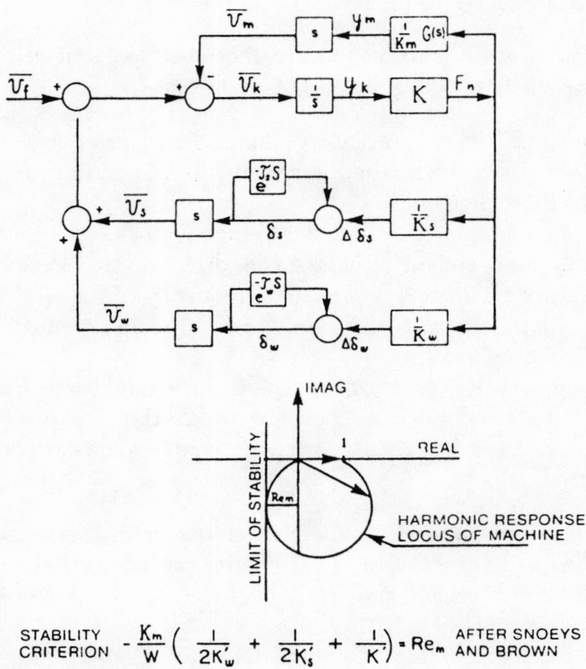

Figure 15.26

the so-called "characteristic equation":

$$-\frac{K_m}{K_s}\frac{1}{(1-e^{-\tau_s s})}-\frac{K_m}{K_{cw}}\frac{1}{(1-e^{-\tau_w s})}-\frac{K_m}{K}=G_m(s). \qquad (15.33)$$

The cutting stiffness K_{cw} is obtained from the slope Λ_w shown on the wheelwork characteristic chart in Fig. 15.9; for example, from Eq. (15.23)

$$K_{cw}=\frac{F_n}{h_w}=\frac{V_w w}{\Lambda_w}.$$

Thus, the cutting stiffness of the workpiece is proportional to the work surface speed V_w and the width of cut w, and inversely as Λ_w. Similarly, the wear stiffness of the grinding wheel is

$$K_s=\frac{F_n}{h_s}=\frac{V_s w}{\Lambda_s}.$$

The wheelwork contact stiffness K is also a function of normal load and is given by Snoeys and Brown[9] as:

$$K = K_0 D_e^{-0.25} (F_n')^{0.75}.$$

The roots of the characteristic equation (15.33) determine whether the system is stable or unstable. Most practical grinding operations are carried out in the unstable region. This means that some of the roots of the above equation have positive real parts; i.e., $\alpha > 0$, where $(\alpha + j\omega)$ is the form of a root and, after a length of time, a vibration of sufficient amplitude will develop, as governed by

$$X = ae^{(\alpha + j\omega)t}. \tag{15.34}$$

If α is very small and positive, it will take a long time for the vibration to grow. The threshold of instability corresponds to the condition where $\alpha = 0$. The condition for unconditional stability is[9]

$$-K_m\left(\frac{1}{2K_{cw}} + \frac{1}{2K_s} + \frac{1}{K}\right) = Re_m. \tag{15.35}$$

Since most practical grinding operations lie in the unstable region and since satisfactory grinding can be accomplished as long as the vibration amplitude is less than a certain value, the concept of a chatter-free grind time (CFGT) has been proposed by Lindsay and Navarro.[12] Consequently, the value of α determines the CFGT. The objective in production grinding operations is to select grinding conditions so that the CFGT is sufficiently long to accommodate the cycle time. Redressing the grinding wheel, of course, reconditions the wheel allowing it to start another CFGT. The influence of the various grinding parameters on the magnitude of α and thus CFGT will be discussed below.

Influence of Grinding Parameters on CFGT The CFGT for certain internal grinding operations as measured by Lindsay is plotted in Fig. 15.27 against the stock removal rate Z_w for a short ($K_m = 60{,}000$ lb/in.) and long ($K_m = 16{,}000$ lb/in.) grinding quill. The CFGT was determined when the vibration reached an audible level. The actual limiting vibration amplitude is not known, but is probably of the order of 0.25 μm. It will be seen that the CFGT falls with a -2 slope in a log-log plot with stock removal rate Z_w', and that it was approximately twice as long for the stiffer quill. The beneficial effect of higher machine stiffness is to be expected, and assuming one could solve Eq. (15.33) for the positive real roots α—a formidable task—it is expected that large K_m will decrease α and give longer CFGT. However, the strong dependence of α on the stock removal rate or its equivalent force intensity is completely ignored by Eq. (15.33). Equation (15.33) implies that

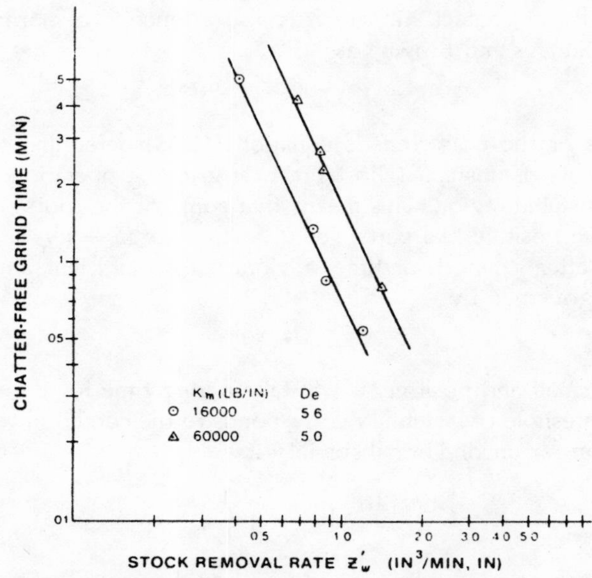

Figure 15.27

the positive real part of the roots will be some function of the system parameters K_s, K, K_m, τ_s, τ_w, and will be independent of the force level in the system. Evidently this is not true, and some of the parameters, notably K_s and K, are nonlinear and change with increasing force levels (or feed rates) as mentioned earlier.

Therefore, the nonlinear "grinding wheelwear law" and the nonlinear contact stiffness K play an important role in determining CFGT. As the preloading force increases, the wheelwear stiffness drops and the contact stiffness increases. This causes large force pulsations to occur for small hardness variations in the wheel. Consequently, a high value of K may tend to increase instability.

As a result of the strong dependence of CFGT on force intensity F'_n, or its equivalent stock removal rate Z'_w, Lindsay has shown that the volume of material removed from the workpiece before the chatter amplitude reaches its limit is also a function of stock removal rate Z. This is called the chatter free volume (CFV). The CFV is given by

$$\text{CFV} = \text{CFGT} \times Z_w. \tag{15.36}$$

Figure 15.28 shows the CFV plotted against stock removal rate. The existence of a maximum CFV and an optimum Z_w is clearly indicated. If the volume to be removed from the workpiece exceeds the maximum CFV, then the

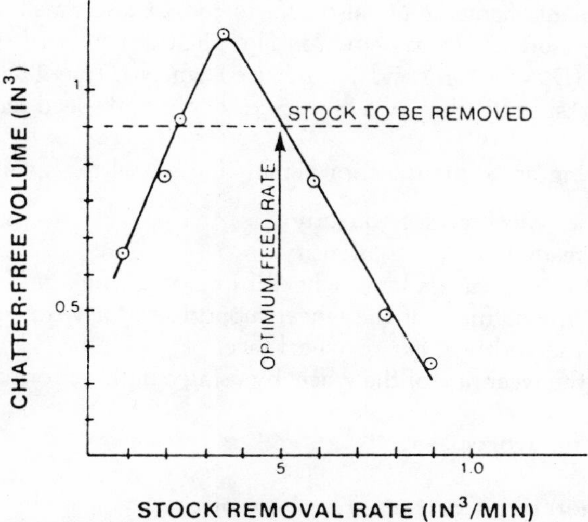

Figure 15.28

wheel will require multiple dressing in order to remove the stock. Otherwise, the fastest rate should be used to get the shortest cycle time.

Anything that reduces the wheelwear rate should tend to increase CFGT or CFV. Figure 15.29 shows a plot of wheelwear rate Z'_s, CFV′, and stock removal rate Z'_w plotted against interface force intensity for grinding with a water-base coolant and a grinding oil. The CFV′ is significantly increased when using oil.

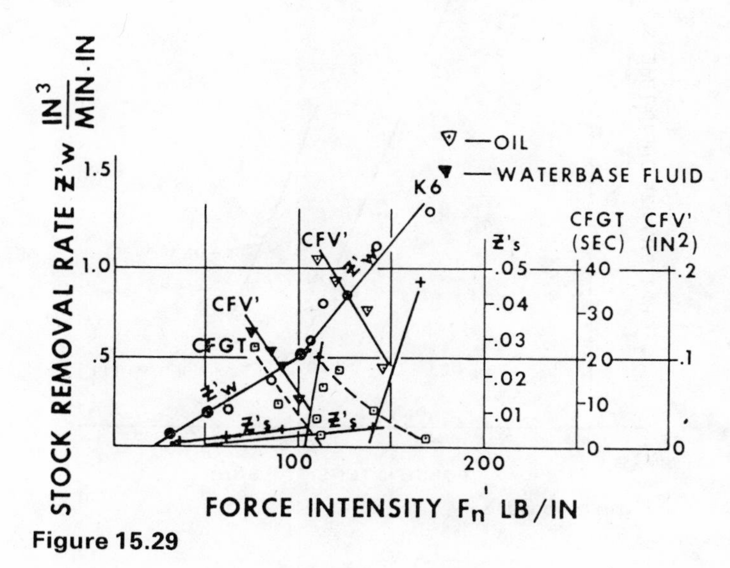

Figure 15.29

The equivalent diameter D_e also affects the CFGT. Small values of D_e tend to cause short CFGT as shown in Fig. 15.30 for internal grinding with a new wheel ($D_e = 5.7$ in.) and with a used wheel ($D_e = 2.38$ in.). The influence of workspeed is shown in Fig. 15.31. High workspeed tends to reduce CFGT.

The following are some rules for eliminating wheel regenerative chatter:

1. Dress the wheel more frequently.
2. Reduce feedrate or force intensity.
3. Increase D_e; i.e., use a large wheel if internal grinding.
4. Increase the stiffness of the wheel support and/or work support.
5. Reduce the width of cut or wheel face.
6. Reduce the wear rate of the wheel by using a high-performance grinding fluid.
7. Reduce the workspeed.

Work Regenerative Chatter This type of chatter tends to occur at high workspeeds. In this case, a small disturbance in the system causes a small transient wave to be ground into the workpiece. One work revolution later this wavy work surface acts as a driving force to cause the system to again

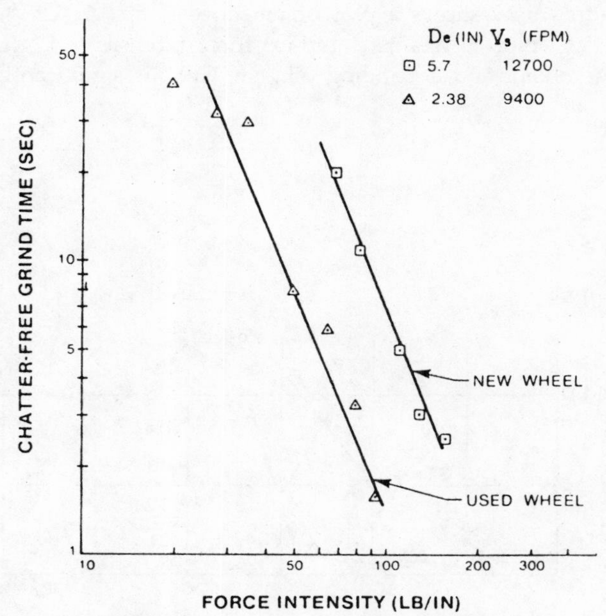

Figure 15.30

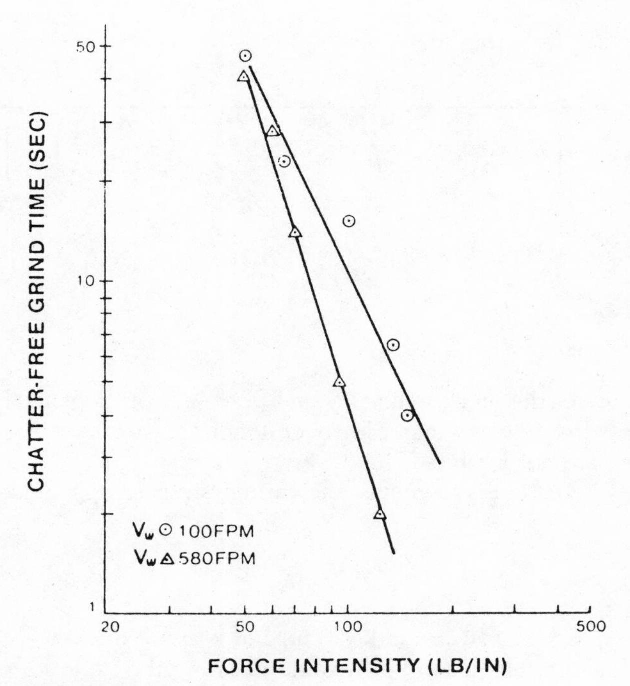

Figure 15.31

vibrate. Under unstable conditions a small wavelet can develop and extend around the work circumference. Equation (15.33) also applies to this type of chatter. Snoeys and Brown[9] have shown that work regenerative chatter tends to be suppressed if the wheelwork contact length L_c is greater than one-half the chatter wavelength λ. Therefore,

$$\lambda = \frac{v_w}{f_b} = 2L_c,$$

where f_b is the "break frequency" above which chatter frequencies cannot occur. The length of contact may be taken as

$$L_c = \frac{0.016 F_n'^{1/3} D_e^{1/3}}{(1.33H + 2.2s - 8)^{0.33}} + \left(\frac{F_n \Lambda_w D_e}{v_w w}\right)^{0.5}, \qquad (15.37)$$

where the first term is due to elastic deflection and the second is due to the wheel depth of cut h, since

$$h = \frac{F_n}{K_{cw}} = \frac{F_n \Lambda_w}{V_w w}. \qquad (15.38)$$

Then the break frequency becomes

$$f_b = \frac{V_w}{2\left[0.016\left(\dfrac{F'_n D_e}{1.33H + 2.2s - 8}\right)^{1/3} + \left(\dfrac{F_n \Lambda_w D_e}{wV_w}\right)^{0.5}\right]}. \quad (15.39)$$

From this equation it can be seen that

High forces F_n
Large D_e
Low workspeed V_w
Soft wheels

all tend to lower the break frequency and prevent work regenerative chatter in the higher-frequency ranges. However, low-frequency structural modes of the machine are not inhibited.

Another alternative is to reduce the cutting stiffness K_{cw}. Since

$$K_{cw} = \frac{V_w w}{\Lambda_w},$$

lower workspeed, narrower width of cut, and keeping the wheel sharp (high Λ_w) would tend to eliminate the chatter. Increasing the static stiffness and/or the damping in the offending structural mode also will tend to eliminate work regenerative chatter.

Wheel Mottle Patterns These random patterns repeat at wheelspeed frequency, but do not have a straight-line character. Figure 15.32 illustrates a typical wheel mottle pattern. These patterns are caused by local hardness and stiffness variations in the grinding wheel and are not the result of mechanical vibration. They are sometimes hardly measurable and only appear as a visual imperfection of the surface finish. Periodic roughness in a Talyrond chart is sometimes due to local hardness variations in the grinding wheel as shown by Hahn and Price.[11] Figure 15.33 shows the Talyronds corresponding to two cross sections in a plunge-ground ball track. For Section 1 there is a strong wheel rotational periodic roughness, while in Section 2 the surface is true. This is caused by a local hard zone in the wheel, not by vibration. Hahn and Price[11] have shown that the quality of the Talyrond trace is related to the degree of stiffness variation in the grinding wheel.

Mottle patterns, sometimes, are very prominent visually when the hard or stiff zone of the wheel operates in the "cutting region" (see wheelwork characteristic chart, Fig. 15.9), while the softer zone operates in the "ploughing region," since the surface finish produced in these two regions is significantly different. The cure in this case would be to change the finish feedrate or

Figure 15.32

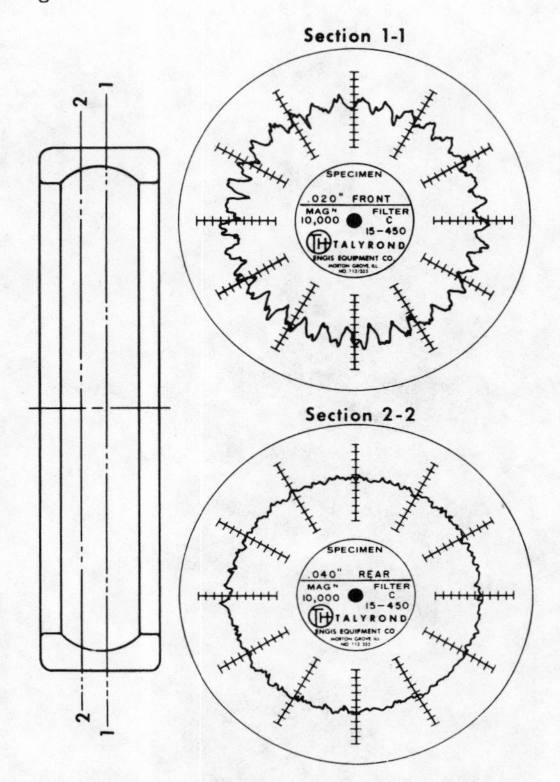

Local Imperfections in Ground Surface

Figure 15.33

sparkout so that the grind terminates with all zones of the wheel operating completely in either the ploughing or the cutting region but not straddling the "plowing–cutting" transition.

Production Grinding Cycles

Grinding Force Profiles

Production grinding machines generally consist of a rotating work-holding device into which workpieces are automatically loaded and unloaded, and a grinding wheelhead mounted on a cross-slide, which is capable of moving the wheel radially into contact with the workpiece. Figure 15.34 shows a record of the radial cross-slide displacement (upper trace) during two consecu-

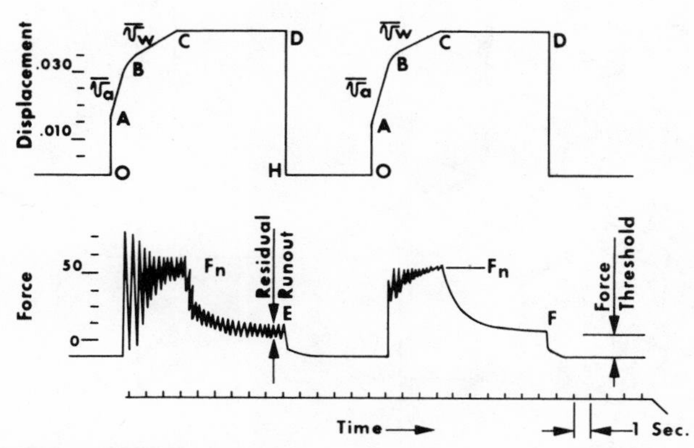

Figure 15.34

tive grinding cycles of a production grinding machine. The lower trace is a record of the normal interface force existing at the wheelwork contact area during the grind cycle as obtained by a force transducer. In a controlled-force grinding machine, the cross-slide is moved at a high velocity from O to A by the applied force F_a (see Fig. 15.35), bringing the wheel close to the unground running-out workpiece. At a point A, the dashpot C is engaged, causing the slide to approach the workpiece at the approach velocity $\bar{v}_a$, which is simply

$$\bar{v}_a = \frac{F_a}{C}. \tag{15.40}$$

As the cross-slide moves from A to B it picks up the running-out workpiece. Large instantaneous force pulsations may occur on the wheel during the early stages of rounding up, as shown by the first force trace. As rounding up takes place, these force pulsations reduce and tend toward an average grinding force F_n according to

$$F_n = F_a - C\bar{v}_w, \tag{15.41}$$

where $\bar{v}_w$ is the plunge-grinding velocity. The majority of the stock is removed and further round up accomplished from B to C. At C, the cross-slide strikes a stop (Fig. 15.35) and comes to rest. However, the grinding wheel continues to grind due to residual spindle deflection and sparks out during the interval CD. At D, the slide is retracted and the workpiece, it is hoped,

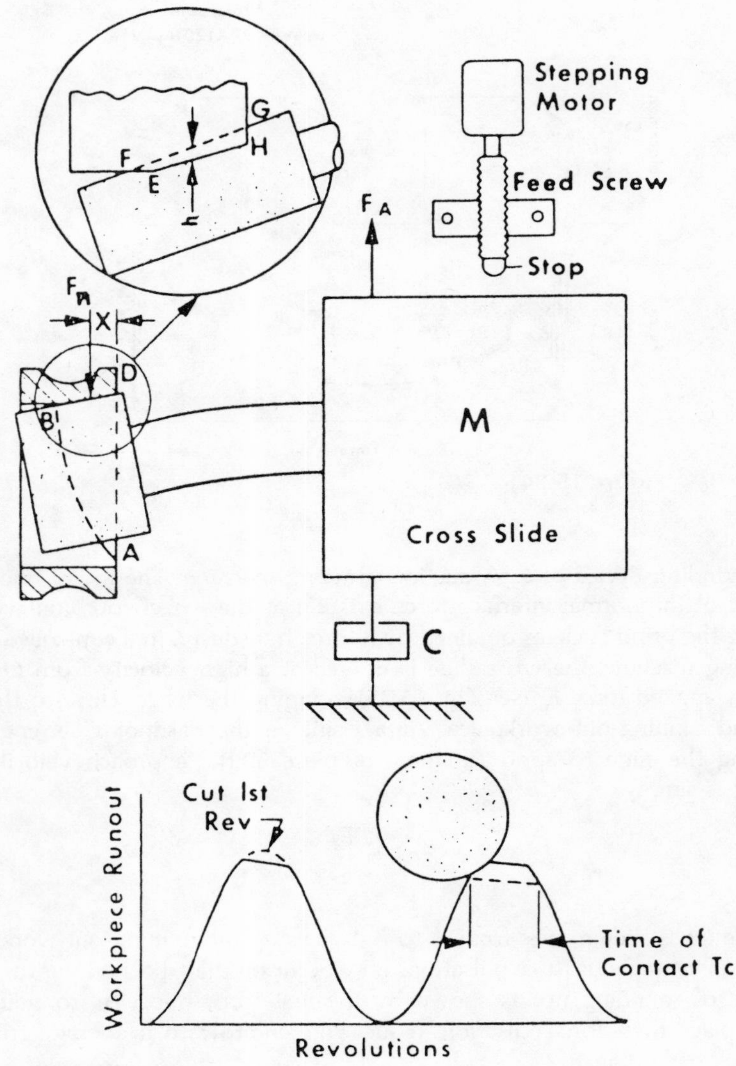

Figure 15.35

is at size with the proper surface finish, roundness, and microprofile. The work is unloaded and a new workpiece reloaded from *H* to *O* in preparation for the next cycle.

Feed-rate grinding machines, in contrast to controlled-force machines, feed the cross-slide at a prescribed feed rate $\bar{v}_f$, gradually developing grinding

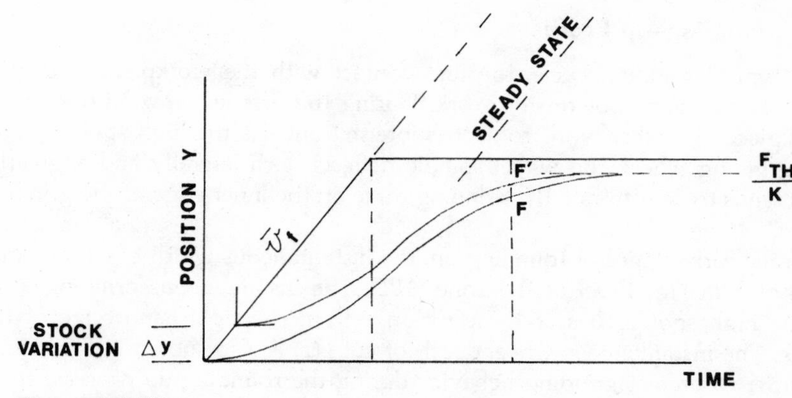

Figure 15.36

force as the deflection of the system increases. Figure 15.36 shows the cross-slide position (line *OAB*) and the position of the cutting surface of the wheel (line *OEF*) during a simple feed and sparkout cycle. The vertical distance between line *OAB* and line *OEF* is the deflection in the system, which is proportional to the grinding force. During the sparkout period *AB*, the force (and deflection) decays to a small but finite threshold force or deflection F_{th}/K_m.

Stock variations of incoming parts often cause size variations in the finished parts if the steady-state force has not been achieved in the rough-grind portion of the cycle, as illustrated in Fig. 15.37, where the first part had more stock than the second part. The value of the force in the system at the termination of the cycle governs the final size and surface finish (unless in-process gaging is used). Consequently, the size envelope produced may depend on initial stock variations and variations in threshold force, which, in turn, depend on the wheel sharpness variations and changes in the conformity D_e.

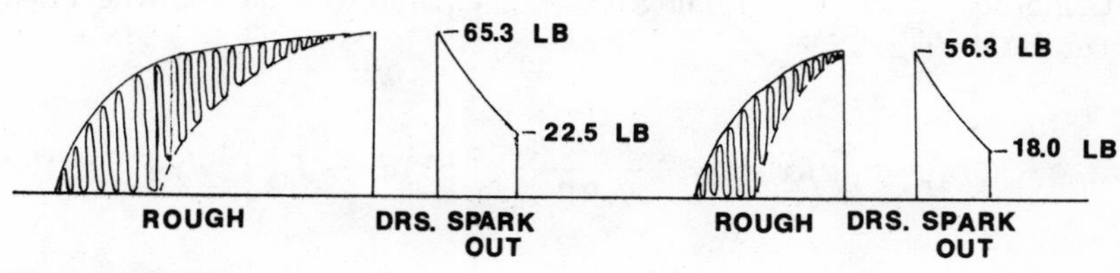

Figure 15.37

The Rounding-Up Process

In a typical grinding cycle, the first contact with the workpiece is usually made at the high spot on the work. During the first few revolutions of the workpiece, the wheelwork contact is intermittent. As the high spot engages the grinding wheel, the wheel spindle deflects both laterally and angularly. This tends to concentrate the grinding force on the inner edge of the grinding wheel.

In the early stages of rounding up, the instantaneous width of cut w varies as shown in Fig. 15.35 as the zone ABD. The instantaneous grinding force at the high spot F_n lies at the position x from the right-hand edge of the work. The instantaneous wheel depth of cut $h(x)$ is also shown in Fig. 15.35. In order to predict grinding behavior during the rounding-up process, equations describing the dynamic behavior of the grinding machine have been combined with grinding-process equations in a computer program called TRUFI.[13] This program prints out for each work revolution the progressively diminishing workpiece runout, the peak-force intensity existing on the grinding wheel at the high point, the peak force itself, and the amount of wheelwear. Also given is the amount of stock required to true-up the workpiece to a given accuracy, and the time required to round up the workpiece.

The grinding process Eqs. (15.16) and (15.19) can be adapted to give instantaneous values of wheel depth of cut and wheelwear over contact regions with nonuniform force intensity distributions as follows.

Dividing both sides of Eq. (5.16) by the work speed N_w gives the wheel depth of cut $h(x)$ as

$$h(x) \equiv \frac{\bar{v}_w}{N_w} = \frac{\text{WRP}}{\pi D_w 60 N_w} [f(x) - F'_{th}], \tag{15.42}$$

where $f(x)$ = local-force intensity,
F'_{th} = threshold force intensity assumed here equal to ploughing-cutting force intensity I_{pc},
x = the position along wheelwork contact zone.

Using Eq. (15.42), the grinding force F_n required to make the wheel depth of cut $EFGH$ (Fig. 15.35) is

$$F_n = \int_0^w f(x)\,dx = \frac{\pi D_w 60 N_w}{\text{WRP}} \int_0^w h(x)\,dx + F'_{th} \int_0^w dx. \tag{15.43}$$

Equating this grinding force to the elastic force required to hold the wheelhead spindle at a given deflection forms the computational basis that enables the computer program to calculate peak force intensities and wheel depth of cuts on each successive work revolution.

The wheelwear (ww) is obtained from Eq. (15.19):

$$\text{ww} \equiv \bar{v}_s t_c = \frac{\text{WWP}(F_n')^2 t_c}{\pi D_s} \tag{15.44}$$

where t_c is time of grinding contact during a work revolution (see Fig. 15.35).

In order to verify the computer program, a production grinder was equipped with instrumentation to record the instantaneous peak grinding force and the instantaneous workpiece runout during the grinding of No. 3920 taper roller-bearing cups (4.4375 in. OD or 112.5 mm). Figure 15.38 shows peak grinding forces, which reach a maximum of 300 lb (1130 N) after 1 sec followed by a decline to around 100 lb (444 N) after 6 sec. The instantaneous runout is shown in the lower part of the figure, where an initial runout of 0.008 in. (0.2 mm) is reduced to less than 0.001 in. (0.025 mm) in about 6 sec. Measured values of runout and force as well as TRUFI computed values are plotted on semilog paper for comparison in Figure 15.39. It will be seen that the agreement between computed and measured values is reasonable.

Workpieces flowing to a production grinder usually have a joint random distribution of stock allowance and runout. A sampling of workpieces is

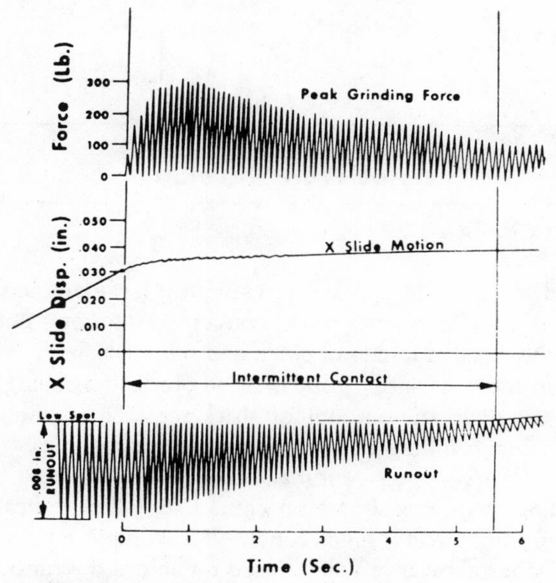

Figure 15.38

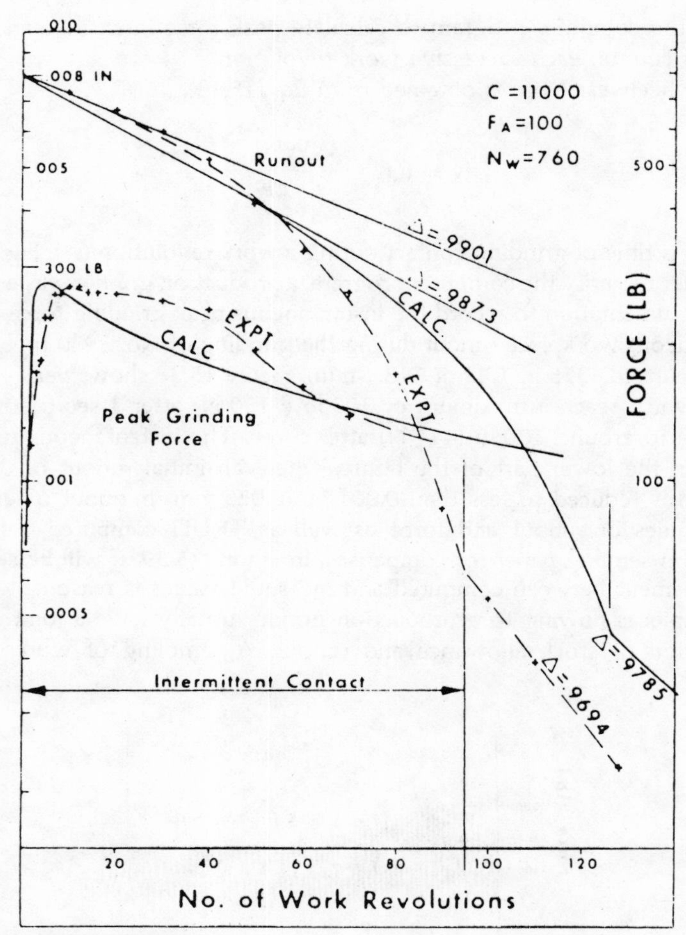

Figure 15.39

shown plotted in Fig. 15.40. Workpieces with a large amount of stock and small runout are easy to round up and convert into a good finished product. Workpieces with small amounts of stock and relatively large runout are difficult to round up and make into a good finished product. By using the rounding-up computer program, the amount of stock required to round up a given initial runout when grinding under a given force or feed rate can be found. Doing this for several values of initial runout, one can draw the force lines or corresponding feed rates shown on Fig. 15.40, which indicate the amount of stock required to correct a given runout. If a workpiece falls below a given force line, the stock allowance will be used up before the runout is corrected. For the case shown in Fig. 15.40, all workpieces will be rounded up if ground under 90 lb (400 N) of force or less. For 100 lb (44 N) grinding force, there

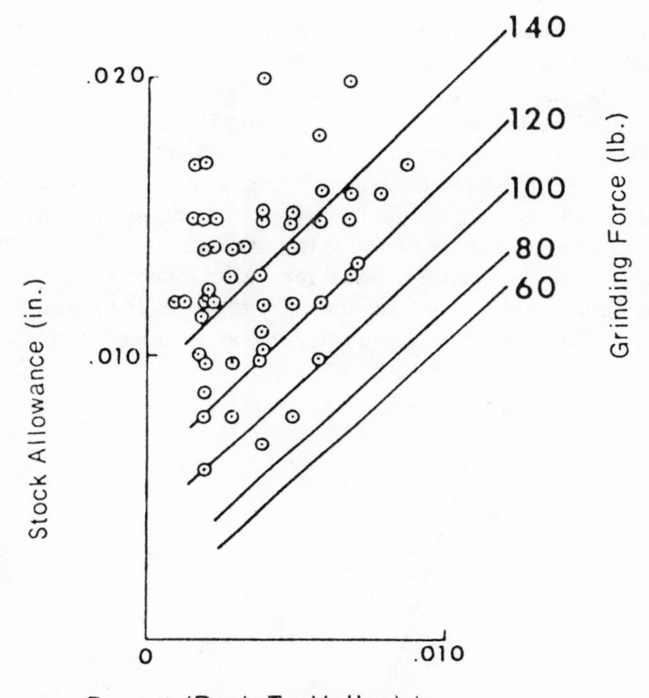

Figure 15.40

will be two parts (the two points below the 100-lb line) that will not round up. Charts such as shown in Fig. 15.40 are useful in selecting the highest permissible grinding force or feed rate.

The computer program TRUFI also outputs the amount of wheelwear occurring during the round-up process. This is an important variable since it determines the amount of infeed, or so-called "compensation," of the diamond required to true the grinding wheel. If the "compensation" is set too large, the wheel will be rapidly consumed and grinding cost per part will rise because of increased abrasive cost. If the "compensation" is set too low, failure to dress will occur, and surface finish, size, and microprofile will deteriorate. If the grinding force or feed rate is set too low in order to protect and conserve the wheel, cycle times tend to be long and, again, grinding costs go up. Consequently, it is important to select the grinding force/feed rate so that cycle times are as short as possible without causing excessive wheel breakdown.

The Sparkout Process

After the grinding wheel has engaged the "black" workpiece and has rounded up the initial runout and removed the majority of the stock, it is necessary to bring the workpiece to the required size and taper tolerance, surface finish, microprofile, and surface integrity, and to do this in the least possible time. At the point C (Fig. 15.34) the cross-slide strikes a stop (Fig. 15.35), which may or may not be fed as desired. On striking the stop, the cross-slide velocity suddenly drops to zero, or, if desired, to a small finish feed velocity $\bar{v}_{ff}$. The velocity of the grinding wheel $\bar{v}_w$ gradually begins to decrease as the spindle deflection and grinding force F_n start to decay. The grinding force decays down to the ploughing-cutting transition through the cutting region (see Fig. 15.9) with the "cutting" metal removal parameter Λ_{wc}, or WRP, for the cutting regime. When the force intensity drops to the ploughing-cutting transition F'_{pc}, the decay rate of the force becomes slower, corresponding to

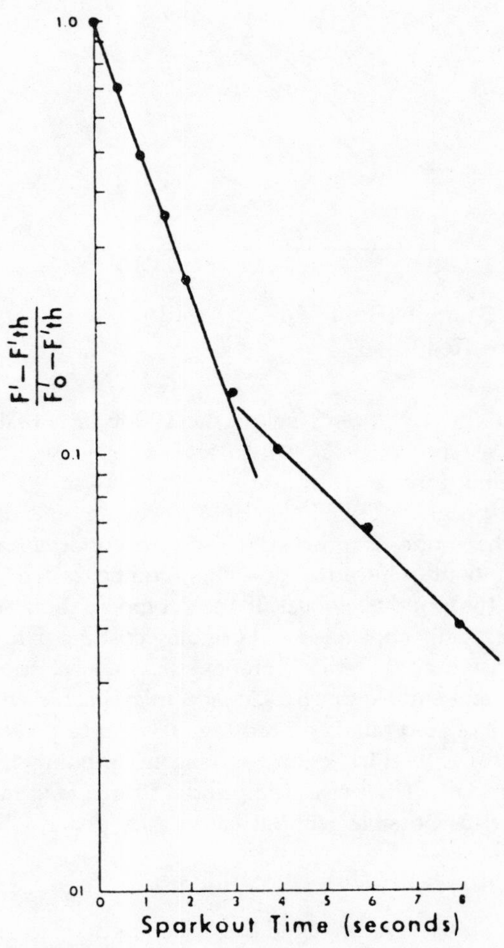

Figure 15.41

the smaller "ploughing" metal removal parameter Λ_{wp}. Further decay of the force intensity reaches the threshold force intensity F'_{th}. At this point, material removal ceases, and the wheel simply rubs on the work. The normalized grinding force intensity during a sparkout is shown plotted on semilog paper in Fig. 15.41. It will be seen that a kinked straight line results, corresponding to the two values Λ_{wc} and Λ_{wp} shown in Fig. 15.42. It will be appreciated that the rounding-up process is most rapid when Λ_w is large. Therefore, it is very desirable to have rounded up the workpiece before the wheel enters the ploughing regime.

It has been shown that the sparkout process follows an exponential decay with a time constant[14] τ_0 given by

$$\tau_0 = \frac{w}{K_m\left(\dfrac{\Lambda_w}{60\pi D_w} + \dfrac{\Lambda_s}{60\pi D_s}\right)}. \tag{15.45}$$

Thus, two time constants occur corresponding to Λ_{wc} and Λ_{wp}, where distinct ploughing and cutting regions exist.

On grinding cycles where size, taper, and surface finish are determined by a sparkout of fixed time, several factors affect the size, taper, and finish

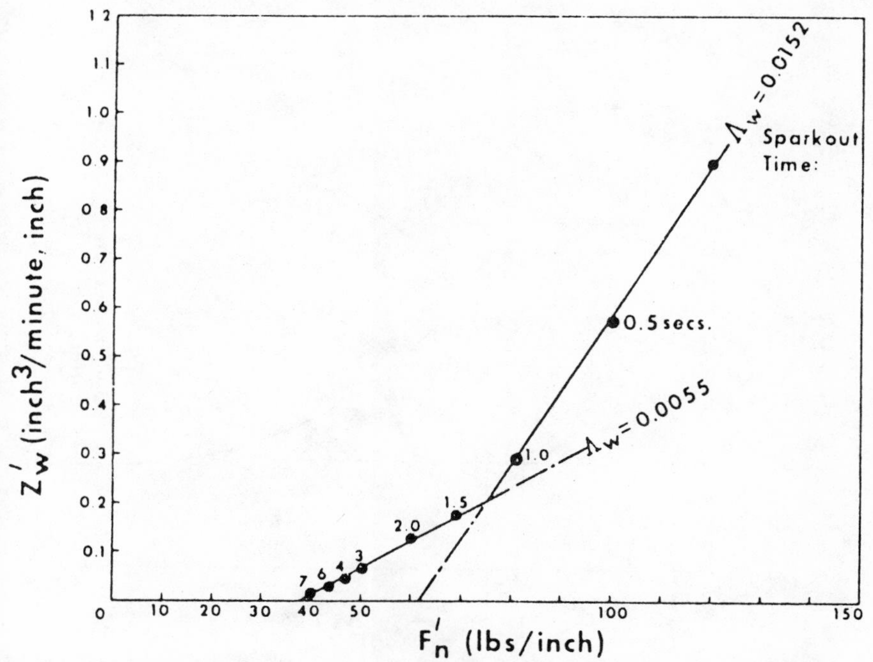

Figure 15.42 Wheelwork characteristic chart corresponding to sparkout in Fig. 15.41.

variations. First, for cycles with a long sparkout time, the influence of the initial condition or force level from which sparkout began is completely lost. Size variations under these conditions are determined solely by variations in the threshold force F_{th}. For shorter sparkout times, errors in size may be caused by both variations in the initial force level, variations in Λ_{wc} and Λ_{wp}, and variations in the threshold force. For the case where the ploughing regime is negligible, a single sparkout time constant results, and the quill deflection during a sparkout can be plotted as illustrated in Fig. 15.43. The solid line represents a standard cycle or, for instance, grinding with a new wheel that has been freshly dressed with a sharp diamond. The dotted line represents the situation where: (1) the initial force level has changed from F_n to $(F_n + \Delta F_n)$, (2) the time constant has changed from τ_0 to τ_1 due to changes in Λ_w in Eq. (15.45), and (3) the threshold force or corresponding quill deflection has changed by ΔF_{th}.

The influence of system rigidity K_m on size-holding ability at the end of a 6-sec sparkout (about three time constants) is illustrated in the accompanying table in the grinding of the two parallel ball tracks (Fig. 15.44) in automotive water pump bearings. In one setup the rigidity of the grinding quill and wheelhead was increased from 23,000 to 50,000 lb/in. (4180 to 9100 N/mm)

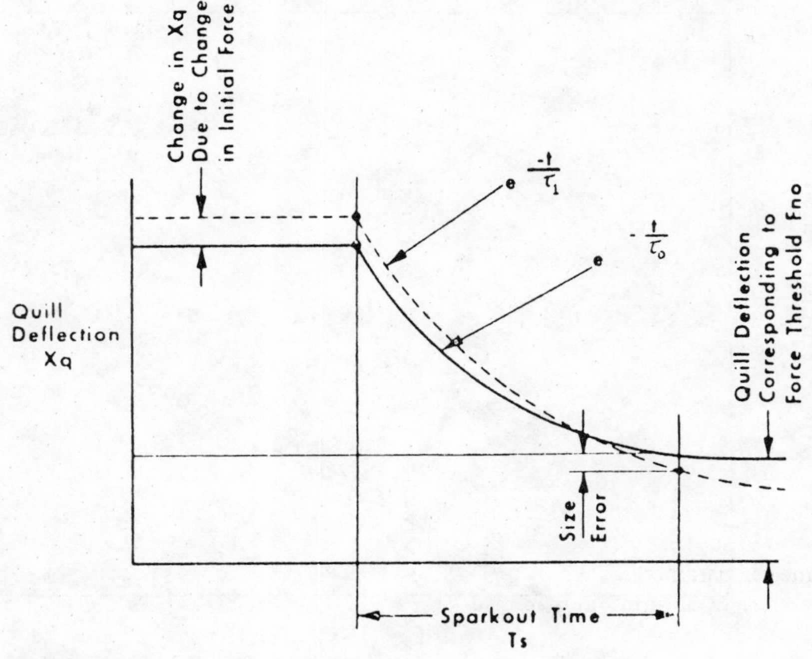

Figure 15.43

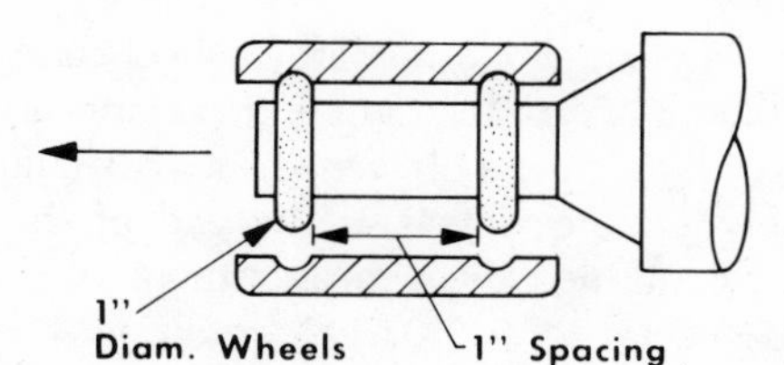

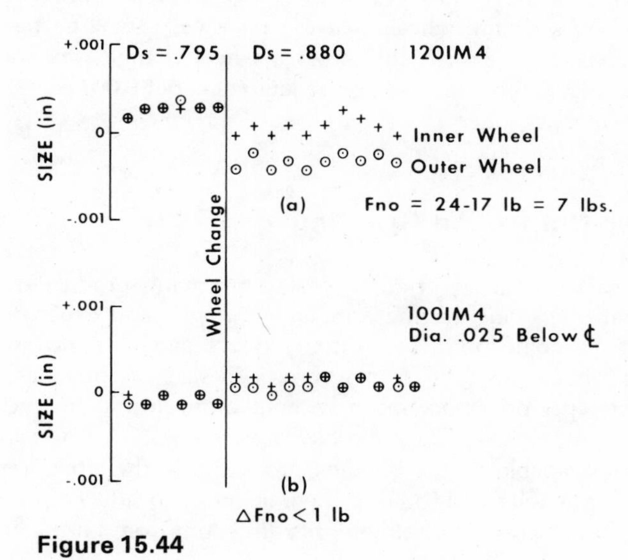

Figure 15.44

at the outer wheel, all other factors remaining the same.

System Rigidity (lb/in.)	Size-Holding Ability Standard Deviation, σ (in.)
23,000	110×10^{-6}
50,000	32×10^{-6}

It is seen that the more rigid system gives greater precision. This is due primarily to reduction of the variation in threshold quill deflection.

Variation of the threshold force F_{th} and the cutting and ploughing metal removal parameters, Λ_{wc} and Λ_{wp}, as the grinding wheel becomes smaller can cause a disruption in size at wheel change in production internal grinders. This is illustrated by the size plot of the ball tracks in the double-groove automotive water pump bearing shown in Fig. 15.44a. The bore size is shown plotted just before wheel change and just after wheel change. In Fig. 15.44a, a size drop of about 0.0006 in. (0.015 mm) and the appearance of 0.0004 in. (0.01 mm) size difference or taper between the two simultaneously ground ball tracks indicates higher force levels at the end of the sparkout for the large wheel. This was confirmed by a measurement of the threshold force for the new and old wheels, 24 lb (107 N) for the new large wheel, 17 lb (75 N) for the used small wheel. This illustrates the use of wheelwork characteristic charts as shown in Fig. 5.9 for explaining and improving grinding performance. By selecting wheels that exhibit a very small change in F_{th} as wheel size changes, the size plot across a wheel change was improved to that shown in Fig. 5.44b where a positive jump of about 0.001 in. (0.0025 mm) occurred at wheel change.

Computer-Controlled Grinding

The problems of reducing grinding costs, improving product quality, and operating with personnel having a minimum of grinding expertise confront many plants. Grinding operations often produce parts that do not conform precisely to what is desired. Holding precision size, or roundness and concentricity, or taper or surface finish without burn, in a short cycle time, is often difficult to achieve. These problems arise because of the inability to control certain variables in the grinding process. It is, therefore, important to identify those variables and to bring them under control to ensure consistent size, taper, surface finish, surface integrity (freedom from burn), and product quality.

On conventional grinding machines, the feedrate is controlled. As the grinding wheel engages the workpiece, forces are induced between wheel and work—the higher the force, the faster the stock removal. The induced force also governs the surface finish, the deflection in the machine, and the onset of thermal damage. Therefore, the induced force is one of the important variables that is uncontrolled in conventional feedrate grinding machines.

The ability of the cutting surface of the grinding wheel to remove stock, called the wheel sharpness, is the second extremely important variable in the grinding process. In feed-rate grinding, as the wheel sharpness drops (the wheel becomes dull or glazed), the induced force rises, resulting in increased deflection and, sometimes, thermal damage.

The size-holding ability of feed-rate "sizematic" grinders is directly related to their ability to maintain the same force between wheel and work at the instant of retraction. If the induced force fluctuates in value at the termination point in the cycle, the system deflection will also fluctuate, and a size error will result unless in-process gaging is used. Even with in-process gaging, taper and surface finish fluctuations will occur. Variations in the induced force are caused by stock variations, wheel sharpness variations, and workpiece hardness variations.

Computer-Controlled Force Grinding

In contrast to feedrate grinding machines, better control of the grinding process can be obtained by computer-controlled force grinding, where the grinding forces are directly under the control of the computer.[8] A force sensor, or load cell, is introduced in the machine that can be calibrated in terms of grinding force. The computer can read the load cell at any time and can control the grinding cycle appropriately. With this system, faster, more efficient grinding cycles can be obtained. On fast production cycles, parts with more stock cause higher forces to be induced, and, correspondingly, more deflection. This leaves more stock for the finish grind or sparkout, and can result in a size, surface finish, or taper error. With computer control, the same force will occur regardless of the amount of rough stock. At the end of the sparkout, the grinding force generally decays down to some small but finite value. If the cycle is terminated at various force levels, size errors occur corresponding to the varying deflection in the wheelwork system at the instant of termination. With computer-controlled force grinding these errors are eliminated.

Variation in the cutting ability of the wheel or wheel sharpness is the most troublesome variable in the grinding process. At a given feed rate, this variable can cause changes in induced force by as much as three or four times. With computer control, the wheel sharpness can be monitored and not permitted to drop below a prescribed value. This avoids thermal damage that might otherwise take place. Size errors due to variations in wheel sharpness are also eliminated as described above. Accordingly, with computer-controlled force grinding two important variables—the system deflection and the wheel sharpness—are brought under control to give more consistent grinding results.

In production grinding it is important to have a fast cycle. Some grind cycles are limited by wheelhead power, by wheel breakdown, by work driving or holding ability, or by the stock runout race. In the stock runout race the stock allowance may be consumed before the part has been rounded up. This means a slower cycle has to be set to give sufficient time to round up. If only a small percentage of all parts run out excessively, the cycle has to be set slow enough to accommodate these parts thereby penalizing the true-running parts. With computer control, the initial runout can be sensed and the cycle slowed down only on those parts with excessive runout. In addition, the round-up process is faster with computer control. As a result, faster cycles can be obtained in those cases where cycle time is limited by the stock runout race. This is especially true in grinding the bore and seat of fuel injection nozzles.

Some production grinding operations are required to hold size, surface finish, etc., during the time the wheel wears down from "NEW" wheel to

"USED" wheel. As the wheel becomes smaller, the surface speed and wheel-work conformity change, which results in changing cutting characteristics. Generally, a fixed cycle is set, which gives satisfactory results only for a limited amount of wheelwear. With computer control, the cycle can be varied with wheel size so as to maintain size and surface finish over a wide range of wheelwear, resulting in more parts per wheel and less downtime for wheel change.

The P.B. Panel Control shown in Figure 15.45 used with a DEC PDP 11/03 standard microcomputer interfaced to a load cell and the machine slides performs computer-controlled force grinding with the advantages outlined above. The load cell can be retrofitted into existing wheelheads in many cases. This control monitors the stock, stock runout, and wheel sharpness on each and every part and eliminates size, taper, and surface finish variations due to these disturbing influences to give exceptionally consistent results without in-process gaging over a wide range of wheelwear. However, slow size drift errors due to thermal expansion or dresser diamond wear have to be fed back to the control either manually or automatically from a postprocess gage. The P.B. Panel Control also can control the postprocess gaging station as indicated in Fig. 15.46 where each and every part is automatically gaged to give a high degree of quality control. During the process of gaging each part, the control accumulates job performance data as shown in Fig. 15.46. These performance data can be displayed on a terminal or transmitted to a "host" computer in the neighborhood.

Figure 15.45

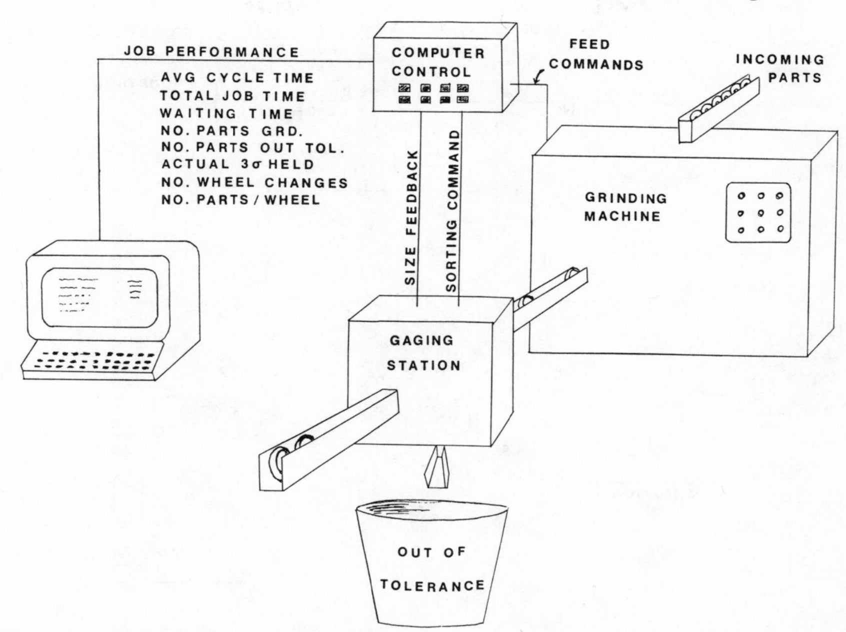

Figure 15.46 Computer-integrated manufacture of ID ground parts.

Multioperation Grinding

Frequently, many parts require more than one grinding operation. This often means that a number of setups one one or more grinding machines are required. Significant savings can sometimes be achieved by performing a series of grinding operations in one setup. By mounting two wheelheads on the cross-slide, as shown in Fig. 15.47, several operations can be carried out with one head followed by other operations with the other head with wheels and cycles being appropriately selected. This reduces the number of setups required, the loading and unloading time for separately executed operations, and the number of machines required, and ensures perfect concentricity and squareness.

The use of compound wheels with multiple cutting surfaces, combined with the ability to feed in the X or Z or vector directions, greatly increases the opportunities for Multioperation Grinding as suggested in Fig. 15.2.

The control shown in Fig. 15.45 generates its own grinding cycle parameters directly from part-print data. The control sets the rough and finish feed rates, the wheelwear compensation, the amount of finish stock, and the sparkout time. These values can be overridden if desired. The computer provides aggressive grinding cycles that are consistent with wheelhead power, work stall torque, and wheel breakdown, thereby reducing the need for personnel with extensive grinding expertise.

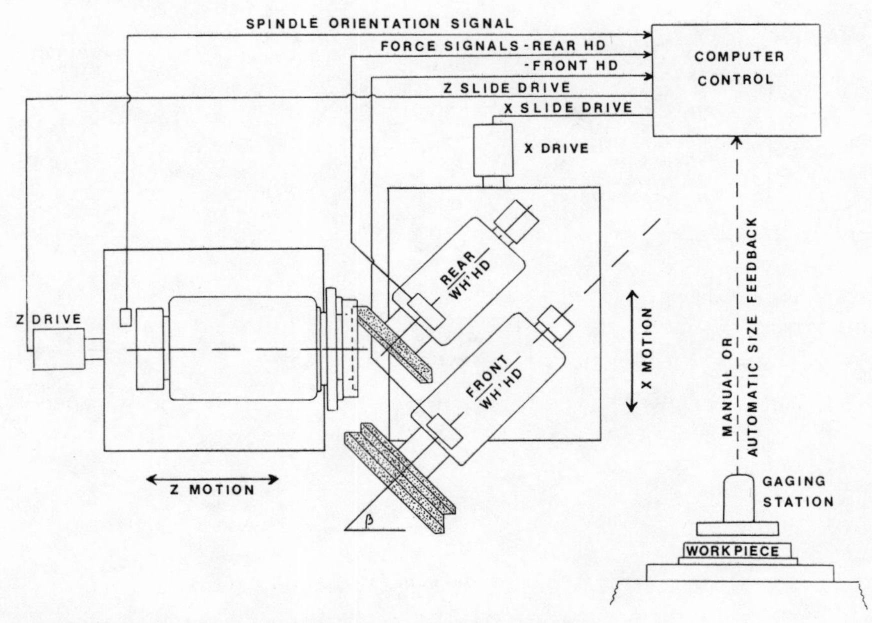

Figure 15.47 Two-axis, two-headed grinding system.

Computer controls can be used on a stand-alone basis or as satellite controls that communicate with a host computer in the department, as illustrated in Fig. 15.48. In this case, the floppy disc drive and the console are shared among several machines, considerably reducing the control cost per machine.

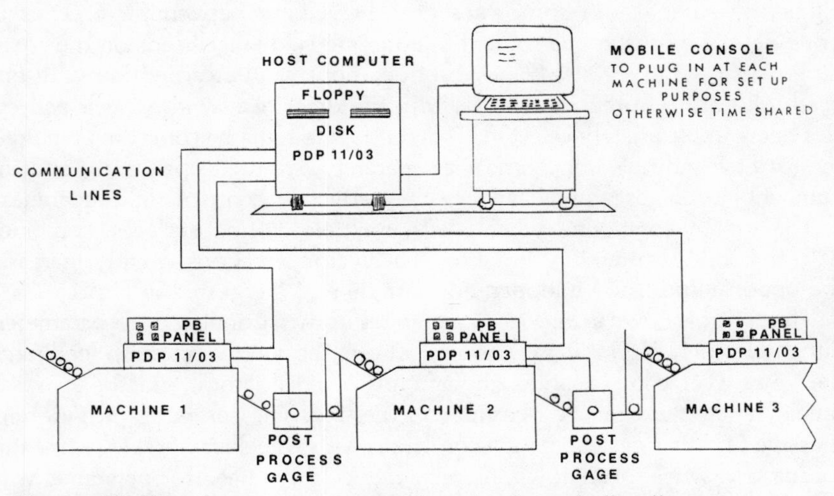

Figure 15.48 Computer-integrated production grinding lines.

Untended Grinding Systems

Significant increases in manufacturing productivity can be realized by increasing machine utilization. Since many machine tools are actually cutting or grinding less than one-third of the time, large improvements in efficiency can be achieved by employing untended grinding systems, where a group of machines operate, essentially, without human attention during the second and third shifts, thereby attaining a large increase in machine utilization and a substantial reduction in the machine tools required. However, owing to the temperamental nature of grinding machines, improved control over the grinding process is required before they can operate without constant human supervision. The following items are prerequisites for untended operation:

1. Auto Loading: Workpieces must be automatically loaded and unloaded—possible with current technology.
2. One Shift Wheel Life: The grinding wheel should last for a complete shift—possible with CBN wheels.
3. Consistent Size-Holding Ability—possible with computer-deflection-compensated sizing in conjunction with postprocess gaging.
4. Comprehensive Computer-Controlled Postprocess Gaging: Gaging for size, taper, roundness, surface finish, waviness, surface profile, chatter—possible with current technology.
5. Computer-Controlled Force Grinding Cycles: Capable of adapting to excessive stock runout, stock variations, hardness variations, changes in wheel size—possible with computer-controlled force grinding.
6. Wheel Sharpness Monitoring: Ability to monitor and adapt to changes in wheel sharpness and to request wheel dressing as required to avoid thermal damage or burn—possible with computer-controlled force grinding.
7. Dresser Diamond Sharpness: Ability to monitor diamond sharpness; can be inferred indirectly by monitoring wheel sharpness and when the dressing operation does not restore wheel sharpness to a prescribed level, the computer concludes the dresser diamond is dull—possible with computer-controlled force grinding.

Nomenclature

V_w = Work surface speed (m/sec)

N_w = Work speed (rpm)

V_s = Wheel surface speed (m/sec)

N_s = Wheel speed (rpm)

V_t = Traverse speed (m/sec)

D_w = Work diameter (mm)
D_s = Wheel diameter (mm)
l = Dress lead (μm/rev)
$c = 2 \times$ diamond depth of dress (μm)
d = Grain diameter (μm)
D_e = Equivalent diameter (cm)
vol = $1.33H + 2.2S - 8$ (vol% of bond in wheel):
$H = 0, 1, 2, 3, \ldots$, for $H, I, J, K, \ldots$, hardness;
S = wheel structure number
F'_n = Normal force intensity (N/mm)
F_n = Normal force (N)
F_t = Tangential force (N)
$F'_{th} = F'_{n0}$ = Threshold force intensity (N/mm)
F'_{pc} = Ploughing-cutting transition force intensity (N/mm)
F'_{bd} = Wheel breakdown force intensity (N/mm)
h = Wheel depth of cut (μm)
α = Dimensionless machining–elasticity number
P = Grinding power (W)
P_s = Specific power (J/mm^3)
τ_0 = Time constant
L_c = Wheelwork contact length
R_c = Rockwell hardness C scale
WRP = Λ_w, work removal parameter (mm^3/min$\cdot$N)
K_m = System stiffness (N/μ)
E = Modulus of elasticity (N/mm^2)
I = Moment of inertia (mm^4)
l_e = Length of cantilever (mm)
K = Angular stiffness (N/rad)
W = Width of wheelwork contact (mm)
$\bar{v}_f$ = Feedrate (μm/sec)
$\bar{v}_w$ = Penetration velocity of wheel into work (μm/sec)
$\bar{v}_s$ = Radial wheelwear velocity (μm/sec)
Z_w = Volumetric rate of stock removal (mm^3/sec)
Z'_w = Stock removal rate per unit width (mm^2/sec)
Z_s = Volumetric wheelwear rate (mm^3/sec)

$Z_s' =$ Wheelwear rate per unit width (mm^2/sec)
$S =$ Wheel sharpness (m^2/N)
WWP = Wheelwear parameter ($mm^4/min \cdot N^2$)
$K_s = V_s w/\Lambda_s =$ Wearing stiffness of grinding wheel
$K_{cw} = V_w w/\Lambda_w =$ Cutting stiffness of workpiece (N/mm)
$K =$ Wheelwork contact stiffness
$\tau_s = 1/N_s =$ Period of wheel rotation speed
$\tau_w = 1/N_w =$ Period of work rotation speed
$G_m(j\omega) =$ Harmonic response locus of wheelwork support
$Re_m =$ Maximum negative real part of $G_m(j\omega)$
$\Lambda_s =$ Wheelwear rate ($\Lambda_s = WWP \times F_n'$)
$h = \bar{v}_w/N_w =$ Wheel depth of cut

References

1. Hahn, R. S., "On the Nature of the Grinding Process," *Proceedings of the Third MTDR Conference*, Pergamon Press, Elmsford, N.Y., 1961, pp. 129–154.

2. Lindsay, R. P., "On Metal Removal and Wheel Removal Parameters, Surface Finish, Geometry and Thermal Damage in Precision Grinding," PhD dissertation, Worcester Mass., Polytechnic Institute, 1971.

3. Hahn, R. S. and R. P. Lindsay, "Principles of Grinding," *Machinery*, July, August, September, October, and November 1971.

4. Lindsay, R. P. "Variables Affecting Metal Removal and Specific Horsepower in Precision Grinding," SME Paper No. MR71-269.

5. Lindsay, R. P., "On the Surface Finish–Metal Removal Relationship in Precision Grinding," ASME Paper No. 72WA/Prod-13.

6. Hahn, R. S. and R. P. Lindsay, "The Production of Fine Surfaces while Maintaining Good Surface Integrity at High Production Rates by Grinding," Proc. International Conf. on Surface Technology 1973, SME, Dearborn, Michigan.

7. Opitz, H. and K. Guhring, "High Speed Grinding," *Annals of C.I.R.P.*, 1967.

8. Hahn, R. S., "High Performance Production Grinding Systems," SME Paper No. MR82-230 (1982).

9. Snoeys, R. and D. Brown, "Dominating Parameters in Grinding Wheel and Workpiece Regenerative Chatter," Proc. 10th International MTDR Conf., Pergamon Press, Elmsford, N.Y., 1969, pp. 325–348.

10. Inasaki, I. and S. Yonetsu, "Regenerative Chatter in Grinding," Proc. 18th International MTDR Conf., Pergamon Press, Elmsford, N.Y., 1977, pp. 423–429.

11. Hahn, R. S. and R. L. Price, "A Nondestructive Method of Measuring Local Hardness Variations in Grinding Wheels," *Annals of C.I.R.P.*, Pergamon Press, Elmsford, N.Y., 1968, Vol. XVI, pp. 19–30.

12. Lindsay, R. P. and N. P. Navarro, "Principles of Grinding with Borazon, Part 2, Difficult-to-Grind Alloys," *Machinery*, July/August. 1973, pp. 33–36.

13. Hahn, R. S. and R. P. Lindsay, "On the Rounding-Up Process in High Production Internal Grinding Machines by Digital Computer Simulation," Proc. of the 12th MTDR Conf., 1971.

14. Hahn, R. S. and R. P. Lindsay, "The Influence of Process Variables on Material Removal, Surface Integrity, Surface Finish, and Vibration in Grinding," Proc. of the 10th MTDR Conf., Pergamon Press, Elmsford, N.Y., 1969, pp. 95–117.

Part Six

Laser Applications

CHAPTER 16

Laser Applications

Stephen M. Copley
Departments of Materials Science
and Mechanical Engineering
University of Southern California

This chapter covers the application of lasers to the shaping of materials. Two approaches are described: laser-assisted machining (LAM), in which the laser heats material as it is sheared by a single-point cutting tool with the objective of improving its machinability; and laser machining (LM), in which the laser heats and vaporizes material. The former is applied to metallic materials, while the latter is applied to nitrides and carbides. Current interest in these approaches is prompted by the development of high-power CO_2 and Nd-YAG infrared lasers with sufficient ruggedness, reliability, and simplicity of operation for use as directed-energy heat sources in manufacturing facilities.

Laser-Assisted Machining (LAM)

The idea of heating a material to improve its machinability is not a new one. The use of gas-torch and induction heating to assist in the turning of metals was first studied in the United States in the 1940s by Tour and Fletcher.[1] Concurrently, Schmidt investigated the use of gas-torch heating in milling.[2] Although many advantages have been reported, such as reduction in force and machining power, increased tool life, and improved surface finish, hot machining has not been adopted as a practical and technologically feasible method by industry. Recently, however, with the development of more-intense directed-energy heat sources, such as the plasma arc[3] and the laser, hot machining has become a subject of renewed interest and investigation.

LAM Facility

A facility for LAM is shown in Fig. 16.1. A CW laser beam is directed by turning mirror (*M*1) along a path parallel to the turning axis of the lathe. The beam is then reflected from carriage mirror (*M*2), cross-slide mirror (*M*3), and workpiece mirror (*M*4), and is finally focused by a lens (*L*) onto the work-

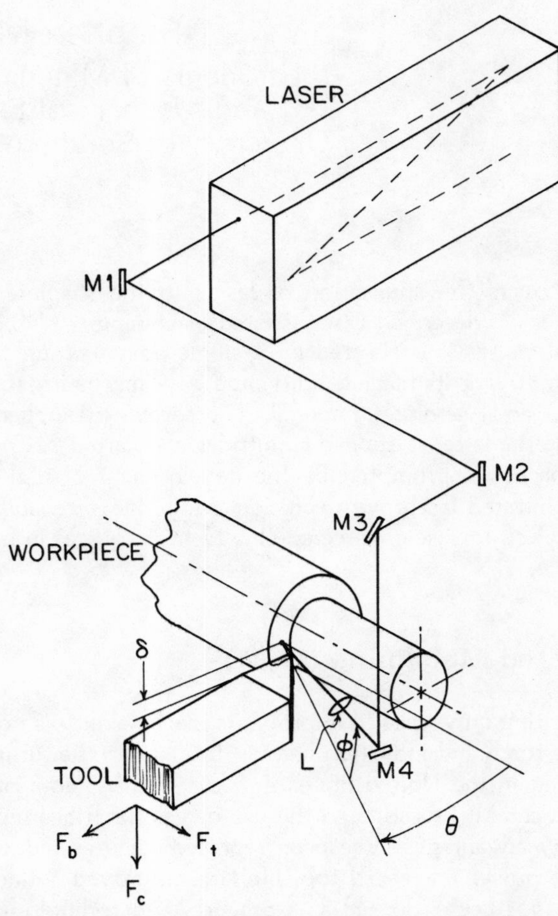

Fig. 16.1 Experimental arrangement for laser-assisted machining.

piece a distance δ in front of the edge of the cutting tool. Because the direction of the carriage motion is parallel to the turning axis and that of the cross-slide motion is perpendicular to the turning axis, the beam retains its alignment with respect to the mirrors during operation of the lathe.

The absorption of the beam is maximized if it is oriented perpendicular to the surface it is heating. Thus in turning, where it is desired to heat the shoulder of the workpiece, the angle θ is set equal to $\pi/2 - \kappa_r$, where κ_r is the major cutting edge angle. On the other hand, in facing, θ is set equal to κ_r. If it is desired to heat the cylindrical surface of the workpiece, then θ is set equal to $\pi/2$.

In our work, the beam is produced by a 1400-W, CW CO_2 laser operating in the TEM_{00} mode (Gaussian spatial distribution of intensity). The mirrors are highly polished, water-cooled copper flats and the lens is ZnSe. Investigations at the Illinois Institute of Technology Research Institute have employed a very-high-intensity beam.[4] In this case, the lens is replaced by a focusing mirror.

The three components of tool force (F_c, F_t, and F_b) are defined in Fig. 16.1. They are measured by a piezoelectric dynamometer mounted in the base of the tool holder.

Figure 16.2 shows the relationship of the laser beam to the workpiece, chip, and tool. The laser beam moves at a constant speed v in the positive x direction.

Laser Heating

The Rosenthal solution predicts the temperature distribution due to a point heat source moving in the positive x direction at a constant velocity.[5] The

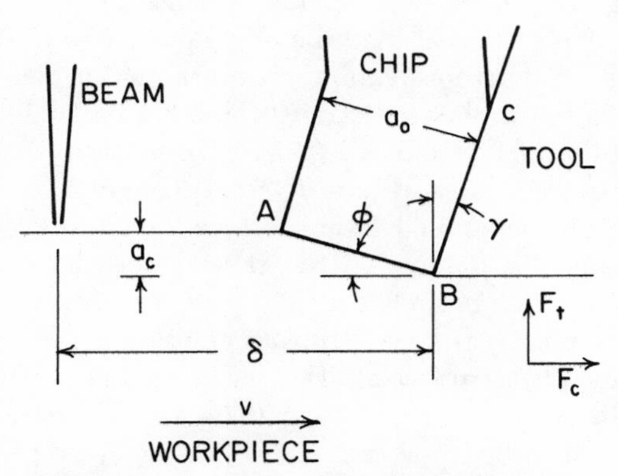

Fig. 16.2 Schematic diagram defining the parameters used to describe laser-assisted machining.

temperature is given in closed form by the equation

$$T = \frac{P}{C_p D 2\pi r} \exp\left(-\frac{v(r + x)}{2D}\right), \tag{16.1}$$

where

$$r = (x^2 + y^2 + z^2)^{1/2} \tag{16.2}$$

and P is the absorbed beam power. Although laser beams cannot be focused to a point, the Rosenthal solution does give a good description of the temperature distribution due to laser heating for distances large compared to the beam diameter. For smaller distances, the spatial distribution of intensity of the beam must be taken into account, which varies with the operating mode of the laser.

Cline and Anthony have calculated the temperature distribution due to a Gaussian heat source moving in the positive x direction at a constant velocity.[6] They give the temperature distribution by the equation

$$T(x, y, z, v) = P(C_p DR)^{-1} f(X, Y, Z, V) \tag{16.3}$$

where R is the beam radius; $X = x/R$, $Y = y/R$ and $Z = z/R$ are dimensionless distances; $V = Rv/D$ is a dimensionless velocity; and $f = f(x, y, z, v)$ is a dimensionless temperature, which must be evaluated by numerical integration. Maruo et al. have given solutions similar to those developed by Cline and Anthony for a Gaussian beam with elliptical cross section.[7]

Figure 16.3 shows isotherms on the shear plane calculated by the Cline–Anthony solution for a Gaussian beam moving at a speed of 50 cm s^{-1} (1.65 fps). The beam diameter is 0.10 cm (0.04 in.) and the absorbed beam power is 1000 W. The material is Inconel 718 with volumetric specific heat equal to 4.28 J cm^{-3} $°C^{-1}$ and thermal diffusivity equal to 0.0466 cm^2 s^{-1}. These are average values for the temperature range 38–1204°C (100-2200°F). The shear plane angle ϕ is 20° and the feed a_L is 0.050 cm (0.02 in.). Figure 16.3a shows the temperature distribution on the shear plane if the distance from the beam to the tool, δ, is 1 cm (0.4 in.). Figure 16.3b shows the temperature distribution if $\delta = 0.6$ cm (0.24 in.). Decreasing δ considerably increases the shear plane area where $T \geq 550$°C (1022°F); however, it has almost no effect on the amount of area where $T \geq 250$°C (482°F).

The temperature distribution on the shear plane can be changed if the spatial distribution of the beam is altered. Figure 16.4 shows isotherms on the shear plane for a Gaussian beam with an elliptical cross section calculated from the solution of Maruo et al. The major axis of the ellipse is 0.4 cm (0.16 in.) and the minor axis is 0.1 cm (0.04 in.). The speed is 75 cm s^{-1} (2.5 fps) and the absorbed power is 2000 W. All other parameters are the same as for Fig. 16.3. In Fig. 16.4a, the major axis of the ellipse was oriented perpendicular to the direction of motion of the beam. This produces a very broad but shallow heated zone. In Fig. 16.4b, the major axis of the ellipse

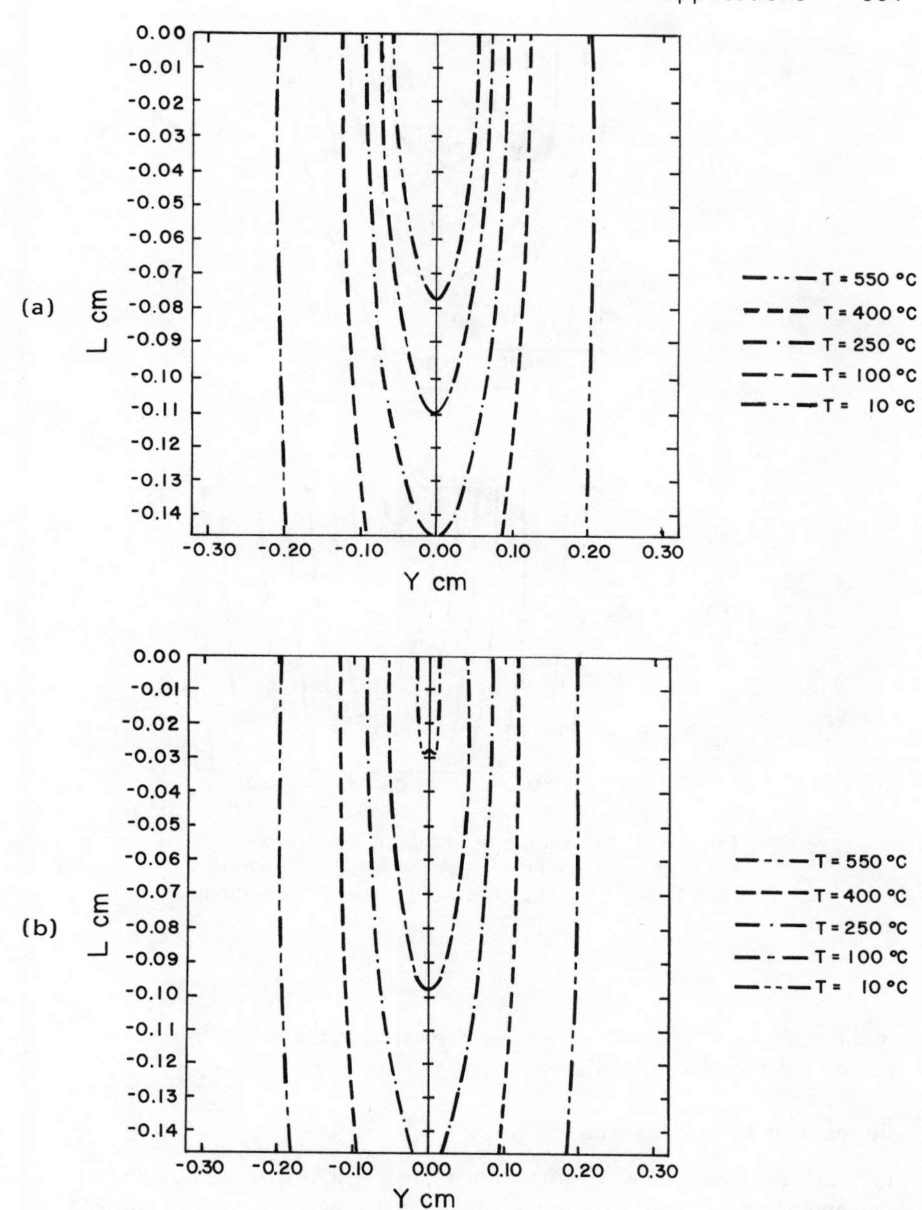

Fig. 16.3 Shear plane isotherms for Inconel 718: (a) $\delta = 1$ cm (0.4 in.); (b) $\delta = 0.6$ cm (0.24 in.).

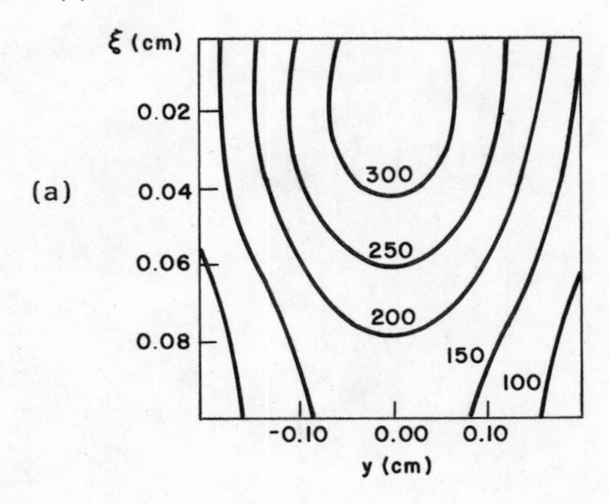

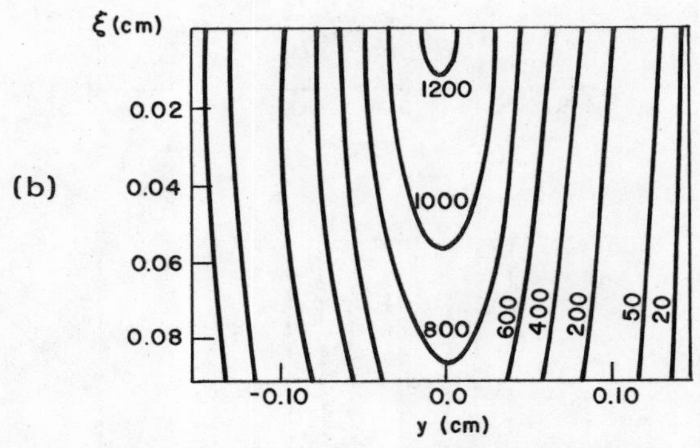

Fig. 16.4 Shear plane isotherms for Inconel 718; (a) major axis of ellipse is perpendicular to the direction of motion of the beam; (b) major axis of ellipse is parallel to the direction of motion of the beam.

was oriented parallel to the direction of motion of the beam. In this case, the heated zone is narrow and deep.

Benefits of Laser Heating

In LAM, the laser is employed as a heat source; it is used to heat the material on the shear plane as the chip is being formed. The advantage of the laser is its capability of heating material on most of the shear plane as the chip is being formed without heating significantly the material that contacts the edge or face of the cutting tool. This, of course, is a consequence of the heat flow associated with laser heating. Heating the material on the shear plane may result in benefits such as decreased cutting forces, increased material removal rate, increased tool life, and improved surface conditions such as smoothness and residual stress or flaw distribution. However, heating the material that contacts the tool is likely to decrease tool life.

In LAM, the laser scan speed must equal the speed at which the cutting tool passes over the workpiece. Heating the material on the shear plane may change the mode of the chip formation from discontinuous to continuous or decrease the tendency to form a built-up edge. Such changes affect the smoothness and flaw distribution of the machined surface. One of the most important reasons for heating material on the shear plane is to produce a decrease in cutting force. Such a decrease in anticipated in precipitation-hardened alloys such as Inconel 718, if they can be heated by combined laser heating and shear plane heating into the temperature range where the yield stress decreases markedly with increasing temperature; see Fig. 16.5. We will present evidence for this effect in Inconel 718 and also Ti-6Al-4V. Such a decrease in machining forces may result in direct benefits, such as making possible the machining of a workpiece that would deflect elastically too much to maintain tolerances or, perhaps, even plastically deform under conventional machining conditions. Such a decrease in machining forces might also produce an increase in tool life, if the concomitant small increase in temperature of the material contacting the tool due to laser heating does not offset this effect. Finally, such a decrease may make possible an indirect but very important benefit, namely, an increase in material removal rate and thus a decrease in machining cost. A discussion of this possibility will now be given.

Consider the cost of turning a unit volume of material. If C_0 is the labor and overhead rate, C_t is the tool cost per edge, T is the tool life, and t_m is

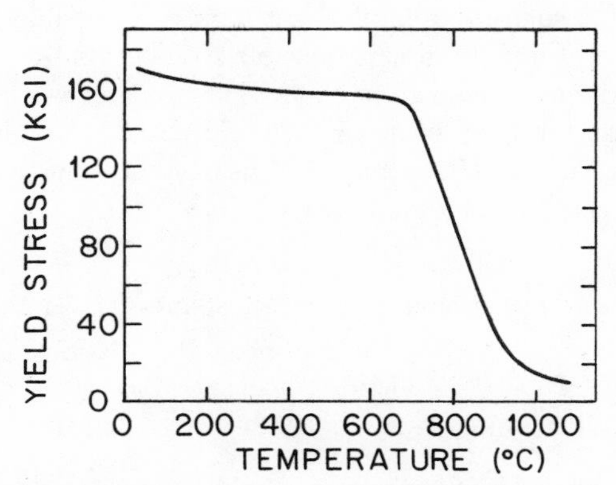

Fig. 16.5 Yield stress versus temperature curve for Inconel 718.

the machining time, then the cost is given by the equation

$$U = C_0 t_m + C_t \frac{t_m}{T}. \tag{16.4}$$

Nonproductive time such as loading and unloading time and tool change time have been omitted in deriving this equation because they depend on the specific workpiece and the shop conditions. For machining a unit volume, the machining time is given by

$$t_m = 1/Z, \tag{16.5}$$

where Z is the material removal rate. Thus,

$$U = \frac{1}{Z}\left(C_0 + \frac{C_t}{T}\right). \tag{16.6}$$

One approach to decreasing the cost is to increase tool life; however, even for an infinite tool life the cost of machining a unit volume of material is C_0/Z. The more effective approach is to increase the material removal rate.

In single-point turning, the material removal rate is given by the equation

$$Z = v a_c a_p \tag{16.7}$$

where v is the cutting speed; a_c is the undeformed chip thickness, which in single-point turning is equal to the feed; and a_p is the back engagement, which is also known as the depth of cut. The magnitude of the cutting force is approximately proportional to the undeformed chip cross section ($a_c a_p$). For roughing and semiroughing operations, commercial practice dictates selecting as large an undeformed chip cross section as possible taking into consideration the rigidity of the tool–workpiece system and the power of the lathe. Current limits for preturned Inconel 718 and Ti-6Al-4V are given in Table 16.1.[8] If, for such force-limited cutting, it is possible by laser heating to decrease cutting force, then it is possible to increase the undeformed chip cross section and thus the material removal rate. The use of a laser would, of course, increase the magnitude of C_0 in Eq. (16.3). Also it might result in a decrease in tool life T. Ultimately, the use of the laser would have to be justified on the basis of decreased cost.

Table 16.1 Current Machining Limits for Semirough Turning

Alloy	Tool	Cutting speed [cm sec^{-1} (fps)]	Feed [cm (in.)]	Depth of Cut [cm (in.)]
Inconel 718	Carbide (C-3)	50	0.0250	0.152–0.318
		(1.64)	(0.01)	(0.06–0.126)
Ti-6Al-4V	Carbide (C-2)	70	0.0250	0.152–0.318
		(2.3)	(0.01)	(0.06–0.126)

Results

At this time, LAM is under active investigation in several laboratories, including our own. Although there are data available supporting many of the benefits mentioned previously, insufficient data are available to draw firm conclusions regarding the technological feasibility of the LAM process. In this section, we give results for Inconel 718 and Ti-6Al-4V alloys where the shear plane was laser heated during chip formation.

Figure 16.6 shows the results of an experiment designed to illustrate the effect of velocity on force drop. A workpiece was machined in the facing configuration so that the cutting velocity continuously decreased as the cutting tool approached the turning axis. During a facing cut the laser was turned on and off several times and the force recorded continuously. Turning the laser on was observed to produce a force drop. The magnitude of the force drop increases with decreasing velocity. The experimental conditions for this experiment were 600-W incident beam power, beam diameter $(1/e) =$ 212 μm (0.008 in.), $\delta = 5$ mm (0.2 in.), $a_c = 0.0075$ cm (0.003 in.), and $a_p =$ 0.0375 cm (0.015 in.).

Figure 16.7 shows the effect of varying the distance δ between the impingement point of the laser and the cutting tool edge on the reduction in cutting force for two cutting speeds, 10 and 20 cm s^{-1} (0.33 and 0.66 fps). The other conditions for the experiment were: incident beam power = 600 W; beam diameter $(1/e) = 212$ μm (0.008 in.); $a_c = 0.0075$ cm (0.003 in.); and $a_p =$ 0.050 cm (0.02 in.). At both cutting speeds, the reduction cutting force first increases and then decreases with increasing distance δ. The maximum in cutting force reduction occurs in the range $\delta = 3$–4 mm (0.12–0.16 in.). At distances less than this range, chip interference is thought to account for the decrease in force reduction. At distances greater than this range, the material has time to cool, so the average temperature of the shear plane with the laser

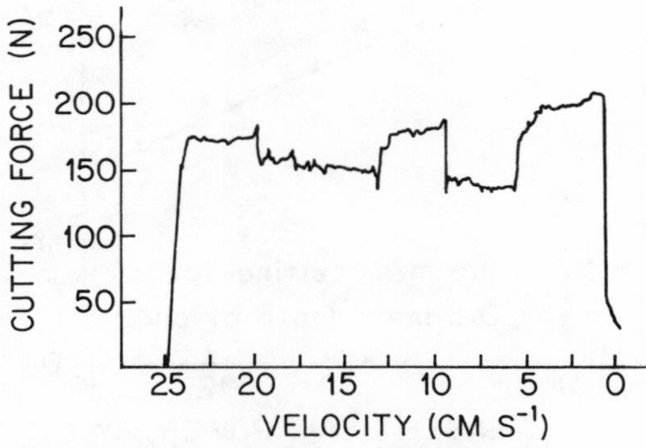

Fig. 16.6 Effect of cutting speed on force decrease in LAM of Inconel 718.

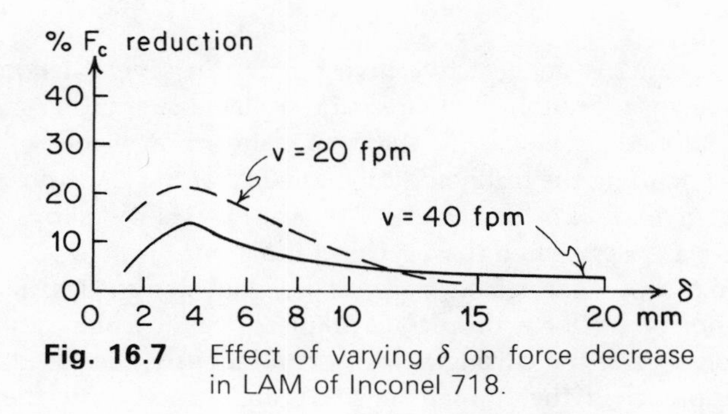

Fig. 16.7 Effect of varying δ on force decrease in LAM of Inconel 718.

turned on decreases; see Fig. 16.3. At the 20 cm s^{-1} (0.66 fps) cutting speed, there is less time for absorbed power of the laser beam to heat the material on the shear plane than at 10 cm s^{-1} (0.33 fps). Thus, the reduction in force observed at highest velocity is less than that observed at the lowest velocity.

To obtain significant decreases in force for cutting conditions approaching the current machining limits given in Table 16.1, it is necessary to employ a

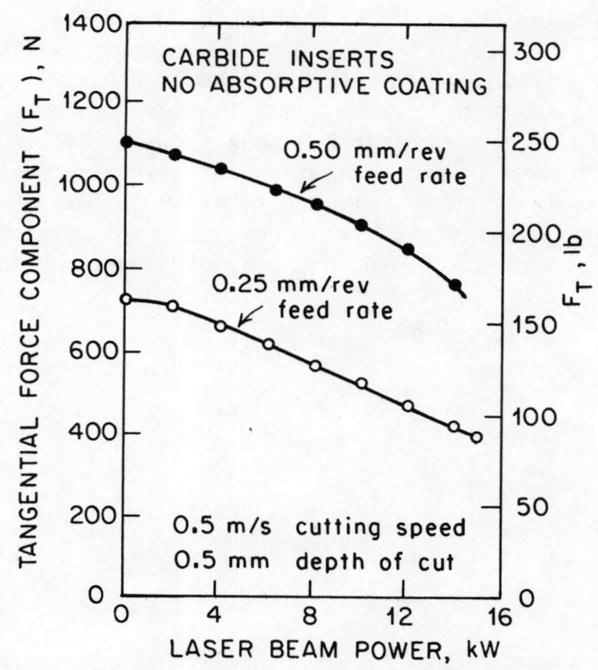

Fig. 16.8 Cutting force versus incident beam power for machining of Inconel 718 at 5 cm s^{-1} (1.64 fps).

laser with beam power considerably greater than 600 W. Figure 16.8 shows a plot of cutting force versus incident beam power for Inconel 718 obtained by Rajagopal et al.[9] The data were obtained at: constant cutting speed, 50 cm s^{-1} (1.64 fps); constant feed, 0.025 cm rev^{-1} (0.01 in. rev^{-1}); and, constant depth of cut, 0.050 cm (0.02 in.). After recording the cutting force with no laser heating, the laser was turned on at 2 kW and then increased in 2-kW steps up to an incident beam power of 14 kW. The cutting time at each power step was 5 sec. Tool wear effects were taken into account by repeating the experiment at decreasing power levels, again changing the power in steps of 2 kW. The value plotted at each power setting was found by averaging the values obtained for increasing power and decreasing power. In this experiment, the distance δ was 1.2 cm (0.5 in.) and the focused beam was elliptical in cross section with a major axis of 0.25 cm (0.1 in.) and a minor axis of 0.05 cm (0.02 in.). The major axis was oriented parallel to the direction of workpiece motion in order to maximize the depth of heat penetration.

Referring to Fig. 16.8, it is clear that sufficient shear plane heating occurred at the higher incident powers to produce a significant force drop. By heating with a 14-kW beam, it is possible to double the feed thereby doubling the material removal rate with only a slight increase in cutting force. Although the cutting speed and feed [0.025 cm rev^{-1} (0.01 in. rev^{-1})] used to obtain the data shown in Fig. 16.8 were equal to the current machining limits indicated in Table 16.1, the depth of cut was much less than that normally employed in semiroughing operations. Rajagopal et al. restricted the depth of cut to 0.050 cm (0.02 in.) so that the incident beam would cover the workpiece shoulder.

Rajagopal et al. observed severe tool wear in their experiments, which suggested that the heat from the laser beam was penetrating too deeply. The results shown in Fig. 16.4 indicate that the penetration can be decreased and the depth of cut can be increased, while maintaining beam coverage of the shoulder, by orienting the major axis of the beam cross section perpendicular to the direction of workpiece motion. This approach was tried in subsequent work by Rajagopal and Plankenhorn.[10] They found that a twofold increase in material removal rate was readily attainable by LAM at 12 kW without any increase in flank wear rate. For example, the feed was increased from 0.05 to 1.00 mm rev^{-1} (0.002 to 0.04 in. rev^{-1}) by laser heating for a cut taken at a speed of 70 cm s^{-1} (2.3 fps) and a depth of cut of 0.25 cm (0.1 in.) using a C-2 carbide insert while maintaining a 0.75 mm min^{-1} (0.03 in. min^{-1}) flank wear rate.

Theoretical Understanding

Shear plane heating and laser heating contribute additively to determine the temperature distribution on the shear plane during LAM. The larger the power absorbed from the laser beam for a given set of machining conditions, the higher is the average temperature on the shear plane. If a sufficiently high average temperature is attained, then the dynamic shear yield stress and thus the cutting force may be decreased. In the case of a force-limited cut, this makes possible an increase in material removal rate. In this section, we will show how the analyses of heat flow due to shear plane heating and laser heating may be employed to predict the amount of absorbed beam power required for a specific increase in material removal rate.

According to Weiner, the temperature distribution on the shear plane due to shear plane heating can be calculated given: the undeformed chip thickness (a_c); the back engagement (a_p); the cutting velocity (v); the tool geometry; the machining forces (F_c, F_t, and F_b); and the thermal properties.[11] In applying the Weiner solution to turning, we assume orthogonal cutting equations can be applied. This assumption, which greatly simplifies the calculation, is reasonable because $a_p \gg a_c$.

Parameters used in our calculations are based on the following machining parameters and measured force data: rake angle, $\gamma = 8°$; $v = 50 \text{ cm s}^{-1}$ (1.64 fps); $a_c = 0.0150$ cm (0.006 in.); $a_p = 0.0875$ cm (0.034 in.); shear plane angle; $\phi = 20°$; $F_c = 380$ N (85.5 lb); and, $F_t = 175$ N (39.4 lb).

We shall calculate the temperature distribution on the shear plane due to shear plane heating for Inconel 718 machined under conditions approaching the current machining limit: namely, $v = 50 \text{ cm s}^{-1}$ (1.64 fps); $a_c = 0.025$ cm (0.01 in.); and $a_p = 0.125$ cm (0.05 in.). It is assumed that F_c and F_t are proportional to the undeformed chip cross section. Thus, for $a_c \times a_p = 0.025$ cm (0.01 in.) $\times$ 0.125 cm (0.05 in.), $F_c = 905$ N (204 lb) and $F_t = 417$ N (94 lb). Details of the calculation are as follows: The machining power is

$$P_M = F_c v = 452 \text{ W}. \tag{16.8}$$

The force parallel to the rake face of the tool, F_R, is

$$F_R = F_c \sin \gamma + F_t \cos \gamma = 539 \text{ N (121.3 lb)}. \tag{16.9}$$

The velocity parallel to the rake face, v_0, is

$$v_0 = \frac{v \sin \phi}{\cos(\phi - \gamma)} = 17.5 \text{ cm s}^{-1} \text{ (0.6 fps)}. \tag{16.10}$$

Thus, the power dissipated at the rake face, P_R, is

$$P_R = F_R v_0 = 94.3 \text{ W}. \tag{16.11}$$

The power dissipated on the shear plane, P_s, is

$$P_s = P_M - P_R = 358 \text{ W}. \tag{16.12}$$

The area of the shear plane, A_s, is

$$A_s = \frac{a_c a_p}{\sin \phi} = 9.14 \times 10^{-3}\ \text{cm}^2\ (1.4 \times 10^{-3}\ \text{in.}^2). \quad (16.13)$$

The heat liberated on the shear plane per unit area associated with the formation of the chip, q, is

$$q = \frac{P_s}{A_s} = 3.91 \times 10^4\ \text{W cm}^{-2}. \quad (16.14)$$

Weiner (Ref. 11, Fig. 3) gives a plot of $T \sin \phi$ as a function of $\sqrt{y}$, where T is the dimensionless shear plane temperature and y is the dimensionless distance measured along the shear plane from the workpiece surface. The parameter y is given by the equation

$$\sqrt{y} = \frac{v\psi^2 \cos \phi}{4D} \xi^{1/2} = 5.78\xi^{1/2}. \quad (16.15)$$

The parameter ψ equals $\tan \phi$ and ξ is the distance from the leading edge of the shear plane measured along the shear plane. The shear plane temperature, u, is

$$u = \frac{1}{\sin \phi} \frac{q}{vC_p} (T \sin \phi) = 534(T \sin \phi). \quad (16.16)$$

The results of this calculation are given in Table 16.2.

Let us calculate the absorbed beam power required to effect a 25% increase in material removal rate beyond the current machining limit for Inconel 718. Such an increase can be attained by increasing the feed from 0.0250 to 0.03125 cm (0.01 to 0.012 in.), while keeping the cutting speed and depth of cut the same. Ordinarily, such an increase would result in an increase in

Table 16.2 Temperature Distribution on the Shear Plane Due to Shear Plane Heating by Plastic Deformation at $v = 50$ cm sec^{-1} (1.64 fps)

ξ	$\sqrt{y}$	$T \sin \phi$	u [°C (°F)]
0.01	0.578	0.78	416 (781)
0.02	0.817	0.90	481 (898)
0.04	1.15	0.96	513 (955)
0.06	1.42	0.99	529 (984)
0.08	1.64	1.00	534 (993)
0.10	—	1.00	534 (993)
0.12	—	1.00	534 (993)
0.14	—	1.00	534 (993)
0.16	—	1.00	534 (993)

cutting force; however, by laser-heating material on the shear plane, so as to decrease its average dynamic shear yield stress, it is possible to increase the feed while keeping the cutting force the same. For the purpose of this calculation, we will assume that the ratio of the dynamic shear yield stress to the static yield stress is constant for the temperature range of interest, and that the variation of static yield stress with temperature is given by Fig. 16.5.

To calculate the decrease in the average dynamic shear yield stress required to keep the cutting force the same, we consider the equation

$$F_c = \frac{1}{\cos(\phi - \gamma)} (F_s \cos \gamma + F_R \sin \phi). \tag{16.17}$$

The force acting parallel to the shear plane, F_s, is given by the equation

$$F_s = \bar{\tau}_s \frac{a_c a_p}{\sin \phi}, \tag{16.18}$$

where

$$\bar{\tau}_s = \frac{1}{A_s} \iint \tau_s \, dA. \tag{16.19}$$

The force acting parallel to the rake face of the cutting tool, F_R, is given by the equation

$$F_R = \bar{\tau}_R L a_p, \tag{16.20}$$

where

$$\bar{\tau}_R = \frac{1}{L a_p} \iint \tau_R \, dA \tag{16.21}$$

and L is the chip contact length along the rake face of the tool. Thus,

$$F_c = \frac{\cos \gamma}{\cos(\phi - \gamma) \sin \phi} \bar{\tau}_s (a_c a_p) + \frac{\sin \phi}{\cos(\phi - \gamma)} \bar{\tau}_R L a_p. \tag{16.22}$$

We previously assumed that F_c was proportional to $a_c a_p$. This behavior follows from Eq. (16.22), if we set $L = L_0 a_c$, where L_0 is a constant. Setting $\bar{\tau}_R L_0 a_c a_p = \bar{\tau}_R L_0 (0.0250)(0.125) = F_R = 539$ N (121.3 lb) [see Eq. (16.9)], we find that $\bar{\tau}_R L_0 = 1.725 \times 10^5$ N cm^{-2} (2.5×10^5 psi). Thus if $\phi = 20°$ and $\gamma = 8°$, then $F_c = 2.95 \bar{\tau}_s (a_c a_p) + 6.04 \times 10^4 (a_c a_p)$. Setting $F_c = 905$ N (204 lb), $a_c = 0.0250$ cm (0.01 in.), and $a_p = 0.1250$ cm (0.05 in.), we find that $\bar{\tau}_s = 7.77 \times 10^4$ N cm^{-2} = 777 MPa (11.3×10^4 lb). Thus without laser heating the ratio of the average dynamic shear yield stress to the average static yield stress is $\bar{\tau}_s / \sigma_y$ = 777 MPa/1069 MPa = 0.727.

If we keep $F_c = 905$ N (204 lb) when we increase the feed (a_c) to 0.03125 cm (0.0123 in.) and ϕ, $\bar{\tau}_R$ and L_0 do not change, then we must decrease $\bar{\tau}_s$ to 5.8×10^4 N cm^{-2} = 580 MPa (8.4×10^4 psi). This corresponds to a 25% decrease in average dynamic shear yield stress. The average static yield stress is 580 MPa/0.727 = 798 MPa or 115,800 psi. If ϕ decreases due to laser heating, then $\bar{\tau}_s < 580$ MPa. Although this behavior has not been documented, it would not be surprising because decreasing ϕ moves the shear plane closer to the laser beam causing the average shear plane temperature to increase. If L_0 increases due to laser heating, then $\bar{\tau}_s < 580$ MPa. If $\bar{\tau}_R$ decreases due to laser heating, then $\bar{\tau}_s > 580$ MPa. One of the basic features of laser heating is, however, minimal heating of material contacting the rake face of the tool. Thus, significant changes in L_0 or $\bar{\tau}_R$ are not considered to be likely.

In order to attain an average dynamic shear yield stress in the shear plane of 580 MPa, which corresponds to an average static yield stress in the shear plane of 798 MPa (115,800 psi), we heat with the laser. The temperature increment due to laser heating added to that due to shear plane heating must give a temperature distribution such that

$$\sigma_y = \frac{1}{A_s} \iint \sigma_y \, dA = 115{,}800 \text{ psi}, \tag{16.23}$$

where values of σ_y corresponding to a particular temperature are obtained from Fig. 16.5.

If the average dynamic shear yield stress is 580 MPa, then $q = \bar{\tau}_s v_s = 5.8 \times 10^4 \times 0.51 = 2.94 \times 10^4$ W cm^{-2}, where $v_s = v \cos\gamma \cos^{-1}(\phi - \gamma)$ is the change in velocity parallel to the shear plane as the metal crosses the plane. According to Eq. (16.16), the temperature increment due to shear plane heating equals $404(T \sin\phi)$°C. The absorbed laser beam power, which must be found by trial and error, must provide a sufficient temperature increment so that Eq. (16.23) is satisfied.

Figure 16.9 shows the shear plane divided into elements for the calculation of temperature increments and the average static yield stress. Figure 16.10 shows the temperature increment due to laser heating (300 W absorbed power), the temperature increment due to shear plane heating, and the resulting shear plane temperature distribution. Figure 16.11 shows the resulting distribution of static yield stresses obtained from Figs. 16.5 and 16.10, which give an average value satisfying Eq. (16.23). The predicted laser beam power must be regarded as a minimum value. If ϕ decreases or L_0 increases due to laser heating, more power would be required unless a compensating change in $\bar{\tau}_R$ occurs.

Table 16.3 gives the absorbed power calculated for 25%, 50%, 75%, and 100% increases in material removal rate. If it is assumed that the loss in the optical train is 20% and the absorption constant of the workpiece is 15%, then the laser output power required for a 25%, 50%, 75%, and 100% increase in material removal rate is 2.5, 5.0, 10.0 and 15.8 kW, respectively.

Summary

The calculations presented in the previous section indicate that for a point heat source 15.8 kW of incident power would be required to produce a 100% increase in material removal rate without increasing the cutting force. A large,

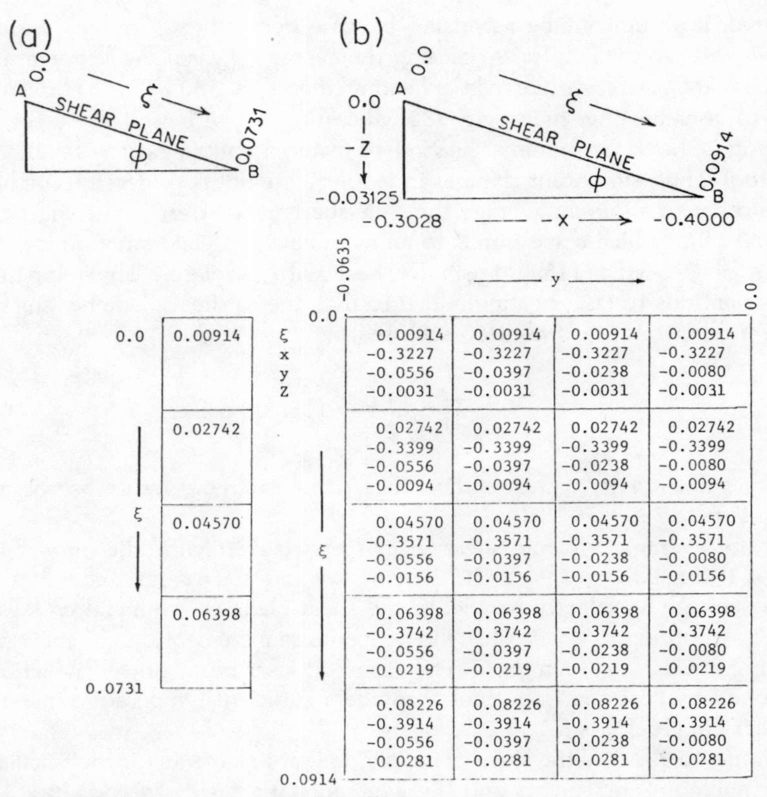

Fig. 16.9 Shear plane divided into elements for calculation of temperature increments and the average static yield stress.

LASER HEATING INCREMENT

83	223	484	724
84	214	445	651
81	196	390	558
76	172	327	459
69	147	267	366

+

SHEAR PLANE HEATING INCREMENT

311	311	311	311
380	380	380	380
396	396	396	396
402	402	402	402
404	404	404	404

=

SUM

394	534	795	1035
464	594	825	1031
477	592	786	954
478	574	729	861
473	551	671	770

Fig. 16.10 Temperature distribution on shear plane due to laser heating at 300 W absorbed power and shear plane heating (temperature given in degrees Celsius)

156	156	92	12
156	156	75	12
156	156	98	21
156	156	130	58
156	156	154	108

Fig. 16.11 Static yield stress distribution on shear plane after laser heat after laser heating. The average static yield stress on the shear plane is 116 ksi.

Table 16.3 Absorbed Power Required for Specific Increase in Material Removal Rate

Increase in Z (%)	a_c [cm (in.)]	Absorbed Power (W)
0	0.02500 (0.01)	0
25	0.03125 (0.0123)	300
50	0.03750 (0.0148)	600
75	0.04375 (0.0172)	1200
100	0.05000 (0.02)	1900

expensive laser would be needed to produce a beam with this incident power. The amount of incident power required might be decreased, however, by changing δ, beam shape, and orientation by using absorption-enchancing coatings and by reducing optical train losses. Theoretical and experimental research is necessary to evaluate these approaches. The economic benefits of increased material removal rate must exceed the increased costs associated with laser heating if LAM is to be adopted. Further tool wear studies and fatigue studies on laser machined specimens are needed to complete the technological evaluation of the LAM process.

Laser Machining (LM)

Laser machining (LM) is based on the formation of a groove in the surface of the workpiece by vaporization owing to heating by a moving laser beam. By overlapping grooves it is possible to remove layers of material and thus shape workpieces. LM has been applied to carbides and nitrides including SiC, Si_3N_4, and SiAlON, which are very hard and normally are shaped by diamond grinding. In the case of SiC, the groove is produced by selective oxidation of the carbon to form CO_2 and liquid Si. Silicon nitride decomposes to form N_2 and liquid Si. In both cases, the Si liquid is either expelled from the grooves as droplets or remains on the grooves' walls.

The arrangement for LM is similar to that employed in LAM, however, no cutting tool is used; see Fig. 16.1. After reflection by the cross-slide mirrors, the laser beam is focused by a ZnSe lens onto the workpiece along a radial direction for turning or parallel to the turning axis for facing.

Figure 16.12 shows a laser-machined piece of SiAlON, which originally had the square cross section remaining near the center of the piece. The maximum peak to valley roughness of the cylindrically turned surface is 7.5 μm. The thread is $\frac{1}{2}$ in. × 13 threads per inch.

Beam Polarization

Laser beams are often partially polarized due to the use of internal Brewster windows and turning mirrors. Wallace et al. have shown that the angular relationship between the polarization direction of the beam and the translation direction of the workpiece determines the shape of the groove cross section.[12]

Figure 16.13 shows the cross section of grooves in NC-132 Si_3N_4 (Norton Co., Worchester, MA), laser machined at an incident power of 560 W and a translation speed of 5 cm s^{-1} (0.16 fps). The angle θ is the angle between the electric vector of the polarized beam and the direction of translation of the specimen. When $\theta = 0°$, the groove is straight, narrow, and the deepest

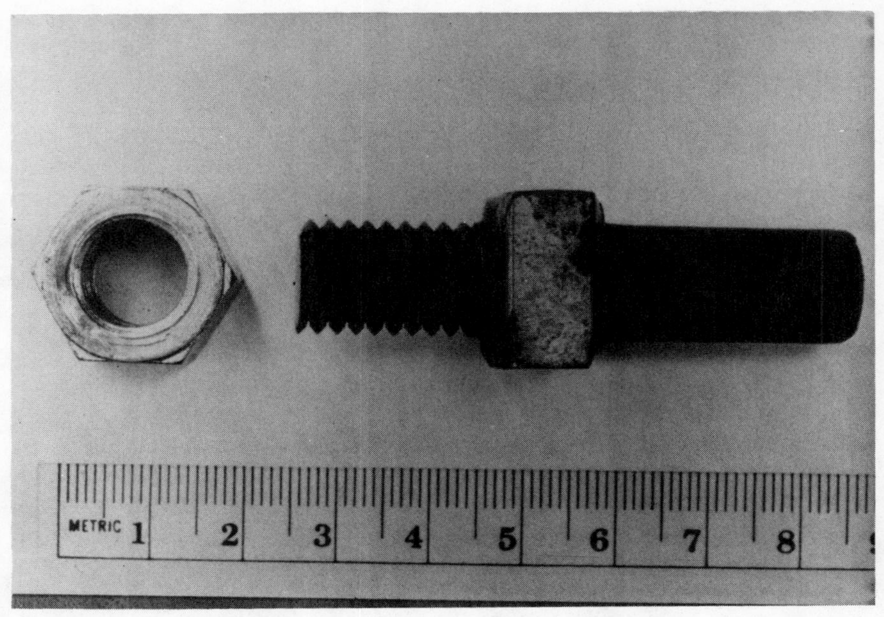

Fig. 16.12 Laser-machined $\frac{1}{2}$ in. × 13 screw thread in SiAlON.

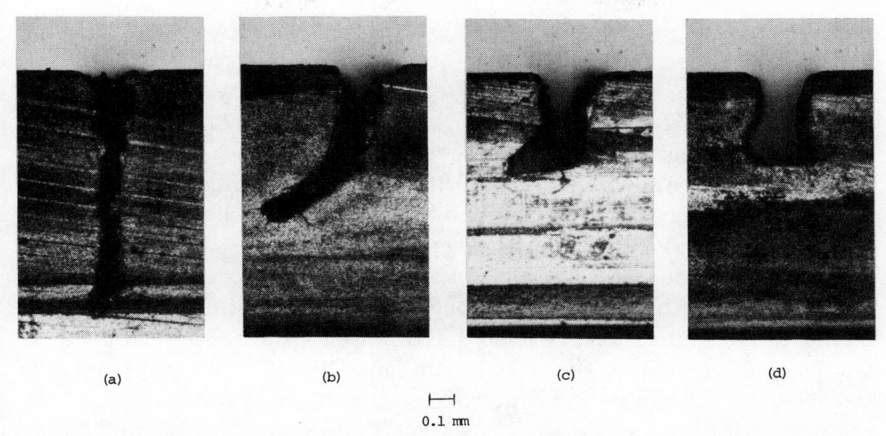

Fig. 16.13 Shape of groove cross sections as a function of θ: (a) $\theta = 0°$; (b) $\theta = 42°$; (c) $\theta = 72°$; and (d) $\theta = 90°$.

observed at any angle. When $\theta = 90°$, the groove is also straight but is wider and shallower than the $\theta = 0°$ groove. For angles between 0° and 90°, the groove is curved.

The mechanism, which is believed to explain the polarization-induced curvature, is illustrated in Fig. 16.14. The diagram shows the angular relation-

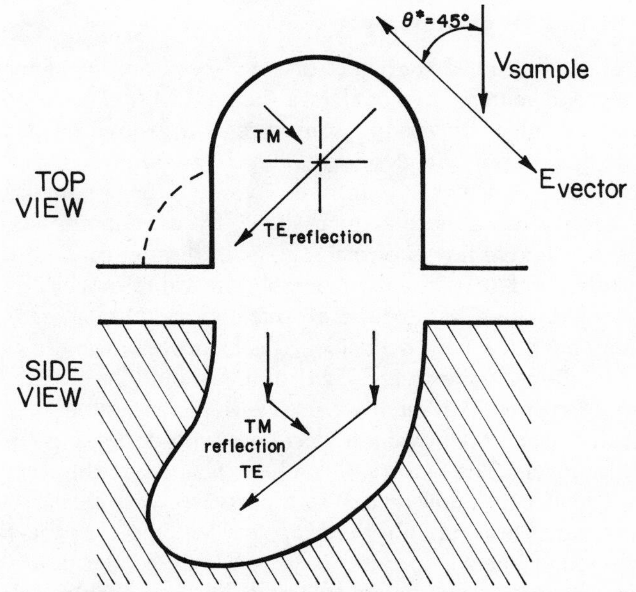

Fig. 16.14 Schematic diagram showing the effect of beam polarization of groove shape.

ship between the translation direction and the electric vectors designating the TM and TE reflections. The TM reflection is produced when the incident beam's electric vector is parallel to the plane defined by the incident and reflected beams; the TE, when it is perpendicular. For steady-state cutting represented in Fig. 16.14, the focused beam will interact with the slanted wall of the groove and be reflected. Although the whole front surface reflects the light, for simplicity only the two reflections mentioned previously are shown, and because the velocity vector is oriented 45° from the electric vector, the direction of the reflected beam is symmetric around the velocity vector as shown in Fig. 16.14. Although the direction of reflections is symmetric, the magnitude of the TE and TM reflections are not equal. Fresnel's law predicts, for any angle of incidence larger than 0°, the reflectivity for the TE reflection will be the largest. For large angles of incidence, e.g., 80°, the reflectivity for the TM ray is 0.65 and that for the TE ray is 0.97. This larger magnitude of the TE reflection is represented schematically by the longer length of the TE vector in Fig. 16.14. After the initial reflections both the TE and TM rays are directed down and across to the opposing wall where the energy will be absorbed. But since the intensity of the light reflected as TE rays is larger than that reflected as TM rays, more energy will be deposited where the TM rays are reflected and therefore more material will be removed in this area. The uneven reflection and subsequent asymmetric energy deposition creates the curved cross section.

Beam Power and Translation Speed

Wallace et al. have studied the effect of scan speed and incident power on groove shape and material removal rate during LM.[13] Figures 16.15a and 16.15b show the effect of varying scan speed on groove cross section for 560 W incident beam power and for $\theta = 0°$ and $\theta = 90°$, respectively. At high speeds, the $\theta = 0°$ and $\theta = 90°$ groove cross sections are similar, although the $\theta = 0°$ groove has a more angular shape. By overlapping such grooves it is possible to remove layers of material, as in turning on a lathe.

Figures 16.16a and 16.16b show material removal rate, which is the product of the cross-sectional area of the groove and the beam speed, as a function of beam speed for various incident beam powers, for the $\theta = 0°$ and $\theta = 90°$ orientations, respectively. Two trends should be noted: (1) at high speeds, the material removal rate decreases with increasing beam speed; (2) at low speeds, the material removal rate decreases with decreasing speed, particularly at high powers. The first trend can be explained on the basis of the total energy deposited per unit area, which decreases with beam speed. If the sample is translated fast enough, the irradiated volume near the surface will not absorb enough energy to raise its temperature to the point where the vaporization reaction becomes rapid enough for appreciable material to be removed. The second trend is thought to result from two other factors. The

Fig. 16.15 Cross sections of laser-machined grooves for different velocities: (a) $\theta = 0°$; (b) $\theta = 90°$.

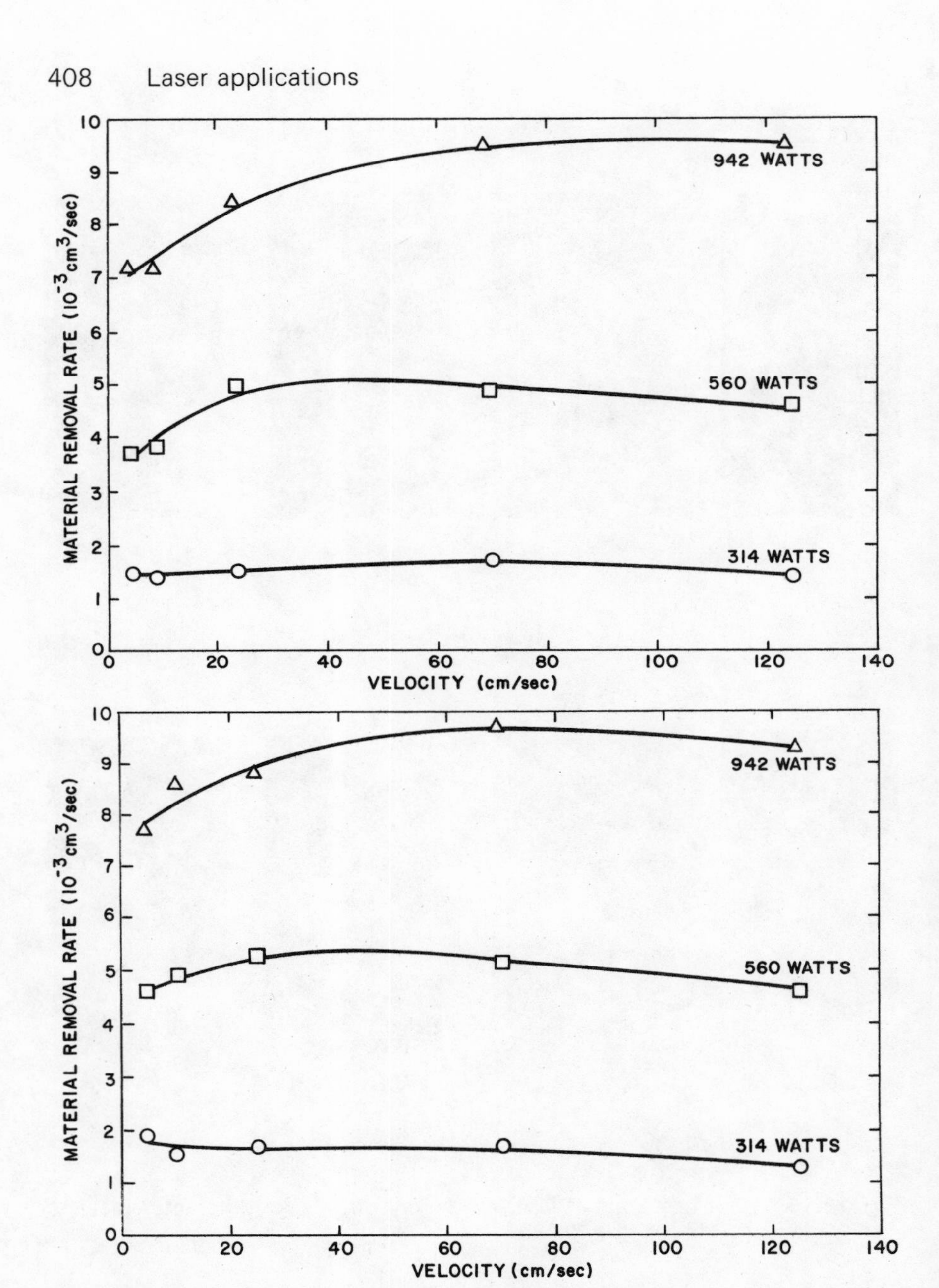

Fig. 16.16 Material removal rate versus translation speed for various incident beam powers: (a) $\theta = 0°$; 'b) $\theta = 90°$.

first factor is the blocking of the incoming beam by ejecta. As the groove becomes deeper at lower velocities, silicon droplets formed near the bottom of the groove are blown up into the incoming beam preventing the beam's energy from reaching the bottom of the groove. The deeper the groove, the longer will be the path that the incoming beam must penetrate. Consequently, less energy is absorbed by the sample resulting in a lower material removal rate. The second factor is conductive loss. As already discussed, the removal of material requires that the temperature be raised to a critical value. If the sample is translated slowly, there is time for energy to be conducted away from the volume being irradiated and, because of this loss, more incident energy is needed to remove the same amount of material.

Feed

Wallace et al. have also investigated the shaping of Si_3N_4 by overlapping grooves.[14] Figure 16.17 shows the arithmetical mean roughness and material removal rate versus feed calculated by overlapping the cross sections of single grooves. The material removal rate increases with increasing feed, approaching the single groove value, because, as the feed is increased, less groove overlap occurs. The roughness increases because the spacing of groove bottoms becomes greater with increasing feed, resulting in a more pronounced scalloping effect.

The observed behavior, however, is not so straightforward. Light-guiding effects create two types of surfaces. The first is characterized by an increase in roughness with decreasing feed as predicted by the overlap theory. The second is characterized by deep initial grooves followed by a region where the grooves become shallow as the end of the cut is approached. Both these effects are illustrated in Fig. 16.18, which shows multiple-pass groove cross sections for a series of different feeds and an incident power of 560 W. The grooves, which were formed at a low speed [9.1 cm s^{-1} (0.3 fps)] and the

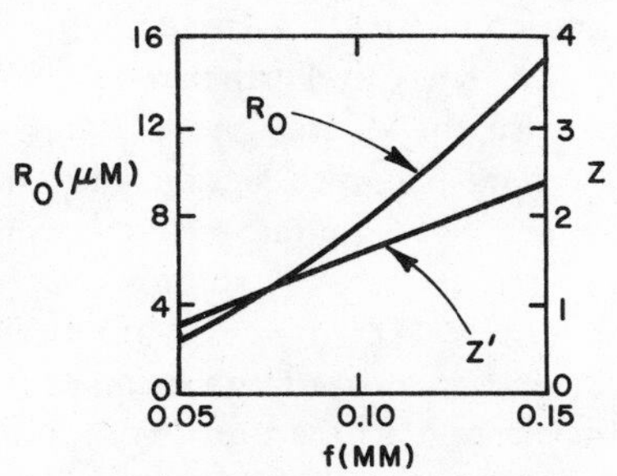

Fig. 16.17 Roughness and effective material removal rate versus feed.

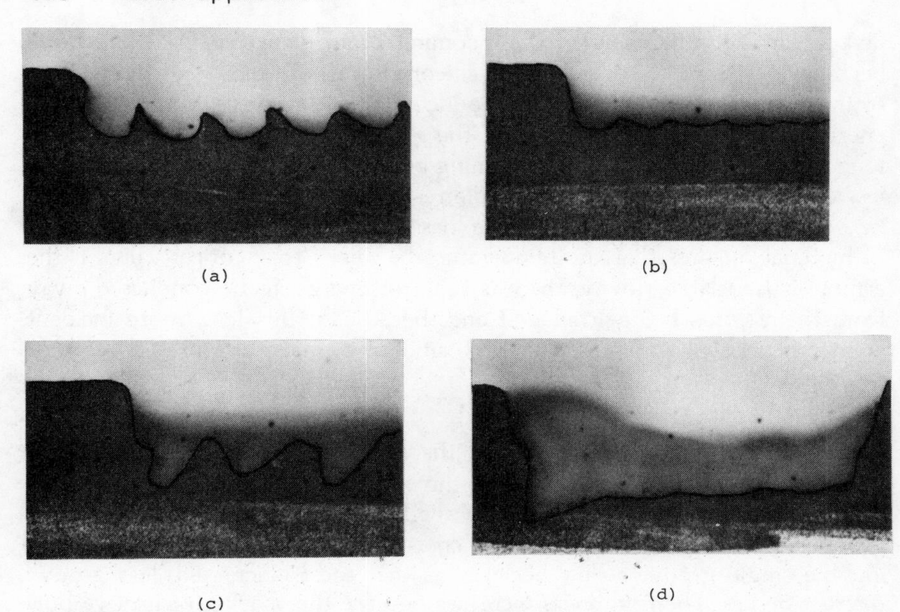

Fig. 16.18 Cross sections of overlapped grooves for various feeds: (a) 0.022 cm rev^{-1} (0.009 in. rev^{-1}); (b) 0.0178 cm rev^{-1} (0.007 in. rev^{-1}); (c) 0.015 cm rev^{-1} (0.006 in. rev^{-1}); and (d) 0.010 cm rev^{-1} (0.004 in. rev^{-1})

$\phi = 0°$ orientation, are similar to those seen at higher velocities and also in the $\phi = 0°$ orientation.

Figure 16.18a shows the bottom surface of a multiple-pass laser-machined sample, machined at a feed of 0.022 cm (0.009 in.). At this feed, a scalloped bottom surface, as would be expected from the repetition of the single-pass groove, was observed. As the feed was reduced to 0.0178 cm (0.007 in.), Fig. 16.18b, the scallops seen in Fig. 16.18a were almost completely removed and the surface became very smooth. Further reduction of the feed would be expected to further increase the degree of smoothness; however, this was not observed. Once the feed becomes significantly smaller than the width of a single-pass groove, the light-guiding effect becomes important. Figure 16.18c shows an example of increasing surface roughness with decreasing feed. A feed reduction from 0.0178 cm (0.007 in.) in Fig. 16.18b is 0.015 cm (0.006 in.) in Fig. 16.18c and creates a large increase in roughness.

In Fig. 16.18c, the first laser-machined groove is on the left. Only a small ridge from the first pass remains. During the second pass, the groove spacing is sufficiently small to cause a large portion of the laser light to be guided to the left by the walls of the first groove. During the third pass the effective groove spacing has been increased, because the beam was guided to the left during the second pass. Thus the third pass groove is not guided by the previously machined surface and is similar to a single-pass groove. The fourth pass is then guided similarly to the second pass, and the process repeats giving the highly contoured profile cross section shown in Fig. 16.18c. It should be noted that this gives a repetitive groove shape the spacing of which is not equal to the feed.

The second type of multiple-pass surface obtained with a feed of 0.010 cm (0.004 in.) varied in depth as shown in Fig. 16.18d. This surface is typical of that produced by laser machining where the feed is reduced to the point that every groove after the first is always guided to the left by the previous grooves. This creates a depression of the type shown in Fig. 16.18d. As one moves to the right from the initial depression, there is a region in which the depth remains constant. From this region the bottom surface gradually arises. This gradual rise occurs over a distance corresponding to many passes.

The removal rate was previously calculated for single-pass grooves. This was done by multiplying the area of the single groove, as measured from a photomicrograph with polar planimeter, by the sample translation speed. For multiple overlapping grooves, the average material removal rate equals the total area of material removed times the sample translation speed divided by the number of passes. The results are shown in Fig. 16.19, where the material removal rate is plotted for different feeds and velocities and an incident power

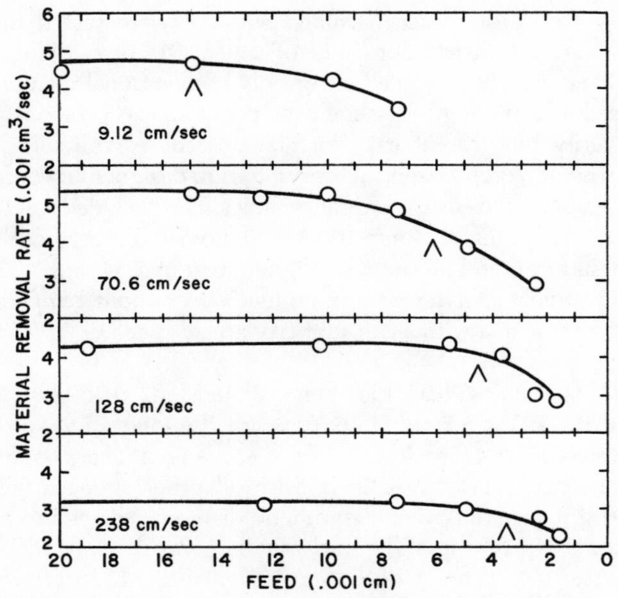

Fig. 16.19 Material removal rate versus feed for various velocities and incident power of 560 W.

of 560 W for the $\phi = 0^\circ$ orientation. Both $\phi = 0^\circ$ and $\phi = 90^\circ$ orientations gave similar results.

In Fig. 16.19, one can see that laser machining of Si_3N_4 produces a constant material removal rate over a wide range of feeds. It was found that this constant rate was equal to the material removal rate for a single groove for the same power, speed, and beam diameter. For each curve corresponding to a different velocity, there is a point beyond which the material removal rate begins to decrease with decreasing feed. The feed corresponding to the onset of this decrease decreases with increasing velocity. For the 9.12 cm sec^{-1} (0.3 fps) velocity curve, the onset is 11×10^{-3} cm (4.33×10^{-4} in.). This value is reduced to approximately 3×10^{-3} cm (1.2×10^{-4} in.) for the 238 cm sec^{-1} (7.8 fps) translation speed curve. This decrease in material removal rate is not related to the decrease predicted by the machining analysis previously proposed, which assumed the shape of a single-pass groove would be repeated during multiple overlapping. Because of light guiding by the already machined walls, the incident energy is deposited over a larger area at small feeds than at large feeds and, therefore, is not as effective in removing material.

Flexural Strength

Specimens with various laser-machined surfaces were tested in four-point bending.[14] The test matrix consisted of seven sets of samples, *A* through *G*. The first set, *A*, was produced using only conventional diamond grinding and was used as a standard that could be compared to previously published results to verify the present experimental procedures. All sets were tested using a four-point flexural strength testing fixture designed to conform to proposed military standards for structural ceramics. This included steel knife edges rounded to a radius of 3.18 mm (0.13 in.), provisions for the alignment of the bearing surfaces, and an outer and inner span of 2.54 and 1.27 cm (1 and 0.5 in.), respectively. All tests were conducted at room temperature in air using a testing machine with a constant cross-head speed of 0.0508 cm min^{-1} (0.02 in. min^{-1}).

All bend specimens had the same dimension, 0.069 cm (0.03 in.) × 0.417 cm (0.164 in.) × 3.5 cm (1.38 in.), with the long edges of the tension surface beveled 0.0794 cm (0.03 in.) at a 45° angle. The samples only had their tension surface laser machined, with all other surfaces being ground using a 250 grit resinoid-bonded synthetic diamond wheel and an infeed of 0.000762 cm (3×10^{-4} in.). The final grinding was done parallel to the long axis of the specimen. Because of the sensitivity of NC-132 to flaws and stress raisers, the smoothest possible laser-machined surface was used. This was obtained with a scan speed of 237 cm sec^{-1} (7.77 fps) and an incident power of 560 W. For this scan speed, a groove overlap of 0.00378 cm (0.00149 in.) was chosen so that the laser machining would not be influenced by the light guiding effect previously discussed.

Sets *B* and *C* were laser machined using the $\phi = 90°$ orientation, while sets *D, E, F,* and *G* used the $\phi = 0°$ orientation. Sets *C* and *D* were longitudinally machined, i.e., their grooves were parallel to the long axis of the sample. Sets *F* and *G* were used to determine if any strength loss due to laser machining could be recovered with a finishing cut by diamond grinding. The laser-machined surface of set *F* was diamond ground as previously described until the features of the laser-machined surface became invisible to the unaided eye. Set *G* was machined in a similar manner but with one extra finishing cut of 0.0008 cm (0.0003 in.) to remove microcracks, which might have been created by laser machining.

A summary of the four-point flexure strength measurements is presented in Table 16.4, which lists the number of samples tested, average bend strengths, standard deviations, and Weibull distribution characteristics.

Table 16.4 indicates the average strength of the laser-machined sample as compared to that of set *A*, 694 MN/m^2, was reduced by 30.6–41.9%. Sets *B* and *D*, which were laser machined in the transverse direction, showed the biggest decrease. Set *B* ($\phi = 90°$) had an average flexure strength of 423 MN/m^2, while set *D* ($\phi = 0°$) had an average strength of 403 MN/m^2. Sets *C* and *E* were laser machined in the longitudinal direction. Set *C* was machined using the $\phi = 90°$ orientation, while set *E* had the $\phi = 0°$ orientation. Set *C* had a strength of 462 MN/m^2, while set *E* had a strength of 482 MN/m^2. The closeness of these values suggest that at the speed of 237 cm sec^{-1} (7.77 fps), used in this investigation, the resultant groove shapes of the two orientations were very similar. The average strength values of sets *F* and *G* listed in Table 16.4 show that the reduction due to laser machining can be recovered by diamond grinding. Not only is the average strength increased, but also the standard deviation is increased to the value observed in set *A* and by previous investigators. The standard deviations of the laser-machined sets in Table 16.4 show a marked decrease in comparison to the diamond-machined sets. The previous investigations and sets *A, F,* and *G* have standard deviations of 9.4%, 8.0%, and 10.3% of their average strengths, respectively, while sets *B, C, D,* and *E* have standard deviations that range from 2.7% to 3.3% with an average of 2.9%. Correspondingly, the Weibull slope, which is a measure of the variability of strength of the material with a larger value being associated with the least variability, also varied significantly. Sets *PI* and *A* both had slopes of 12.6, while the laser-machined sets *B, C, D,* and *E* had values of 34.4, 33.0, 39.2, and 36.1, respectively. This along with the observed reduction in strength suggests that the laser-machined samples had a narrow distribution of flaw sizes, but that the flaws were larger than in the diamond-ground samples.

Table 16.4 Four-Point Flexure Strength of NC-132 Si_3N_4

	Sets							
	PI[a]	*A*	*B*	*C*	*D*	*E*	*F*	*G*
Machining Tool								
D—Diamond (Grit)	D (320)	D (520)	L	L	L	L	L	L
L—Laser							D (520)	D (520)
Machining Direction								
T—Transverse	L	L	T	L	T	L	T	T
L—Longitudinal								
O—Orientation			90°	90°	0°	0°	0°	0°
Number of samples	50	10	10	10	10	10	9	9
Average strength (MN/m^2)	632	694	423	462	403	482	643	683
Standard deviation (MN/m^2)	59	56	12	15	11	14	52	70
Weibull strength (MN/m^2)	658	720	430	469	408	489	668	715
Weibull slope	12.6	12.6	34.4	33.0	39.0	36.0	12.5	9.8

[a] Previous investigation.

Optical and SEM examinations were not able to reveal exact origins of fracture, but did show that the fracture origins were associated with the surface and were not close to the chamfered edges. SEM photographs also showed that the "wetted" silicon had very shallow cracks. Although these cracks were somewhat random in direction, the majority were perpendicular to the groove direction and extended across the width of the "wetted" silicon.

Summary

The feasibility of laser-machining ceramic materials by vaporization and expulsion of reaction products has been demonstrated. High material removal rates have been attained. The mechanical properties of laser-machined articles are sufficiently good to encourage application of this technique.

Acknowledgments

The author is pleased to acknowledge support of this research by the Defense Advance Research Projects Agency with the Air Force Wright Aeronautical Laboratories/MLTM, Air Force Systems Command, Wright-Patterson Air Force Base under Contract F33615-79-C-5119 and with the Department of the Army Defense Supply Service under Contract MDA 903-80-C-0436.

References

1. Tour, S. and L. S. Fletcher, "Hot Spot Machining," *Iron Age*, July 1949.

2. Schmidt, A. O., "Hot Milling," *Iron Age*, April 1949.

3. Moore, A. E. W., "Hot Machining for Single Point Turning—A Breakthrough," *Tooling and Production Magazine*, November 1977.

4. Plankenhorn, D. J., V. L. Hill, and S. Rajagopal, "Design, Construction and Operation of an Experimental Facility for Laser-Assisted Machining," in *Annual Technical Report: Advanced Machining Research Program*, D. G. Flom (ed.), DARPA Contract No. F 33615-79-C-5119, Chap. 17, 16 August, 1980.

5. Rosenthal, D., "The Theory of Moving Sources of Heat and Its Application to Metal Treatments," *Trans. ASME*, Vol. 68, 1946, pp. 849–866.

6. Cline, H. E. and T. R. Anthony, "Heat Treating and Melting Materials with a Scanning Laser or Electron Beam," *J. Appl. Phys.*, Vol. 48, 1977, pp. 3895–3900.

7. Maruo, H., I. Miyamoto, T. Ishida, and Y. Arata, "Investigation of Laser Hardening," *J. JWS*, Vol. 50, 1981, p. 208.

8. Gorsler, F. W., "Opportunities for Laser Assisted Machining," in *Semiannual Technical Report: Advanced Machining Research Program*, D. G. Flom (ed.), DARPA Contract No. F 33615-79-C-5119, Chap. 5, 15 February, 1980.

9. Rajagopal, S., V. L. Hill, and D. J. Plankenhorn, "Laser Assisted Machining of Inconel 718," in *Annual Technical Report: Advanced Machining Research Program,* D. G. Flom (ed.), DARPA Contract No. F-33615-79-C-5119, Chap. 18, 17 August, 1980.

10. Rajagopal, S. and D. J. Plankenhorn, "Laser Assisted Machining of Ti-6Al-4V," in *Annual Technical Report: Advanced Machining Research Program,* D. G. Flom (ed.), DARPA Contract No. F 33615-79-C-5119, Chap. 9, 17 August, 1981.

11. Weiner, J. H., "Shear Plane Temperature Distribution in Orthogonal Cutting," *Trans. ASME,* Vol. 77, 1331–1341 (1955).

12. Wallace, R. J., M. Bass, and S. M. Copley, "Effect of Beam Polarization of Shape of Grooves in Si_3N_4 Produced by Laser Machining" (unpublished results).

13. Wallace, R. J., M. Bass, and S. M. Copley, "Laser Machining of Si_3N_4: I. Energetics," unpublished results.

14. Wallace, R. J., M. Bass, and S. M. Copley, "Laser Machining of Si_3N_4: II. Shaping and Mechanical Properties" (unpublished results).

Part Seven

General Management Considerations

CHAPTER 17

General Management Considerations

William V. Burgess
University of San Francisco

As American industry turns to the 21st century, the major thrust is to come from the application of technology. In a maturing society, the gross national product (GNP) from the output of physical labor obeys the economic law of diminishing returns. For each added increment of labor, there is a reduced increment of production. The increase in GNP will, of necessity, come from added efficiencies of technological capacity.

Managing and controlling those technological inputs is the duty and responsibility of administration, that branch of organizational operations directly concerned with the direction and guidance of human productivity.

But what is the source of direction and guidance for the administration? Where can they turn for inspiration?

The authors of the following chapters in this section offer some suggestions. Stewart proposes a system of factory modeling through computer simulations. Tipnis gives an economic model for direction. Rieker and Burgess concentrate on the contributions of people to the production phase. Each has a message worthy of management consideration.

Stewart offers a system of factory modeling via computer simulations that will be of assistance in the design, development, and utilization of manufacturing processes.

Models serve the purpose of representing in symbolic form the full operations of a factory, including schedules of parts to be processed, the capacities of machines, rates, costs, etc. There are, however, disadvantages to the use of models owing to the necessity of assumptions, idealizations, and simulations of the real world. There can be, for example, no truly accurate prediction of human behavior in unusual circumstances, as witnessed by the failure of the Three Mile Island nuclear accident team to predict the ability of engineers to deny the advent of an "incident" and their capacity for denial of the seriousness of an actual "incident" in the face of catastrophic occurrences. Mathematical programming alone will be an incomplete model.

It is possible, declares Stewart, to assess the shortcomings of a model in comparison with the real process when automatic input and update data are established along with a reporting system that is understood by factory personnel.

The utility of the model is dependent on its ability to provide feedback to management and on their ability to provide "what if" programming to manufacturing personnel.

Tipnis proposes a further aid to management by utilization of economic models for process development, including sequencing, waiting time, transportation, and possible penalty costs. Macroeconomic and microeconomic models are presented for sensitivity analysis within the formulas of economic feasibility.

Rieker turns the topic of management considerations from the purely technical to the ultimately human aspects of production in his discussion of Quality Control Circles. He stresses the need for consideration of all aspects of a quality control problem, including the "blue sky" proposals of the QC committee as they address their problems. The human facilitation of the process is emphasized so as to lend support to the idea of voluntary participation (of course, this participation is real and meaningful) by staff, management, and on-line workers.

Several examples from Japan and the United States are given with positive results as to type of problem solved and monetary savings. Rieker closes with cautions as to use of QC circles as a panacea and expresses the opinion that there are numerous potential benefits from employee participation in production matters.

Burgess closes this section of the text by detailing the process of leadership development, its distinction from purely managerial roles, and a forecast of the style of leadership required to meet the needs of a future work force.

The philosophy expressed in each of these last two chapters is one of faith in the ability of humankind to address and solve problems in a constructive fashion. There is an egalitarianism in the tone of the message of these two authors that is symptomatic of a Jeffersonian democracy where competence is expected, found, and rewarded in all strata of society.

The decline in productivity of American industry is due not so much to a decline in the American work ethic as to a foreshortened view of the capacity of all employees to contribute to productive efficiency, say these authors. By focusing on technology alone, management has lost the utility of expertise that resides in the operator closest to the production line. Rieker and Burgess offer the evidence that human inputs affect the productive output in a positive sense if they are positively used and in a negative sense if negatively employed.

Long-term goals, such as the quality of work life, satisfaction with the job, health of workers, benefits to the community, consumer satisfaction, and lasting improvements for coming generations, exceed the importance of production quotas and quarterly profit/loss statements. Indeed, if the long-term benefits are made a standard of operation, the production and profits accrue as a natural result of good management. There can be no more Love Canal burials of hazardous production.

An organization's value to society ought to be measured first by the positive aspects it contributes to the general welfare, then by its balance sheet reports to stockholders. If there is an assessment that the contributions to societal good exceed the technical requirements of the task undertaken, the organization becomes an integral part of society and is not subject to the whims of the marketplace. The status of indispensability is achieved only by those organizations that serve first the needs of the social system. If a production technique or a company is but a tool for the accomplishment of a specific task, it has no value beyond the task. But when a production technique or an organization serves a human goal, it becomes essential and is assured of survival rates of production.

Mere technology is not enough. An organization must also stand for something. If it can be identified with a sense of responsibility to its employees, to its community, and with a sense of stewardship in relation to cultural values, it may count on long-term success.

Etzioni has mentioned the multiplicity of goals in American organizations, but the most successful—the Fortune 500—have identifiable central themes that contribute to their standing in the marketplace. This, too, is a management consideration.

The ability to say, "This is what we stand for" and to have that verified by public opinion is the hallmark of organizational success. A reputation for innovation in technology is a strong position in the American industrial marketplace, but a reputation for serving society well is the ultimate recognition.

In summary, wise leaders will attend to the human aspects of productivity, particularly as those human aspects relate to our own organizational styles and cultural values. There is a dysfunctional aspect in transplanting directly from outside the corporate structure any system that has not been evaluated by those inside the corporate structure. The American tendency to seek instant answers is exemplified by our obsession with Japanese forms of management currently. The successful adaptation of Japanese management forms with their strong cultural biases for paternalism, lifetime employment, and "slow-track" paths to promotion (Pascale and Athos, 1981) is difficult, but does have several American models, those Theory Z firms (Ouchi, 1981) that fit the requirements of strong corporate themes of societal service, quality of work life, and cooperation among levels in the organizational structure.

The prognosis for survival of such transplants is good, exceeding perhaps the projected survival of management methods that rely solely on technology or solely on human organization. The combination of carefully considered technological innovation and carefully considered human resources development may be hard to beat.

Selected Readings

Blake, R. R., and J. S. Mouton, *Productivity, The Human Side*, AMACOM, New York, 1981.

Cribbin, J. J., *Effective Managerial Leadership*, AMACOM, New York, 1972.

Drucker, P. F., *The Changing World of the Executive*, Truman Talley Books, New York, 1982.

Gellerman, S. W., *Motivation and Productivity*, AMACOM, New York, 1963.

Ouchi, W., *Theory Z: How American Business Can Meet the Japanese Challenge*. Addison-Wesley, Reading, MA, 1981.

Pascale, R. T. and A. C. Athos. *The Art of Japanese Management*, Simon & Schuster, New York, 1981.

CHAPTER 18

Factory Models

Richard C. Stewart
S&D Systems

Factory Models

In this chapter, a factory model is defined as a representation of the flow of parts through a manufacturing facility by use of a digital computer. A factory is considered a manufacturing facility that has an environment which is subject to a great deal of change. The sources of change include manufacturing engineering changes in the process flow variables, product changes due to marketing variables, product changes due to product engineering design variables, and changes brought about by the fact that people/machine systems are subject to the wide variety of unpredictable events inherent in the environment of the real world. From an analytical point of view this atmosphere of rapid change means that a solution to the problem of how to "best" plan and/or schedule a factory can be correct only for a very short period of time.

Because of the nature of the factory world many human decisions must be made daily concerning the next part to be worked, the next purchase order to be issued, the next customer order to be released to the factory, etc. In the extreme this means that the information necessary to make "the best" decision cannot be readily obtained, and thus the factory operates out of control in the sense of feedback/control systems. This situation of operating under uncertainty has given rise to two basic approaches in the study of factory models: (1) to represent the uncertainties as probabilities and determine "optimum" decision rules for operating under these conditions; (2) to develop "what if" models that essentially leave the optimizing function to human management and are tied into a real world information system for rapid updating.

Probabilistic simulation certainly has its place in the realm of systems analysis and has been well covered in the literature. This section is designed to present another approach to factory management: the idea of using the interaction of a human/computer systems approach where the computer is used more as a decision-support system rather than a normative optimizer.

Specific examples of support system uses could be:

1. Provide feedback for productivity studies. Feedback is a necessary element in increasing productivity. Productivity cannot be measured without a feedback system. A model can aid in determining what should have been accomplished in order to compare expected with actual results.

2. Analyze (measure) capacity. Determining the feasibility in terms of capacity of a factory to produce a given slate of products is of constant concern in factories that produce for a market with uncertain demand. Even factories that produce to more or less fixed demand are concerned with matching capacities to schedule requirements. Models can be used to explore alternative manpower assignments and work flows in order to adjust in-plant capacities to meet requirements.

3. Analyze (evaluate) capital investments. Engineering changes and technological improvements should be evaluated in terms of the economic and productivity improvements that they offer. Models are readily adaptable to investigate alternative processing schemes and different machine configurations.

4. Analyze QA/QC function. The question of when, where, and how much testing and inspection should be carried out in order to meet production specifications can become analytically impractical from a statistical methods point of view. A scheduling model with built-in factors based on the relationships of testing/inspecting and yield/rework can be used to evaluate the efficacy of various QA/QC programs.

5. Management planning and control. Planning is a function of management that determines what the desired level of production is for some period of time in the future. Control is a function of management that measures how well the plan is being fullfilled. Factory models, at various levels of detail, can provide the necessary information to allow management to perform these vital functions. The models are not sufficient to accomplish these functions for various reasons cited subsequently, but serve as an estimating mechanism for evaluating alternative courses of action.

6. Track production and inventory. In order to provide a feedback system to the model it is necessary to establish a tracking system that provides information on the actual flow of material in the factory, including inventory. By updating the model, say daily, very close control can be established that permits revising plans. By far the most tedious and important element in a factory model is the establishment of the data collection system. Most of the literature on this subject negates this aspect of developing models, but in our experience data collection is at least 80% of the effort.

7. Simulate forecasts. A plan is a forecast. The capability to forecast the results of factory operation is the most important feature of the model.

There are a couple of disadvantages in such uses, namely, assumptions and simplifications:

1. Assumptions. The real world, especially that part of the real world that involves people and machines, is not completely predictable. Thus models deal with forecasts and not with predictions. Such things as machine failure, employee absenteeism, and order cancellation must be dealt with on a probability basis if at all.

2. Simplifications. Models require simplifications that would be considered unreasonable in a real factory. For example, people can be assigned to machines in very small time increments that would not be practicable in practice.

Systems Analysis

A factory model is part of an overall management planning and control system, which includes:

Manufacturing
Inventory and Purchasing
Accounting and Finance
Marketing and Distribution
Personnel and Payroll

The factory model contains the basic functions of the factory, that is, the machines, the raw materials, and the work force. However, to be practicable the model must also account for the parts of the business that interact with the factory. Thus the current status of the inventory of parts and raw materials must be known as well as the anticipated deliveries of additional material. The accumulation of costs attributable to various modes of operation must be determinable, since cost is a major criterion of efficient operation. Marketing and distribution are needed to supply the demands placed on the production facility and the capacity of the system to deliver finished products. Personnel and payroll are the sources of labor costs and availability.

Another key factor in systems analysis is discipline—those responsible for data entry must perform. Ensuring that the files are maintained in an accurate and timely manner is a major responsibility of first line management. This is not readily understood in the workplace. Two reasons for this lack of appreciation for the importance of data files are that (1) it is not an important factor in the early stages of business development and (2) there is rarely a reward system associated with record keeping. In a complex modern manufacturing environment it is essential for planning and controlling that the location of every item in the factory be known with a high degree of reliability. Whether or not to have real time tracking systems is a function of the size of the plant, the number of parts in progress, the yields (scrap rate), the order rate, and the possible alternatives available to management.

Another key factor is participation—management must be an active participant. It is becoming more apparent that the introduction of any new technology must be accompanied by a participative posture by management. Change is something that is avoided by all systems that have evolved into whatever state that exists. When change is introduced, the system will exert force to resist the change. Unless there is some counter force to make resisting the change undesirable, the system will deteriorate, since the force exerted to resist the change will reduce the prevailing level of efficiency.

A third key factor is acceptance—users must be part of development. Wight has established the aphorism that "a person would rather live with a problem that he cannot solve than have a solution that he cannot understand." Unless the ultimate user of a model is involved with the development phase it will be difficult for the user to assume "ownership" and support the model.

The fourth key factor is scheduling models into planning models. Scheduling models require a level of detail that takes into account the sequences that the parts should take as they progress through the various machines. It is obvious that in an environment as disruptive as a factory the schedules projected by a model cannot be maintained over a long period of time.

Planning models are needed for forecasting requirements in terms of cash, material, and labor over longer periods of time than a scheduling model's limited horizon. Since planning models are a much rougher cut than scheduling models, it is more efficient to smooth over the sequencing problem and use an estimate of the average span over the total operations of a part. The span of a part through the factory is a function of the batch size, the mix of other parts present, the schedule, and the capacities of the machines. Thus care must be taken to ensure that the average spans used remain accurate.

The final factor is determining boundaries for higher-level models. By repeated use of scheduling models it is possible to establish parameters for a region in which the spans for a given part will approximate the expected time required to complete a given level of production. Thus if the planning model does not violate these feasible regions, it may be assumed that the planning model is valid. Otherwise, it is necessary to verify the feasibility of the planning model solution by again setting parameters with the scheduling model.

Building the Model

Defining the Process

The process to be modeled must first be defined in terms of the boundaries for the model. In a factory model this usually means deciding how to handle the functions of inventory and procurement, production requirements, and capacities. Figure 18.1 shows this.

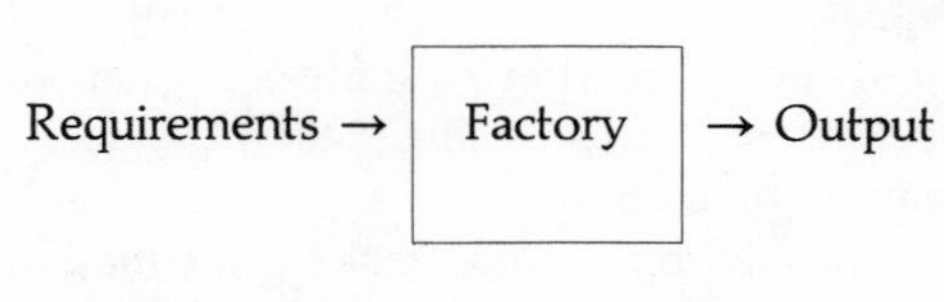

Figure 18.1

Within the factory the processes must be defined in terms of machine and human capacities, and machine and human interchangeability, according to four elements:

1. The objective(s). The objective of the model may be expressed in different ways. For example, we may fix (hold constant) the requirements, the inventory and procurement schedules, and the factory processes, and determine the resulting output. Alternatively, we may open the inventory and procurement schedules and allow infinite inventory. It is also practical to vary the machine configuration within the factory to determine the value of new investments or different working hours.

2. The operating characteristics. The operating characteristics include the paths taken by each part to be manufactured, the setup and run times required at each operation, and any travel time required to get from one operation to the next.

3. The constraints. The constraints include the restrictions imposed by the designer to keep the model within the realm of reality. Physical limitations on inventory, machine performance factors, and different skill levels for the workers are examples.

4. Data source analysis (DSA). Data source analysis is a technique or, better, a set of procedures for determining and establishing the sources of data within an organization and for an efficient method of gathering these data into a rational set of files. DSA also involves establishing the discipline for the collection of the data, i.e., the responsibilities involved.

A management information system can be thought of as a set of computer files accessible to a set of programs (software) such that the file data can be converted into useful information. DSA has historically been the domain of the data processing department, but, if technologists are to become users of models and information systems, the responsibility for designing the data files and the disciplines necessary for the collection of the data must be borne by the user.

The basic philosophy we have found to be the key to effective DSA is that the originator of the data should be responsible for getting the data into a file. Users can be remote from the source of the data necessary for their needs, e.g., accountants from inventory cost data. The drawback to this philosophy is that the originator has no application for some of the data he or she generates and thus has no apparent motivation to maintain these data. This is a management problem that must be overcome for an efficient operation. Obviously, incentives can be established that would make the data gathering function a rewarding one. Another difficult determination in DSA is deciding on what data to keep in the files. There is no way of being sure that any data left out of the files will not be required at a later date. Obviously, it is easier to carry some unnecessary data than to omit some necessary data, but, in the extreme, unnecessary data can become an expensive burden. In practice, this is not considered a major problem, and our rule of thumb is: when in doubt, file the data.

The other major problem in DSA is avoiding redundancy. Files must be relational, but information should not be repeated in another file.

An example of a shop simulator program follows on the next pages.

References

Church, J., "Simulation for the Computer Novice," Illinois Institute of Technology Research Institute, Proceedings of the Winter Simulation Conference, IEEE, New York, 1978, pp. 354–359.

Cox, J. F. and F. P. Adams "Manufacturing Resource Planning an Integrated Decision-Support System," *Simulation*, Vol. 35, No. 3, 1980, pp. 73–79.

Crane, G. A. and D. L. Thompson, "GASP IV Replaces Rules of Thumb," *Ind. Eng.*, Vol. 11, No. 5, 1979, pp. 48–52.

Orlicky, J., *Material Requirements Planning*, McGraw-Hill, New York, 1975.

Phillips, D. T., M. Handwerker, and G. L. Hogg, "OEMS: Generalized Manufacturing Simulator," *Comput. Ind. Eng.*, Vol. 3, No. 3, 1979, pp. 225–233.

Sixth Annual ICAM Days (Integrated Computer-Aided Manufacturing) Proceedings, January 1982, Air Force Wright Aeronautical Laboratories, Materials Laboratory, Manufacturing Technology Division, Ohio.

Example

```
FILE: SHOPSIM1 FORTRAN  A1                 VM/SP CONVERSATIONAL MONITOR SYSTEM


C SHOP SIMULATOR PROGRAM---PROGRAMMER DENNIS L. KRATZ

      DIMENSION IBATCH(20),XMND(20),IQUANT(20),IPOS(20),PRTCLK(20),
     1DTIME(20),STDATE(20),TMLEAD(20),IPART(20),MAXPOS(20),COMPDT(20),
     1SPAN(20),TOTQ(20),AVGQ(20)
      DIMENSION NXLCI(5,5),STDS(5,5),SETUP(5,5),SHFT(15)
      DIMENSION CCKLK(15),LCINUM(15),CAP(15)
      DIMENSION EXTLCI(15,100),BCHAVL(15,100),DWNTME(15,100),
     1NPOS(15,100),RUNTME(15,100),NBATCH(15,100),ENTLCI(15,100),
     1TTLQUE(15,100),NPART(15,100),CCAVL(15,100),NQUANT(15,100),
     1STUP(15,100)
C NXLCI IS THE MATRIX OF PATHS TAKEN BY EACH BATCH OF PART NUMBERS
      DATA NXLCI/1001,1002,1003,1004,1005,1002,1003,1004,1005,1001,
     11003,1004,1005,1001,1002,1004,1005,1001,1002,1003,1005,1001,1002,
     11003,1004/
C STDS ARE THE STANDARD RUN TIMES OF EACH PART IN EACH LOADCENTER
      DATA STDS/25*1.0/
C SETUP ARE THE SETUP TIMES FOR EACH BATCH IN EACH LOAD CENTER
      DATA SETUP/25*.0/
C SHFT ARE THE HOURS PER DAY WORKED IN EACH LOAD CENTER
      DATA SHFT/5*24.0/
C IPART ARE THE PART NUMBER IDENTIFICATIONS
      DATA IPART/1111111,2222222,3333333,4444444,5555555/
C IBATCH ARE THE IDENTIFICATIONS OF THE BATCHES(SOME BATCHES MAY
C HAVE THE SAME PART NUMBER)
      DATA IBATCH/5*1/
C XMND ARE THE EXPECTED COMPLETION DATES OF THE BATCHES
      DATA XMND/50.,50.,50.,50.,50./
C IQUANT ARE THE NUMBER OF PARTS IN EACH BATCH
      DATA IQUANT/5*10/
C CAP IS THE CAPACITY OF THE LCI TO PERFORM AGAINST THE HOURS WORKED
      DATA CAP/5*1.0/
```

```
FILE: SHOPSIM1 FORTRAN  A1                VM/SP CONVERSATIONAL MONITOR SYSTEM

C TRVL IS THE TRAVEL TIME ALLOWED BETWEEN ALL PAIRS OF LCI'S
      DATA TRVL/.0/
C PERFRM IS THE PERFORMANCE FACTOR FOR THE WHOLE SHOP
      DATA PERFRM/1.0/
C LCINUM IS THE LCI IDENTIFICATION
      DATA LCINUM/1001,1002,1003,1004,1005/
C MAXPOS IS THE MAXIMUM DIMENSION OF THE MATRIX OF PATHS
      DATA MAXPOS/5*5/
C JJMAX IS THE COUNTER FOR DAYS
      JJMAX=1
C KKMAX IS NUMBER OF BATCHES
      KKMAX=5
C LLMAX AND MMMAX ARE PROGRAMMING CONSTANTS
      LLMAX=5
      MMMAX=100
C NNMAX IS THE NUMBER OF LCI'S
      NNMAX=5
      START=999.
C SET PART CLOCKS TO ZERO
      DO 111 KK=  1,KKMAX
      PRTCLK(KK)=0.
  111 CONTINUE
C SET LCI CLOCKS TO ZERO
      DO 222 NN=1,NNMAX
      CCKLK(NN)=0
  222 CONTINUE
C SET INITIAL BATCH POSITION TO ONE
      DO 333 KK=1,KKMAX
      IPOS(KK)=1
  333 CONTINUE
      DO 150 KK=1,KKMAX
      IF(XMND(KK).GT.START)GO TO 150
      START=STDATE(KK)
  150 CONTINUE
      DAYCLK=START+1.
```

```
FILE: SHOPSIM1 FORTRAN  A1                    VM/SP CONVERSATIONAL MONITOR SYSTEM

C MAIN PROGRAM
C START NEW DAY
      DO 4001 JJ=1,JJMAX
C RESET NSTAT AND SELECT LCI
  501 CONTINUE
      NSTAT=0
      DO 2222 NN=1,NNMAX
C RESET TEST AND NEXT PART
      TESTA=999
      TESTB=999
      TESTC=999
      NEXTA=0
      NEXTB=0
      NEXTC=0
      NEXTP=0
C SELECT NEXT PART TO BE PROCESSED
      DO 7005 KK=1,KKMAX
      IF(KK.EQ.II)PRTCLK(KK)=TMPCLK
      IF(NXLCI(KK,IPOS(KK)).EQ.0.OR.NXLCI(KK,IPOS(KK)).NE.
     1LCINUM(NN)) GO TO 7005
      IF(PRTCLK(KK).EQ.0)PRTCLK(KK)=STDATE(KK)
      IF(PRTCLK(KK).GE.DAYCLK.OR.CCKLK(NN).GE.DAYCLK)GO TO 7005
      DTIME(KK)=PRTCLK(KK)-CCKLK(NN)
      IF(IPOS(KK).EQ.1.OR.CCKLK(NN).EQ.0)DTIME(KK)=0
      IF(DTIME(KK))7006,7007,7008
 7006 IF(XMND(KK).GT.TESTA)GO TO 7005
      TESTA=XMND(KK)
      NEXTA=KK
      GO TO 7005
 7007 IF(XMND(KK).GT.TESTB)GO TO 7005
      TESTB=XMND(KK)
      NEXTB=KK
      GO TO 7005
 7008 IF(XMND(KK).GT.TESTC)GO TO 7005
      TESTC=XMND(KK)
      NEXTC=KK
```

```
FILE: SHOPSIM1 FORTRAN  A1               VM/SP CONVERSATIONAL MONITOR SYSTEM

 7005 CONTINUE
      NEXTP=NEXTA
       IF(NEXTA.EQ.O)NEXTP=NEXTB
       IF(NEXTA.EQ.O.AND.NEXTB.EQ.O)NEXTP=NEXTC
       IF(NEXTP.EQ.O)NSTAT=NSTAT+1
       IF(NSTAT.GE.NNMAX)GO TO 4002
       IF(NEXTP.EQ.O)GO TO 2222
      KK=NEXTP
      II=KK
      DO 401 MM=1,MMMAX
      IF(NBATCH(NN,MM).EQ.O) GO TO 402
  401 CONTINUE
  402 CONTINUE
C PROCESS PART AND UPDATE CLOCKS
      BCHAVL(NN,MM)=PRTCLK(KK)
      CCAVL(NN,MM)=CCKLK(NN)
      IF(CCKLK(NN).EQ.O)CCAVL(NN,MM)=BCHAVL(NN,MM)
      TTLQUE(NN,MM)=CCAVL(NN,MM)-BCHAVL(NN,MM)
      IF(TTLQUE(NN,MM).LT.O)TTLQUE(NN,MM)=O
      TOTQ(KK)=TTLQUE(NN,MM)
      DWNTME(NN,MM)=BCHAVL(NN,MM)-CCAVL(NN,MM)
      IF(DWNTME(NN,MM).LT.O)DWNTME(NN,MM)=O
      IF(CCKLK(NN).LT.PRTCLK(KK))CCKLK(NN)=PRTCLK(KK)
      S=SETUP(KK,IPOS(KK))
      P=STDS(KK,IPOS(KK))*IQUANT(KK)*(1./24.)*CAP(NN)*SHFT(NN)
      NBATCH(NN,MM)=IBATCH(KK)
      NQUANT(NN,MM)=IQUANT(KK)
      ENTLCI(NN,MM)=CCKLK(NN)
      EXTLCI(NN,MM)=ENTLCI(NN,MM)+S+P
      NPOS(NN,MM)=IPOS(KK)
      IF(NPOS(NN,MM).EQ.MAXPOS(KK))COMPDT(KK)=EXTLCI(NN,MM)
      NPART(NN,MM)=IPART(KK)
      CCKLK(NN)=EXTLCI(NN,MM)
```

```
FILE: SHOPSIM1 FORTRAN  A1                    VM/SP CONVERSATIONAL MONITOR SYSTEM

      TMPCLK=EXTLCI(NN,MM)+TRVL
      IPOS(KK)=IPOS(KK)+1
      RUNTME(NN,MM)=P
      IF(NPOS(NN,MM).EQ.MAXPOS(KK))ISTAT=ISTAT+1
      IF(ISTAT.EQ.KKMAX)GO TO 1001
C SELECT NEXT LCI
 2222 CONTINUE
      GO TO 501
C UPDATE MASTER CLOCKS AND START NEW DAY
 4002 JJMAX=JJMAX+1.
      DAYCLK=DAYCLK+1.
      START=START+1.
 4001 CONTINUE
C
C WRITE RESULTS
 1001 DO 997 NN=1,NNMAX
      WRITE (6,998)LCINUM(NN)
  998 FORMAT('0','LCI',2X,I5)
      WRITE(6,1212)
      WRITE(6,1213)
 1212 FORMAT(T14,'BCH',1X,'BCH',7X,'PART',5X,'LCI',4X,'TOTAL',3X,
     1'DOWN',4X,'ENTER',10X,'EXIT')
 1213 FORMAT(1X,'PART NUMBER',1X,'NO.',1X,'SIZ',1X,'POS',3X,'AVAIL',3X,
     1'AVAIL',4X,'QUE',4X,'TIME',5X,'LCI',4X,'RUN',5X,'LCI')
      DO 990 MM=1,MMMAX
      IF(NPOS(NN,MM).EQ.0) GO TO 997
      WRITE(6,999)NPART(NN,MM),NBATCH(NN,MM),NQUANT(NN,MM),NPOS(NN,MM),
     1BCHAVL(NN,MM),CCAVL(NN,MM),TTLQUE(NN,M),DWNTME(NN,MM),ENTLCI
     1(NN,MM),RUNTME(NN,MM),EXTLCI(NN,MM)
  999 FORMAT(1X,I8,4X,I3,1X,I3,1X,I3,2X,F6.2,2X,F6.2,2X,F6.2,2X,F6.2,
     12X,F6.2,1X,F6.3,2X,F6.2)
  990 CONTINUE
  997 CONTINUE
  881 DO 885 LL=1,LLMAX
      IF(IPART(LL).EQ.ISTAT)GO TO 885
      IF(IPART(LL).EQ.0)GO TO 885
```

```
FILE: SHOPSIM1 FORTRAN  A1            VM/SP CONVERSATIONAL MONITOR SYSTEM

      ISTAT=IPART(LL)
      TQ=0
      KSTAT=0
      ASPAN=0
      WRITE (6,882)
  882 FORMAT('0',T28,'PART NUMBER SUMMARY')
      WRITE(6,883)
  883 FORMAT('0',T17,'BCH',4X,'STRT',4X,'COMP',12X,'LEAD',4X,'TOT',5X,
     1'AVG',5X,'BCH')
      WRITE(6,884)
  884 FORMAT(1X,'PART NUMBER',4X,'NUM',4X,'DATE',4X,'DATE',4X,'SPAN',
     14X,'TIME',4X,'QUE',5X,'QUE',5X,'QTY')
      DO 888 KK=1,KKMAX
      IF(IPART(KK).NE.ISTAT)GO TO 888
      SPAN(KK)=COMPDT(KK)-STDATE(KK)
      ASPAN=ASPAN+SPAN(KK)
      AVGQ(KK)=TOTQ(KK)/MAXPOS(KK)
      TQ=TQ+TOTQ(KK)
      KSTAT=KSTAT+1
      WRITE(6,887)IPART(KK),IBATCH(KK),STDATE(KK),COMPDT(KK),SPAN(KK),
     1TMLEAD(KK),TOTQ(KK),AVGQ(KK),IQUANT(KK)
  887 FORMAT(1X,I8,7X,I3,4X,F4.0,4X,F4.0,4X,F4.0,4X,F4.0,4X,F4.0,4X,
     1F4.0,4X,I3)
      IPART(KK)=0
  888 CONTINUE
      AQ=TQ/KSTAT
      AVGSPN=ASPAN/KSTAT
      WRITE(6,886)TQ,AQ,AVGSPN
  886 FORMAT('0','TOTAL QUE=',F4.0,4X,'AVG QUE=',F4.0,4X,'AVG SPAN=',
     1F4.0)
  885 CONTINUE
      WRITE(6,581)
  581 FORMAT('0',T21,'LCI SUMMARY')
      WRITE(6,579)
  579 FORMAT('0',8X,'TOT',4X,'TOT',10X,'TOT')
      WRITE(6,578)
```

```
FILE: SHOPSIM1 FORTRAN  A1                      VM/SP CONVERSATIONAL MONITOR SYSTEM

  578 FORMAT(1X,8X,'HRS',4X,'HRS',10X,'DOWN',10X,'TOT',3X,'AVG')
      WRITE(6,577)
  577 FORMAT(1X,'LCI',5X,'AVL',4X,'USED',5X,'%',3X,'TIME',5X,'%',
     14X,'QUE',3X,'QUE')
      DO 675 NN=1,NNMAX
      USED=0
      DNTM=0
      TLQ=0
      DO 575 MM=1,MMMAX
      IF(NPOS(NN,MM).EQ.0)GO TO 676
      USED=USED+RUNTME(NN,MM)+STUP(NN,MM)
      DNTM=DNTM+DWNTME(NN,MM)
      TLQ=TLQ+TTLQUE(NN,MM)
  575 CONTINUE
  676 MM=MM-1
      HRSAV=(EXTLCI(NN,MM)-CCAVL(NN,1))*24.
      NHRSAV=HRSAV+.5
      USED=USED*24.+.5
      NUSED=USED
      IPCT=USED/HRSAV*100
      DNTM=DNTM*24.
      NDNTM=DNTM
      JPCT=DNTM/HRSAV*100
      TLQ=TLQ*24.
      NTLQ=TLQ
      KPCT=TLQ/HRSAV*100
      AVQ=TLQ/MM
      NAVQ=AVQ
      WRITE(6,677)LCINUM(NN),NHRSAV,NUSED,IPCT,NDNTM,JPCT,NTLQ,
     1NAVQ
  677 FORMAT(1X,I4,4X,I4,3X,I4,3X,I3,3X,I4,3X,I3,3X,I4,
     13X,I3)
  675 CONTINUE
      END
```

CHAPTER 19

Economic Models for Process Development

Vijay H. Tipnis
Tipnis Associates

Introduction

Economics of Machining Operations

The economics of conventional machining operations such as turning, drilling, milling, and grinding has been determined through time and cost models developed since the early 1900s. Traditionally, the time and cost models are expressed in terms of speed, feed, depth, tool life, tool change time, setup time, and overhead and tool usage costs. During the mid-1970s, a generalized economic model applicable to any machining operation was introduced.[1] Since the structure of this model is conducive to trade-off and sensitivity analysis, it is used as the basis for formulating economic models for process development.

Economic Model for Process Development: Overall Framework

To establish the overall framework, we must begin with the definition of process and how it relates to the economics of machining. During the past two to three decades, several attempts have been made to formulate economic models that go beyond the unit machining operation—i.e., they consider a sequence of cuts, setups, waiting, transportation, and penalty costs as well as product mix. Each of these attempts involves different sets of assumptions peculiar to the specific problem considered. Consequently, a consistent model incorporating all the aspects necessary for process development is not available.

Process

By process we mean a discrete parts manufacturing process concerned with the conversion of input material into a prescribed output configuration of desired shape, size, and accuracy. The process requires input energy to effect this conversion. Typically, a discrete parts manufacturing system involves a sequence of such processing units (also called workstations) through which the input material is progressively transformed until it reaches the configuration, shape, size, and accuracy necessary for it to be accepted for subassembly or assembly where it performs a designed function (see Fig. 19.1). Any process performed at any processing unit in the discrete parts manufacturing system has a characteristic rate that is a function of the operating variables, work material, and tool or energy employed. The process requires resetting or adjustment at a certain frequency (e.g., tool change) depending on the same variables that affect the rate as well as others such as tool geometry. The operating rate and adjustment frequency are known as process models. The limits of accuracy and safe performance (such as tool breakage and machine tool horsepower, speed and feed range, force limits) are the constraints

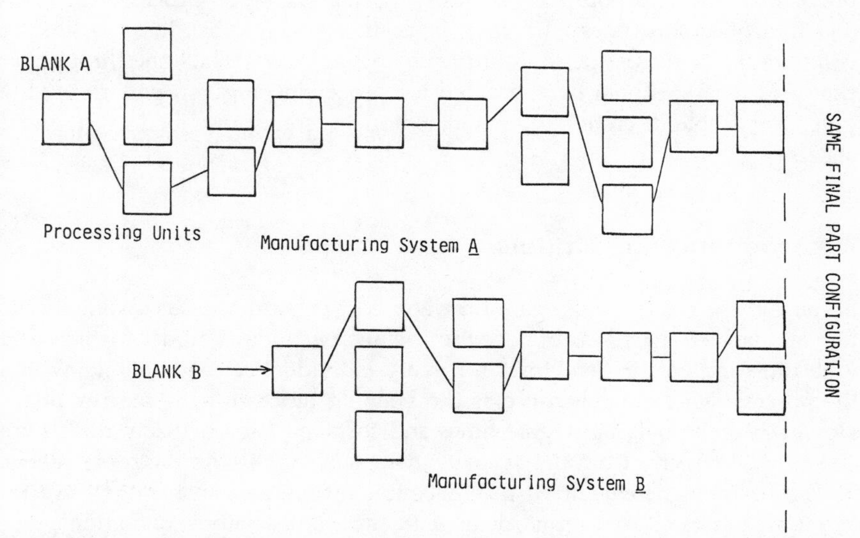

Figure 19.1 Manufacturing systems for discrete parts manufacturing process.

that circumscribe the region of process operating variables within which the process models are applicable and the process operable.

Time and Cost Expenditures

Each discrete parts manufacturing process requires an initial setup for each new input material configuration before the process can be executed. Thereafter, prior to process execution, only the operations of loading/unloading and clamping/unclamping are required for each subsequent piece of input material in the same lot.

The time for processing plus operations such as setup, load and unload, and any slack allowed is defined as the "floor-to-floor time" at that processing unit. The throughput time is defined as the total time including the floor-to-floor time at each processing unit in the sequence, the waiting time, and the transportation time between all the processing units. In computing the times, it is necessary to make sure that if there are any overlapping operations that are conducted simultaneously on the same input material, then only the longest of those overlapping times are counted.

Costs at each processing unit are due to the operating expenses, including labor, machine, and support cost, plus the costs of consumable materials such as cutting tools, cutting fluids, etc. The hourly cost rates are usually established for machine and support activity on the basis of a target yearly expenditure divided by anticipated actual hours the unit is to be available for processing. After appropriately multiplying the waiting (in-process inventory) and transportation times with their respective hourly rates, it is possible to compute the total cost per output part by aggregation. Thus, the throughput time and total cost can be computed for processing a given part through a given applicable discrete parts manufacturing system prescribed in the route sheets.

Macroeconomic Models

Because these models aggregate time and cost expenditures as stated above, they are labeled macro- as opposed to the microeconomic models, which deal with time and cost expenditures for every individual cut and activity within the process. Micromodels serve as the building blocks for the macromodels. However, when only aggregate times and costs are known, macromodels are constructed directly from the available data, such as that on the route sheets.

The following, generalized macroeconomic time and cost models are introduced based on the terms defined in the nomenclature, definitions, and Fig. 19.2.

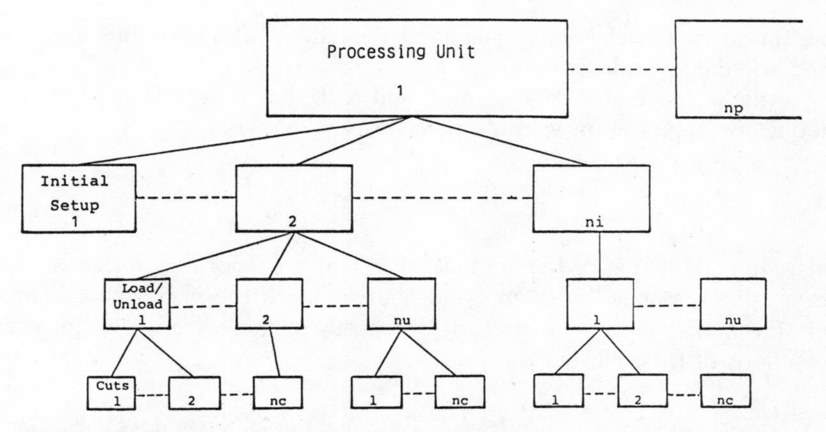

Figure 19.2 Processing sequence for macroeconomic models.

Total throughput time per part, T:

$$T = \sum_{np} \left(\sum_{ni} \sum_{nu} \sum_{nc} tc + \sum_{ni} \sum_{nu} tu + (1/N) \sum_{ni} ti \right) + \sum_{np} tw + \sum_{np} tt + \sum_{np} tr. \qquad (19.1)$$

Total cost per part, C:

$$C = \sum_{np} \left(\sum_{ni} \sum_{nu} \sum_{nc} tc(Cr) + \sum_{ni} \sum_{nu} tu(Ur) + (1/N) \sum_{ni} ti(Ir) \right) + \sum_{np} tw(cw) + \sum_{np} (tt)(ct) + \sum_{np} (tr)(cr) + cm + cs, \qquad (19.2)$$

where sums np, ni, nu, nc are taken over all processing units, all setups within a processing unit, all load/unloads within a setup, and all cuts within a load/unload, respectively. As stated in the previous section, the summing for throughput time must be done with due consideration for all overlapping activities.

Macroeconomic Models Applied to Route Sheet Data

Route sheets generally contain aggregate estimates on processing and set-up times, labor and machine rates, and total work material and consumable expenditures per lot. Also, based on the shop performance, a ratio between the actual time and planned time from labor vouchers is generally known. Data on part waiting, transportation, and rework times as well as processing unit idle times are not readily available because these data depend on the job shop scheduling and performance.

For the purpose of estimating time and cost, the Eqs. (19.1) and (19.2) are modified for application to route sheet data as follows:

$$\text{Time/part} = \sum_{np} tp + (1/N) \sum_{np} ts \tag{19.3}$$

where *tp* is the larger of the machine time, planned labor time, or actual processing time obtained after applying allowances to the planned labor time; *ts* is the initial set-up time at each processing unit. Figure 19.3 shows the definitions of these times.

$$\text{Cost/part} = \left(\sum_{np} tp + (1/N) \sum_{np} ts \right) (LMS) + cm + cs, \tag{19.4}$$

where (*LMS*) is the combined labor, machine, and support cost rate (\$/hr); (*LMS*) can be broken down into separate cost rates, if the time expenditures for labor, machine, and support are different. Also, the set-up time, *ts*, can be multiplied only by labor cost rate, if the machine did not remain idle during set-up.

Microeconomic Models

Microeconomic models are concerned with time and cost expenditures within each individual cut. The following generalized microeconomic time and cost

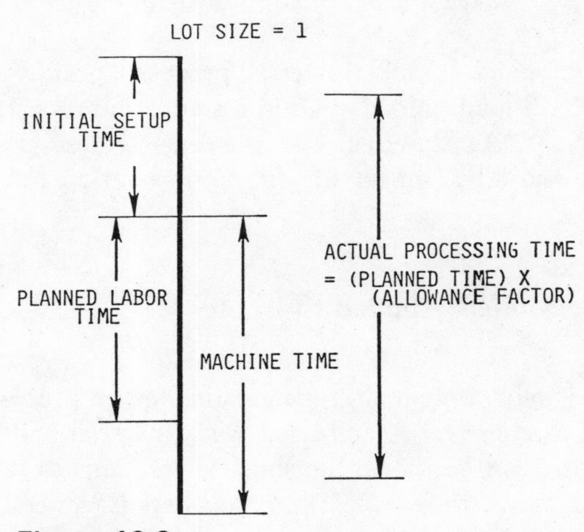

Figure 19.3

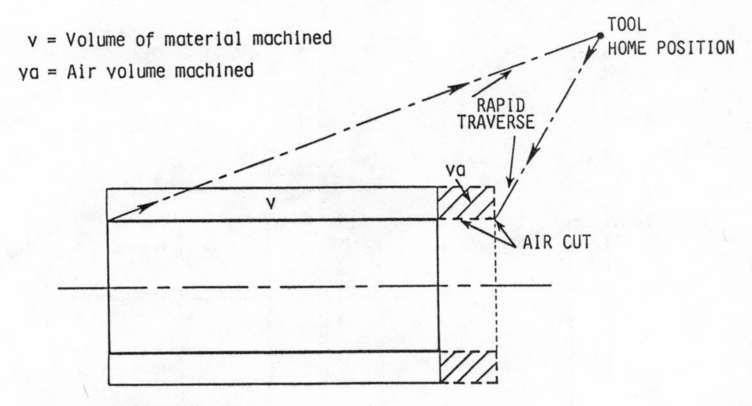

Figure 19.4

models are introduced; the terms are defined in the nomenclature and Fig. 19.4. Although a simple turning example is shown in the figure, these microeconomic models are applicable to any machining operation:

Time/cut

$$tc = (1/r) + (v + va)/R + (v/RT)td. \tag{19.5}$$

Cost/cut

$$cc = \{(1/r) + (v + va)/R + (v/RT)td\}Cr + (v/RT)Y. \tag{19.6}$$

Referring to Fig. 19.4, $(1/r)$ is the time spent by tool in rapid traverse; $(v + va)/R$, in cutting metal and in tool advance and retreat through the air volume, va, at feed; $(v/RT)td$, in prorated down time for tool change for the cut. In addition, $(tc)(Cr)$, cost per cut, contains the additional term $(v/RT)Y$, the prorated tool usage cost for the cut.

Process Model

The above microeconomic models include a process model as defined by T, R and the working region circumscribed by the constraints within which the process can be operated (a) without violating the dimensional accuracy, surface finish, surface and functional integrity; (b) without breaking the cutters or damaging the machine tool fixture setup; and (c) without exceeding the safety limits for noise, toxicity, electrical shock, and harmful radiation. Without going into a detailed mathematical statement of the process working region at this time, note that within the working region, T and R are functions of the operating variables, and assume unique relationships only at their tradeoff, known as the cutting rate–tool life function (R-T-F). It is proven that the

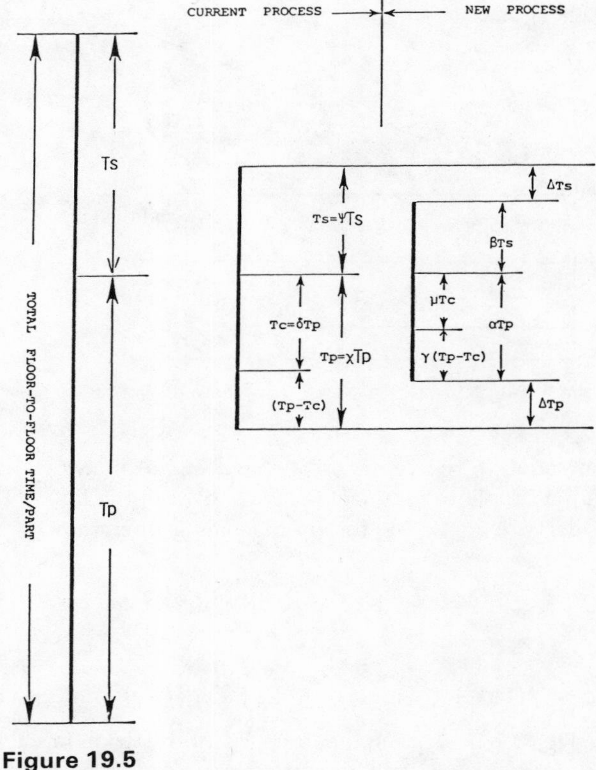

Figure 19.5

economic optima (minimum cost, maximum production rate, or profit rate) must lie on the *R-T-F* or as close to *R-T-F* as possible, if the *R-T-F* is beyond the working region.[1] Ideally, the new and existing processes should be compared at their respective optima. At least, premature rejection of a new process should be avoided until its optima are established. Also, acceptance of a new process should be postponed until it is able to exceed the optima of the existing process.

Economic Feasibility of a Process

Traditionally, the economic feasibility of a new process is determined on the basis of operating cost savings; increased capacity due to time savings is generally computed separately. The new criteria proposed below combine the two savings.

Let

$$T = \text{total time with the current process (hr/part)}$$
$$T' = \text{total time with the new process (hr/part)}$$
$$\Delta T = (T - T'), \text{ total time savings/part (hr/part)}$$
$$\Delta C = \text{total cost savings/part (\$/part)}$$
$$\Delta I = \text{added investment for the new process (\$)}$$
$$(oc) = \text{(farm-out rate) − (in-house rate) (\$/hr)}$$
$$\Delta \text{ROI} = \text{contribution to return on investment by changing to the new process}$$
$$P = \text{number of parts produced in a year} \leqq \text{(capacity of the processing unit with the current process)}$$
$$P(T/T') = \text{number of parts produced in a year} \leqq \text{(capacity of the processing unit with the new process)}$$

The *necessary condition* for economic feasibility of a new process is that the cost savings per part plus the value of the time savings per part must be greater than zero:

$$\Delta C + \Delta T(\text{oc}) > 0. \tag{19.7}$$

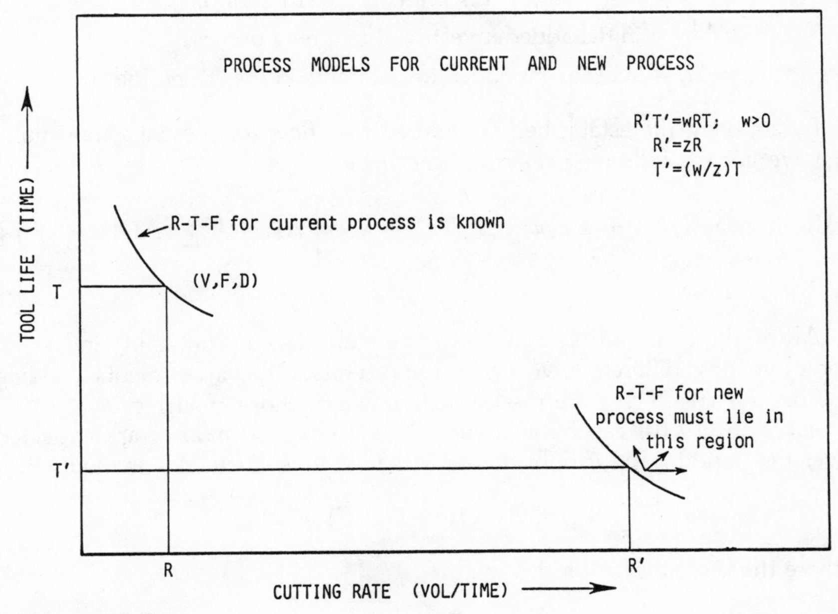

Figure 19.6

Note that either ΔC or ΔT may be negative as long as the criterion given by inequality (19.7) is satisfied; clearly if both ΔC and ΔT are negative, the necessary condition is violated. Also, the extra capacity ΔT has value only as long as there is a finite opportunity cost (*oc*).

The *sufficiency condition* for economic feasibility requires that ΔROI must meet a certain prescribe level to justify the added investment, ΔI:

$$\Delta \text{ROI} = [\{\Delta C + \Delta T(oc)\}P(Y/T')/\Delta I] > 1, \tag{19.8}$$

and

$$\text{payback period} < 1/\Delta \text{ROI}. \tag{19.9}$$

The *sufficiency conditions* given by inequalities (19.8) and (19.9) assume uniform cost savings and undiscounted cash inflows.

A more rigorous sufficiency criterion is obtained by applying the discounted cash flow/present value methods that recognize the time value of money.

Let

$\{\Delta C + \Delta T(oc)\}$ be the net savings over the tth period

P = number of parts to be produced in the tth period

$P_t(T_t/T_t')$ = number of parts produced in the tth period with the new process $\leqq$ (capacity of the processing unit)

k = discount (interest) rate in the tth period.

ΔI_0 = initial added investment in a new process

ΔI_t = subsequent added investments in the tth period

By applying the established discounted cash flow (or present value analysis), we get the following *sufficiency condition*:

$$\Delta \text{ROI} = \sum_{t=1}^{n} [\{\Delta C + \Delta T(oc)\}P_t T_t/T_t'(1+k)^t] \Bigg/ \left[\Delta I_0 + \left(\sum_{t=1}^{n} \Delta I_t/(1+k)^t\right)\right] > 1. \tag{19.10}$$

ΔROI, the contribution to ROI, can be viewed as a profitability index (*PI*) for evaluating different investment opportunities; the investments yielding the highest $\Delta \text{ROI} \leqq 1$ are the most desirable of those available.

Equation (19.10) can be simplified, if uniform cash inflows are expected over the period t and if only the initial added investment ΔI_0 is required:

$$\Phi\{\Delta C + \Delta T(oc)\}P(T/T') > \Delta I_0, \tag{19.11}$$

where the discount factor

$$\Phi = \sum_{t=1}^{n} 1/(1+k)^t.$$

The sufficiency conditions stated in Eqs. (19.10) and (19.11) can be modified to include *the effect of monetary inflation* simply by incorporating the inflation factor ρ during the time period t, into the discounting factor. For example,

$$\Phi = \sum_{t=1}^{n} 1/(1 + k + \rho)^t. \tag{19.12}$$

Another refinement for the sufficiency conditions stated in Eqs. (19.10) and (19.11) is obtained by incorporating *the effect of learning curve* on the time and cost savings obtained after the xth lot is processed. In most instances, the current and the new processes will be compared at the xth copy of the part; the time and cost of making the first unit is computed by applying the well-known learning curve formula:

$$Y_x = Kx^n, \tag{19.13}$$

where Y_x = time to make xth unit

K = time to make first unit

$n = \log b/\log 2$

b = learning factor $0 \leqq b < 1$

For the new process, the cost equivalent of time over period x less the steady-state cost equivalent of time is obtained by

$$K \int_0^x x\,dx - xY_x. \tag{19.14}$$

The amount represented by Eq. (19.14) can be treated either as initial expenditure or as additional investment and incorporated into Eq. (19.10) appropriately.

The methodology presented above provides a basis for establishing economic feasibility of a new process. For economic feasibility, the necessary as well as the sufficiency conditions stated by Eqs. (19.7) and (19.10), respectively, must be satisfied. These conditions and their variants or refinements are used for establishing economic feasibility of LAM, HSM, and UHSM on a continuing basis.

Sensitivity Analyses for Macroeconomic and Microeconomic Models

Sensitivity analysis is useful in evaluating the impact of addition/deletion of a variable or constraint, and in a change in a coefficient of an objective function or constraints in the model. A parametric sensitivity analysis approach is adapted for the sensitivity analysis presented here. In this approach, the effect of change in the parameter under given specific conditions can be computed. Also, assuming continuous functions, partial derivatives of the output with respect to the parameters can be examined.

The basic questions that can be investigated by the sensitivity analysis are:

1. What is the necessary and sufficient degree of improvement in processing time or material removal rate a new process must demonstrate to justify the added investment in the new process?
2. What are the accompanying input parameter ranges that define the acceptable performance envelope?

Thus, the sensitivity analysis can provide specific direction for process development and identify "opportunity windows" for a new process whose technological feasibility is under investigation. A detailed discussion of sensitivity analysis techniques is covered in the references.

CHAPTER 20

Quality Control Circles: The Key to Employee Performance Improvement

Wayne S. Rieker
President, Quality Control Circles, Inc.

What is a QC Circle?

A QC Circle is the heart of a total quality program. It is a group composed of members of the normal organizational work crew and their supervisor. Membership in the Circle is purely *voluntary*. This is probably the single, most unique feature of the program, one which is absolutely necessary, as it assures workers that this is not just "another management program." These members are supported in their efforts by management, union officials, and the Facilitator (whose role, along with that of the Circle Leader, will be discussed later). Since it is true that voluntarism generates enthusiasm, it is likewise important that the participation of this support group also be voluntary.

It is necessary to emphasize from the start that QC Circles does not envision any need to create new bureaucratic structures, nor does it encourage any by-passing of existing structures.

Can we give a precise definition of a QC Circle? Many people have tried to do so. The difficulty lies in the flexibility of the concept: There can be many variations in the make-up of a QC Circle. Although we cannot give an exact definition, the following points do identify the key elements of the normal QC Circle:

1. Members of the normal organizational *work crew* and their *supervisor*.
2. Meeting on a *voluntary* basis.
3. At *regularly scheduled* periodic meetings.
4. To receive *training* in problem-solving techniques.
5. Then *identifying* and *prioritizing* problems, *investigating* and *analyzing* causes.
6. And *developing* and *implementing solutions* when the authority to do so is within their purview.

Let us pursue some of these elements further. First, a Circle is primarily a *normal work crew*—a group of people who work together to produce a part of a product or a service. This is not a program in which a few are selected from various parts of the organization to solve problems for those who are not present. A foundational principle of QC Circles is that people represent themselves—no one else speaks for them.

Second, there is nothing sacred about the length or the frequency of meetings. But experience has proven that in the beginning, weekly one-hour meetings are the best plan. It is important that the meetings be regularly scheduled and not just held when there is some problem. Remember, a meeting provides the opportunity for members to bring up problems that are not necessarily apparent to the supervisor.

Finally, QC Circles is not just another employee suggestion program, in which employees raise complaints or suggestions for others to investigate and act upon. Rather, it is a process whereby the group identifies problems, prioritizes working on them, finds causes, proposes solutions, and, where possible, implements them. *The key is involvement—involving the members in every feasible aspect.*

Forming and Operating QC Circles

With the above as background, let us now move on to how to start and operate a QC Circle program.

How to Form a Circle

After volunteer leaders have had their training, a meeting is scheduled with the employees of that supervisory group. These employees are then told about the program and are asked to consider volunteering to become a part of a QC Circle. This must all be done in a very nonthreatening atmosphere; the employees must be confident that they are not being forced into anything by management.

Quite often the question arises as to what happens if not enough employees volunteer to make up a Circle. The answer is—you just do not start a Circle at that time. However, you really need not worry about it. In literally hundreds of situations, I have encountered this lack of response only one time. On every other occasion a sufficient number of people have volunteered. This is an opportunity for the employees and, properly explained, they will see it as such.

Development of Sub-Circles

More frequently than not, the problem is one of *too many* volunteers for just one Circle. More than 10 persons seem to make a Circle so large that it is unwieldy and does not operate effectively.

What do you do in that event? You must find some mechanism for choosing the members of the first Circle from among the volunteers. Then you set about training another leader for another Circle for those who are left out. Earlier it was stated that the official supervisor was the Circle leader, but when you set up more than one Circle (sub-Circles) under one supervisor, you need to choose another leader. The reason for this second leader is that it is too much of a burden on a supervisor's time to lead more than one Circle. The choice of a leader for the sub-Circle can be done in any reasonable fashion. One rule to keep in mind is that since these sub-Circle members still work for the supervisor, the sub-Circle is still the responsibility of that supervisor although he/she need not be active in that sub-Circle.

Where Circles Operate

You probably understand by now that Circles are usually started in the line manufacturing or operations part of the company. This is normally the case; but Circles can be set up anywhere in the organization where a group of people must work together to produce a common product or service. One point to emphasize: QC Circles do not involve just the *quality* organization. Any organization having a need to improve its operation will find QC Circles valuable.

How Circles Operate

The focus of Circle activity is problem identification and solution. The first step in solving problems is the choice of a problem to work on. Where do the ideas for problem selection come from? The answer is anywhere—the management, other organizations, and the members themselves. However, the main point to remember is that the final decision in selecting a problem to work on is made by the Circle members alone. This is very important—the Circle members must repeatedly receive assurance through just such decisions that this is their program and that they make the decisions about it.

Quite often during the investigation of a problem, a Circle will need the expertise of someone outside the Circle. For instance, it may be necessary to seek out the assistance of the technical people in the manufacturing, engineering, or quality engineering parts of the company to help solve a particular problem. When this occurs, it is important that these experts know about QC Circles and how they operate. They must be cautioned that they do not allow their superior technical knowledge to lead to a take-over or a monopolization of the Circle. Properly handled, the technically trained engineer, who possesses special expertise, and the Circle members, who know the job best, can become a powerful team for solving problems.

It was emphasized previously that the QC Circle program is run by the members. However, it must also be stressed that any solutions to problems proposed by the Circle must have management's approval if the Circle does not have the authority under normal procedures to make the changes proposed. There is nothing about the QC Circle program which usurps the authority of management to control the operation of the business. The Circle must be told in advance that not all of their proposed solutions will be accepted by management. There may not be a budget allocated to cover it, or management may disagree with the solution, or whatever. However, it is also important for management not to turn down proposed solutions without a reason, and it is very helpful if management explains to a Circle the reasons for any rejection. For if the Circle finds out that management is playing a game with them—that it has no intention of accepting any of their solutions, regardless of how good they are—then the Circle will see this and will undoubtedly give up and revert to pre-QC Circle performance.

Where to Hold Circle Meetings

In Japan a visitor will generally see areas throughout a shop specifically set aside for Circle meetings. This is not the case in the United States. Consequently, from the very beginning Circle meetings have tended to be held in conference rooms. There has been a surprise benefit to this. Circle members have been heard to comment that this was the first time in their worklife that they had been invited to a conference room to sit down at a meeting. They had begun to believe that maybe management did feel that they were important and that the QC Circle program was important.

Not all Circle meetings have to be held in conference rooms. All that is being suggested is that the area be quiet enough for people to concentrate. And most importantly, the Circle members should be treated the same as would be the company's professional people.

How Circle Meetings Are Conducted

One of the keys to the success of QC Circles is the training received by the leaders and members on how to conduct QC Circle meetings. They are taught the importance of separating meeting "process" from meeting "content." More specifically, they learn to understand, through paying attention to the "process" and the "roles" that need to be filled, how the effectiveness of the meeting "content" can be greatly enhanced. As for content, the early meetings focus on training in the problem-solving techniques. This provides a basis for creating the team spirit under which QC Circles operate. However, we have found that Circle members soon tire of just studying techniques; so it is the responsibility of the facilitator and leader to recognize when it is time to move on to actual problem-solving.

Meetings on Paid Time

If QC Circles are ever to become a normal and widely used method of operating, it is important that a way be found to hold the meetings as a part of the normal work week. Otherwise, if they are held mostly on overtime, management may become conscious only of the cost and might be motivated to reduce Circles rather than to increase their use.

Union Participation

From the very beginning of my involvement in QC Circles, the need for support from union leadership has been clear. On my first study tour to Japan, the president of the union at Lockheed was invited to go along; but he was not able.

There is still some fear in unions that QC Circles will be a "speed-up" campaign. Time usually eases that fear, though the unions understandably remain vigilant. In reality, some of the most active members are union stewards. Experience shows that when a union indicates some reluctance, it is most valuable to invite full union participation in the leader training course. This fully informs the union of the value of QC Circles to their people.

QC Circle Advisory Committees

Many organizations choose to have a steering committee to administer QC Circles. This is certainly their prerogative. However, my negative experience with such committees and my belief that wherever possible Quality Control Circles should utilize the existing organizational structure cause me to advise that careful consideration be given to the advantages and disadvantages before deciding to set up such a committee. For example:

A. Some Benefits
- Broad "ownership" of the QC Circles program is provided.
- The "executive" workload in the planning for implementation is spread out.
- Early integration of QC Circles with company practices, policies, and procedures is facilitated.

- Opportunity to share planning and implementation with union leadership is offered.

B. Some Concerns

- Another "committee" is created to get a job done that existing organizational structure can do.
- Committees seem to have a "limited lifespan" of effectiveness.
- Strong committees may usurp responsibilities from line management.

In any event, should an Advisory Committee be established, one should consider phasing it out after the program has taken root and is running well.

Facilitator's Role

The role of facilitator was created in the United States to do precisely what the title implies—to "make easy" the operation of the QC Circle program, i.e., to enable it to go and to grow. It seems this process is new and unfamiliar enough that it needs the help of someone whose job it is to be the in-plant expert on QC Circles and to take whatever action appears necessary to keep things moving. Experience has taught that the facilitator is one of the most important elements to a successful program. What does a facilitator do? This will vary, of course, but one primary function is teaching—teaching leaders and then helping leaders to teach the techniques to the members. In addition, the facilitator provides whatever other assistance a leader needs: contacting other organizations, interfacing with management, encouraging nonmembers to join, scheduling management presentations. Moreover, it is also the facilitator's responsibility to promote and publicize the program, and to measure and evaluate its success. The initiative and ingenuity of the facilitator can go a long way toward creating a successful program.

What attributes should an effective facilitator possess? Primarily, this individual should be a well-organized self-starter; should care about people and enjoy working with shop personnel as well as management; must believe in participative management and be willing to give credit to others.

It is also very important that this person know the company and its operations, and possess some technical background, e.g., engineering, statistics, accounting. The point here is that some dimension other than the behavioral sciences be present. Since the facilitator's role involves a large amount of teaching, some experience in this field is desirable. Finally, since the task is one of working with volunteers, some involvement with the dynamics of volunteer groups would also prove helpful.

Although the facilitator needs to be a multidimensional and multitalented individual, such a person is normally already present in an organization. One need not, and should not, hire an outside Doctor of Psychology, nor does one want an impersonal statistician. What is needed is a person who blends the personal and technical, as the latter applies to your particular organization.

How many facilitators are needed? A program can start with one facilitator. Then, as the first Circles mature, they need less of the facilitator's time (at this stage, a facilitator can handle probably between 15 and 20 Circles). In fact, over a period of years, as your program expands and supervisors use the QC Circle problem-solving techniques as a normal way of operating, the need for facilitators will probably greatly diminish. One other point: Some organizations have chosen to have a member of management fill some portion of the facilitator's role and to have another person fill the working-level role. As long as the program's needs are met, it makes little difference how this is accomplished.

Leader's Role

The QC Circle leader is usually the official foreman or supervisor. This person is chosen because the philosophy of QC Circles strives to maintain the normal organizational structure, while operating in a participative manner that enables everyone in the group to voice an opinion. Other persons can be leaders, though. For example, in Japan, Circle leadership is often determined by vote, or may even be rotated among the members. It was believed, however, that in the United States supervisors would not be comfortable with such an arrangement. Nevertheless, experience has surfaced some supervisors who saw the leader's duties as a burden they did not wish to add to their already overtaxing jobs. Given this reality and the desire to stay within the official structure, and given the fundamental principle that the acceptance of leadership is voluntary, every effort has been made to keep the duties of a leader such that they would not discourage supervisors from volunteering.

The role of the successful Circle leader does require the acceptance of responsibility. Such a person must adhere to two basic principles: *offer sincere support* and *guide without dominating*. The leader is also expected to perform certain functions: preside at meetings; involve all members; see that adequate records are kept, assignments made, and the meetings are fruitful, businesslike discussions. Although the leader is encouraged to do the teaching of the other Circle members, often this task will be undertaken by the facilitator. Finally, a Circle leader should regularly communicate with management about Circle activities and help the facilitator keep nonmembers informed.

Member's Role

Obviously, the primary role of the member is to participate in the Circle meetings. To do this effectively, the member must learn the techniques that will be used to identify and select problems, to collect and analyze data, and to discover possible solutions. Once corrective actions have been decided upon, it is necessary that each member follow the plan of action and record the results. The members will also be involved in the preparation and delivery of management presentations. Through all this, it is important that members keep in touch with nonmembers so that no new "clique" be seen as being formed. After all, what members are trying to do is improve communication and teamwork, not create a barrier between Circle members and nonmembers.

When asked why a worker would join a QC Circle, the most ready answer is certainly "not for the money!" This program is not based on monetary rewards. Some of the answers are provided by a survey taken at Westinghouse. People receive (and are likely to desire) a voice in decision-making, increased job satisfaction, enhanced job security, a better quality of work life, and—perhaps most importantly—recognition.

Recognition—Management Presentations

After a QC Circle has solved a family of problems (often referred to as a "theme"), the time has arrived for them to receive recognition for their accomplishments. Originally, it was not felt that this element of the program was as important as some of the others. Experience has shown otherwise. Workers desperately need to be recognized for what they are accomplishing. Although a great deal of satisfaction can come from the internal realization that long-lasting problems have been solved, people have a real need to tell management about it. (In Japan, many companies set aside one or two days each year for their Board of Directors to listen to the best Circles tell their stories.)

In Japan one can view many very elaborate management presentations. I did not believe American workers would do this; how wrong I was! Many Circles take great care to impress management that they can organize and coordinate a proper business meeting. Moreover, among normally casually attired floor workers, one sees women who have their hair done for the occasion and supervisors in a coat and tie—indications of the importance members place on this presentation. However, not all Circles will be so formal. I witnessed one Circle of machinists who used grease pencil for their charts, had some misspelled words, etc. We must recognize that what is important is the content of their problem-solving, not how well they speak or how neat their charts look.

Management's Role

It is because of this need for recognition that management's role is so important. In fact, lack of middle management support is probably the biggest threat to the success of Circle programs today. Management at all levels must understand the concept and operation of the program and provide sincere and active support. This support can take several forms: offer suggestions and feedback, and use Circles to achieve goals; attend management presentations and implement Circle solutions when feasible; communicate and stay informed. Above all, management must provide recognition for achievements. If time is taken to provide these kinds of support, the program will have a high probability of success.

QC Circle Conference

In Japan, October is *Quality Month*. Every year *The Deming Prize* is awarded to a company in recognition of its excellence in quality. And the Union of Japanese Scientists and Engineers (JUSE) sponsors a National QC Circle Conference which features the best Circle presentations.

Some companies in the United States have begun to have annual gatherings to acknowledge the efforts of their Circles. A small start has been made at the yearly conferences of the American Society for Quality Control (ASQC) and the International Association of Quality Circles (IAQC) to spotlight the work of individual Circles. But we still have a long way to go to match the recognition given by the Japanese. My own experience of the first "All QC Circle Conference" at Lockheed convinced me of the value of Circles exchanging information with one another and with management. Management is duly impressed and Circle members become convinced that their accomplishments are being recognized. It is the type of rewarding experience we all need to have!

Time-Tested Characteristics of Successful QC Circles

QC Circles can and will be different in every company that uses them, but there are 10 elements necessary to ensure success:

1. *People building.* A QC Circle program will not work unless there is a sincere desire on the part of management to help their employees to grow and develop through QC Circles. Any company whose only goals are selfish gains for management is advised not to bother to try QC Circles. Such an effort will be seen for just what it would be: another attempt at manipulation by management; a program begun on this premise will fail.
2. *Voluntary.* This is the second most important element of the program and one which seems difficult for management to accept, or at least, to deal with. This is the visible proof to the members that the program is for their benefit. They are completely free to take advantage of it or to not participate.

3. *Leader gets participation from everyone.* QC Circles are a participative program; so the leader must see that the quiet, more introverted persons have a chance to say what is on their minds.
4. *Members help each other to develop.* Since all members will not be equally able to understand and use the techniques, it is important that all the members help in everyone else's development. It is not only the leader's job to see this happens: every member must look out for the development and growth of the others.
5. *Projects are Circle efforts—not individual efforts.* A QC Circle is a group, a team. Everything a Circle does, then, should be a team effort. The projects chosen to work on should be of interest and value to all the members. Also, the Circle, as a whole, should receive recognition for any achievements it has accomplished.
6. *Training is provided to workers and management.* For workers to find solutions to their problems it is not enough to simply turn them loose to proceed in any unstructured manner. They need to know effective techniques, or they will become frustrated at their ineptitude. Management must also receive training for the role they are to play: a role of support without domination.
7. *Creativity is encouraged.* A nonthreatening environment for ideas must be created. People will not let go and risk suggesting a "half-thought-through" idea if they feel they will be ridiculed or rejected. Remember, from seemingly "wild" ideas often come practical solutions.
8. *Theme selection.* The projects that Circles undertake must pertain to their own work, not the work of others; and they must not involve non-work-related subjects. Members are experts at what they do, but not at what other people do.
9. *Management that is supportive.* Unless there is someone in management willing to give QC Circles some time, some advice, and some commitment in the beginning, QC Circles will not have the nourishment it needs to grow into a mature, fruitful program.
10. *Quality and improvement consciousness develops.* All of the above will be useless unless it achieves an awareness on the part of members that they must always be thinking of ways to improve quality and reduce errors.

What Subjects Do QC Circles Address?

Because of the name, "Quality Control Circles," there is a tendency for people to think that this program is only for solving what we label "quality problems." Not so! My original interest in QC Circles came about because I was looking for a way to motivate my work force. As a manufacturing manager, I found myself in the plant on weekends and began to wonder why I was the only one who seemed worried about the problems. How could I get everyone concerned? These were not just quality problems; rather they encompassed the whole spectrum of a manufacturing manager's responsibility, i.e., schedule, quality, cost, productivity, motivation, etc.

Experience clearly indicates that QC Circles' techniques are effective in solving any kind of problem. In fact, in Japan current data show that more and more Circles are now working on methods improvements or cost reduction problems. Obviously, it is a company's decision about what kinds of problems Circles should undertake. However, a word of advice: Circle members' activities should be limited to work-related items, not such things as the cafeteria food, the parking lot, working hours, etc.

Why Call It QC Circles?

As time passes and more experience is gained in the United States with QC Circle programs, it becomes more evident that the name given this program has considerable importance in giving direction and showing clearly the program's intent. Although there are many companies who have given their programs other names—for example, American Airlines' "Participative Action Circles," Buick's "Employee Participation Circles," Verbatim's "Employee Communication Circles"—the name *Quality Control Circles* or *Quality Circles* tells everyone precisely what the focus of the program is: the control of quality. That is its primary intent. The rejection of the words "Quality Control" in the United States seems to be based on our idea that "quality" is an attribute of interest to only a few people or to the Quality Control department of an organization. A quite different notion prevails in Japan, where "Quality Control" is understood to be everyone's responsibility. That means that everyone is responsible for *whatever* is related to the tasks, attitudes, etc., which are a part of the process that produces a *quality* product of whatever kind. More specifically, each person is directly responsible for his/her *own part* in that process. Our efforts should be aimed at changing our perception of who is responsible for quality, as well as altering our understanding of the breadth of the concept. To do these things makes far more sense, and holds much more for our future, than changing the name of the program.

The Roots of QC Circles

Quality Control Circles is founded in the teachings of United States' behavioral scientists such as Abraham Maslow, Douglas McGregor, and Frederick Herzberg, which have been married to the practices of the Quality

Assurance profession as represented by Drs. W. Edwards Deming, J. M. Juran, and Kaoru Ishikawa. For those not familiar with quality control personalities, Dr. Deming was invited to Japan by General MacArthur after World War II, and he taught the Japanese Statistical Quality Control (SQC). The Deming Award is the most valued prize for good quality in Japan today; companies go to great lengths to receive this recognition. Dr. Juran is probably the most renowned quality professional in the world today. He is responsible for the *Handbook of Quality Control* as well as other texts. He has spent a good deal of time in Japan and was the one who taught Total Quality Control (TQC) concepts to the Japanese. He has spoken highly of the QC Circle program in Japan. Dr. Ishikawa, President, Musashi Institute of Technology and an Executive Director, QC Circle Headquarters, Tokyo, is credited with beginning QC Circles by having started workshop discussion groups using the quality control statistical techniques as problem-solving tools.

QC Circles, then, is a style of supervising, or managing, which uses McGregor's "theory Y" management principle that employees have their intelligence to offer to their jobs (and not just their hands) and Herzberg's teachings that the motivating elements to a job are involvement, responsibility, recognition, communication, and feedback (and not just money.)

References

American Society for Quality Control, *Annual Technical Conference Transactions,* Milwaukee, 1975, 1978, 1979, 1980, 1981, 1982.

Cole, R. E., "Will QC Circles Work in the U.S.?" *Quality Progress* July 1980.

Gryna, F. M., Jr., *Quality Circles—A Team Approach to Problem Solving,* American Management Association, New York, 1981.

International Association of Quality Circles, *Annual International Conference Transactions,* Midwest City, OK, 1979, 1980, 1981, 1982.

Ishikawa, K., ed., *QC Circles Koryo: General Principles of the QC Circle,* Union of Japanese Scientists and Engineers (JUSE), Tokyo, 1980.

Rieker, W. S., "Quality Control Circles—Development and Implementation," *29th Annual Technical Conference Transactions,* American Society for Quality Control. Milwaukee, 1975.

Rieker, W. S., "The QC Circle Phenomenon—An Update," *33rd Annual Technical Conference Transactions,* American Society for Quality Control, Milwaukee, 1979.

Rieker, W. S., "Introduction of QC Circles Into Service Industries," *Transactions International Association of Quality Circles,* Midwest City, OK, 1982.

Rieker, W. S., and S. J. Sullivan, "QC Circles' Trend in the U.S.—A Concern," *36th Annual Conference Transactions,* American Society for Quality Control, Milwaukee, 1982.

CHAPTER 21

Leadership: The "Right Stuff" for Management of Human Productivity

William V. Burgess
University of San Francisco

Introduction

The implementation of nontraditional technology is usually only one step in a long sequence of actions taken by the most effective managers. Before a technological system can become fully functional, thoughtful administrators will have considered at least two principal issues: the central function of their organization and the characteristics of the organization's employees (managers and subordinates) in relation to technological, economic, and social expectations.

It is a complex process, and one that requires leadership as opposed to management, synthesis as opposed to analysis, and the foresight of prophecy as opposed to the hindsight of tradition.

This chapter is organized around the notions that leadership qualities are different than managerial qualities, that people are a major factor of consideration in the advancement of any technological development, and that the future is a powerful source of influence for improvement in leadership.

Leadership

What is it about leadership that sets it apart from management? Why can't a good manager be a leader?

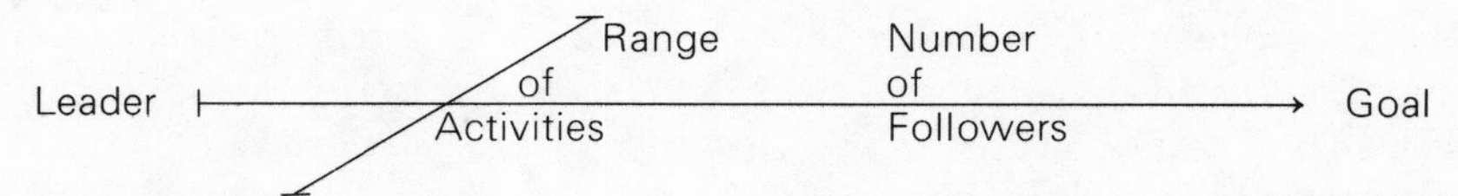

Fig. 21.1 Model of leader influence on followers and range of activities in goal orientation.

To answer the second question first, we would say that a good manager can indeed be a leader. In fact, it is harder to be a leader without managerial skills than it is to be a manager without leadership skills. A good manager can definitely be a leader, but not simply by being a manager. There is a difference.

First, leadership is defined by this writer as a "goal-centered activity that generates a following set of activities among others whose intentions and commitments adhere to the path indicated by the initial activity" (see Fig. 21.1).

While it is acknowledged that leadership does require certain personal characteristics such as initiative, stamina, intellect, etc., the view depicted in the model is that it is the nature of the responses given to the influence attempts of the leader that marks leadership ability. Therefore, we will pursue the discussion of leadership from an interactive base rather than from a personality base.

In further analysis, when leadership is studied in relation to management, the comparisons and contrasts shown in Table 21.1 emerge.

Table 21.1

Leadership	Management
Goals to achieve	Standards to maintain
Followers to influence	Subordinates to command
Ideas to generate	Rules to follow
Risks to take	Perils to avoid
Power to activate	Authority to stabilize

Leadership is seen to be challenging, stimulating, actively changing, while management is seen to be controlling, directive, and more operational than ideational.

Selznick[1] said that the task of leadership is to select key values and create a social structure that embodies those values by giving the organization a sense of mission and purpose. He was careful to note, however, that leadership is not constantly in operation within an organization but may rise and fall with the need to prepare critical decisions or to protect institutional integrity.

Zalesnik[2] explained the rarity of leaders in modern organizations by pointing out that forces within the corporation operate against the exercise of leadership. The need to hedge against loss of control and to provide a balance of power within the organization has led to a reduction of opportunity for the imaginative and creative thinking activities associated with leadership. The inability of most organizations to foster an air of "special consideration" for selected persons or groups of persons stymies the development of potential leaders.

Styles of Leadership

The popularity of recent books with titles such as *How to Be the Toughest Manager of Them All, Intimidating Your Way to Success,* or *Winning Every Time (At Work or Play)* is evidence of the appeal that the two-fisted, hard-nosed, rock-'em, sock-'em approach to management has for so many people. The killer instinct is thought to be necessary in the scramble to the top of the career ladder.

I am reminded of a couple of other little sayings about organizational life that seem to give some balance to the techniques implied and expressed by such an approach. One of them is to "be careful whose toes you step on in climbing the ladder for you will meet the same people on your way down." The other comes from the psychotherapist Victor Frankl who said, "We are pushed by our drives, but pulled by our values." Very few of us value being kicked and shoved, ridiculed, or castigated. Given that fact, I cannot imagine how the valid leadership behaviors such as aggressiveness in solving problems, commitment to goals, and forcefulness, energy, and striving were ever translated into an apparent necessity of being an "S.O.B." in order to be an effective manager. There is no sense to it.

Neither is there good research to support such a contention. The research does support the no-nonsense, directive approach *under certain specific conditions;* but, in the main, research over the last 30 years shows that higher production is attained when the manager and subordinates behave democratically and nonpunitively toward each other.

Lippit and White[3] reported that democratic leadership was most favored by Boys Club members. Under that form of leadership, constructive activities continued even during the supervisor's absence.

In several replications of their experiment, I have found that adult workers respond to the different styles of leadership in the same way as Lippit and White reported for their subjects. The preponderance of approximately 50 replications shows that adults assigned to a task of specific duration in a competitive arena will temporarily accept a dictatorial style of leadership. They will carry out tasks under the pressure of a deadline but will not accept any continuance of authoritarianism beyond the specified time limits (under the experimental conditions I have used time blocks of 15 min and 30 min). In addition, when the task is completed, the members of the work group repudiate the product of their efforts. They disavow any sense of ownership, pride in their accomplishment, or satisfaction with the product.

Work groups under a democratic style of leadership, however, working under the same experimental conditions, express satisfaction with their product, themselves, and their designated leader. They tend to take longer to get started, but they do finish on schedule.

Likert[4] describes four patterns of management that range from exploitative–authoritarian (system 1) through benevolent–authoritarian (system 2), consultative (system 3), to participative (system 4). The ideal condition, system 4, is characterized by confidence and trust among members in all matters, a sense of responsibility and of satisfaction, accurate communication in all directions, and extremely high goals.

Bachman et al.[5] studied five different organizational settings, public and private, white collar and blue collar, to answer questions about why employees comply with supervision and how this relates to organizational effectiveness and employee satisfaction. Bachman and his associates found that the reasons given for compliance were mostly in response to the manager's administrative rights of office and to the manager's expertise. The more professional–technical the type of organization, the more that expertise was given as the reason for compliance. Expertise was also uniformly related to satisfaction and performance while administrative prerogative as a reason for compliance was *not* significantly related to organizational effectiveness and/or employee satisfaction.

The value of expert leadership in sustaining high goals and high productivity is clearly expressed in the recent book *The Soul of a New Machine*,[6] which represents a case study of the development of a new line of computer by a modern high technology corporation.

In short, people will work very well for those in whom they can place confidence and trust and those they believe to be qualified. Thus, the leader is not expected to display a power to command but to display competence in the situation at hand by choosing the best alternative for solving the problem.

How does a manager become a leader? How can a manager know what will be effective in influencing others, gaining a commitment and motivating them toward higher achievement?

The implications from the research cited above led Tannenbaum and Schmidt[7] to suggest that there are three sets of forces that warrant the managers special consideration before embarking on a course of action:

1. Forces in the manager.
2. Forces in the subordinates.
3. Forces in the situation.

The first two of these represent value systems in people as individuals or as small work groups. The third represents organizational constraints emanating from societal conditions at large. Each will be discussed in turn in the following sections.

Value Systems

It used to be thought that our values, beliefs, or conscience, so to speak, were formed early in life and not subject to much modification in later years except under the pressure of prolonged influence of sharp, immediate, highly emotional events. The Biblical proverb "As the twig is bent, so the tree will grow" is an example of that line of thought and also demonstrates how long such thinking has guided our understanding of individual behavior. "It's the way I was brought up" is a contemporary version of the same concept.

We have, in a similar manner, followed the idea that only powerful motivators could alter that early pattern of values. Events such as a family tragedy, a business failure, or a religious conversion have demonstrated such powers of alteration. These influential events are highly individualistic, though, and like falling in love, relatively private modifications of manner.

Those values which guide the activities of people in the work place are more public and, consequently, more subject to public influence. Our individual desire to belong to a supportive group is strong. It is also intelligent and evaluative, allowing us to make some conscious changes in what we believe and what we do.

Social psychologists following the work of Lewin in the 1940s and 1950s found it was possible to influence expressed attitudes toward others by placing them in a situation where they had to interact by sharing ideas and opinions, and where they could learn how others felt about things through direct experience with other kinds of people. Prejudice, for example, was modified and behavior was significantly changed among members of military units exposed to such interaction.

In other studies, several researchers found that work group norms of production were directly determined by what the work group believed to be "right," not by physical ability to perform the tasks. People who were interested in belonging to a support group and gaining the personal satisfaction that the work environment could provide adopted the same codes and modes of performance as their peers and simultaneously internalized a set of values different than those held prior to the influence of the peer group.

This is not to say that early values are not carried with us throughout our lives. Nor does it imply that we are so malleable as to lose lifelong beliefs at the first presentation of contrary opinion. It does suggest that in the process of our identifying with others those early values can be held up for investigation and may be subject to change under conditions other than traumatic events or prolonged impact.

Thank goodness for that! Otherwise, we should not be able to learn and would forever be victim (rather than beneficiary) of new developments.

Management Models of Value Systems

Four separate, distinct models of managerial types have been developed by Maccoby[10] from composites of several ($N = 250$) managers in 12 major companies located throughout the United States. As a part of his research, Maccoby administered the Rorschach Test (a projective test of personality) and conducted interviews of three hours or more about personal beliefs and behaviors. The composite types are described as:

1. Craftsman
2. Jungle Fighter
3. Company Man
4. Gamesman

The Craftsman This type refers to the traditional builder, the farmer, the artisan of the 18th and 19th centuries who holds values of a productive and/or hoarding character. Maccoby says they are motivated by an appeal to greed and ambition in that they want to do well and make money but what they really like to do is solve problems such as putting things together and making them run with mechanical perfection. In his survey, Maccoby found the Craftsman most often in the lower level of the organization, since this type preferred to remain with the technical operations system. For that reason, and because they lacked sensitivity to different interests and values, they are unlikely to rise higher in the organization. They are not seen as capable leaders of a complex and changing organization. An example is the university research scientist.

The Jungle Fighter This type is the empire builder of history and the current entrepreneur. Representatives of this type need to subdue or outsmart others constantly and they do it by brute force or by devious unscrupulous behavior toward peers, subordinates, and colleagues. The Jungle Fighter's goal is power but most of them cannot enjoy it when they achieve it because of their innate suspicion and distrust of others. Needless to say, they create so much hostility within an organization that they have no staying power. Many of them leave because they have become a liability in all their manipulations except in such short-term service as the company hatchet man.

The Company Man Everybody recognizes this type because everybody is known by the Company Man. His interests are in people, in the organization, in rules and regulations, and in his own personal security. Effective representatives of this type serve as organizers and as a stimulant to cooperation. They also protect the company from moving too fast into areas of risk. The Company Man is often the one who will catch the corporate cheat or stop the unprincipled Jungle Fighter by applying regulatory force. The type lacks the self-assurance and the ability to accept risk that is part of the make-up of those managers who are promoted to chief executive officer.

The Gamesman This is the most recently evolved type of manager, the product of the 1960s style of corporate life and the subject of "game theory" research studies of the past decade. The Gamesman enjoys competition, heads himself for the "fast track," and fits in well in those companies where innovation and change are the rule, where deadlines and schedules are highly flexible. His goal is to win through teamwork, but the Gamesman has to be recognized as team captain or Most Valuable Player. Essentially a liberal, nonbigoted, nonidealogical person, the Gamesman will play according to the known rules with anybody who can keep up such an adventurous pace.

Large organizations have needed all types, says Maccoby, excluding possibly the Jungle Fighter. These leader types and their value systems met the needs of the organizations that have employed them in the past decades. They did fit the social character of their institutions during those times. Those times may be coming to an end.

Certainly in an era of high-speed mass production, the Craftsman will be too painstakingly slow. The Jungle Fighter will not be long tolerated in any outfit, so his effectiveness is temporary. The Company Man resists change too much and the Gamesman burns up scarce resources too swiftly. To fit the social character of organizational life over the next couple of generations, a new managerial value system is needed—one that can adapt to a changing technology, to a changing work force, and to a changing organizational environment.

Mentoring for Model Management

The new managerial value system is most likely to be a combination of organizational values and public attitudes. In an era of increasing watchfulness by the public and decreasing resources for industry, organizations are limited in their ability to ignore either influence. Consequently, the new leadership is likely to develop from a sensitivity to public opinion (not just shareholder's opinions) and from a sense of institutional purpose.

Managers who have risen by office politicking or ruthless drive for profit will be passed over in the next generation in favor of leaders who been molded by a sense of "mission" for the organization in response to an environmental expectation. Such leaders are developed through careful mentoring strategies, not through "sink or swim" survival strategies.

In a 1978 interview with J. F. Lunding, G. L. Clements, and D. S. Perkins, the three successive chief executives of Jewel Tea Company, the interviewers concluded that no one ever makes it to the top without having gone through a mentoring process where they are allowed to bring in new ideas under the support and direction of an experienced sponsor who will teach them the central traditional values of the organization and allow them to challenge that system in the name of growth and development.

Levinson[12] sees that process as essential for executive growth. The younger managers identify with the older managers, model their values and behaviors, then, just as in father–son relationships, try to outdo their mentors. The presence of those models, those mentors, those sponsors who will support and encourage while allowing for mistakes is a source of energy and a source of learning that serves the individual and the organization well in adaptations to societal and other environmental pressures. As the younger executive participates in the firm's activities, he brings new insights and values from his world into synthesis with those of the ongoing system to create a vitality similar to what geneticists call "hybrid vigor." The wrenching and tearing of forced change in response to crisis is converted by the mentoring process into a mechanism for continuous executive and organizational performance at a higher level of productivity.

The New Work Force

Social forces contribute to technological productivity in many ways. Attitudes of investors regarding the general (and specific) economic system affect the availability of monetary resources. Movements centered on environmental health and safety can stimulate or constrain technological development. A "law-suit" mentality of consumer groups and of communities can result in legislation and regulation that limits productivity or channels energies away from the central manufacturing process. All of these attitudes are carried to the work place by the members of the work force and there concentrated so that productivity can be affected. It behooves the capable leader, then, to examine this work force, find out what it is like, and make management decisions based on a solid interpretation of what these characteristics imply for the organization's performance.

Conclusion

Research studies often seem to confirm those judgments that insightful, intelligent people have made and in other cases seem to emphasize the common sense feeling of what "ought to be." In this country, for example, where democracy is a knee-jerk reflex and equality is so fiercely sought as a natural right, why do we need any research to tell us that participatory management and democratic leadership yield higher levels of productivity, more pride of accomplishment, and less employee turnover? I would surmise that it is because we see people in the political sense as integral parts of the system and we see people in the production sense as inventory items.

With the growing influence of informed public opinion as part of the organizational environment, though, the successful leaders will be those who can anticipate societal changes, translate them into production needs, and use them as a source of energy for the operation of an integrated sociotechnical system. Then they will have the "Right Stuff."

References

1. Selznick, P., *Leadership in Administration: A Sociological Interpretation*, Harper & Row, New York, 1957.

2. Zalesnik, A., "Managers and Leaders: Are They Different?," *Harvard Business Review*, 68–78, May–June (1977).

3. White, R. and R. Lippitt, "Leader Behavior and Member Reaction in Three 'Social Climates,'" in D. Cartwright and A. Zander, eds., *Group Dynamics*, 3d. ed. Harper & Row, New York, 1968, pp. 318–335.

4. Likert, R., *The Human Organization: Its Management and Value*, McGraw-Hill, New York, 1967.

5. Bachman, J., et al., "Bases of Supervisory Power: A Comparative Study in Five Organizational Settings," in A. S. Tannenbaum, ed., *Control in Organizations*, McGraw-Hill, New York, 1968, pp. 229–235.

6. Kidder, J. T., *The Soul of a New Machine*, Little, Brown & Co., Boston, 1981.

7. Tannenbaum, R. and W. H. Schmidt, "How to Choose a Leadership Pattern," *Harvard Business Review*, May–June 1973, pp. 162–171.

8. Coch, L. and J. R. P. French, Jr., "Overcoming Resistance to Change," *Human Relations*, 1948, No. 11, pp. 512–532.

9. Morris Massey, *What You Are Is Where You Were When*. Farmington Hills, Miss.: Magnetic Video Library, 1975.

10. Maccoby, M., *The Gamesman*, Simon & Schuster, New York, 1976.

11. Collins, E. G. C. and P. Scott, "Everyone Who Makes It Has a Mentor," *Harvard Business Review*, July–August 1978, pp. 89–91.

12. Levinson, H., "A Psychologist Looks at Executive Development," *Harvard Business Review*, September–October 1962, pp. 69–75.

Index